Grundlehren der mathematischen Wissenschaften 309

A Series of Comprehensive Studies in Mathematics

Alejandro Adem R. James Milgram

Cohomology of Finite Groups

Springer-Verlag
Berlin Heidelberg New York
London Paris Tokyo
Hong Kong Barcelona
Budapest

Alejandro Adem
Department of Mathematics
University of Wisconsin
Madison, WI 53706, USA

R. James Milgram
Department of Applied Homotopy
Stanford University
Stanford, CA 94305-9701, USA

Mathematics Subject Classification (1991): 20J05, 20J06, 20J10, 55R35, 55R40, 57S17, 18G10, 18G15, 18G20, 18G40

ISBN 3-540-57025-X Springer-Verlag Berlin Heidelberg New York
ISBN 0-387-57025-X Springer-Verlag New York Berlin Heidelberg

Library of Congress Cataloging-in-Publication Data
Adem, Alejandro. Cohomology of finite groups/
Alejandro Adem, Richard James Milgram.
p. cm. – (Grundlehren der mathematischen Wissenschaften; 309)
Includes bibliographical references and index.
ISBN 0-387-57025-X
1. Finite groups. 2. Homology theory. I. Milgram, R. James. II. Title. III. Series.
QA177.A34 1995 512'.55–dc20 94-13318 CIP

Printed in Germany

Typesetting: Camera-ready copy produced by the authors' output file using a Springer TEX macro package
SPIN 10078665 41/3140-5 4 3 2 1 0 Printed on acid-free paper

Table of Contents

Chapter III. Modular Invariant Theory

Chapter IV. Spectral Sequences and Detection Theorems

Chapter V. G–Complexes and Equivariant Cohomology

Chapter VI. The Cohomology of Symmetric Groups

Chapter VII. Finite Groups of Lie Type

Chapter VIII. Cohomology of Sporadic Simple Groups

Chapter IX. The Plus Construction and Applications

Chapter X. The Schur Subgroup of the Brauer Group

Introduction

Some Historical Background

This book deals with the cohomology of groups, particularly finite ones. Historically, the subject has been one of significant interaction between algebra and topology and has directly led to the creation of such important areas of mathematics as homological algebra and algebraic K-theory. It arose primarily in the 1920's and 1930's independently in number theory and topology. In topology the main focus was on the work of H. Hopf, but B. Eckmann, S. Eilenberg, and S. MacLane (among others) made significant contributions. The main thrust of the early work here was to try to understand the meanings of the low dimensional homology groups of a space X. For example, if the universal cover of X was three connected, it was known that $H_2(X;\mathbb{A})$ depends only on the fundamental group of X. Group cohomology initially appeared to explain this dependence.

In number theory, group cohomology arose as a natural device for describing the main theorems of class field theory and, in particular, for describing and analyzing the Brauer group of a field. It also arose naturally in the study of group extensions, $N \triangleleft E \longrightarrow G$, where $H^3(G;C)$ carries the obstruction to the existence of any extension at all, $H^2(G;C)$ counts the number of distinct extensions. Here C is the center of N.

In these original algebraic applications the emphasis was on cohomology with *twisted* coefficients, i.e., where the group acts non-trivially on the coefficient group. For example, the action in $H^*(G;C)$ is that induced from the action of E on N by conjugation. On the other hand, in topology the emphasis was almost exclusively on the situation where the action is trivial. These basic applications are discussed in Chapter I which is, to a large degree, a historical introduction to the subject, concentrating on the extension problem for groups and the description of the Brauer group in terms of group cohomology. In particular it develops the extension theory for groups and central simple algebras to motivate the definition of group cohomology.

Finite groups and their cohomology also arise in myriad other contexts in topology and algebra. One of the classic successes in the area was the proof (due to P. Smith) that any finite group which acts freely on a sphere must be periodic (equivalently, have Krull dimension one, i.e., have all its abelian subgroups cyclic). Such groups have been classified. They are discussed in

(IV.6). The simplest of them is $(\mathbb{Z}/2)$ and its classifying space, the infinite real projective space $\mathbb{RP}^\infty$, was the main tool used by J.F. Adams in solving the problem of vector fields on spheres. Likewise, Quillen used the structure of the cohomology of the finite groups of Lie type (Chapter VII) to prove the Adams conjecture identifying the $im(J)$ groups as direct summands of the stable homotopy groups of spheres. More recently, with the proof by G. Carlsson of the Segal conjecture, it is evident that the dominant influence on the structure of stable homotopy theory is contained in the structure of the finite symmetric groups.

From a more algebraic point of view the cohomology ring $H^*(G;\mathbb{K})$ of a finite group with coefficients in a finite field is connected via work of D. Quillen, J. Alperin, L. Evens, J. Carlson, and D. Benson to the structure of the modular representations of G. This connection and its ramifications have provided a vast increase in interest in the subject among algebraists. The theoretical underpinnings here have been discussed at length in the excellent books of D. Benson, [Be], and L. Evens, [Ev2], so we do not repeat them here. What we do is to supply the techniques and examples needed to flesh out the theory. For example see Chapter VIII where we discuss the cohomology of some of the sporadic groups, notably the Mathieu groups M_{11}, M_{12}, M_{22} and the group $O'N$, and Chapter VII where we discuss the finite Chevalley groups.

In topology the main source of examples and test spaces are classifying spaces of groups, various natural subspaces, and maps induced by homomorphisms of the groups. This is hardly suprising since the Kan-Thurston theorem and Quillen's plus construction (both discussed in Chapter IX) show that any simply connected space can be constructed from the classifying spaces of groups in very simple ways. For example the plus construction on the classifying space of the infinite symmetric group, $\mathcal{S}_\infty$, is identified with the space $\lim_n \Omega^n S^n$, the infinite loop space of the infinite sphere. Also, the groups of Lie type over finite fields lead to models for B_O, B_U, B_{Sp}, etc.

Related to this, and a major motivation for the study of group cohomology in algebraic topology was Steenrod's construction of cohomology operations in arbitrary topological spaces using the cohomology of the symmetric groups. (This construction is reviewed in (IV.7) and the cohomology of the symmetric groups is discussed in Chapter VI.)

Another reason for the significance of group cohomology stems from its direct relationship with both group actions and homotopy theory. Methods developed by P. Smith to study finite group actions uncovered substantial cohomological restrictions in transformation groups (Chapter V) which led A. Borel to develop a systematic method for analyzing group actions.

Objectives

Our main purposes in writing this book were to collect and make available the theory and its major applications, and provide the backround and techniques necessary for serious calculations. Even today we still hear it said by otherwise knowledgeable mathematicians that group cohomology is a theory without examples (and even if there are any, they are too complex to understand). With Chapters VI, VII, VIII, and X we hope we are able to lay this view to rest. There we discuss the cohomology rings of the symmetric and alternating group, most of the finite Chevalley groups and some of the sporadic groups. Additionally we show how (Chapter X) the theory can be used effectively, even with twisted coefficients, to determine those division algebras which occur in the rational representation rings of finite groups.

The role of classical modular invariant theory is emphasized throughout as a major means of explaining the cohomology classes which occur, and hints are given as to the role in topology of many of the cohomology groups and classes. Space and time prevented us from going into this further, but it should be possible for the interested reader to understand the genesis of most of the cohomology classes for the groups above.

We have tried to lay out the subject to enable us to get to the foundational ideas and techniques as quickly as possible. After Chapter I which, as we have indicated, is primarily a historical introduction to the subject, we turn to its modern development. The basic theory is contained in Chapters II, III, and IV, where we develop the three building blocks of the subject: classifying spaces, invariant theory, and spectral sequences. We have tried throughout, and particularly in these chapters, to provide useful but accessible examples to help clarify the material.

After these core chapters we deal with the areas of interaction previously described, keeping track of their relationship to the cohomology of finite groups. We describe applications to group actions in Chapter V. We also analyze the symmetric groups (Chapter VI) and the general linear groups over a finite field (Chapter VII) in some detail, as they are key groups for stable homotopy theory and algebraic K–theory respectively. Chapter VIII is devoted to the sporadic simple groups, where we describe recent work by the authors aimed at trying to understand the role of some of these groups in topology. In Chapter IX we provide a description of Quillen's plus construction with applications, as well as a proof of the Kan–Thurston Theorem, which states that the cohomology of any topological space is the cohomology of a group. Finally in Chapter X we use cohomological methods to present a solution to the Schur subgroup problem, which classifies all the division algebras that can occur in the rational group rings of finite groups.

Prerequisites

The book has come about from a series of seminars and occasional courses by the second author over a large number of years at Stanford, Aarhus Universitet, and the University of Minnesota, and above all from a two year continuing seminar at Stanford that the authors ran jointly. The first author has also used parts of the book for a second year graduate course at the University of Wisconsin. Overall, the book only assumes that the reader has the backround of a beginning second year graduate student, that is, a one year graduate course in algebra and at least a half year of algebraic topology. This is certainly all that is needed for Chapter I. Chapter II requires more topology however. Familiarity with the first half of the classic text of Milnor and Stasheff, [MS], should be more than sufficient. From Chapter III on the book is largely self contained, though at a few points (IV.7) and (VI.1), some familiarity with the book of Steenrod and Epstein, [SE], is useful, and in Chapter X where we discuss the interplay between group cohomology and classical representation theory useful texts would be Serre's book on local fields, [Se4], and Pierce's book on Associative algebras, [P]. For supplementary material on homological algebra we recommend the text by Cartan and Eilenberg [CE], which has an excellent chapter on Tate cohomology. Likewise the text by K. Brown [Brown] is a nice introduction to group cohomology and given its emphasis on infinite groups we recommend it as a complement to our text. We would like to thank the students at Stanford for their remarks, as well as Su Han Chan for pointing out several typos in an early draft.

Notation and Closing Remarks

The only peculiarity of notation is with regard to coefficients. As abelian groups and even as rings $\mathbb{Z}/p$ and $\mathbb{F}_p$ are the same, though the second notation emphasizes that this ring is the finite field with p-elements. At times throughout the text the reader will see cohomology with coefficients $H^*(G;\mathbb{Z}/p)$ or $H^*(G;\mathbb{F}_p)$. When $(\mathbb{Z}/p)$ is the coefficients we regard $\mathbb{Z}/p$ as an additive abelian group, possibly acted on by G. This occurs when, for example, we are studying group extensions. When $\mathbb{F}_p$ occurs we mean to emphasize the ring structure of the coefficients. Typically, when this occurs the action is trivial and we are interested in the ring structure of the resulting cohomology groups.

Currently the cohomology of groups is a fertile common ground for exchange of ideas between areas such as homotopy theory, modular representation theory, group actions, number theory, etc. We hope that this text will provide a solid background for any person interested in probing deeper into a subject which has been so influential in the development of contemporary algebra and topology.

We would like to acknowledge being partially supported by NSF grants while working on this project. The first author was also supported by an

NSF Young Investigator Award (NYI) and the University of Wisconsin Research Foundation. We would like to thank Beno Eckmann for his advice and encouragement. The first author would like to express his sincere appreciation to Jim Milgram for inviting him to participate in this project, thereby sharing his unique insight, wealth of knowledge and enthusiasm for group cohomology.

Finally, we would like to thank Melania and Judy for their patience, support and love while we were fighting through the writing of this book. It could not have been done without them. We dedicate the book to them.

Chapter I. Group Extensions, Simple Algebras and Cohomology

0. Introduction

In this chapter we will study the structure of extensions of groups and the structure of central simple algebras over a field $\mathbb{F}$. The theory of group extensions,

$$N \triangleleft E \longrightarrow G\,,$$

their existence and classification, will be reduced to two questions about low dimensional cohomology groups. Specifically, we will associate to G and the center C of N, abelian groups $H^2_\phi(G;\ C)$ and $H^3_\phi(G;\ C)$, depending only on G, C, and the action ϕ of G on C. The second group will contain an element, unambiguously defined for each triple $(N,\ G,\ \tau\colon\ G \longrightarrow \mathrm{Out}(N))$ so that τ induces ϕ on restriction to C, and a necessary and sufficient condition for the existence of an extension with the data $(N,\ G,\ \tau)$ will be that that element be zero. Then, once it is known that some extension exists, the elements in the first group will count the number of distinct extensions which are possible up to an appropriate notion of isomorphism. These results are due to S. Eilenberg and S. MacLane, and first appeared in a paper in the Annals of Mathematics in 1947, [EM].

Our exposition of these results begins in §1 with a preliminary discussion of extensions. We associate to the extension a homomorphism $G \longrightarrow \mathrm{Out}(N)$, and begin to study the question of when an extension exists with a given homomorphism, and if so, how many. Then, the next two sections give important examples of how such extensions occur. In §2 we study extensions arising from the unit quaternions, thought of as a non-trivial extension by $(\mathbb{Z}/2)$ of the special orthogonal group SO(3). Specializing to finite groups this leads to the binary icosahedral, tetrahedral, and octahedral groups. In §3 we consider the extensions by $(\mathbb{Z})$ of the fundamental groups of 2-dimensional surfaces, particularly the torus. We see that they correspond to bundles over the surfaces with the circle S^1 as fiber.

Then we turn to the general situation. We first study a special case, the pull-back construction. With this example in hand we go on to the general case. Then, in §7, we use certain explicit extensions associated to free groups to show that every element in $H^3_\phi(G;\ C)$ actually occurs as an obstruction to the existence of an extension for some N with center C.

After this, §8 is devoted to the discussion of how similar cohomological techniques enter into, and indeed, completely determine, the structure of finite dimensional division algebras with center a given field $\mathbb{F}$. Here, we first review the standard theory of simple algebras, the Wedderburn and Noether-Skolem theorems, and then construct factor sets associated to maximal subfields of the algebras. These are then connected to the cohomology groups $H^2_\phi(G; C)$ which arise naturally in the classification of group extensions. Finally we describe the Brauer group in terms of these H^2_ϕ's.

This chapter is written so that it can be read with a minimum of backround, a first year graduate course in algebra and a half year course in topology should be sufficient. It is mainly historical in nature, introducing the concepts and main properties of these low dimensional cohomology groups in very much the manner they originally appeared. In chapter II we begin again with a modern approach to their definition through "resolutions" and "classifying spaces".

1. Group Extensions

Let $N \lhd E$ be a normal subgroup, then there is a group $G = E/N$ and a sequence $N \lhd E \longrightarrow E/N = G$. An important question in group theory is how many distinct E are there having N as a normal subgroup with G as the quotient E/N.

As an example, consider the case where both N and G are simple groups (that is they have no non-trivial normal subgroups). Group theorists working very hard over the last 100 or so years have been able to classify all finite simple groups. Then any non-simple group E must have a normal subgroup N and a quotient group G which is simple. Similiarly N is either itself simple or has a simple quotient, and in order to understand all finite groups from our knowledge of simple groups we should first understand how to proceed when N, G are simple.

The most elementary simple groups are the cyclic groups of prime order $\mathbb{Z}/p$. So as a beginning we might ask to understand all finite groups E having a normal subgroup $N = \mathbb{Z}/p$ and having quotient group $\mathbb{Z}/q$ with p and q primes, not necessarily distinct. To be explicit, the extension has the form

$$N = \mathbb{Z}/p \lhd E \longrightarrow E/N = \mathbb{Z}/q \,.$$

We proceed as follows. First, since $N \lhd E$, then given $g \in E$ the map $\tau_g \colon N \longrightarrow N$, defined on points by $\tau_g(n) = gng^{-1}$ is an automorphism of N. Moreover, $\tau_{gh}(n) = ghn(gh)^{-1} = ghnh^{-1}g^{-1} = \tau_g(\tau_h(n))$ or equivalently $\tau_{gh} = \tau_g\tau_h$ in $\mathrm{Aut}(N)$. Thus the map $\phi\colon E \longrightarrow \mathrm{Aut}(N)$ defined by the rule $g \mapsto \tau_g$ is a homomorphism. ϕ restricted to N takes the elements of N into inner automorphisms of N, since $\tau_n(m) = nmn^{-1}$, for all $m \in N$, is just the inner automorphism *conjugation by* n for $n \in N$. Of course, for $g \notin N$ there is no reason that τ_g should be inner. Indeed, in our example, $\mathbb{Z}/p \lhd E\mathbb{Z}/q$, the inner automorphism group $\mathrm{Inn}(\mathbb{Z}/p) = 1$, so any non-trivial automorphism is outer.

Lemma 1.1 *Let $f: E \longrightarrow H$ be a homomorphism and suppose there are normal subgroups $N \lhd E$ and $W \lhd H$ so that $f(N) \subset W$, then f induces a well defined homomorphism $\hat{f}: E/N \longrightarrow H/W$.*

Proof. Define $\hat{f}\{g\} = \pi f(g)$ where $\pi: H \longrightarrow H/W$ is the projection. Then if g' also represents $\{g\}$ we have $g' = gn$ and $\pi f(gn) = \pi(f(g)f(n)) = \pi f(g)$ since $f(n) \in W$. □

Thus since $\mathrm{Inn}(N)$ is a normal subgroup of $\mathrm{Aut}(N)$ we have a well defined homomorphism

$$\hat{\phi}: E/N \longrightarrow \mathrm{Out}(N) = \mathrm{Aut}(N)/\mathrm{Inn}(N) .$$

Example 1.2 The simplest extension is the Cartesian product $E = N \times G$ with group structure $(n, g)(n', g') = (nn', gg')$. Here $\hat{\phi}: G \longrightarrow \mathrm{Out}(N)$ is the trivial map $\hat{\phi}(g) = 1$ all $g \in G$.

Theorem 1.3 *Let $\lambda: G \longrightarrow \mathrm{Aut}(N)$ be a homomorphism, then there is a group E with N as a normal subgroup and $E/N = G$. Moreover, the associated map, $\hat{\phi}: G \longrightarrow \mathrm{Out}(N)$, is just $\pi\lambda$; where $\pi: \mathrm{Aut}(N) \longrightarrow \mathrm{Out}(N)$ is the projection.*

Proof. On the Cartesian product $N \times G$ define a new multiplication by the rule

$$(n, g)(n', g') = (n\lambda(g)(n'), gg') .$$

It is direct to show that this multiplication gives the product the structure of a group with N as a normal subgroup and G as the quotient by N. However, to give the reader experience with such formal arguments we give the details of the verification now.

The definition shows that $(1, 1)(n', g') = (\lambda(1)(n'), g')$ and this, in turn is (n', g') since $\lambda(1) = \mathrm{id}$. Likewise, $(n, g)(1, 1) = (n\lambda(g)(1), g) = (n, g)$ since $\lambda(g)$ is a homomorphism $N \longrightarrow N$. Hence $(1, 1)$ is a unit. Also,

$$\begin{aligned} (n, g)(\lambda(g^{-1})(n^{-1}), g^{-1}) &= (n\lambda(g)(\lambda(g^{-1})(n^{-1})), 1) \\ &= (n\lambda(1)(n^{-1}), 1) \\ &= (1, 1) \end{aligned}$$

while

$$\begin{aligned}(\lambda(g^{-1})(n^{-1}),g^{-1})(n,g) &= (\lambda(g^{-1})(n^{-1})\lambda(g^{-1})(n),1)\\ &= (\lambda(g^{-1})(n^{-1}n),1)\\ &= (\lambda(g^{-1})(1),1)\\ &= (1,1)\,.\end{aligned}$$

Thus $(\lambda(g^{-1})(n^{-1}), g^{-1})$ is a 2-sided inverse to (n, g).

We now show that the product is associative.

$$\begin{aligned}(n,g)((n',g')(n'',g'')) &= (n,g)(n'\lambda(g')(n''),g'g'')\\ &= (n\lambda(g)(n'\lambda(g')(n'')),g(g'g''))\\ &= (n(\lambda(g)(n')\lambda(g)\lambda(g')(n'')),g(g'g''))\\ &= (n\lambda(g)(n')\lambda(gg')(n''),g(g'g''))\\ &= (n\lambda(g)(n'),gg')(n'',g'')\\ &= ((n,g)(n',g'))(n'',g'').\end{aligned}$$

Hence $N \times G$ with the new multiplication is a group.

It remains to evaluate the map $\hat{\phi}: G \longrightarrow \mathrm{Out}(N)$. We have $G \hookrightarrow N \times G$ as the set of pairs $(1, g)$. Associated to $g \in G$ we obtain $(1,g)(n,1)(1,g^{-1}) = (\lambda(g)(n), 1)$. Thus the map $\phi: E \longrightarrow \mathrm{Aut}(N)$ restricted to G is our original map (λ). □

Examples 1.4 $\mathrm{Aut}(\mathbb{Z}/nm) = \mathrm{Aut}(\mathbb{Z}/n) \times \mathrm{Aut}(\mathbb{Z}/m)$ when n and m are relatively prime. Also, $\mathrm{Aut}(\mathbb{Z}/2^r) = \mathbb{Z}/2 \times \mathbb{Z}/2^{r-2}$ for $r \geq 3$ with generators, multiplication by -1 giving the element of order two, and multiplication by 5 or -3 giving the element of order 2^{r-2}. (Note that $(1+4k)^2 = 1+8k(1+2k)$, so $(1+4)^{2^{r-2}} \equiv 1 \bmod (2^r)$ while $(1+4)^{2^{r-3}} \not\equiv 1 \bmod (2^r)$.)

In general, if p is prime, then $\mathrm{Aut}(\mathbb{Z}/p^r)$ is given by multiplication by the elements of $\mathbb{Z}/p^r$ which are not divisible by p, and there are $p^{r-1}(p-1)$ of them, which is thus the order of $\mathrm{Aut}(\mathbb{Z}/p^r)$. If p is odd then $1+p$ has order p^{r-1}, and hence generates the p-Sylow subgroup. The projection $\mathrm{Aut}(\mathbb{Z}/p^r) \longrightarrow \mathrm{Aut}(\mathbb{Z}/p)$ induced from the surjection $\mathbb{Z}/p^r \longrightarrow \mathbb{Z}/p$ is onto, but $\mathbb{Z}/p = \mathbb{F}_p$ is a field so its multiplicative group, being finite, consists of roots of unity, and hence is cyclic and isomorphic to $\mathbb{Z}/(p-1)$. From this $\mathrm{Aut}(\mathbb{Z}/p^r) = \mathbb{Z}/(p-1) \times \mathbb{Z}/p^{r-1}$ for p an odd prime.

As a particular example, $\mathrm{Aut}(\mathbb{Z}/7) = \mathbb{Z}/6$ with generator, multiplication by 3. Explicitly, the automorphism is given by $n \mapsto 3n \bmod 7$, so its square is $n \mapsto 9n \equiv 2n$ which has order 3. Hence there is a homomorphism $\mathbb{Z}/3 \longrightarrow \mathrm{Aut}(\mathbb{Z}/7)$ sending the generator T to multiplication by 2. Consequently (1.3) gives a group extension

$$\mathbb{Z}/7 \triangleleft E \longrightarrow \mathbb{Z}/3$$

with $\phi: \mathbb{Z}/3 \longrightarrow \mathrm{Aut}(\mathbb{Z}/7) = \mathrm{Out}(\mathbb{Z}/7)$ generated by $\phi(1) = \times 2$.

Similarly $\mathbb{Z}/2 \hookrightarrow \mathrm{Aut}(\mathbb{Z}/n)$ is just the map $\begin{cases} 1 \mapsto 1 \\ T \mapsto u \end{cases}$ where $u(\alpha) = \alpha^{-1}$ all $\alpha \in \mathbb{Z}/n$. (Written additively this is just $\alpha \rightarrow -\alpha$ so it is multiplication by -1.)

Then the associated extension $\mathbb{Z}/(n) \longrightarrow E \longrightarrow \mathbb{Z}/2$ given by the construction in the theorem is the dihedral group of order $2n$, which we generally write D_{2n}.

More generally, the types of groups arising above, extensions of cyclic groups, $\mathbb{Z}/n$, by abelian groups, A, where the multiplication is twisted by a homomorphism $\varphi: A \longrightarrow \mathrm{Aut}(\mathbb{Z}/n)$ are called *metabelian groups* and play a key role in the study of periodic groups in (IV.6) and of important subgroups of the Brauer groups discussed in Chapter X and I.8.

Definition 1.5 Given a triple (N, G, λ) where N and G are groups and $\lambda: G \longrightarrow \mathrm{Aut}(N)$ is a homomorphism, then the group $N \times G$ with multiplication

$$(n, g)(n', g') \mapsto (n\lambda(g)[n'], gg')$$

constructed in the proof of (1.3) is called the *semi-direct product* of N and G twisted by λ, and is written $N \times_\lambda G$.

Let us take a slightly different point of view now. We have

Definition 1.6 Let $N \triangleleft E$ with quotient group $E/N = G$. A *section* $v: G \longrightarrow E$ is a map $v: G \longrightarrow E$, which selects, for each $g \in G$, an element $v(g) \in \pi^{-1}(g)$.

A section $v: G \longrightarrow E$ gives us a way of writing E as a product, $E = Nv(G)$. That is to say, every element in E has a unique expression $nv(g)$.

In particular, suppose N is abelian and $\tau: G \longrightarrow \mathrm{Out}(N) = \mathrm{Aut}(N)$ is known. Then we have $v(g)n = v(g)nv(g)^{-1}v(g) = \tau(g)(n)v(g)$. This implies that in the multiplication table for E we know how to multiply all elements of N with elements in E. What we *do not quite know* are the products

$$v(g_1)v(g_2) = n(g_1, g_2, v)v(g_1g_2)\ .$$

Indeed, this last formula shows that our uncertainty corresponds to a map

$$\lambda_v:\ G \times G \longrightarrow N$$

Exactly what the conditions on this map might be is not clear at this point but will be discussed later. It measures the degree to which τ fails to determine the extension E completely.

We remark at this point that the semi-direct product is the group which results when it is assumed that it is possible to pick a section v so that $\lambda_v(g_1, g_2) = 1$ for all $g_1, g_2 \in G$. An extension is called *split* when there exists a section λ_v which is a homomorphism, and any split extension is isomorphic to the semi-direct product.

Example 1.7 Consider the set of distinct extensions of the form $(\mathbb{Z}/4, \mathbb{Z}/2, \tau)$ where $\tau: \mathbb{Z}/2 \longrightarrow \mathrm{Aut}(\mathbb{Z}/4) = \mathbb{Z}/2$ is the non-trivial map, i.e. $\tau(T) = \times(-1)$, $\tau(1) = \mathrm{id}$. Given an extension of this form,

$$\mathbb{Z}/4 \triangleleft E \longrightarrow \mathbb{Z}/2 ,$$

consider a lift $v: \mathbb{Z}/2 \longrightarrow E$. There are 2 cosets of N in E. Pick v so that $v(1) = 1_E$ and $v(T)$ is some element in the other coset. We have $v(T)^2 \in N$; and

$$\tau(T)[v(T)^2] = v(T)v(T)^2v(T)^{-1} = v(T)^2 .$$

Hence $v(T)^2$ is fixed under $\times(-1)$. Moreover, given any other choice for $v(T)$, $v'(T) = v(T)e$ we have $v'(T)^2 = v(T)^2e^{-1}e = v(T)^2$, so this class is independent of the choice of lifting. Writing the elements of $\mathbb{Z}/4$ as $(1, e, e^2, e^3)$ only 1 and e^2 are fixed under $\times(-1)$ so these are the only choices for $v(T)^2$. If $v(T)^2 = 1$, then we get the dihedral group D_8,

$$D_8 = \{a, b \mid a^4 = b^2 = 1,\ bab = a^3\} .$$

But if $v(T)^2 = e^2$, we obtain the quaternion group Q_8,

$$Q_8 = \{a, b \mid a^4 = 1,\ b^2 = a^2,\ bab^{-1} = a^3\} ,$$

which is not isomorphic to D_8.

2. Extensions Associated to the Quaternions

In this section we introduce the quaternions. Then we show that the unit quaternions can be regarded as an extension of the special orthogonal group SO(3), indeed, the group $N = \mathbb{Z}/2$ in this case. We then restrict to certain subgroups of SO(3), obtaining extensions like the binary tetrahedral group as an extension of the alternating group on 4 letters $\mathcal{A}_4$ by $(\mathbb{Z}/2)$, and the generalized quaternion groups of order $4n$ as extensions by $(\mathbb{Z}/2)$ of the dihedral groups D_{2n}. These results are classical, and a modern but different exposition of them can be found in [Wo], §3.5.

We write $\mathbb{R}$ for the real numbers, $\mathbb{C}$ for the complexes, and represent the quaternions as $\mathbb{H} = \mathbb{C} \oplus \mathbb{C}$, i.e., pairs of complex numbers $z_1 + z_2 j$, with $\mathbb{R}$-linear multiplication given by the formula

$$(2.1) \qquad (z_1 + z_2 j) \cdot (z_3 + z_4 j) = (z_1 z_3 - z_2 \bar{z}_4) + (z_1 z_4 + z_2 \bar{z}_3) j$$

(Here if $z = a + bi$ then $\bar{z} = a - bi$ is the usual conjugation.) The following theorem describes the basic properties of the quaternions.

Theorem 2.2
1. *The multiplication in (2.1) is associative.*

2. *The multiplication in (2.1) has a two sided identity,* $\mathbf{1} = (1 + 0j)$.
3. *There is a conjugation operation* $\omega\colon \mathbb{H} \longrightarrow \mathbb{H}$ *defined by* $\omega(z_1 + z_2 j) = \bar{z}_1 - z_2 j$ *and* $(z_1 + z_2 j)\omega(z_1 + z_2 j) = \omega(z_1 + z_2 j)(z_1 + z_2 j) = z_1\bar{z}_1 + z_2\bar{z}_2 \geq 0$, *and* $\omega(z_1 + z_2 j)(z_1 + z_2 j) = 0$ *if and only if both* z_1 *and* z_2 *are 0. Moreover, for any* $\alpha \in \mathbb{H}$, $\alpha\omega(\alpha)$ *is contained in the center of* $\mathbb{H}$ *which is just* $\mathbb{R}$.
4. $\omega(\alpha\beta) = \omega(\beta)\omega(\alpha)$.

Proof. Again, as in (1.3), the proof consists of a sequence of formal expansions. We write down the details for (2.2(1)) as an example.

$$\begin{aligned}(z_1 + z_2 j) \cdot ((z_3 + z_4 j) \cdot (z_5 + z_6 j)) = &\ (z_1 + z_2 j) \cdot (z_3 z_5 - z_4 \bar{z}_6 \\ &+ (z_3 z_6 + z_4 \bar{z}_5) j) \\ = &\ z_1 z_3 z_5 - z_1 z_4 \bar{z}_6 \\ &- z_2 \bar{z}_3 \bar{z}_6 - z_2 \bar{z}_4 z_5 \\ &+ [z_2(\bar{z}_3 \bar{z}_5 - \bar{z}_4 z_6) \\ &+ z_1(z_3 z_6 + z_4 \bar{z}_5)] j\end{aligned}$$

Similarly

$$\begin{aligned}((z_1 + z_2 j) \cdot (z_3 + z_4 j)) \cdot (z_5 + z_6 j) = &\ (z_1 z_3 - z_2 \bar{z}_4 \\ &+ (z_1 z_4 + z_2 \bar{z}_3) j) \cdot (z_5 + z_6 j) \\ = &\ (z_1 z_3 - z_2 \bar{z}_4) z_5 \\ &- (z_1 z_4 + z_2 \bar{z}_3) \bar{z}_6 \\ &+ (z_1 z_3 z_6 - z_2 \bar{z}_4 z_6 \\ &+ (z_1 z_4 + z_2 \bar{z}_3) \bar{z}_5) j\end{aligned}$$

and comparing coefficients these expressions are the same. □

Corollary 2.3 $\mathbb{H} - \{0\}$ *is a multiplicative group under the quaternion multiplication, (2.1).*

(The inverse is given by $\alpha^{-1} = w(\alpha)/\alpha w(\alpha.)$

It is a special kind of group, though, one in which the multiplication is continuous with respect to the usual topology on $\mathbb{R}$ and the Cartesian product topology on $\mathbb{H}$.

Definition 2.4 The map Norm: $\mathbb{H} \longrightarrow \bar{\mathbb{R}}_+$, which maps $\mathbb{H}$ into the non-negative reals, is defined as $\text{Norm}(\alpha) = \omega(\alpha)\alpha$.

Corollary 2.5 Norm *is a homomorphism onto the multiplicative group of positive reals,*

$$\text{Norm:}\ \mathbb{H} - \{0\} \longrightarrow \mathbb{R}_+ .$$

Proof.

$$\begin{aligned}\mathrm{Norm}(\alpha\beta) &= \alpha\beta w(\alpha\beta) = \alpha\beta w(\beta)w(\alpha)\\ &= \alpha\mathrm{Norm}(\beta)w(\alpha) = \mathrm{Norm}(\beta)\alpha w(\alpha)\\ &= \mathrm{Norm}(\beta)\mathrm{Norm}(\alpha) = \mathrm{Norm}(\alpha)\mathrm{Norm}(\beta).\end{aligned}$$ □

$\mathrm{Norm}^{-1}\{+1\} = S^3 \subset \mathbb{H} - \{0\}$, the subgroup of unit quaternions, which is thus the kernel of Norm.

We may embed $\mathbf{R}_+ \rightarrow \mathbb{H} - \{0\}$ by setting $\lambda(r) = r\mathbf{1}$. Then the composition is the map $\lambda \rightarrow \lambda^2$ which is an isomorphism $\mathbb{R}_+ \rightarrow \mathbb{R}_+$. Also, the image, $\mathrm{im}(\lambda) \subset \mathrm{center}\mathbb{H} - \{0\}$, and we have a map

$$\tau\colon \mathbf{R}_+ \times S^3 \longrightarrow \mathbb{H} - \{0\}$$

defined by $\tau(r, \alpha) = \lambda(r) \cdot \alpha$. Thus we obtain the following structure theorem which describes the multiplicative group $\mathbb{H} - \{0\}$.

Theorem 2.6 *τ is an isomorphism.*

Proof. First note that τ is a homomorphism

$$\begin{aligned}\tau((r_1, \alpha_1)(r_2, \alpha_2)) &= \tau(r_1 r_2, \alpha_1\alpha_2)\\ &= \lambda(r_1)\lambda(r_2)(\alpha_1\alpha_2)\\ &= \lambda(r_1)\alpha_1\lambda(r_2)\alpha_2\\ &= \tau(r_1, \alpha_1)\tau(r_2, \alpha_2).\end{aligned}$$

Now note that $\tau(r, 1)$ and $\tau(1, \alpha)$ are both injections, hence the kernel of τ is determined by the intersection $\tau(r, 1) \cap \tau(1, \alpha)$ which is just the single element $\{1\}$! Finally, suppose $\beta \in \mathbb{H} - \{0\}$, then, setting $v = \mathrm{Norm}(\beta)$, we have $\lambda(v)^{-1}\beta \in S^3$. Hence $\beta = \lambda(v)(\lambda(v)^{-1}\beta)$. □

The Group of Unit Quaternions and SO(3)

We now study the group S^3 of unit quaternions. We will construct a surjective homomorphism $S^3 \rightarrow \mathrm{SO}(3)$ with kernel $\mathbb{Z}/2$, a basic non-split extension which allows us to construct many other non-trivial extensions by specializing to subgroups.

Note that S^3 can be characterized as those $\alpha \in \mathbb{H} - \{0\}$ for which $\omega(\alpha) = \alpha^{-1}$.

Let $W^3 \subset \mathbb{H}$ be the (-1)-eigenspace of ω, the "pure imaginary" subspace of vectors $ri + z_2 j$ with r real. Thus, for $\beta \in W$, $\gamma \in S^3$ we have

$$w(\gamma\beta\gamma^{-1}) = w(\gamma^{-1})w(\beta)w(\gamma) = \gamma(-\beta)\gamma^{-1} = -\gamma\beta\gamma^{-1}$$

and it follows that $\gamma\beta\gamma^{-1} \in W$ if $\beta \in W$.

Also note that

$$\gamma^{-1}(r\beta + s\beta')\gamma = r\gamma^{-1}\beta\gamma + s\gamma^{-1}\beta'\gamma$$

so the conjugation action of S^3 on W is a map $\phi: S^3 \rightarrow \mathrm{GL}_3(\mathbb{R})$, where $\mathrm{GL}_3(\mathbb{R})$ is the group of $\mathbb{R}$-linear isomorphisms of W. But also, $\mathrm{Norm}(\gamma^{-1}\beta\gamma) = \mathrm{Norm}(\beta)$, and, for $\beta \in W$ so β has the form

$$\begin{aligned} v &= ri + sj + tij \\ &= v_i i + v_j j + v_{ij} ij, \end{aligned}$$

we see that $\mathrm{Norm}(v) = r^2 + s^2 + t^2$. Thus, the linear transformation of W is length preserving in the usual metric on W, and consequently $\mathrm{image}(\phi) \subset \mathrm{O}(3)$, the subgroup of orthogonal transformations.

Lemma 2.7 *$\phi: S^3 \rightarrow \mathrm{O}(3)$ is a homomorphism with kernel $\mathbb{Z}/2 = \{\pm \mathbf{1}\} \subset S^3$. Moreover, the image of ϕ is exactly the component of $\mathbf{1}$, $\mathrm{SO}(3) \subset \mathrm{O}(3)$.*

Proof. Conjugation by $\alpha\beta$ is conjugation first by α and then by β. This shows that ϕ is a homomorphism.

We now determine the kernel of ϕ. Consider a non-zero $m \in W$, $\omega(m) = -m$ so $m^2 = -\mathrm{Norm}(m)$. Also, suppose $v \perp m$, i.e., $\langle v, m\rangle = v_i m_i + v_j m_j + v_{ij} m_{ij} = 0$, then $vm \in W$ since the coefficient of 1 in vm is $-\langle v, m\rangle$, and

$$-vm \;=\; \omega(vm) \;=\; \omega(m)\omega(v) \;=\; m\beta\,.$$

But the converse is also true. We have

Proposition 2.8 *Let $m, v \in W$ be non-zero, then $\langle m, v\rangle = 0$ if and only if $mv = -vm$.*

Proof of 2.8. It suffices to show $\omega(mv) = -mv$. But $\omega(mv) = w(v)\omega(m) = vm = -mv\,.$ □

Corollary 2.9 *If $m, v \in W$ are non-zero and orthogonal, then m, v, mv form a mutually orthogonal basis for W.*

Proof of 2.9. $mmv = m^2 v$, $mvm = -m^2 v$, so $m \perp mv$. Also $v(mv) = (vm)v = -mv^2$ so $v \perp mv$. Moreover since $\mathrm{Norm}(mv) = \mathrm{Norm}(m)\mathrm{Norm}(v) \neq 0$ it follows that $mv \neq 0$. □

Now, consider an element $\gamma = A + m \in S^3$ with $m \in W$, A real. Then $\gamma^{-1} = A - m$, and we have

$$\begin{aligned} (A+m)m(A-m) &= (Am - \mathrm{Norm}(m))(A-m) \\ &= (A^2 + \mathrm{Norm}(m))m \\ &= m. \end{aligned}$$

Moreover, if $z \perp m$, then

$$(A+m)z(A-m) = (Az - zm)(A-m) = z(A^2+m^2) - 2zmA\,.$$

Now, noting that $A^2 - m^2 = 1$ implies we can write $A = cos(\theta)$, $\mathrm{Norm}(m)^{1/2} = sin(\theta)$, so the previous proposition shows that we are dealing with a rotation through an angle 2θ in the plane perpendicular to m.

If $A + m \in$ Kernel, then m must be zero or $A^2 + m^2 = 1$. However, since, by assumption $A^2 - m^2 = 1$, this implies again that $m^2 = 0$, so $m = 0$, $A = \pm 1$ and $\mathrm{Kernel}(\phi) = \mathbb{Z}/2$.

To show the third statement of (2.7) consider $\alpha \in \mathrm{O}(3)$ with $det(\alpha) = 1$. Then its characteristic polynomial $det(xI - \alpha) = x^3 - \mathrm{tr}(\alpha)x^2 + a_2 x - 1$ has a real root λ, and λ is not 0 (cubics with real coefficients always have real roots). Hence, there is a non-zero vector $v \in \mathbf{R}^3$ so that $\alpha(v) = \lambda v$, and, since α preserves length, $\lambda = \pm 1$. v gives a splitting $\mathbf{R}^3 = \langle v \rangle \perp \mathbf{R}^2$, and since α preserves angle, $\alpha(\mathbf{R}^2) = \mathbf{R}^2$. Moreover α restricted to $\mathbf{R}^2$ is also length preserving, hence, α is either a reflection or a rotation. But $det(\alpha) = 1 = \lambda det(\alpha)$ on $\mathbf{R}^2$. Hence, if $\lambda = 1$, then α restricted to $\mathbf{R}^2$ is a rotation, and we have already seen that all such α are in the image.

It remains to check the case $\lambda = -1$. Here α on $\mathbf{R}^2$ is a reflection, and so has an eigenvector v' with eigenvalue $+1$. We now choose v' instead of v and use the above argument. □

Summarizing, we have shown that there is an extension

$$\mathbb{Z}/2 \triangleleft S^3 \xrightarrow{\pi} \mathrm{SO}(3)\,.$$

We claim this extension is not a semi-direct product $\mathbb{Z}/2 \times_\lambda \mathrm{SO}(3)$. Indeed, if it were, since $\mathrm{Aut}(\mathbb{Z}/2) = 1$, it would, of necessity, be a product. In particular, if we take the matrix

$$\alpha = \begin{pmatrix} -1 & 0 & 0 \\ 0 & 1 & 0 \\ 0 & 0 & -1 \end{pmatrix} \in \mathrm{SO}(3)$$

we have $\alpha^2 = Id$, so one of the two elements in $\pi^{-1}(\alpha)$ would also satisfy this relation. However, $\pi^{-1}(\alpha) = \pm j$ and $(\pm j)^2 = -1$, so the splitting is not possible!

The Generalized Quaternion Groups and Binary Tetrahedral Group

We now consider some special subgroups of SO(3).

Lemma 2.11 *Let $N \triangleleft G \xrightarrow{\pi} G/N$ be given with N finite. Let $V \subset G/N$ be a subgroup, then there is an extension*

$$N \triangleleft \pi^{-1}(V) \xrightarrow{\pi} V .$$

In particular, if V is finite, then so is $\pi^{-1}(V) \subset G$. Indeed, $|\pi^{-1}(V)| = |N||V|$.

(This is clear.)

Example 2.12 $\mathbb{Z}/2 \times \mathbb{Z}/2 \subset \mathrm{SO}(3)$ as the subgroup generated by

$$\alpha = \begin{pmatrix} 1 & 0 & 0 \\ 0 & -1 & 0 \\ 0 & 0 & -1 \end{pmatrix} \text{ and } \beta = \begin{pmatrix} -1 & 0 & 0 \\ 0 & 1 & 0 \\ 0 & 0 & -1 \end{pmatrix} .$$

We thus have an extension $G = \pi^{-1}(\mathbb{Z}/2 \times \mathbb{Z}/2)$,

$$\mathbb{Z}/2 \triangleleft \pi^{-1}(\mathbb{Z}/2 \times \mathbb{Z}/2) \longrightarrow \mathbb{Z}/2 \times \mathbb{Z}/2 .$$

$|G| = 8$ and, since $i \longrightarrow \alpha$, $j \longrightarrow \beta$, the elements in G are ± 1, $\pm i$, $\pm j$, $\pm ij$. The squares of each of the last 6 elements are -1, and this G is just the quaternion group Q_8 discussed briefly in 1.7.

Example 2.13 Let

$$\alpha = \begin{pmatrix} 1 & 0 & 0 \\ 0 & \cos(\theta) & -\sin(\theta) \\ 0 & \sin(\theta) & \cos(\theta) \end{pmatrix}, \quad \beta = \begin{pmatrix} -1 & 0 & 0 \\ 0 & 1 & 0 \\ 0 & 0 & -1 \end{pmatrix}$$

with $\theta = 2\pi/k$, for some integer k. Then the subgroup $H(k)$ generated by α, β in $\mathrm{SO}(3)$ has order $2k$. In fact it is the dihedral group, as we can easily check that $\beta\alpha\beta^{-1} = \alpha^{-1}$, $\alpha^k = 1$, and $\beta^2 = 1$. Then the extension group $\pi^{-1}(H(k)) \subset S^3$ has order $4k$. It is called the *generalized quaternion group* of order $4k$, and is written Q_{4k}.

Example 2.14 Let σ^3 be the regular tetrahedron with center at the origin. Let T_3 be the subgroup of $\mathrm{SO}(3)$ consisting of rigid motions which take σ^3 to itself. For each vertex there are 3 rotations (including the identity) fixing that vertex. Moreover, if $\alpha \in T_3$ fixes v_1, then it must be one of those 3 rotations.

Also, the only rigid motion fixing 2 or more vertices must be the identity. Thus the homomorphism

$$\phi \colon T_3 \longrightarrow S_4 ,$$

taking α to the permutation on the 4 vertices which it generates must be an *injection.* We claim that $\mathrm{im}(\phi)$ is the subgroup of S_4 of order 12 consisting of exactly the even permutations.

Denote the four vertices of σ^3 by the symbols $1, 2, 3, 4$. Note that by our remarks every permutation of the form (i, j, k) of 3 of the 4 vertices (fixing the fourth) is in the image of ϕ. There are 8 of these. Then note that the product $(i, j, k)(i, j, l) = (i, l)(j, k)$, and there are 3 new elements obtained in this way. Finally, there is the identity element, making 12 in all. The remaining elements in the group fix no vertices or fix 2 vertices. But if one of the remaining elements fixing no vertices is in the image of ϕ, then an element fixing 2 vertices is in the image, and this is impossible.

Thus, we have identified T_3 as the subgroup of even permutations, the alternating group $\mathcal{A}_4 \subset \mathcal{S}_4$.

Now, the inverse image $\pi^{-1}(T_3)$ is called the binary tetrahedral group, and has order 24. S_4 also has order 24, but we will see later that the two groups are distinct.

Remark 2.15 There are two other groups which may be obtained in this way, the binary octahedral group (from the rigid motions of the cube) and the binary icosahedral group (from the rigid motions of the icosahedron). See (IV.6) for a more complete discussion of these groups.

3. Central Extensions and S^1 Bundles on the Torus T^2

In this section we give another class of examples of extensions which occur quite naturally in algebraic geometry and topology. They are connected with complex line bundles over Riemann surfaces since each line bundle is uniquely associated with a circle bundle over the surface, and these circle bundles all arise in the manner described here. As it is most direct we concentrate on the case where the surface is a torus.

We consider the subgroup of $\mathrm{GL}_3(\mathbb{R})$ consisting of the upper triangular matrices,

$$UT_3(\mathbb{R}) = \left\{ \begin{pmatrix} 1 & a & b \\ 0 & 1 & c \\ 0 & 0 & 1 \end{pmatrix} \middle| \; a,\ b,\ c,\ \in \mathbb{R} \right\} .$$

$UT_3(\mathbb{R})$ has the form of an extension

$$\mathbb{R} \triangleleft UT_3(\mathbb{R}) \longrightarrow \mathbb{R} \oplus \mathbb{R} ,$$

where the normal subgroup is the set of matrices

$$N = \left\{ \begin{pmatrix} 1 & 0 & b \\ 0 & 1 & 0 \\ 0 & 0 & 1 \end{pmatrix} \middle| \; b \in \mathbb{R} \right\} .$$

Let G_n, n a positive integer, be the subgroup of $UT_3(\mathbb{R})$ generated by the three matrices

$$A = \begin{pmatrix} 1 & 1 & 0 \\ 0 & 1 & 0 \\ 0 & 0 & 1 \end{pmatrix}, \quad B_n = \begin{pmatrix} 1 & 0 & 0 \\ 0 & 1 & n \\ 0 & 0 & 1 \end{pmatrix}, \quad C = \begin{pmatrix} 1 & 0 & 1 \\ 0 & 1 & 0 \\ 0 & 0 & 1 \end{pmatrix} .$$

Then G_n is given as a (non-split) central extension

$$\mathbb{Z} \triangleleft G_n \longrightarrow \mathbb{Z} \oplus \mathbb{Z}$$

where the defining relations are $ABA^{-1}B^{-1} = C^n$, $AC = CA$, $BC = CB$.

Note that since G_n is a subgroup of $UT_3(\mathbb{R})$ it acts freely on $UT_3(\mathbb{R})$ with quotient a compact three dimensional manifold M^3. Factoring out by the image of N we get $\mathbb{R}^2/\mathbb{Z}^2 = T^2$, the two dimensional torus, and the associated projection $\pi: M^3 \rightarrow T^2$ is a fibering with fiber the circle S^1.

We now show that these fiberings are topologically distinct.

Lemma 3.1 *The first integral homology group of the quotient $UT_3(\mathbb{R})/G_n$, the group $H_1(UT_3(\mathbb{R})/G_n;\mathbb{Z})$, is isomorphic to $\mathbb{Z}/n \oplus \mathbb{Z}^2$. Consequently, the quotients for different n are topologically distinct.*

Proof. Recall that $H_1(X;\mathbb{Z}) = \pi_1(X)/[\pi_1(X), \pi_1(X)]$ where $[G,G]$ denotes the commutator subgroup of G. We now determine $\pi_1(UT_3(\mathbb{R})/G_n)$.

Proposition 3.2 $\pi_1(UT_3(\mathbb{R})/G_n) \cong G_n$.

Proof of 3.2 $UT_3(\mathbb{R})$ is contractible, with a contraction being given by

$$H: \left(t, \begin{pmatrix} 1 & a & b \\ 0 & 1 & c \\ 0 & 0 & 1 \end{pmatrix}\right) \mapsto \begin{pmatrix} 1 & ta & tb \\ 0 & 1 & tc \\ 0 & 0 & 1 \end{pmatrix}.$$

Consequently $\pi_1(UT_3(\mathbb{R})) = 0$ and $UT_3(\mathbb{R})$ is the universal cover of

$$UT_3(\mathbb{R})/G_n$$

which thus has π_1 equal to the group of covering transformations, and this is G_n. □

It remains to determine $G_n/[G_n, G_n]$. We write the defining relations for G_n *additively.* $A + B - A - B = nC$, $A + C = C + A$, $B + C = C + B$, so we see that the commutator quotient is presented as

$$G_n/[G_n, G_n] \;=\; \{A, B, C \mid nC = 0\} \;=\; \mathbb{Z}/n \oplus \mathbb{Z}^2$$

as required. □

Remark 3.3 Using the results of Chapter II we will be able to show that these groups G_n are the only non-trivial central extensions of $\mathbb{Z}^2$ by $\mathbb{Z}$.

Remark 3.4 The fundamental group of the general Riemann surface M_g^2 of genus g has a presentation

$$\pi_1(M_g^2) \;=\; \{a_1, b_1, a_2, b_2, \ldots, a_g, b_g \mid [a_1, b_1][a_2, b_2] \cdots [a_g, b_g] = 1\}$$

where $[a, b] = aba^{-1}b^{-1}$ is the commutator. For each integer $n \in \mathbb{Z}_+$ there is a central extension $\mathbb{Z} \triangleleft G_n(g) \longrightarrow \pi_1(M_g^2)$ where the extension data is $[a_1, b_1][a_2, b_2] \cdots [a_g, b_g] = e^n$ and e generates the central $\mathbb{Z}$. As in the case of the torus, these are all the central extensions of $\pi_1(M_g^2)$ by $\mathbb{Z}$. Again, they are all topologically distinct.

4. The Pull-back Construction and Extensions

In our previous discussion of group extensions we considered triples $(N, G, \phi:)$ where ϕ is a homomorphism $\phi: G \to \mathrm{Aut}(N))$. Using this data we constructed the semi-direct product $N \times_\phi G$ as an example of a group extension $N \triangleleft N \times_\phi G \to G$. However, the more general situation $N \triangleleft E \to G$ gives rise to a map $\phi: G \longrightarrow \mathrm{Out}(N)$, not into $\mathrm{Aut}(N)$, and the next three sections discuss this more general situation. It turns out that the critical consideration for defining an extension from a map $G \to Out(N)$ is the structure of the center $C(N)$. When $C(N) = \{1\}$ there is no difficulty, and that is the situation we discuss in this section. Then in §5 we begin the analysis for the case $C(N) \neq \{1\}$. Here there are two questions, existence, discussed in §5, and the number of distinct extensions which we analyze in §6.

The key construction we need in this section is given as follows.

Definition 4.1 Let G_1, G_2, G_3 be groups, $f_1: G_1 \to G_3$, $f_2: G_2 \to G_3$ homomorphisms then the pull-back $G_1 \times_{G_3} G_2 \subset G_1 \times G_2$ is defined as the set of all pairs (g_1, g_2) satisfying the condition $f_1(g_1) = f_2(g_2)$ in G_3, $\{(g_1, g_2) \mid f_1(g_1) = f_2(g_2) \in G_3\}$.

The two coordinate projections $p_i: G_1 \times G_2 \to G_i$ restrict to give maps $p_i: G_1 \times_{G_3} G_2 \to G_i$ and the following diagram commutes

$$\begin{array}{ccc} G_1 \times_{G_3} G_2 & \xrightarrow{p_1} & G_1 \\ \downarrow{\scriptstyle p_2} & & \downarrow{\scriptstyle f_1} \\ G_2 & \xrightarrow{f_2} & G_3 \end{array}$$

Lemma 4.2

a. *$G_1 \times_{G_3} G_2$ is a group and p_1, p_2 above are homomorphisms.*

b. *$G_1 \times_{G_3} G_2$ is the unique group satisfying the universal property that if L is a group and there are homomorphisms $h_i: L \to G_i$, $i = 1, 2$ so that $f_1 h_1 = f_2 h_2$, then there is a unique homomorphism*

$$h_1 \times h_1 = e(h_1, h_2): \ L \longrightarrow G_1 \times_{G_3} G_2$$

so that $p_i \cdot e(h_1, h_2) = h_i$, $i = 1, 2$.

Proof. (a) If $(g_1, g_2), (g_1', g_2') \in G_1 \times_{G_3} G_2$ then, since both f_1, f_2 are homomorphisms it follows that $(g_1 g_1', g_2 g_2'), (g_1^{-1}, g_2^{-1}) \in G_1 \times_{G_3} G_2$. (a) follows. (b) h_1 and h_2 induce the homomorphism $h_1 \times h_2 \colon L \longrightarrow G_1 \times G_2$ defined on elements by $h_1 \times h_2(l) = (h_1(l), h_2(l))$. Since $f_1 h_1 = f_2 h_2$ we have $\mathrm{im}(h_1 \times h_2) \subset G_1 \times_{G_3} G_2$. Then $e(h_1, h_2)$ is $h_1 \times h_2$ regarded as a homomorphism to $G_1 \times_{G_3} G_2$. It is clearly unique. □

Example 4.3 If $G_1 = \{1\}$ then $G_1 \times_{G_3} G_2 \cong \mathrm{Ker}(f_2)$.

Example 4.4 Let $G_1 = \mathbb{Z}/4$, $G_3 = \{1, T\} \cong \mathbb{Z}/2$, and

$$G_2 = D_8 = \{a, b \mid a^2 = b^4 = 1, aba^{-1} = b^{-1}\}.$$

Let f_1 be the projection and define f_2 by setting $f_2(a) = T$, $f_2(b) = 1$, where T generates G_3. The generators of $G_1 \times_{G_3} G_2$ are the pairs $(1, b)$, $(T, a) \in G_1 \times G_2$, and we have relations $(T, a)^4 = 1$, $(T, a)(1, b)(T^3, a) = (1, b^3)$. Consequently we have a presentation for the pull-back given as

$$G_1 \times_{G_3} G_2 = \{a, b \mid a^4 = b^4 = 1, aba^{-1} = b^{-1}\}.$$

In particular $|G_1 \times_{G_3} G_2| = 16$, and $p_2 \colon G_1 \times_{G_3} G_2 \longrightarrow G_2$ is onto, so we can think of $G_1 \times_{G_3} G_2$ as an extension of $\mathbb{Z}/2$ by G_2.

This second example illustrates the following principle:

Theorem 4.5 *Let $f_1 \colon G_1 \longrightarrow G_3$ be surjective, with kernel N, then $p_2 \colon G_1 \times_{G_3} G_2 \longrightarrow G_2$ is surjective with kernel N' isomorphic to N. In other words we have an extension*

$$N' \lhd G_1 \times_{G_3} G_2 \longrightarrow G_2 = (G_1 \times_{G_3} G_2)/N'$$

where $N' = (N, 1)$. Moreover, if $\phi \colon G_3 \longrightarrow \mathrm{Out}(N)$ is associated to G_1, then the corresponding ϕ' for $G_1 \times_{G_3} G_2 \longrightarrow G_2$ is $\phi f_2 \colon G_2 \longrightarrow \mathrm{Out}(N)$.

Proof. For $g_2 \in G_2$, we can always find $g_1 \in G_1$ so that $f_2(g_2) = f_1(g_1)$ since f_1 is onto. Thus p_2 is onto. The kernel of p_2 is the set of pairs $(n, 1)$ with $f_1(n) = 1 \in G_3$. But this is precisely N. The remaining assertion follows when we observe that the action of (g_1, g_2) on $(n, 1)$ is precisely the action of g_1, so

$$\phi' p_2(g_1, g_2) = \phi f_1(g_1) = \phi f_2(g_2).$$ □

Example 4.6 Let $G_1 = \mathrm{Aut}(G)$, $G_3 = \mathrm{Out}(G)$ and $\pi = f_1 \colon \mathrm{Aut}(G) \longrightarrow \mathrm{Out}(G)$ is the usual projection with kernel $\mathrm{Inn}(G)$. Let $f_2 \colon G_2 \longrightarrow \mathrm{Out}(G)$ be any homomorphism, then we have an extension $\mathrm{Inn}(G) \lhd \mathrm{Aut}(G) \times_{\mathrm{Out}(G)} G_2 \longrightarrow G_2$. Moreover, the twisting data is given as follows:

Lemma 4.7 *The usual surjection* $c: G \longrightarrow \mathrm{Inn}(G)$ *with kernel* $Z(G)$*, the center of* G*, induces homomorphisms*

$$\mu:\ \mathrm{Aut}(G) \longrightarrow \mathrm{Aut}(\mathrm{Inn}(G))$$

$$\hat{\mu}:\ \mathrm{Out}(G) \longrightarrow \mathrm{Out}(\mathrm{Inn}(G))$$

so that the diagram

$$\begin{array}{ccc} \mathrm{Aut}(G) & \xrightarrow{\mu} & \mathrm{Aut}(\mathrm{Inn}(G)) \\ \Big\downarrow{\scriptstyle \pi} & & \Big\downarrow{\scriptstyle \pi} \\ \mathrm{Out}(G) & \xrightarrow{\hat{\mu}} & \mathrm{Out}(\mathrm{Inn}(G)) \end{array}$$

commutes, and in the extension for $\mathrm{Aut}(G) \times_{\mathrm{Out}(G)} G_2$*, the twisting map*

$$\phi = \hat{\mu} \cdot f_2:\ G_2 \longrightarrow \mathrm{Out}(\mathrm{Inn}(G))\,.$$

Proof. Any automorphism $\alpha: G \longrightarrow G$ must take the center of G to itself. Hence it induces a well defined automorphism $\hat{\alpha}: \mathrm{Inn}(G) \longrightarrow \mathrm{Inn}(G)$. Then μ is defined by $\mu(\alpha) = \hat{\alpha}$. If $\alpha(g) = kgk^{-1}$ for $k \in G$, then $\hat{\alpha}(g) = \{k\}\{g\}\{k^{-1}\}$ for all $\{g\} \in \mathrm{Inn}(G)$, so μ takes $\mathrm{Inn}(G)$ into $\mathrm{Inn}(\mathrm{Inn}(G))$, hence $\hat{\mu}$ exists. Moreover, for $\lambda \in \mathrm{Aut}(G)$, $\{n\} \in \mathrm{Inn}(G)$ we have $\lambda\{n\}\lambda^{-1} = \{\lambda(n)\}$ so $\hat{\mu}$ can be identified with the twisting map for the extension $\mathrm{Inn}(G) \triangleleft \mathrm{Aut}(G) \longrightarrow \mathrm{Out}(G)$. □

This example shows that the existence of an extension $G \triangleleft E \longrightarrow G_2$ with given twisting data is made difficult only when G has a non-trivial center. In particular we have

Theorem 4.8 *If* $Z(G) = \{1\}$*, then given any triple* $(G,\ G_2,\ \phi: G_2 \longrightarrow \mathrm{Out}(G))$ *there is a* **unique** *extension* $G \triangleleft E \longrightarrow G_2$ *with twisting map* ϕ*.*

Proof. Since $\mathrm{Inn}(G) = G$ it follows that $\hat{\mu}$ above is the identity map. Hence the extension $G \triangleleft \mathrm{Aut}(G) \times_{\mathrm{Out}(G)} G_2 \longrightarrow G_2$ has ϕ as twisting data. But also, suppose there is a second such extension $G \triangleleft E' \xrightarrow{p} G_2$. Then we have a commutative diagram

$$\begin{array}{ccc} E' & \xrightarrow{\lambda} & \mathrm{Aut}(G) \\ \Big\downarrow{\scriptstyle p} & & \Big\downarrow{\scriptstyle \pi} \\ G_2 & \xrightarrow{\phi} & \mathrm{Out}(G) \end{array}$$

and consequently a homomorphism $e(p, \lambda): E' \longrightarrow \mathrm{Aut}(G) \times_{\mathrm{Out}(G)} G_2$. Note that λ is the identity on $G = \mathrm{Inn}(G)$, and p is the composition $p_2 e(p, \lambda)$

implies $p_2 e(p, \lambda)$ is surjective onto G_2. Thus $e(p, \lambda)$ is an isomorphism preserving G, G_2 and ϕ. □

Example 4.9 If G is a non-abelian simple group then given any homomorphism, ϕ, from G_2 to $\mathrm{Out}(G)$, $\phi: G_2 \longrightarrow \mathrm{Out}(G)$, there is one and only one extension $G \triangleleft E \longrightarrow G_2$ with twisting map ϕ.

Example 4.10 If $G = S_n$ for $n \neq 6$ then $\mathrm{Aut}(S_n) = \mathrm{Inn}(S_n)$ so $\mathrm{Out}(S_n) = 1$ and we have

Corollary 4.11 *If $S_n \triangleleft G$ with quotient H and $n \neq 6$ then $G = S_n \times H$.*

Remark 4.12 A group G with trivial center and $\mathrm{Out}(G) = \{1\}$ is called *complete.* We will show in the next sections that these are *precisely* the groups for which any extension with G as normal subgroup must be a product.

In the next section we consider the problem of constructing extensions when N has a non-trivial center.

5. The Obstruction to Extension When the Center Is Non-Trivial

We now assume N has a non-trivial center C, we assume given a homomorphism $\phi: G \longrightarrow \mathrm{Out}(N)$ and wish to analyze the obstruction to constructing an extension $N \triangleleft E \longrightarrow G$ with twisting map ϕ. We will construct a map $\mu: G \times G \times G \longrightarrow C$ depending on ϕ which vanishes if and only if an extension exists, (5.4). In turn μ will be used to motivate the definition of the third cohomology group $H^3_\phi(G; C)$, (5.7).

Choose a lifting $L: G \longrightarrow \mathrm{Aut}(N)$, i.e. any map so that $\pi \cdot L = \phi$. We can assume that L is normalized so that $L(1) = 1$. There is no reason to assume L is a homomorphism. Indeed, if it were possible to choose L so as to be a homomorphism then the semi-direct product $N \times_L G$ would solve our problem.

Hence, we construct a two variable map $f: G \times G \longrightarrow \mathrm{Aut}(N)$, which measures the obstruction to finding an L which is a homomorphism.

Definition 5.1 Let $\phi: G \longrightarrow \mathrm{Out}(N)$ be a homomorphism and $L: G \longrightarrow \mathrm{Aut}(N)$ any lifting of ϕ. Then the *obstruction* to L being a homomorphism is the two variable map, $f: G \times G \longrightarrow \mathrm{Aut}(N)$ defined on elements by the formula

$$f(g_1, g_2) = L(g_2)^{-1} L(g_1)^{-1} L(g_1 g_2) \,.$$

The map f has the following properties

1. $f(1,g) = f(g,1) = 1$ for all $g \in G$,
2. $f(g_1,g_2) = 1$ for all $g_1,g_2 \in G$ if and only if L is a homomorphism
3. $f(g_1,g_2) \in \mathrm{Inn}(N)$ for all $g_1,g_2 \in G$.

We see that (3) is true since ϕ is a homomorphism, and $\pi \cdot L = \phi$, so

$$\begin{aligned}\pi(f(g_1,\ g_2)) &= \pi(L(g_2)^{-1})\pi(L(g_1)^{-1})\pi(L(g_1g_2))\\ &= \phi(g_2^{-1})\phi(g_1)^{-1}\phi(g_1g_2)\\ &= 1.\end{aligned}$$

The following is less obvious.

Proposition 5.2 *For any triple $(g_1,g_2,g_3) \in G^3$ we have the relation*

$$1 = (L(g_3))^{-1}f(g_1,g_2)L(g_3)f(g_1g_2,g_3)f(g_1,g_2g_3)^{-1}f(g_2,g_3)^{-1}.$$

Proof. Expanding out we have

$$\begin{aligned}f(g_1,g_2) &= L(g_2)^{-1}L(g_1)^{-1}L(g_1g_2)\\ f(g_1g_2,g_3) &= L(g_3)^{-1}L(g_1g_2)^{-1}L(g_1g_2g_3)\\ f(g_1,g_2g_3)^{-1} &= \left(L(g_2g_3)^{-1}L(g_1)^{-1}L(g_1g_2g_3)\right)^{-1}\\ f(g_2,g_3)^{-1} &= \left(L(g_3)^{-1}L(g_2)^{-1}L(g_1g_2)\right)^{-1}\end{aligned}$$

and (5.2) follows by direct substitution. □

Since $\mathrm{im}(f) \subset \mathrm{Inn}(N)$ we can lift f to $f'\colon G\times G\longrightarrow N$. The indeterminacy of f' is C and it is here that C enters into the discussion. For convenience choose f' so that if $f(g_1,g_2) = 1$ then $f'(g_1,g_2) = 1$ as well. Thus we have that the class

$$\mu(g_1,g_2,g_3) = L(g_3)^{-1}[f'(g_1,g_2)]f'(g_1g_2,g_3)(f'(g_1,g_2g_3))^{-1}(f'(g_2,g_3))^{-1} \tag{5.3}$$

is contained in the center of N, C.

Remark. Herc we have introduced the notational convention that we use throughout the remainder of this chapter, $\alpha[h] = \alpha h \alpha^{-1}$, but see the proof of (5.4) for more details.

It follows that we have defined a map

$$\mu\colon\ G\times G\times G\longrightarrow C$$

depending on ϕ, our choice of L lifting ϕ and our choice of f' lifting f.

Theorem 5.4 *An extension $N \triangleleft E \longrightarrow G$ with twisting map $\phi: G \longrightarrow \mathrm{Out}(N)$ exists if and only if it is possible to choose L, f' above so that $\mu: G \times G \times G \longrightarrow C$ is the trivial map, $\mu(g_1, g_2, g_3) = 1$, for all triples $(g_1, g_2, g_3) \in G^3$.*

Proof. Suppose E exists, then we have the commutative diagram

$$\begin{array}{ccc} E & \xrightarrow{\lambda} & \mathrm{Aut}(N) \\ \downarrow p & & \downarrow \pi \\ G & \xrightarrow{\phi} & \mathrm{Out}(N) \end{array}$$

Pick for $g \in G$ some $l_g \in p^{-1}(g)$ and define $L(g) = \lambda[l_g]$. Then $L(g)(n) = l_g n l_g^{-1}$, and

$$f(g_1, g_2)(n) = l_{g_2}^{-1} l_{g_1}^{-1} l_{g_1 g_2} n l_{g_1 g_2}^{-1} l_{g_1} l_{g_2}$$

so we can choose $f'(g_1, g_2) = l_{g_2}^{-1} l_{g_1}^{-1} l_{g_1 g_2}$. But the proposition showing the four fold product $= 1$ in $\mathrm{Inn}(N)$ is really a proposition about functions in 2 variables which can be expanded as $l_{g_2}^{-1} l_{g_1}^{-1} l_{g_1 g_2}$. Hence

$$l_{g_3}^{-1} f'(g_1, g_2) l_{g_3} f'(g_1 g_2, g_3) f'(g_1, g_2 g_3)^{-1} f'(g_2, g_3)^{-1} = 1$$

and we have shown the first part of the theorem.

Let us now assume that there exist liftings $L: G \longrightarrow \mathrm{Aut}(N)$ and $f': G \times G \longrightarrow N$ so that the above formula holds. A multiplication, ω is defined on the Cartesian product $G \times N$, by setting

$$\omega((g, n), (g', n')) = (gg', f'^{-1}(g, g') L(g')^{-1}[n] n') .$$

We verify that ω defines a group structure on $G \times N$. First, since $f'(1, g) =- f'(g, 1) = 1$ and $L(1) = 1$, we have $(1, 1)(g, n) = (g, n)(1, 1) = (g, n)$. Likewise, setting

$$m = L(g)[f'(g^{-1}, g) n^{-1}]$$

we have $(g^{-1}, m)(g, n) = (1, 1)$. Thus it suffices to show the multiplication above is associative. We have

$$\begin{aligned} (g, n)((g', n')(g'', n'')) &= (g, n)(g'g'', f'^{-1}(g', g'') L(g'')^{-1}[n'] n'') \\ &= (gg'g'', f'^{-1}(g, g'g'') L(g'g'')^{-1}[n] f'^{-1}(g', g'') L(g'')^{-1}[n'] n'') \\ &= (gg'g'', f'^{-1}(g, g'g'') f'^{-1}(g', g'') L(g'')^{-1} L(g')^{-1}[n] L(g'')^{-1}[n'] n'') \end{aligned}$$

On the other hand

$$((g, n)(g', n'))(g'', n'') = (gg', f'^{-1}(g, g') L(g')^{-1}[n] n')(g'', n'') =$$
$$(gg'g'', f'^{-1}(gg', g'') L(g'')^{-1} f'^{-1}(g, g') L(g'')^{-1}(L(g')^{-1}[n]) L(g'')^{-1}[n'] n'')$$

and these two expressions are equal if and only if $\mu(g_1, g_2, g_3) \equiv 1$ for all $(g_1, g_2, g_3) \in G^3$. $\square$

The Dependence of $\mu(g_1, g_2, g_3)$ on f' and the Lifting L

Note that any two liftings f', $\bar{f}'$ associated to the same L differ by elements in C. Precisely $\bar{f}'(g_1, g_2)f'^{-1}(g_1, g_2) \in C$ for each pair $(g_1, g_2) \in G^2$. Or equivalently, $\bar{f}'(g_1, g_2) = f'(g_1, g_2)h(g_1, g_2)$, where $h: G \times G \to C$ is an arbitrary function only constrained by the requirement $h(1, g) = h(g, 1) = 1$. We have

Lemma 5.5 *Let (L, f'), $(L, \bar{f}')$ be two liftings associated to the same ϕ. Then there is a two variable map $h: G \times G \to C$ with $h(1, g) = h(g, 1) = 1$ so that*

$$\bar{\mu}(g_1, g_2, g_3) = \mu(g_1, g_2, g_3)L(g_3)^{-1}[h(g_1, g_2)]h(g_1g_2, g_3)h(g_1, g_2g_3)^{-1}(h(g_2, g_3))^{-1}$$

where μ is associated to (L, f) and $\bar{\mu}$ corresponds to $(L, \bar{f}')$.

Next consider the effect of varying the lifting $L(g)$, so that $\bar{L}(g) =$- $L(g)\langle n(g)\rangle$, where $\langle n(g)\rangle$ represents the inner automorphism on N induced by the element $n(g) \in N$. Then we easily calculate

$$\bar{f}'(g_1, g_2) = n(g_2)^{-1}(L(g_2)^{-1}[n(g_1)]f'(g_1, g_2)n(g_1g_2) .$$

Now, expanding out $\bar{\mu}(g_1, g_2, g_3)$ and doing some obvious cancelling we obtain

$$\bar{\mu}(g_1, g_2, g_3) = L(g_3)^{-1}[L(g_2)^{-1}(n(g_1)^{-1})]L(g_3)^{-1}[f'(g_1, g_2)] f'(g_1g_2, g_3)f'^{-1}(g_1, g_2g_3)[n(g_1)]f'^{-1}(g_2, g_3)$$

which can be rewritten as

$$\mu(g_1, g_2, g_3)L(g_3)^{-1}[L(g_2)^{-1}(n(g_1)^{-1})]f'(g_2, g_3)L(g_2g_3)^{-1}[n(g_1)]f'^{-1}(g_2, g_3).$$

But this just reduces to $\mu(g_1, g_2, g_3)$. Thus the total variation of $\mu(g_1, g_2, g_3)$ arises from arbitrary functions $h: G \times G \to C$ according to the formula in lemma (5.5).

Remark 5.6 The set of maps $\mathcal{C}^3 = \{\mu: G^3 \to C\}$ where C is an abelian group is itself an abelian group when we set

$$(\mu\nu)(g_1, g_2, g_3) = \mu(g_1, g_2, g_3)\nu(g_1, g_2, g_3) .$$

Thus we have defined the obstruction as a well determined coset in $\mathcal{C}^3/B^3$ with $B^3 \subset \mathcal{C}^3$ the set of maps f where

$$f(g_1, g_2, g_3) = L(g_3)^{-1}[h(g_1, g_2)] + h(g_1g_2, g_3) - h(g_1, g_2g_3) - h(g_2, g_3) .$$

In (7.2) we will show that μ is not an arbitrary element in C^3, but rather lies in a certain subgroup $Z^3 \subset C^3$ and thus the triple (N, G, ϕ) gives a well defined coset $\theta(N, G, \phi)$ contained in the quotient $\mathbf{Z}^3/B^3$. Moreover, this coset is zero if and only if (N, G, ϕ) comes from an extension $N \triangleleft E \longrightarrow G$ with ϕ as the twisting map. Anticipating (7.2) we make a definition.

Definition 5.7 The quotient group $\mathbf{Z}^3/B^3$ is called the 3rd *cohomology group of G with coefficients in C and twisted by* $\bar{\phi}$. It is written $H^3_{\bar{\phi}}(G;\ C)$.

(Here $\bar{\phi}$ is given by the action of G by ϕ after restricting to C.)

Remark 5.8 $H^3_{\bar{\phi}}(G;\ C)$ is part of a very important family of invariants which we can associate to G, the cohomology groups of G with (twisted) coefficients C which, from Chapter II on form the main topic of this book.

However, before we specify the group $\mathbf{Z}^3$ exactly we turn to the question of how many distinct extensions we can construct if the class $\theta(N, G, \phi) = 0$.

Remark 5.9 The constructions above and (7.1), (7.2), which show that $\mathbf{Z}^3$ is defined only in terms of C and the action of G on C, shows that as we vary N over all groups with center C, (without varying the effect of restricting ϕ to its action on C) we will always obtain the same group $H^3_{\bar{\phi}}(G;\ C)$ as our obstruction set, but (7.7) shows that the obstructions $\theta(N, G, \phi)$ will run over *all the elements* in $H^3_{\bar{\phi}}(G;\ C)$ provided that $G \neq \mathbb{Z}/2$. In particular when $H^3_{\bar{\phi}}(G;\ C) \neq 0$ and $G \neq \mathbb{Z}/2$ there are triples (N, G, ϕ) which correspond to no extension at all! The simplest cases where this happens are $G = \mathbb{Z}/3$, $C = \mathbb{Z}/3$, where, as we will see in Chapter II, $H^3(G;\ C) = \mathbb{Z}/3$ and $G = (\mathbb{Z}/2)^2$, $C = \mathbb{Z}/2$ where $H^3(G;\ C) = (\mathbb{Z}/2)^4$.

6. Counting the Number of Extensions

In 6.2 we define a very natural notion of equivalence classes of extensions, which amounts to isomorphism of the extension groups together with the extension data. Then in 6.8 we identify the set of equivalence classes of extensions with the quotient of a certain naturally occuring abelian group by an appropriate subgroup. Finally, in 6.11 we use this identification to define the second cohomology group $H^2_{\bar{\phi}}(G; C)$.

We suppose that (N, G, ϕ) as well as $L: G \longrightarrow \mathrm{Aut}(N)$, $f: G \times G \longrightarrow N$ are given with $\mu(g_1, g_2, g_3) = 1$ for all $(g_1, g_2, g_3) \in G^3$. Then this gives an extension group which as a set is $(G \times N)$ but with a twisted multiplication $u(L, f)((g, n), (g', n')) = (gg', f^{-1}(g, g')L^{-1}(g')[n]n')$.

Definition 6.1 We say that the pair (L, f) is *equivalent* to the pair (L', f') if there is an isomorphism

$$\tau\colon\ (G \times N, \mu_{(L,f)}) \longrightarrow (G \times N, \mu_{(L',f')})$$

so that $\tau(1 \times N)$ is the identity and the induced map $\hat{\tau}\colon G \to G$ is also the identity.

Definition 6.2 Two extensions E, E' are *equivalent* if and only if there is an isomorphism $\tau\colon E \longrightarrow E'$ so that the diagram

$$\begin{array}{ccccc} N & \longrightarrow & E & \longrightarrow & G \\ \downarrow{\scriptstyle 1} & & \downarrow{\scriptstyle \tau} & & \downarrow{\scriptstyle 1} \\ N & \longrightarrow & E' & \longrightarrow & G \end{array}$$

commutes.

Remark 6.3 If E is an extension, then $E = G \times N$ as a set, and choosing an element g in each coset we have $(g,1)(g',1) = (gg', f^{-1}(g,g'))$. Moreover, we also have

$$(g,n)(g',n') = (gg', f^{-1}(g,g')(g'^{-1}ng')n') = (gg', f^{-1}(g,g')L(g')^{-1}[n]n')$$

and we see that the two definitions of equivalence above are actually the same.

We now introduce a series of algebraic constructions designed to give computable invariants which distinguish extensions. At this stage they will appear somewhat ad hoc. However, after the *bar construction* is introduced in II.3.4, they will appear in a natural context.

Proposition 6.4 *Any homomorphism* $\tau\colon (G \times N, \mu_{(L,f)}) \to (G \times N, \mu_{(L',f')})$ *satisfying the condition that* $\tau|_N = \mathrm{id}$, *and the quotient map* $\hat{\tau}\colon G \to G$ *is also the identity must have the form* $\tau(g,n) = (g, \kappa(g)n)$ *where* $\kappa(g) = c(g)n(g)^{-1}$ *with* $c(g) \in C$, $L(g)\langle n(g)\rangle = L'(g)$.

Proof. Certainly $(1,n) \to (1,n)$ and $(g,1) \to (g,\kappa(g))$ for some $\kappa\colon G \to N$. It follows that $(1,n)(g,1) \to$ has image $(1,n)(g,\kappa(g))$ which can be rewritten as $(g, L'^{-1}(g)[n]\kappa(g))$. But $(1,n)(g,1) = (g, L(g)^{-1}[n]) = (g,1)(1, L(g)^{-1}[n])$ which maps to $(g,\kappa(g))(1, L(g)^{-1}[n]) = (g, \kappa(g)L(g)^{-1}(n))$. Comparing, we must have $\kappa(g)L(g)^{-1}[n] = L'(g)^{-1}[n]\kappa(g)$. □

Lemma 6.5 *Suppose that*

$$f'(g,g') = n(g')^{-1}L(g')^{-1}[n(g)^{-1}]f(g,g')n(gg')^{-1},$$

then, setting $\kappa(g) = n(g)^{-1}$ *we have an equivalence* $(G \times N, \mu_{(L,f)}) \approx (G \times N, \mu_{(L',f')})$ *where* $L'(g) = L(g)\langle n(g)\rangle$.

Proof. It is formal to check that if we set $\tau(g,1) = (g, n(g)^{-1})$, then $\tau((g,1)(g',1)) = \tau(gg',1) = \tau(g,1)\tau(g',1)$. But this is all that is needed. □

Consequently we can assume that $L = L'$ for our classification.

Theorem 6.6 $(G \times N, \mu_{(L,f)}) \sim (G \times N, \mu_{(L,f')})$ *if and only if* $f'(g_1, g_2)$ *can be written as a product* $f(g_1, g_2)k(g_1, g_2)$ *where* $k\colon G\times G \longrightarrow C$ *satisfies the property that there exists* $\lambda\colon G \longrightarrow C$ *and* $k(g_1, g_2) = L(g_2)^{-1}(\lambda(g_1))\lambda(g_1g_2)^{-1}\lambda(g_2)$.

Proof. Suppose τ exists then

$$\begin{aligned}\tau((g,1)(g',1)) &= \tau(gg', f^{-1}(g,g')) \\ &= (gg', \kappa(gg')f^{-1}(g,g')),\end{aligned}$$

while

$$\begin{aligned}\tau(g,1)\tau(g',1) &= (g,\kappa(g))(g',\kappa(g')) \\ &= (gg', f'^{-1}(g,g')L(g')^{-1}[\kappa(g)]\kappa(g')).\end{aligned}$$

On the other hand, since $L = L'$ and $\kappa(g) \in C$ for all $g \in G$, so, comparing, we have $f'(g,g') = L(g')^{-1}[\kappa(g)]\kappa(g')\kappa(gg')^{-1}f(g,g')$. Then, setting $\lambda(g) = \kappa(g)$ gives the result in one direction. On the other hand if we define $f'(g,g')$ by the above equation we easily check that the μ associated to f' is 1. □

Let $\mathcal{C}^2$ be the set of all maps

$$h\colon G \times G \longrightarrow C\,.$$

(As before $\mathcal{C}^2$ is an abelian group).

Definition 6.7 Let $\mathbf{Z}^2(G,C)$ be the kernel of the homomorphism $\delta\colon \mathcal{C}^2 \longrightarrow \mathcal{C}^3$, where

$$\delta(h)(g_1,g_2,g_3) = L^{-1}(g_3)[h(g_1,g_2)]h(g_1,g_2g_3)^{-1}h(g_1g_2,g_3)h(g_2,g_3)^{-1}\,.$$

We also denote by $\mathbf{B}^2(G,C)$ the set of all

$$h(g_1,g_2) = (g_2)^{-1}(\kappa(g_1))\kappa(g_1g_2)^{-1}\kappa(g_2)\,.$$

With these definitions of $\mathbf{Z}^2(G,C)$ and $\mathbf{B}^2(G,C)$ we have

Theorem 6.8 $\mathbf{B}^2(G,C) \subset \mathbf{Z}^2(G,C)$ *and the set of equivalence classes of extensions is isomorphic to* $\mathbf{Z}^2/B^2$.

Proof. $h \in \mathbf{Z}^2(G,C)$ if and only if, setting $f'(g_1,g_2) = f(g_1,g_2)h(g_1,g_2)$ then the μ associated to f' is identically 1. But these f' represent the set of possible extensions. Now, the previous theorem completes the proof. □

Example 6.9 Consider the case $N = Q_8 = \{a, b \mid a^2 = b^2 = (ab)^2\}$, $G = \mathbb{Z}/2$. To begin the analysis of the extensions we need to note that $\mathrm{Aut}(Q_8) = S_4$ where the inner automorphisms of Q_8 correspond to the Klein group, so $\mathrm{Out}(Q_8) = S_3$. Here, the outer automorphism, $\mathfrak{t}$, which switches generators, $\mathfrak{t}\colon a \leftrightarrow b$, is represented by a transposition, while the automorphism, t, which is given on generators by $t(a) = (b)$, $t(b) = ab$, represents an element of order 3. It follows that any extension data is conjugate to either $(Q_8, \mathbb{Z}/2, 1)$ or $(Q_8, \mathbb{Z}/2, \mathfrak{t})$. There certainly exist extensions here, namely the Cartesian product in the first case and the semi-direct product in the second, $E = \mathbb{Z}/2 \times Q_8$ or $E = Q_8 \times_{\mathfrak{t}} \mathbb{Z}/2$. To evaluate the number of other extensions we need only determine

$$f(\mathfrak{t}, \mathfrak{t}) \in \mathbb{Z}/2 = \langle b^2, 1\rangle = C(Q_8)\,.$$

Hence $f(\mathfrak{t}, \mathfrak{t}) = 1$ (giving the product or semi-direct product) or $f(\mathfrak{t}, \mathfrak{t}) = b^2$. We check the associated μ. In general

$$\mu(\mathfrak{t}^{\epsilon_1}, \mathfrak{t}^{\epsilon_2}, \mathfrak{t}^{\epsilon_3}) = f(\mathfrak{t}^{\epsilon_1}, \mathfrak{t}^{\epsilon_2}) f(\mathfrak{t}^{\epsilon_2}, \mathfrak{t}^{\epsilon_3})^{-1} f(\mathfrak{t}^{\epsilon_1}\mathfrak{t}^{\epsilon_2}, \mathfrak{t}^{\epsilon_3}) f(\mathfrak{t}^{\epsilon_1}, \mathfrak{t}^{\epsilon_2}\mathfrak{t}^{\epsilon_3})^{-1}$$

and if $f(\mathfrak{t}, \mathfrak{t}) = b^2$, it is easily seen that $\mu \equiv 1$. On the other hand, given $h'(\mathfrak{t}) = b^2$, then $h'(\mathfrak{t}, \mathfrak{t}) = h(\mathfrak{t})h^{-1}(1)h(\mathfrak{t}) = 1$ and we see that $\mathbf{Z}^2(G, C)/\mathbf{B}^2(G, C) = \mathbb{Z}/2$ with $f(\mathfrak{t}, \mathfrak{t}) = b^2$ as representative for the non-trivial element.

The second extension group in the case of the trivial twisting is given as the quotient

$$\frac{\mathbb{Z}/4 \times Q_8}{\mathbb{Z}/2}$$

which is an example of a central product, a quotient obtained by identifying isomorphic subgroups of the centers of two groups. In this case the generator of the $\mathbb{Z}/2$ maps to $(2, b^2)$. This group has center $\mathbb{Z}/4$ whereas $\mathbf{Q}_8 \times \mathbb{Z}/2$ has center $\mathbb{Z}/2 \times \mathbb{Z}/2$, and these groups are not only not isomorphic as extensions, they are *not isomorphic.*

Remark. We will see in (II.3.8) that $H^2(\mathbb{Z}/4;\ \mathbb{Z}/4) = \mathbb{Z}/4$ where the action of $\mathbb{Z}/4$ on $\mathbb{Z}/4$ is trivial. However, as isomorphism classes of groups there are only two extensions of the form $\mathbb{Z}/4 \triangleleft E \longrightarrow \mathbb{Z}/4$, $(\mathbb{Z}/4)^2$ and $\mathbb{Z}/16$. Indeed, the extension is determined by $x^4 = y^i$ where x generates the quotient $\mathbb{Z}/4$ and y the central $\mathbb{Z}/4$, and for $\mathbb{Z}/16$ the two distinct extension classes are $x^4 = y$ and $x^4 = y^3$.

In the non-trivial case the extension is given as the quotient

$$\frac{Q_8 \times_T \mathbb{Z}/4}{\mathbb{Z}/2}\,,$$

the $\mathbb{Z}/2$ generated by the element $(b^2, 2)$. This group is not isomorphic to any of the others. Indeed, we have

$$(xa)^4 = (b^3a)^2 = b^2 \neq 0$$

and so xa is of order 8. In fact, this group is just the quaternion group of order 16, since $xxax^3 = xb$, and $xaxb = 1$. Consequently, its center is $\mathbb{Z}/2$, so it is not isomorphic to either of the two groups above. In the semi-direct product we have again that xa has order 8. But here, x is a non-central element of order 2, and one easily checks that Q_{16} has no elements of order 2 except for its center.

Corollary 6.10 *There are exactly 4 distinct groups H of order 16 which contain Q_8 as a subgroup.*

Proof. Q_8 must be normal in H so H is one of the extensions above. □

Remark 6.11 As before, for given G, C and ϕ restricted to C, define

$$H^2_\phi(G,C) = \mathbf{Z}^2(G,C)/\mathbf{B}^2(G,C)\ .$$

This is called the second cohomology group of G with coefficients in C twisted by ϕ. This group counts the number of distinct extensions of N by G with twisting ϕ (on C) if, indeed, there are any at all.

Example 6.12 It is direct (but tedious at this stage) to calculate that

$$H^2_{\text{id}}(\mathbb{Z}/2 \times \mathbb{Z}/2; \mathbb{Z}/2) = (\mathbb{Z}/2)^3\ .$$

Efficient techniques for finding groups of this type will be given in Chapter II, see especially (II.4.3). In terms of the elements in $\mathcal{C}^2$ the generators for $H^2_{\text{id}}(\mathbb{Z}/2 \times \mathbb{Z}/2; \mathbb{Z}/2)$ are given as

$$ef:\quad \begin{array}{r} (T,\tau) \mapsto 1 \\ (T\tau,T\tau) \mapsto 1 \\ (T\tau,\tau) \mapsto 1 \\ (T,T\tau) \mapsto 1 \end{array}$$

(the remaining generators go to zero)

$$b_T:\quad \begin{array}{r} (T,T) \mapsto 1 \\ (T\tau,T\tau) \mapsto 1 \\ (T\tau,T) \mapsto 1 \\ (T,T\tau) \mapsto 1 \end{array}$$

(the remaining generators go to zero)

$$b_\tau:\quad \begin{array}{r} (\tau,\tau) \mapsto 1 \\ (T\tau,\tau) \mapsto 1 \\ (\tau,T\tau) \mapsto 1 \\ (T\tau,T\tau) \mapsto 1 \end{array}$$

(the remaining generators go to zero).

Here $\mathbf{Q}_8$ is the extension group associated to the class $b_T + b_\tau + ef$. The dihedral group D_8 is associated to each of the following three classes ef, $ef + b_T$, $ef + b_\tau$. $\mathbb{Z}/4 \times \mathbb{Z}/2$ is associated to each of the three classes b_τ, b_T, $b_\tau + b_T$, and $\mathbb{Z}/2 \times \mathbb{Z}/2 \times \mathbb{Z}/2$ is associated to 0.

7. The Relation Satisfied by $\mu(g_1, g_2, g_3)$

In this section we return to the remaining question, the obstruction to the existence of an extension $N \triangleleft E \longrightarrow G$ for a given map $\phi: G \longrightarrow \mathrm{Out}(N)$. The class μ of (5.3) is shown to define a specific class in an abelian group $H^3_\varphi(G; C)$ where φ is the action of G on C induced from the map ϕ, and H^3_φ only depends on G, C, and φ. This class is 0 if and only if an extension corresponding to ϕ exists. Then we show that for each element $\alpha \in H^3_\varphi(G; C)$ there is an N with center C, and a map $\phi: G \longrightarrow \mathrm{Out}(N)$ so ϕ induces φ, so that the obstruction associated to ϕ is α. In other words, this group contains all the obstructions to the existence of an extension.

Let $\mathcal{C}^3$ be, as in (5.3), the set of all maps $G \times G \times G \longrightarrow C$, and $\mathcal{C}^4$ be the set of all maps $G \times G \times G \times G \longrightarrow C$. Define a *coboundary map* $\delta^3: \mathcal{C}^3 \longrightarrow \mathcal{C}^4$ by setting

$$\begin{aligned}(\delta^3 f)&(g_1, g_2, g_3, g_4)\\ &= L(g_4)^{-1}[f(g_1, g_2, g_3)] f(g_1, g_2, g_3 g_4)^{-1}\\ &\quad f(g_1, g_2 g_3, g_4) f(g_1 g_2, g_3, g_4)^{-1} f(g_2, g_3, g_4),\end{aligned}$$

and, as in (5.5), recall the similar map which defined the variation in μ, $\delta^2: \mathcal{C}^2 \longrightarrow \mathcal{C}^3$, given by the formula

$$(\delta^2 f)(g_1, g_2, g_3) = L(g_3)^{-1}[f(g_1, g_2)] f(g_1 g_2, g_3) f(g_1, g_2 g_3)^{-1} f(g_2, g_3)^{-1}.$$

Remark. Generalizing this we let $\mathcal{C}^n$ be the set of all maps $G^n \longrightarrow C$ and define a map

$$\begin{aligned}&\delta(f(g_1, \ldots, g_n, g_{n+1}) =\\ &\quad L(g_{n+1})^{-1}[f(g_1, \ldots, g_n)] \left(\prod_1^n f(g_1, \ldots, g_i g_{i+1}, \ldots)^{\epsilon_i}\right) f(g_2, \ldots, g_{n+1})^{\epsilon_0}\end{aligned}$$

where $\epsilon_i = (-1)^{n+1-i}$. We have

Lemma 7.1 $\delta^{n+1}\delta^n = 0$, *for all* $n \geq 1$.

(This is a direct calculation.)

Now, recall the element $\mu(g_1, g_2, g_3) \in \mathcal{C}^3$ that we associated to the extension data (N, G, ϕ) in (5.3), where the center of N is C, in (5.3).

$$\mu(g_1,g_2,g_3) = L(g_3)^{-1}[f(g_1,g_2)]f(g_1g_2,g_3)(f(g_1,g_2g_3))^{-1}(f(g_2,g_3))^{-1}.$$

We have

Theorem 7.2 *Let $\mu(g_1,g_2,g_3) \in C^3$ be the obstruction class associated to the homomorphism $\phi: G \rightarrow \mathrm{Out}(N)$ then $(\delta^3\mu)(g_1,g_2,g_3,g_4) = 1$ for all $(g_1,g_2,g_3,g_4) \in G^4$, i.e., $\delta^3(\mu) = 0$ when we write C additively.*

Proof. Write

$$\begin{aligned}\delta\mu(g_1,g_2,g_3,g_4) &= \{L(g_4)^{-1}[\mu(g_1,g_2,g_3)]\mu(g_2,g_3,g_4)\}\\ &\quad\{\mu(g_1,g_2g_3,g_4)\mu(g_1g_2,g_3,g_4)^{-1}\mu(g_1,g_2,g_3g_4)^{-1}\}\end{aligned}$$

Then note that since $\mu(g_1,g_2,g_3) \in C$, and $f(x,y) \in N$ which centralizes C, we have

$$\begin{aligned}&L(g_3)^{-1}[f(g_1,g_2)]f(g_1g_2,g_3)(f(g_1,g_2g_3))^{-1}(f(g_2,g_3))^{-1} =\\ &\qquad (f(g_2,g_3))^{-1}(L(g_3))^{-1}(f(g_1,g_2))f(g_1g_2,g_3)(f(g_1,g_2g_3)^{-1},\end{aligned}$$

i.e. we can cyclically permute the elements in the product without changing its value. Hence, we can write the product above as

7.3

$$\begin{aligned}&L(g_4)^{-1}[\mu(g_1,g_2,g_3)]\,\{f(g_2,g_3g_4)^{-1}f(g_3,g_4)^{-1}L(g_4)^{-1}[f(g_2,g_3)]f(g_2g_3,g_4)\}\\ &\quad\{f(g_2g_3,g_4)^{-1}L(g_4)^{-1}[f(g_1,g_2g_3)]f(g_1g_2g_3,g_4)(f(g_1,g_2g_3g_4))^{-1}\}\\ &\quad\{(f(g_1g_2g_3,g_4))^{-1}L(g_4)^{-1}[f(g_1g_2,g_3)^{-1}]f(g_3,g_4)f(g_1g_2,g_3g_4)\}\\ &\quad\{f(g_2,g_3g_4)f(g_1,g_2g_3g_4)(f(g_1g_2,g_3g_4))^{-1}L(g_3g_4)^{-1}[f(g_1,g_2)^{-1}]\}.\end{aligned}$$

Similarly, we can pass elements of N over elements of C. So, for example, we can move the term $f(g_2,g_3g_4)$ in the last bracket of (7.3) to follow $L(g_4)^{-1}[\mu(g_1,g_2,g_3)]$. Using these shifts we cancel $f(g_2,g_3g_4)$ and $f(g_1,g_2g_3g_4)$. After this the three terms $f(g_1g_2,\ g_3g_4)$, $f(g_1g_2g_3,\ g_4)$ and $f(g_2g_3,\ g_4)$ pair directly with their inverses and so cancel. The result is

$$\begin{aligned}&L(g_4)^{-1}[\mu(g_2,g_3,g_4)]f(g_3,g_4)^{-1}L(g_4)^{-1}[f(g_2,g_3)f(g_1,g_2g_3)f(g_1g_2,g_3)^{-1}]\\ &\qquad f(g_3,g_4)L(g_3g_4)^{-1}[f(g_1,g_2)^{-1}]\end{aligned}$$

At this point recall, from (5.1), that $f(x,y) = L(y)^{-1}L(x)^{-1}L(xy)$. When we replace $f(g_3,g_4)$ in the expression above by this expansion we get

$$L(g_4)^{-1}[\mu(g_1,g_2,g_3)]L(g_3g_4)^{-1}[L(g_3)[\mu(g_1,g_2,g_3)^{-1}]]$$

but since the action restricted to C is obtained via the lifting L, it follows that the second term is $L(g_4)^{-1}[\mu(g_1,g_2,g_3)^{-1}]$, and the result follows. □

Consequently, if we define $\mathbf{Z}^3(G,C) = \mathrm{Ker}(\delta^3) \subset C^3$ then $\mathrm{im}(\delta^2) \subset \mathbf{Z}^3(G,C)$ and $\mu(g_1,\ g_2,\ g_3) \in \mathbf{Z}^3(G,C)$. Thus, the map μ gives a well defined coset

$$\{\mu\} \in \mathbf{Z}^3(G, C)/(\mathrm{im}(\delta^2)) \;=\; H^3_\phi(G; C)$$

which is the zero element if and only if ϕ is associated with an extension.

We now show that the entire group $H^3_\phi(G;\ C)$ is needed, that is, every element occurs as an obstruction to the existence of an extension for some N and $\phi: G \longrightarrow \mathrm{Out}(N)$.

A Certain Universal Extension

Given a group G then for any set of generators $\{g_j \mid j \in J\}$ there is a surjective map, p, of the free group $F(J)$ onto G, $p: F(J) \longrightarrow G$, defined on points by $(p(j) = g_j)$. Write $N(J)$ for the kernel of p, then $N(J) \triangleleft F(J) \longrightarrow G$ is an extension of $N(J)$ by G. Moreover, such an extension is universal in the sense that if $N \triangleleft L \longrightarrow G$ is any extension by G then there is a homomorphism $\phi: F(J) \longrightarrow L$ so ϕ restricted to $N(J)$ has image in N and $\hat{\phi}: G \longrightarrow G$ is the identity. That is, we have a commutative diagram

$$\begin{array}{ccccc} N(J) & \triangleleft & F(J) & \longrightarrow & G \\ \big\downarrow & & \big\downarrow{\scriptstyle \phi} & & \big\downarrow{\scriptstyle =} \\ N & \triangleleft & L & \longrightarrow & G \end{array} .$$

$N(J)$, being a subgroup of a free group is also free.

We now study the particular universal extension obtained when $J = G - \{1\}$ using the technique of §4. The surjection $p: F(J) \longrightarrow G$ is given by $p([g]) = g$ where we denote the generator of the free group corresponding to the element $g \in J$ by $[g]$. Also, for any two elements, g, g' in J, we define the class $|g, g'|$ as the product $[g']^{-1}[g]^{-1}[gg']$. Clearly, $|g, g'| \in \mathrm{Ker}(p)$.

Lemma 7.4 *$N(J)$ is the free group on generators $|g, g'|$, $g, g' \in J$ with action defined by the formula* $[g]^{-1}(|g', g''|) \;=\; |g'', g||g', g''g||g'g'', g|^{-1}$.

Proof. We have

$$\begin{aligned} [g]^{-1}[g', g''][g] &= [g]^{-1}[g'']^{-1}[g']^{-1}[g'g''][g] \\ &= ([g]^{-1}[g'']^{-1}[g''g])([g''g]^{-1}[g']^{-1}[g'g''g])([g'g''g]^{-1}[g'g''][g]) \end{aligned}$$

so the subgoup generated by the $|g, g'|$ is normal in $F(J)$, and the action is as specified. To verify that these elements span $N(J)$, let $M \;=\; \langle \cdots |g, g'| \cdots \rangle$ and consider the cosets $U = M \sqcup_{g \in J} [g]M$. (These cosets are all distinct since $M \subset N(J)$.) We claim $U = F(J)$. Indeed, let $[g_r][g_{r-1}] \cdots [g_1] \in F(J)$ be any word. Then we have

$$\begin{aligned} &[g_1 g_2 \cdots g_r]^{-1}[g_r][g_{r-1}] \cdots [g_1] \;= \\ &\qquad ([g_1 \cdots g_r]^{-1}[g_r][g_1 \cdots g_{r-1}])([g_1 \cdots g_{r-1}]^{-1}[g_{r-1}][g_1 \cdots g_{r-2}]) \\ &\qquad\quad \cdots [g_1 g_2])([g_1 g_2]^{-1}[g_2][g_1]) \end{aligned}$$

and this belongs to M. Hence the original word belongs to the coset $[g_1 \cdots g_r]M$.

It remains to show that the $|g, g'|$ are free generators. To this end consider the space $Y = \bigvee_{j \in J} S^1_j$ which is defined as the union of a collection of disjoint circles obtained by identifying all their base-points to a single point $*$. Now take the (G) covering $X \to Y$ associated to the subgroup M. A maximal tree consists of all the arcs I_j starting at $\hat{*}$ and covering the identification map $I \to S^1_j \subset Y$. Then any other arc in X starts at the end-point of one such arc I_{g_1} and ends at the end point of another I_{g_2}, and is the lift of an $I \to S^1_{g_3}$. But this implies $g_3 g_1 = g_2$ in G. □

Corollary 7.5 *The map $l: G \to \mathrm{Aut}(M)$ defined by*

$$l(g)^{-1}(|g', g''|) = |g'', g||g', g''g||g'g'', g|^{-1}$$

induces an injection $\phi: G \to \mathrm{Out}(M)$ whenever $G \not\cong \mathbb{Z}/2$. For this ϕ, we have $F(J) = \mathrm{Aut}(M) \times_{\mathrm{Out}(M)} G$.

Proof. Any automorphism $\alpha: F(J) \longrightarrow F(J)$ must certainly take the commutator subgroup to itself, hence it induces an automorphism

$$F(J)/F(J)' \longrightarrow F(J)/F(J)'$$

where $F(J)'$ is the commutator subgroup. But $F(J)/F(J)'$ is the free *Abelian* group in the set J, and any *inner automorphism* of $F(J)$ clearly goes to the identity map on $F(J)/F(J)'$.

Now, note that $l(g)^{-1}(|g, g^{-1}|) = |g^{-1}, g|$, hence $l(g)$ on $F(J)/F(J)'$ is not the identity unless, possibly, $g = g^{-1}$, or $g^2 = 1$. Now, suppose, if $g^2 = 1$, that there is an $h \in G$ so that $1, g, h, gh$ are all distinct. Then $l(g)^{-1}(|h, g|) = |g, g||hg, g|^{-1}$ and, again $l(g)$ is not the identity on $F(J)/F(J)'$. □

Example 7.6 When $G = \mathbb{Z}/2$ we have that $J = \{g\}$ and $F(J) = \mathbb{Z}$. In this case the extension is $\mathbb{Z} \xrightarrow{2} \mathbb{Z} \to \mathbb{Z}/2$ and ϕ is the trivial map.

Each Element in $H^3_\phi(G; C)$ Represents an Obstruction

Theorem 7.7 *Assume $G \neq \mathbb{Z}/2$. Given an action $\hat{\phi}: G \to \mathrm{Aut}(C)$, and an element $\{a\} \in H^3_\phi(G; C)$, there is a group N with center C, a homomorphism $\phi: G \to \mathrm{Out}(N)$, inducing $\hat{\phi}$ on restricting to C, and the obstruction to extension for ϕ is $\{a\}$.*

Proof. Let $N' = F(J)$ where $J = (G - \{1\})^2$, and define an action of G on N' as before by the formula

$$(\tilde{l}(g))^{-1}(|g',g''|) = |g'',g||g',g''g||g'g'',g|^{-1}.$$

Since $G \neq \mathbb{Z}/2$, the free group N' is non-commutative and hence has trivial center. Extend this action to the product $N = C \times N'$ by

$$\begin{aligned} L(g)^{-1}[(c,1)] &= \hat{\phi}(g^{-1}(c),1) \\ L(g)^{-1}[(1,|g',g''|)] &= (a(g',g'',g),(\tilde{l}(g))^{-1}(|g',g''|)) \end{aligned}$$

where $a \in \mathbf{Z}^3(G,C)$ represents $\{a\}$. The fact that this is an action is exactly the fact that $(\delta^3(a)(g',g'',g,h) \equiv 1$, since this and the formula for $(\tilde{l}(g))^{-1}$ implies

$$h^{-1}(a(g',g'',g))a(g'',g,h)a(g',g''g,h)a(g'g'',g,h)^{-1} = a(g',g'',gh).$$

Next, note that, as we have seen in our discussion of the universal extension we can choose $f'(g,g') = f(g,g')$ contained in $N' = \mathrm{Inn}(N)$ as $|g,g'|$. But from this

$$\begin{aligned} (L(g_3))^{-1}(f'(g_1,g_2))f'(g_1g_2,g_3)(f'(g_1,g_2g_3))^{-1}(f'(g_2,g_3))^{-1} \\ = (a(g_1,g_2,g_3),1) \end{aligned}$$

and the obstruction class is indeed represented by $\{a\}$. □

8. Associative Algebras and $H^2_\phi(G;C)$

In this section we review another major algebraic question connected to the groups $H^2_\phi(G;C)$ that we have introduced in (6.11), the classification of finite dimensional central simple $\mathbb{F}$-algebras. This section culminates in the definition of the Brauer group, $B(\mathbb{F})$, and its identification with $\lim_{\mathbb{K}} H^2_\phi(\mathrm{Gal}(\mathbb{K}/\mathbb{F});\mathbb{K}^\bullet)$ as $\mathbb{K}$ runs over all finite Galois extensions of $\mathbb{F}$. The major focus of Chapters II - IX will be the structure and some applications of the cohomology of finite groups. But in Chapter X we will combine the two subjects in our solution of the Schur subgroup problem, which classifies all the division algebras which can occur in the rational group rings of finite groups.

Basic Structure Theorems for Central Simple $\mathbb{F}$-Algebras

From now on, in this section, $\mathbb{A}$ will always denote a simple unitary algebra with center $\mathbb{F}$ where $\mathbb{F}$ is a field. We also assume $\mathbb{A}$ is finite dimensional as a $\mathbb{F}$-vector space. The basic structure theory for such algebras $\mathbb{A}$, is well known, see for example, [P], or [Hu, pp. 450-461]. First there is the fundamental Wedderburn theorem.

Theorem 8.1 *$\mathbb{A}$, as above, is a matrix ring $M_n(D)$, where D is a division ring with center $\mathbb{F}$. Moreover, D and n are uniquely associated to $\mathbb{A}$.*

Next, as regards subalgebras of $\mathbb{A}$, the Noether-Skolem theorem gives a great deal of information.

Theorem 8.2 *Let $\mathbb{B}$ and $\mathbb{C}$ be subalgebras of $\mathbb{A}$ that contain the center $\mathbb{F}$. If $\alpha: \mathbb{B} \longrightarrow \mathbb{C}$ is an algebra isomorphism which is the identity when restricted to $\mathbb{F}$ then there is $\beta \in \mathbb{A}$ so that $\alpha(b) = \beta b \beta^{-1}$ for all $b \in \mathbb{B}$.*

Example : Twisted Group Rings .

Suppose that $\mathbb{F} \subset \mathbb{K}$ is a finite Galois extension, so, from the fundamental theorem of Galois theory,

$$\dim_{\mathbb{F}}(\mathbb{K}) = |\mathrm{Gal}(\mathbb{K}/\mathbb{F})| = n < \infty .$$

Define the algebra $\mathbb{K} \odot \mathrm{Gal}(\mathbb{K}/\mathbb{F})$ as the $\mathbb{K}$-vector space with basis $\mathrm{Gal}(\mathbb{K}/\mathbb{F})$ and multiplication on basis elements given by the rule $kgk'g' = kg[k']gg'$ and extend to linear combinations via the distributive law. Then we have

Lemma 8.3 $\mathbb{K} \odot \mathrm{Gal}(\mathbb{K}/\mathbb{F})$ *is a central simple $\mathbb{F}$ algebra.*

Proof. Associativity is formal, as usual. Now we verify that the center of $\mathbb{K} \odot \mathrm{Gal}(\mathbb{K}/\mathbb{F})$ is $\mathbb{F}$. Suppose that $\lambda = \sum k_g g$ is contained in the center with $k_g \neq 0$ for some $g \neq 1$. Then, for all $k \in \mathbb{K}$ we have $k\lambda = \lambda k$ so $kk_g = k_g g[k]$ and so we must have $k = g[k]$ for all $k \in \mathbb{K}^\bullet$. But this is impossible. Hence $k_g = 0$ for $g \neq 1$ and $\lambda \in \mathbb{K}$. Now commuting with $\mathrm{Gal}(\mathbb{K}/\mathbb{F})$ shows that $\lambda \in \mathbb{F}$.

To show that $\mathbb{K} \odot \mathrm{Gal}(\mathbb{K}/\mathbb{F})$ is simple suppose there is a two sided ideal $B \subset \mathbb{K} \odot \mathrm{Gal}(\mathbb{K}/\mathbb{F})$, and let $b \in B$ be a non-zero term, $b = \sum b_g g$, with a minimum number of non-zero coefficients b_g. If this minimal number is 1 then clearly $B = \mathbb{K} \odot \mathrm{Gal}(\mathbb{K}/\mathbb{F})$. On the other hand if it is greater than 1 we can assume that $b = 1 + \sum_{g \neq 1} b_g g$ with the second summand non-zero. Suppose $b_g \neq 0$ in this second summand. Then there is a $k \in \mathbb{K}$ so that $k \cdot g[k]^{-1} \neq 1$. Now consider $w = b - kbk^{-1}$. Clearly $w \neq 0$ and $w = \sum (1 - kg[k]^{-1}) b_g g$ is contained in the ideal B. But the number of non-zero coefficients in w is clearly less than the number for b. □

In fact we have

Theorem 8.4 *Under the assumptions above, the central simple $\mathbb{F}$-algebra $\mathbb{K} \odot \mathrm{Gal}(\mathbb{K}/\mathbb{F})$ is the matrix ring $M_n(\mathbb{F})$.*

Proof. For $\lambda = \sum b_g g \in K \odot \mathrm{Gal}(\mathbb{K}/\mathbb{F})$, define $\lambda: \mathbb{K} \longrightarrow \mathbb{K}$ by $\lambda(k) = \sum b_g g[k]$. Clearly, λ is an $\mathbb{F}$-linear map of $\mathbb{K}$, thought of as a $\mathbb{F}$-vector space, into itself.

This correspondence is non-trivial and defines a unitary algebra homomorphism $\mathbb{K} \odot \mathrm{Gal}(\mathbb{K}/\mathbb{F}) \longrightarrow M_n(\mathbb{F})$. Since $\mathbb{K} \odot \mathrm{Gal}(\mathbb{K}/\mathbb{F})$ is simple this map must be an embedding, and now, a dimension count shows that it is also an isomorphism. □

Example : Cyclic algebras .

Suppose as above that $\mathbb{F} \subset \mathbb{K}$ and $\mathbb{K}$ is a finite Galois extension of $\mathbb{F}$. Also, assume that $\mathrm{Gal}(\mathbb{K}/\mathbb{F}) = \mathbb{Z}/n$, a cyclic group with generator T. Then we can define an algebra $\mathbb{A}(\mathbb{K}, T, \kappa)$ for any non-zero $\kappa \in \mathbb{F}$ as the $\mathbb{K}$-vector space with basis $\mathrm{Gal}(\mathbb{K}/\mathbb{F})$, multiplication given by $Tk = T(k)T$, and also $T^n = \kappa$. For example $\mathbb{A}(\mathbb{C}, T, -1) = \mathbb{H}$, the quaternions, where T is conjugation.

An obvious modification of the proof of (8.3) gives

Lemma 8.5 $\mathbb{A}(\mathbb{K}, T, \kappa)$ *is a central simple* $\mathbb{F}$*-algebra.*

But these examples are, as the quaternions show, not necessarily just matrix rings.

Lemma 8.6 $\mathbb{A}(\mathbb{K}, T, \kappa) \cong \mathbb{A}(\mathbb{K}, T, \kappa')$ *if there is a* $k \in \mathbb{K}$ *so that*

$$\kappa' = (kT[k]T^2[k] \cdots T^{n-1}[k])\kappa \, .$$

(Indeed, if we replace T by kT then $(kT)^n = \kappa'$ where κ' has the form above.)

Remark. The map $N: \mathbb{K}^\bullet \longrightarrow \mathbb{F}^\bullet$ defined by $N(k) = \prod_{i=0}^{n-1} T^i[k]$ is a multiplicative homomorphism and is generally called the *norm* map.

For example, the norm map $N: \mathbb{C} \longrightarrow \mathbb{R}$ has image the set of all positive reals, and hence the number of possible distinct algebras of the type $\mathbb{A}(\mathbb{C}, T, \kappa)$ where T is conjugation is exactly two, with $\mathbb{A}(\mathbb{C}, T, 1) \cong M_2(\mathbb{R})$ and $\mathbb{A}(\mathbb{C}, T, -1) = \mathbb{H}$.

Tensor Products of Central Simple $\mathbb{F}$-Algebras

The tensor product $\mathbb{A} \otimes_{\mathbb{F}} \mathbb{B}$ of two central simple $\mathbb{F}$-algebras is the quotient of the free abelian group generated by the elements of the cartesian product $\mathbb{A} \times \mathbb{B}$ by factoring out the subgroup generated by the relations

$$\left\{ \begin{array}{l|l} (a, b + b') - (a, b) - (a, b') & \\ (a + a', b) - (a, b) - (a', b) & \text{for all } a,\ a' \in \mathbb{A},\ b,\ b' \in \mathbb{B},\ f \in \mathbb{F} \\ (fa, b) - (a, fb) & \end{array} \right\}$$

We write $a \otimes b$ for the equivalence class of (a,b) in $\mathbb{A} \otimes_{\mathbb{F}} \mathbb{B}$ and every element of $\mathbb{A} \otimes_{\mathbb{F}} \mathbb{B}$ can be written as a linear combination $\sum a_i \otimes b_i$. $\mathbb{F}$ acts on $\mathbb{A} \otimes_{\mathbb{F}} \mathbb{B}$ by $f(a \otimes b) = fa \otimes b$. If $a_1, \ldots, a_l$ form an $\mathbb{F}$-basis for $\mathbb{A}$ and $b_1, \ldots, b_r$ are an $\mathbb{F}$-basis for $\mathbb{B}$, then, as an $\mathbb{F}$-vector space $\mathbb{A} \otimes_{\mathbb{F}} \mathbb{B}$, has a basis consisting of the elements $a_i \otimes b_j$, $1 \leq i \leq l$, $1 \leq j \leq s$.

$\mathbb{A} \otimes_{\mathbb{F}} \mathbb{B}$ becomes a ring when we define the multiplication by the rule $(a \otimes b)(c \otimes d) = ac \otimes bd$ on generating elements and extend to sums using the distributative law. It is direct to check that this operation is well defined.

Here are some standard results about tensor products of central simple $\mathbb{F}$-algebras.

Theorem 8.7 *Let $\mathbb{A}$ and $\mathbb{B}$ be central simple $\mathbb{F}$-algebras, then the tensor product algebra $\mathbb{A} \otimes_{\mathbb{F}} \mathbb{B}$ is also a central simple $\mathbb{F}$-algebra.*

Proof. To check that $\mathbb{F}$ is the center of $\mathbb{A} \otimes_{\mathbb{F}} \mathbb{B}$ consider an element $\lambda = \sum a_i \otimes b_i$ in the center. Write it $\lambda = \sum \theta_i \otimes b_i$ where the b_i now run over a fixed basis of $\mathbb{B}$ and commute it with elements of the form $c \otimes 1$. Since the b_i are independent it follows that $c\theta_i = \theta_i c$ for all $c \in \mathbb{A}$ and $\theta_i \in \mathbb{F}$ for each i. Consequently $\lambda = 1 \otimes b$, and commuting with elements of $\mathbb{B}$, we see that b is in the center of $\mathbb{B}$. The verification that $\mathbb{A} \otimes_{\mathbb{F}} \mathbb{B}$ is simple is similar to the argument in (8.3). Write the elements in the ideal $\mathcal{I}$ in the form $\sum \theta_i \otimes b_i$ and look for an element with a minimal number of non-zero entries θ_i. Then, using the fact that $\mathbb{A}$ is simple we can choose such an element so that it has the form

$$1 \otimes b_i + \sum_{j \neq i} \theta_j \otimes b_j \; .$$

Now conjugate by units, $c \otimes 1$ and reduce the number of non-zero θ_j's. □

A mild, but useful variant of (8.7) occurs in (8.8). Its proof is the same as that of (8.7).

Theorem 8.8 *Let $\mathbb{A}$ be a central simple $\mathbb{F}$-algebra, and suppose that $\mathbb{K}$ is a finite field extension of $\mathbb{F}$, then $\mathbb{A} \otimes_{\mathbb{F}} \mathbb{K}$ is a central simple $\mathbb{K}$-algebra.*

The next result involves the structure of maximal subfields of a central simple division algebra, D. These are the basic tool in analyzing the structure of all central simple algebras.

Theorem 8.9 *Let D be a central simple $\mathbb{F}$-division algebra. Let $\mathbb{K}$ be a maximal subfield of D containing $\mathbb{F}$. Then $D \otimes_{\mathbb{F}} \mathbb{K} = M_n(\mathbb{K})$ where $n = \dim_{\mathbb{F}}(\mathbb{K})$. In particular, $\dim_{\mathbb{F}}(D) = n^2$ and, if $\mathbb{F} \subset \mathbb{K} \subset D$ then $\mathbb{K}$ is a maximal subfield of D if and only if $\dim_{\mathbb{F}}(\mathbb{K}) = n$.*

These theorems are discussed in standard textbooks. For example [Hu], where these results and proofs are given on pp. 450-463. Another good source is [P]. A classic reference is [ANT].

Corollary 8.10 *Let D be a central simple division algebra over the field $\mathbb{F}$. Then there is a finite Galois extension $\mathbb{K}$ of $\mathbb{F}$ with Galois group $G = \mathrm{Gal}(\mathbb{K}/\mathbb{F})$ and a representation of $M_m(D)$, for an appropriate m, as an algebra of the form*

$$\mathbb{A} = \mathbb{K}1 \oplus \mathbb{K}\bar{g}_2 \oplus \cdots \oplus \mathbb{K}\bar{g}_n$$

where the $g_i \in G$ and $g_i k = g_i[k]g_i$, but for each relation $R = \prod g_i^{a_i} = 1$ which holds in G, the relation in $\mathbb{A}$ takes the form

$$\prod \bar{g}_i^{a_i} = k_R \in \mathbb{K}$$

for appropriate $k_R \in \mathbb{K}$.

Proof. First find $\mathbb{K}_1$ with $\mathbb{F} \subset \mathbb{K}_1$ and $\mathbb{K}_1$ maximal in D, and a further extension, $\mathbb{K}$ of $\mathbb{K}_1$, $\mathbb{F} \subset \mathbb{K}_1 \subset \mathbb{K}$, so that $\mathbb{K}$ is a finite Galois extension of $\mathbb{F}$. Then, if $m = deg_{\mathbb{K}_1}(\mathbb{K})$ we have that $\mathbb{K} \subset M_m(D)$ is a maximal subfield, and, by the Skolem-Noether theorem, there are elements $g_i \in M_m(D)$ which act on $\mathbb{K}$ as $\mathrm{Gal}(\mathbb{K}/\mathbb{F})$. By the independence of the elements of the Galois group these elements are independent in $M_m(D)$, regarded as a left $\mathbb{K}$-vector space. Hence we have constructed an $(mn_1)^2$ dimensional sub $\mathbb{F}$-vector space in $M_m(D)$, where $n_1 = \dim_{\mathbb{F}}(\mathbb{K}_1)$. By a dimension count this subspace must be the entire algebra. □

The Cohomological Interpretation of Central Simple Division Algebras

Suppose that we have $\mathbb{F} \subset \mathbb{K} \subset \mathbb{A}$ and $\mathbb{K}$ is both maximal and Galois over $\mathbb{F}$. Also suppose that $\mathbb{K}$ is its own centralizer in $\mathbb{A}$. Then we can write

$$\mathbb{A} = \mathbb{K}1 \oplus \mathbb{K}\bar{g}_2 \oplus \cdots \oplus \mathbb{K}\bar{g}_n$$

where $g_i \in \mathrm{Gal}(\mathbb{K}/\mathbb{F}) = G$ as above. We associate to this decomposition a map $\phi_{\mathbb{A}}: G \times G \longrightarrow \mathbb{K}^\bullet$ where $\mathbb{K}^\bullet$ denotes the units in $\mathbb{K}$, by

$$(g_1, g_2) \mapsto \bar{g}_2^{-1}\bar{g}_1^{-1}(\overline{g_1 g_2}) \in \mathbb{A} .$$

But since $g_1^{-1}g_2^{-1}(g_1 g_2)$ centralizes $\mathbb{K}$ it follows that the product is in $\mathbb{K}$ and we have our association.

Lemma 8.11 *The map $\phi_{\mathbb{A}}$ satisfies $\delta(\phi_{\mathbb{A}}) = 1$.*

Proof. The boundary

$$\begin{aligned}\delta(\phi_{\mathbb{A}})(g_1, g_2, g_3) &= \bar{g}_3^{-1}[\phi_{\mathbb{A}}(g_1, g_2)]\phi_{\mathbb{A}}(g_1 g_2, g_3)\phi_{\mathbb{A}}(g_1, g_2 g_3)^{-1}\phi_{\mathbb{A}}(g_2, g_3)^{-1} \\ &= \bar{g}_3^{-1}\{\bar{g}_2^{-1}\bar{g}_1^{-1}(\overline{g_1 g_2})\}\bar{g}_3\{\bar{g}_3^{-1}(\overline{g_1 g_2})^{-1}(\overline{g_1 g_2 g_3})\} \\ &\quad \cdot \{(\overline{g_1 g_2 g_3})^{-1}\bar{g}_1(\overline{g_2 g_3})\}\{(\overline{g_2 g_3})^{-1}\bar{g}_2\bar{g}_3\}\end{aligned}$$

after expanding out, and this directly cancels to 1. □

Definition 8.12 Let Φ: $\mathrm{Gal}(\mathbb{K}/\mathbb{F})^2 \longrightarrow \mathbb{K}^\bullet$ satisfy the two conditions

(8.13)
1. $\Phi(1,g) = \Phi(g,1) = 1$ for all $g \in \mathrm{Gal}(\mathbb{K}/\mathbb{F})$,
2. $\delta(\Phi)(g_1,g_2,g_3) = 1$ for all triples $(g_1,g_2,g_3) \in \mathrm{Gal}(\mathbb{K}/\mathbb{F})^3$.

Then $\mathbb{A}(\mathbb{K},\mathrm{Gal}(\mathbb{K}/\mathbb{F}),\Phi)$ is the vector space $\mathbb{K}1 \oplus \mathbb{K}g_2 \oplus \cdots \oplus \mathbb{K}g_n$ where the g_i run over the elements of $\mathrm{Gal}(\mathbb{K}/\mathbb{F})$. $\mathbb{A}(\mathbb{K},\mathrm{Gal}(\mathbb{K}/\mathbb{F}),\Phi)$ has an $\mathbb{F}$-linear multiplication $\mathbb{A} \otimes_{\mathbb{F}} \mathbb{A} \longrightarrow \mathbb{A}$ defined by the rules

3. $g_i k \;=\; g_i[k] g_i$,
4. $(g_i)(g_j) \;=\; (g_i g_j)\Phi(g_i,g_j)^{-1}$ where (g) is the basis element corresponding to g.

This construction defines a central simple $\mathbb{F}$-algebra as we see from the next result.

Lemma 8.14 *Assume the triple* $(\mathbb{K},\mathrm{Gal}(\mathbb{K}/\mathbb{F}),\Phi)$ *satisfies the conditions of (8.12). Then we have*

1. $\mathbb{A}(\mathbb{K},\mathrm{Gal}(\mathbb{K}/\mathbb{F}),\Phi)$ *is a central simple* $\mathbb{F}$*-algebra of dimension* n^2 *over* $\mathbb{F}$ *where* $n = |\mathrm{Gal}(\mathbb{K}/\mathbb{F})|$ *and the centralizer of* $\mathbb{K}$ *in* $\mathbb{A}$ *is* $\mathbb{K}$,
2. *Any central simple* $\mathbb{F}$*-algebra,* $\mathbb{B}$*, of dimension* n^2 *over* $\mathbb{F}$ *with* $\mathbb{K}$ *as a maximal subfield has the form* $\mathbb{A}(\mathbb{K},\mathrm{Gal}(\mathbb{K}/\mathbb{F}),\Phi)$ *with* Φ *as in (8.12),*
3. $\mathbb{A}(\mathbb{K},\mathrm{Gal}(\mathbb{K}/\mathbb{F}),\Phi) \cong \mathbb{A}(\mathbb{K},\mathrm{Gal}(\mathbb{K}/\mathbb{F}),\Psi)$ *if and only if there is a* $\tau\colon G \longrightarrow \mathbb{K}^\bullet$ *with* $\tau(1) = 1$ *and*

$$\Phi(g_1,g_2)\Psi(g_1,g_2)^{-1} \;=\; g_2^{-1}[\tau(g_1)]\tau(g_2)\tau(g_1g_2)^{-1}$$

for all $(g_1,g_2) \in \mathrm{Gal}(\mathbb{K}/\mathbb{F})^2$.

Proof. (1). (8.12.1) assures that $1 \in \mathrm{Gal}(\mathbb{K}/\mathbb{F})$ is the identity for this multiplication and (8.12.2) assures that the multiplication is associative. Next, using essentially the same argument as in the proof of (8.3) or (8.7) one formally checks that the resulting algebra is a central simple $\mathbb{F}$-algebra. Finally, the same argument can be used to show that $\mathbb{K}$ is its own centralizer in $\mathbb{A}$.
(2). The argument here is very similar to the proof of (8.10). There are elements $\bar{g}_i \in \mathbb{B}$ so that $\bar{g}_i k \bar{g}_i^{-1} = g_i[k]$ for each element $g_i \in \mathrm{Gal}(\mathbb{K}/\mathbb{F})$. Moreover, by the independence of the elements of $\mathrm{Gal}(\mathbb{K}/\mathbb{F})$, the $\bar{g}_i$ are linearly independent in $\mathbb{B}$, thought of as a vector space over $\mathbb{K}$. A dimension count now shows that we have

$$\mathbb{B} \;=\; \mathbb{K}1 \oplus \mathbb{K}\bar{g}_2 \oplus \cdots \oplus \mathbb{K}\bar{g}_n\,.$$

Then the discussion preceeding (8.11) shows that $\Phi \;=\; \phi_{\mathbb{A}}$, and $\mathbb{B} \;=\; \mathbb{A}(\mathbb{K},\mathrm{Gal}(\mathbb{K}/\mathbb{F}),\Phi)$.

(3). The elements $\bar{g}_i$ are not well defined from their action on

$$\mathbb{K} \subset \mathbb{A}(\mathbb{K}, \mathrm{Gal}(\mathbb{K}/\mathbb{F}), \Phi)$$

since $\bar{g}_i k_i$ has exactly the same action on $\mathbb{K}$ as does $\bar{g}_i$. However, any other element, λ, which acts on $\mathbb{K}$ in the same way as $\bar{g}$ satisfies the condition that $\lambda^{-1}\bar{g}$ centralizes $\mathbb{K}$, and hence, since $\mathbb{K}$ is its own centralizer, this is the only indeterminacy which is allowed.

Consequently, given $\mathbb{K}$, two different choices of the elements $g_i \in \mathbb{A}$ lead to the following change in $\phi_{\mathbb{A}}$, $\phi'_{\mathbb{A}} = \phi_{\mathbb{A}} g_2^{-1}[k_{g_1}^{-1}]k_{g_2}^{-1}k_{g_1g_2}$, or the same,

$$\phi_{\mathbb{A}}(g_1, g_2)\phi'_{\mathbb{A}}(g_1, g_2)^{-1} = g_2^{-1}[k_{g_1}^{-1}]k_{g_2}^{-1}k_{(g_1g_2)} .$$

Likewise, if we have $\Phi(g_1, g_2)\Psi(g_1, g_2)^{-1} = g_2^{-1}[\tau(g_1)]\tau(g_2)\tau(g_1g_2)^{-1}$ then the resulting algebras are isomorphic.

Suppose now that $\mathbb{A}(\mathbb{K}, \mathrm{Gal}(\mathbb{K}/\mathbb{F}), \Phi) \cong \mathbb{A}(\mathbb{K}, \mathrm{Gal}(\mathbb{K}/\mathbb{F}), \Psi)$. Then, by the Noether-Skolem theorem there is an isomorphism

$$\varphi\colon \mathbb{A}(\mathbb{K}, \mathrm{Gal}(\mathbb{K}/\mathbb{F}), \Phi) \longrightarrow \mathbb{A}(\mathbb{K}, \mathrm{Gal}(\mathbb{K}/\mathbb{F}), \Psi)$$

so that φ maps $\mathbb{K}$ to $\mathbb{K}$ and is, in fact, the identity when restricted to $\mathbb{K}$. But then, as pointed out above, $\varphi(g_i)$ must have the form $g'_i k_i$ for each $g_i \in \mathrm{Gal}(\mathbb{K}/\mathbb{F})$. □

Example. The class $\phi_{\mathbb{A}}$ for $\mathbb{K} \odot \mathrm{Gal}(\mathbb{K}/\mathbb{F})$ is $\phi_{\mathbb{A}}(g_1, g_2) \equiv 1$ for all pairs (g_1, g_2), and this clearly represents the trivial element in the group $H^2_\phi(\mathrm{Gal}(\mathbb{K}/\mathbb{F}); \mathbb{K}^\bullet)$. Shortly we will introduce the Brauer group $B(\mathbb{F})$ and describe the relation between these groups $H^2_\phi(\mathrm{Gal}(\mathbb{K}/\mathbb{F}), \mathbb{K}^\bullet)$ and $B(\mathbb{F})$. Additionally this twisted cohomology group plays a very basic role in class field theory.

Example. Let $\mathbb{A}$ be central simple over the finite field $\mathbb{F}$. Then any finite extension $\mathbb{K}$ of $\mathbb{F}$ is cyclic. Also, it is well known that for finite fields the norm map $\mathbb{K}^\bullet \longrightarrow \mathbb{F}^\bullet$ is surjective. Indeed, suppose for explicitness that $|\mathbb{F}| = p^r$, $|\mathbb{K}| = p^{rn}$. Then a generator for $\mathrm{Gal}(\mathbb{K}/\mathbb{F})$ is the map $x \mapsto x^{p^r}$. Hence the norm map has the form $x \mapsto x^t$ where $t = 1 + p^r + p^{2r} + \cdots p^{r(n-1)}$. But $p^{rn} - 1 = (p^r - 1)t$, so, in the cyclic group $\mathbb{Z}/(p^{nr} - 1)$ raising elements to the power t is a surjective homomorphism onto the cyclic subgroup $\mathbb{Z}/(p^r - 1)$.

It follows that when $\mathbb{F}$ is a finite field any central simple division algebra D has a maximal subfield $\mathbb{K}$ with cyclic Galois group. But we have already seen that, in this case, $\mathbb{A}(\mathbb{K}, T, \kappa) \simeq \mathbb{A}(\mathbb{K}, T, N(k)\kappa)$ so $\mathbb{A}(\mathbb{K}, T, \kappa) \simeq \mathbb{A}(\mathbb{K}, T, 1) = M_n(\mathbb{F})$ and there are no non-commutative central simple $\mathbb{F}$-algebras. On the other hand the discussion above shows that $H^2_\phi(\mathrm{Gal}(\mathbb{K}/\mathbb{F}); \mathbb{K}^\bullet)$ corresponds to the central simple $\mathbb{F}$ algebras which contain $\mathbb{K}$ as a maximal subfield, so $H^2_\phi(\mathrm{Gal}(\mathbb{K}/\mathbb{F}); \mathbb{K}^\bullet) = 0$ when $\mathbb{F}$ is a finite field.

But for a fixed $\mathbb{A}$ this is as far as we can go. It can well happen that there are non-isomorphic maximal subfields of $\mathbb{A}$ which are Galois and contain $\mathbb{F}$. For example $M_2(\mathbb{Q})$ contains any quadratic extension of $\mathbb{Q}$ as a maximal subfield.

Comparing Different Maximal Subfields, the Brauer Group

To continue further we must compare the algebras $\mathbb{A}(\mathbb{K}, \text{Gal}(\mathbb{K}/\mathbb{F}), \Phi)$ and $\mathbb{A}(\mathbb{K}', \text{Gal}(\mathbb{K}'/\mathbb{F}), \Psi)$ for different $\mathbb{K}$ and $\mathbb{K}'$. To do this we tensor $\mathbb{A}$ with $M_n(\mathbb{F})$. By the Wedderburn theorem, this does not change the underlying division algebra, D. But this enlargement does allow us to compare the distinct $\mathbb{K}$'s since we can use the composite field, $\mathbb{K}\mathbb{K}'$, as a maximal subfield of $\mathbb{A} \otimes_{\mathbb{F}} M_n(\mathbb{F})$ for an appropriate n.

We begin by determining the class $\phi_{\mathbb{A}\otimes M_n(\mathbb{F})}$. Assume that $\mathbb{K}_1$ is a degree n Galois extension of $\mathbb{K}$ so that $\mathbb{K}_1$ is Galois over $\mathbb{F}$. Since $M_n(\mathbb{K}) \subset \mathbb{A}\otimes_{\mathbb{F}} M_n(\mathbb{F})$ we see that $\mathbb{K}_1$ is a maximal subfield in $\mathbb{A} \otimes_{\mathbb{F}} M_n(\mathbb{F})$, and $\text{Gal}(\mathbb{K}_1/\mathbb{F})$ is given as a normal extension

$$\text{Gal}(\mathbb{K}_1/\mathbb{K}) \overset{\triangleleft}{\longrightarrow} \text{Gal}(\mathbb{K}_1/\mathbb{F}) \overset{\pi}{\longrightarrow} \text{Gal}(\mathbb{K}/\mathbb{F}) .$$

Then, we see that $\phi_{\mathbb{A}\otimes M_n(\mathbb{F})}\colon \text{Gal}(\mathbb{K}_1/\mathbb{F})^2 \longrightarrow \mathbb{K}_1^\bullet$ is defined by

$$\phi_{\mathbb{A}\otimes M_n(\mathbb{F})}(g_1, g_2) = \phi_{\mathbb{A}}(\pi(g_1), \pi(g_2))$$

since $g_2^{-1}n_2^{-1}g_1^{-1}n_1^{-1}(n_1g_1n_2g_2) = g_2^{-1}g_1^{-1}(g_1g_2)$ when we make appropriate choices for the lifting classes. We may formulate this in terms of a homomorphism

$$H^2_\phi(\text{Gal}(\mathbb{K}/\mathbb{F}); \mathbb{K}^\bullet) \overset{inf}{\longrightarrow} H^2_\phi(\text{Gal}(\mathbb{K}_1/\mathbb{F}); \mathbb{K}_1^\bullet) \tag{8.15}$$

called the inflation map, which decomposes as the composition of two maps, first the map $H^2_\phi(\text{Gal}(\mathbb{K}/\mathbb{F}); \mathbb{K}^\bullet) \overset{\text{res}}{\longrightarrow} H^2_\phi(\text{Gal}(\mathbb{K}_1/\mathbb{F}; \mathbb{K}^\bullet)$ defined via a map on the $\mathcal{C}^i$'s,

$$\text{res}^{\#}\colon\ \mathcal{C}^i \longrightarrow (\mathcal{C}')^i \text{ where } \text{res}^{\#}(f)(g_1, \ldots, g_i) = f(\pi(g_1), \ldots, \pi(g_i)) .$$

It is routine to check that $\text{res}^{\#}\delta(f) = \delta(\text{res}^{\#}(f))$. Consequently, $\text{res}^{\#}$ maps $\mathbf{Z}^i$ for the first $\mathcal{C}^i$'s to the $\mathbf{Z}^i$'s for the second, and similarly for the $\mathbf{B}^i$'s. Thus we get a well defined map of quotients.

The second map is induced by including $\mathbb{K}^\bullet$ as a multiplicative subgroup of $\mathbb{K}_1^\bullet$ and has the form $i\colon H^2_\phi(\text{Gal}(\mathbb{K}_1/\mathbb{F}); \mathbb{K}^\bullet) \longrightarrow H^2_\phi(\text{Gal}(\mathbb{K}_1/\mathbb{F}); \mathbb{K}_1^\bullet)$.

Now, suppose that $\mathbb{K}$ and $\mathbb{K}'$ are maximal subfields in $\mathbb{A}$ which contain $\mathbb{F}$, both are Galois over $\mathbb{F}$, and both are their own centralizers. Then we have that the composite field $\mathbb{K}\mathbb{K}' = \mathbb{K}_1$ is Galois over $\mathbb{K}$, $\mathbb{K}'$ and $\mathbb{F}$. Hence, setting $n = \dim_{\mathbb{K}}(\mathbb{K}_1)$ we get that the inclusion $\mathbb{A} \subset \mathbb{A}\otimes_{\mathbb{F}} M_n(\mathbb{F})$ allows us to use the argument above on both the representative of $\mathbb{A}$ in $H^2_\phi(\text{Gal}(\mathbb{K}/\mathbb{F}); \mathbb{K}^\bullet)$ and the representative in $H^2_\phi(\text{Gal}(\mathbb{K}'/\mathbb{F}); (\mathbb{K}')^\bullet)$. We get the diagram

$$\begin{array}{ccc} H^2_\phi(\text{Gal}(\mathbb{K}/\mathbb{F}); \mathbb{K}^\bullet) & \overset{inf}{\longrightarrow} & H^2_\phi(\text{Gal}(\mathbb{K}_1/\mathbb{F}); \mathbb{K}_1^\bullet) \\ & & \Big\uparrow {\scriptstyle inf} \\ & & H^2_\phi(\text{Gal}(\mathbb{K}'/\mathbb{F}); (\mathbb{K}')^\bullet) \end{array}$$

and each inflation map inf takes the class corresponding to $\mathbb{A}$ to the class corresponding to $\mathbb{A} \otimes_{\mathbb{F}} M_n(\mathbb{F})$.

Now, note that $M_r(D) \otimes_{\mathbb{F}} M_s(\mathbb{F}) = M_{rs}(D)$, so that passing to limits over inflation maps we get that every central simple division algebra over $\mathbb{F}$ is represented by a unique class in the abelian group

$$B(\mathbb{F}) = \lim_{\mathbb{K}} H^2_{\phi}(\mathrm{Gal}(\mathbb{K}/\mathbb{F}); \mathbb{K}^{\bullet}) .$$

We write this limit as $H^2(\mathbb{F})$, and it is commonly called the second cohomology group of the Galois group of the separable algebraic closure of $\mathbb{F}$ with coefficients in the multiplicative group of this algebraic closure, (separable since we only use Galois extensions). But we emphasize that, in actuality, it is the limit over finite Galois extensions, $\mathbb{F} \subset \mathbb{K}$, of the H^2's above.

Definition 8.16 The group $B(\mathbb{F})$ defined above is called the *Brauer group* of the field $\mathbb{F}$. Its elements are in one to one correspondence with the distinct isomorphism classes of central simple $\mathbb{F}$-division algebras.

Remark. If $\alpha \in B(\mathbb{F})$ corresponds to D and β corresponds to D' then the class $\alpha + \beta$ corresponds to the division algebra in the simple algebra $D \otimes D'$. (Check explicit representative classes for the generating classes $\phi_{\mathbb{B}}$, and note that $M_s(D) \otimes_{\mathbb{F}} M_r(D') = M_{rs}(D \otimes_{\mathbb{F}} D')$.)

Remark. A more leisurely discussion of the topics in this section can be found in the book [P] pp. 250-270.

Chapter II. Classifying Spaces and Group Cohomology

0. Introduction

This is one of the basic chapters in the book. We start in §1, §2, with preliminaries from topology. The results reviewed in §1 on the basic structure and properties of classifying spaces are essential throughout the remainder of the text. The material in §2 on the Steenrod algebra is not needed in the rest of this chapter and is placed here only for continuity. It is used, however, in Chapter III, §3, and, from then on, more and more frequently throughout the book. Here we only review the basic facts and give Steenrod's axiomatic treatment of the pth power operations. However, in Chapter IV, §7 we will use the cohomology of groups to provide the construction of the Steenrod operations. In §3 we give the definition of group cohomology and devote the remainder of this chapter, with the exception of §8 to basic facts and techniques for calculating these cohomology groups, particularly for finite groups. §8 gives a nice application of these ideas to construct non-trivial outer automorphisms for p-groups.

The reader will notice that this chapter is very different from the preceding one, in that it assumes a considerable background in algebraic topology. This is an inevitable consequence of the historical development described in the introduction. However, there are some excellent sources for this material available in the literature, notably [Sp] for general background in homotopy theory and [MS] for bundle theory. Rather than burden the reader with full details on what we need, we have on occasion opted for introducing material as we go along, hoping that the reader will look into it on his own. The main reason for doing this is to afford a quick route into interesting aspects of group cohomology, avoiding potentially soporific generalities whenever possible.

1. Preliminaries on Classifying Spaces

For the purposes of the material in this chapter we will use the following conventions.

1. X a topological space will always mean X is a CW complex, probably a simplicial complex, with the compactly generated topology. (A subset

$U \subset X$ is open in the compactly generated topology if and only if $U \cap C$ is open in C for every compact subset $C \subset X$.)

2. $X \times Y$ will mean the Cartesian product of the spaces X and Y with the compactly generated topology.
3. Let G be a topological group; the unit $e \in G$ will always be the distinguished base point. We will also assume that e is regular, i.e. e has a regular neighborhood, but more, we will assume (G, e) is an NDR pair. This means there is a neighborhood N of e and a homotopy of the identity, H_t, so that $H_t(N) \subset N$, $H_t(e) = e$ for all t, $0 \leq t \leq 1$, and $H_1(N) = e$. This is a technical condition which is trivially satisfied if, for example G is discrete which is the usual condition we consider, or if G is a CW complex.
4. Assuming that G is a CW complex we will also assume that the multiplication $\mu: G \times G \longrightarrow G$ is cellular. Again, this is a technical condition which is trivially satisfied when G is discrete, since then the points of G are the (0) cells of the CW decomposition of G, and there are no higher dimensional cells.
5. Moreover the same conditions will hold for every subgroup $H \subset G$ which we discuss. Specifically, H will always be a sub-CW complex of G, and the NDR structure will preserve $N \cap H$.

Next we review the following facts about fibre bundles:

1. Let $F \longrightarrow E' \longrightarrow I \times B$ be a fibering. Then $E' = I \times E$ for some unique fibering $F \longrightarrow E \longrightarrow B$ and the map to $I \times B$ preserves the t coordinate in I.
2. Given a fibering $F \longrightarrow E \xrightarrow{\pi} B$ and a map $f: X \longrightarrow B$, there is an induced bundle, denoted $f^!(E)$ which is the subspace of $X \times E$ consisting of the set of pairs (x, y) with $\pi(y) = f(x)$, and we have the commutative diagram

$$
\begin{array}{ccc}
F & \longrightarrow & F \\
\downarrow & & \downarrow \\
f^!(E) & \xrightarrow{p_2} & E \\
\downarrow{\scriptstyle p_1} & & \downarrow{\scriptstyle \pi} \\
X & \xrightarrow{f} & B
\end{array}
$$

Given a fibre bundle $F \longrightarrow E \longrightarrow B$ with group $G \hookrightarrow \operatorname{Homeo}(F)$, where $\operatorname{Homeo}(F)$ is the group of homeomorphisms of F with the compact-open topology, we can form the *associated principal bundle*, $\operatorname{Prin}(E)$, by replacing the fiber F by G and using the *same* transition functions as previously in measuring the change of product structures as one moves from chart to chart. (Recall that the group G of the bundle is the subgroup of $\operatorname{Homeo}(F)$ generated by these transition functions.) Since the transition functions act from the right on G, G is free to act from the left, and we have a fiber preserving action

$$G \times \text{Prin}(E) \longrightarrow E$$

which is transitive on fibers. Conversely, given a principal G bundle $G \to E \to B$ and a right action of G on a space F we can form the *associated* F *bundle* $F \times_G E \to B$, $(f, y) \sim (fg, y) \sim (f, gy)$ for all $g \in G$. We have the theorem of Milnor,

Theorem 1.1 *Given a group G there exists a space B_G and a G-bundle with total space E_G,*

$$G \longrightarrow E_G \longrightarrow B_G \ ,$$

so that if $G \to E \to X$ is any principal G-bundle over X there is a unique homotopy class of maps $f: X \to B_G$ so that $f^!(E_G) = E$.

In particular, this implies that B_G is unique up to homotopy type. If B_G and B'_G are classifying spaces for G there must be maps $f: B_G \to B'_G$ inducing E_G and $g: B'_G \to B_G$ inducing E'_G. Then the composite $fg: B_G \to B_G$ induces E_G so must be homotopic to the identity, and similarly for $gf: B'_G \to B'_G$.

This space B_G is called the *classifying space* for G and has the following basic property

$$[X, B_G] = \{\text{the set of isomorphism classes of principal } G \text{ bundles over } X\}$$

known as the Steenrod recognition principle which gives a quite explicit method of constructing and recognizing the classifying space for G.

Theorem 1.2 *$G \to E \to B$ is a classifying space if and only if E is contractible. If $E \to B$ is a principal G bundle with E n-connected, then the classifying map $f: B \to B_G$ is a homotopy equivalence through dimension n.*

Remark 1.3 If G is discrete, the exact homotopy sequence of the fibration shows that

$$\pi_i(B_G) = \begin{cases} 0 & i > 1 \\ G & i = 1 \end{cases} .$$

A space having a non-zero homotopy group in a single dimension is called an Eilenberg-MacLane space. Such a space is unique up to homotopy equivalence and one exists for any (discrete) group G provided the dimension is 0 or 1, or the dimension is greater than one but G is Abelian. If the group is G and the dimension n, the Eilenberg-MacLane space is usually denoted $K(G, n)$. We will shortly introduce an explicit construction of the spaces E_G and B_G which will make the construction of the $K(G, n)$ very easy.

Corollary 1.4 *A model for the classifying space for the product $G \times H$ of two groups, $B_{G \times H}$, can be chosen to be $B_G \times B_H$.*

Proof. A principal $G \times H$ bundle over $B_G \times B_H$ is

$$E_G \times E_H \xrightarrow{\pi_G \times \pi_H} B_G \times B_H$$

but the product of two contractible spaces is itself contractible, so the result follows. □

The set of homotopy classes of maps $X \rightarrow B_G$ which is denoted $[X, B_G]$ is equal to the set of isomorphism classes of principal G-bundles on X when X is a CW-complex, and the isomorphism is explicit, $f^!(E_G)$, the pull-back of the principal G-bundle $E_G \rightarrow B_G$ is defined as the set of pairs $(x, l) \in X \times E_G$ for which $f(x) = p(l)$. This is the bundle over X associated to f.

In the case of the $K(G, n)$ for $n > 1$ the set $[X, K(G, n)] \cong H^n(X; G)$ and the isomorphism is again explicit. Note that $H^n(K(G, n); G) = \mathrm{Hom}(G, G)$ and if we set $\iota = \{\mathrm{id}\}$ then $\{f\} \mapsto f^*(\iota) \in H^n(X; G)$ gives the correspondence. In particular, $K(G, n) \times K(G, n) = K(G \times G, n)$ by (1.4), and $H^n(K(G \times G, n); G) = \mathrm{Hom}(G \times G, G)$ for $n > 1$ so that G is abelian. The element $\iota^+ \in H^n(K(G \times G, n); G)$ is the class corresponding to the sum map $G \times G \xrightarrow{+} G$ and, in the correspondence above between $[X, K(G, n)]$ and $H^n(X; G)$ we have that $\alpha + \beta$ is represented by the composition

1.5 $$X \xrightarrow{\Delta} X \times X \xrightarrow{\alpha \times \beta} K(G, n) \times K(G, n) \xrightarrow{\iota^+} K(G, n).$$

In the case of $K(G, 1)$, $[X, K(G, 1)]$ is the set of conjugacy classes of homomorphisms $\phi: \pi_1(X) \rightarrow G$. In particular $[K(G, 1), K(G, 1)] = \mathrm{Out}(G)$, but generally, for G non-commutative $[X, K(G, 1)]$ is just a set.

Examples 1.6

1. S^1 is the Eilenberg-MacLane space $K(\mathbb{Z}, 1)$, and $E_G \rightarrow S^1$ is the usual map $\mathbb{R} \xrightarrow{exp} S^1$ where $exp(t) = e^{2\pi i t}$.
2. For the group $G = \mathbb{Z}/n$ we have a free action

$$G \times S^{2m-1} \longrightarrow S^{2m-1}$$

defined by $T(z_1, \ldots, z_m) = (\rho_n z_1, \ldots \rho_n z_m)$ where $\rho_n = e^{2\pi i/n}$ is a primitive n^{th}-root of 1. The quotient is the Lens space L_n^{2m-1}. The inclusion $S^{2m-1} \subset S^{2m+1}$ as the set of points $(z_1, \ldots, z_m, 0)$ commutes with the $\mathbb{Z}/n$ action and consequently induces an inclusion of quotients $L_n^{2m-1} \hookrightarrow L_n^{2m+1}$, so, using the Steenrod Recognition theorem and passing to limits we see that $B_{\mathbb{Z}/n} = \lim_{m \mapsto \infty} L_n^{2m-1}$.
3. We have the principal (Hopf)-fiberings $S^1 \rightarrow S^{2m+1} \rightarrow \mathbb{CP}^m$, which, as above, under the inclusion $S^{2m+1} \hookrightarrow S^{2m+3}$ induce inclusions $\mathbb{CP}^m \hookrightarrow \mathbb{CP}^{m+1}$, and passing to limits we have

$$B_{S^1} = \lim_{m \mapsto \infty} \mathbb{CP}^m = \mathbb{CP}^\infty .$$

We now explicitly construct a model for the classifying space B_G. Let σ^n be the usual n dimensional simplex. It is given with barycentric coordinates as the set of all $n+1$-tuples of non-negative real numbers $(t_0, t_1, \ldots, t_n)$ with $\sum_{i=0}^{n} t_1 = 1$. The $n+1$ faces are the sets F_i where $t_i = 0$, $0 \le i \le n$. A more convenient coordinatization for our purposes is given by

Lemma 1.7 $\sigma^n = \{(\tau_1, \ldots, \tau_n) \mid 0 \le \tau_1 \le \tau_2 \le \cdots \le \tau_n \le 1\}$.

Proof. Set $\tau_i = t_0 + t_1 + \cdots + t_{i-1}$. This defines a map from barycentric coordinates to the coordinates in the lemma. Conversely, given $0 \le \tau_1 \le \tau_2 \le \cdots \le \tau_n \le 1$, define $t_0 = \tau_1$, $t_{i-1} = \tau_i - \tau_{i-1}$, $1 < i < n$, and $t_n = 1 - \tau_n$. This defines the inverse map. □

Let $E_G = \coprod_{n=0}^{\infty} G \times \sigma^n \times G^n/(\text{relations})$ where the relations are given as follows.

$$\begin{aligned}
(g, \tau_1, \ldots, \tau_n, g_1, \ldots, g_n) &\sim (g, \tau_1 \ldots, \hat{\tau}_i, \ldots, \hat{g}_i, g_i g_{i+1}, \ldots, g_n) \\
&\quad \text{if } g_i = e \text{ or } \tau_i = \tau_{i+1} \\
&\sim (gg_1, \tau_2, \tau_3, \ldots, \tau_n, g_2, \ldots, g_n) \\
&\quad \text{if } \tau_1 = 0 \\
&\sim (g, \tau_1, \ldots, \tau_{n-1}, g_1, \ldots, g_{n-1}) \\
&\quad \text{if } \tau_n = 1\,.
\end{aligned}$$

Here the $\wedge$ over the coordinate τ_i or g_i means that we simply delete that coordinate. This space E_G admits a free G action, $G \times E_G \rightarrow E_G$, which is defined as

$$(g, (\bar{g}, \tau_1, \ldots, \tau_n, g_1, \ldots, g_n)) \mapsto (g\bar{g}, \tau_1, \ldots, \tau_n, g_1, \ldots, g_n)\,,$$

and then, using the relation, to pass to equivalence classes. It is routine to check that this is well defined and G-free in the quotient. Also E_G is contractible. The contraction is given by

$$H_t((g, \tau_1, \ldots, \tau_n, g_1, \ldots, g_n)) = (e, t, \overline{\tau_1 + t}, \ldots, \overline{\tau_n + t}, g, g_1, \ldots, g_n)$$

on passing to equivalence classes, where $\bar{t} = \begin{cases} t & t \le 1 \\ 1 & t \ge 1\,. \end{cases}$ The quotient space B_G is given explicitly as

$$B_G = \coprod_0^{\infty} \sigma^n \times G^n/(\text{relations})$$

where the relations are those given above, but we set the leading term g equal to a single point, pt. Explicitly

$$\begin{aligned}(\tau_1, \dots, \tau_n, g_1, \dots g_n) \sim\ & (\tau_1 \dots, \hat{\tau}_i, \dots, \hat{g}_i, g_i g_{i+1}, \dots, g_n) \\ & \qquad \text{if } g_i = e \text{ or } \tau_i = \tau_{i+1} \\ & (\tau_2, \tau_3, \dots, \tau_n, g_2, \dots, g_n) \\ & \qquad \text{if } \tau_1 = 0 \\ & (\tau_1, \dots, \tau_{n-1}, g_1, \dots, g_{n-1}) \\ & \qquad \text{if } \tau_n = 1\,.\end{aligned}$$

Of course it must be proved that the quotient map $E_G \longrightarrow B_G$ is a fibration with fiber G, but this is a direct inductive construction. There is a natural filtration $E_G^0 \subset E_G^1 \subset E_G^2 \subset \cdots$ covering a similar filtration $B_G^0 \subset B_G^1 \subset B_G^2 \subset \cdots$ where E_G^i is the image in E_G of $G \times \sigma^i \times G^i$. Using the NDR assumption on $e \in G$ we obtain the existence in B_G^{i+1} of a neighborhood of B_G^i in which B_G^i is a deformation retract $H_t(i)$.

Now, assuming we have product neighborhoods $N_l \in B_G^i$, we easily obtain that in this neighborhood $H_1(i)^{-1}(N_l)$ is also a product neighborhood. On the other hand it is evident that $B_G^{i+1} - B_G^i$ is a product neighborhood (there are no identifications in this region and the original map $G \times \sigma^i \times G^i \longrightarrow \sigma^i \times G^i$ over it was a product). It follows that B_G^{i+1} has an open covering by product neighborhoods, and the inductive step is complete.

The above construction has very good naturality properties. First, if $f: G \longrightarrow H$ is a (continuous) homomorphism then the map E_f defined by $\tilde{E}_f(g, \tau_1, \dots, \tau_n, g_1, \dots, g_n) \mapsto (f(g), \tau_1, \dots, \tau_n, f(g_1), \dots, f(g_n))$ evidently preserves relations and induces maps

$$E_f\colon\ E_G \longrightarrow E_G, \qquad B_f\colon\ B_G \longrightarrow B_H\ ,$$

so that $E_{fg} = E_f E_g$, $B_{fg} = B_f B_g$ for compositions of homomorphisms.

It can also be verified that $B_{G \times H} = B_G \times B_H$, and, if $H \lhd G$ is a normal subgroup then the natural map $B_G \longrightarrow B_{G/H}$ is a fibration with fiber B_H. The following theorem of N. Steenrod gives one of the most important properties which distinguishes this construction of the classifying space from others.

Theorem 1.8 *Let $G = H \times K$, a (compactly generated) product of two groups, then the two projections p_1 and p_2 give a homeomorphism $B_{p_1} \times B_{p_2}: B_G \longrightarrow B_H \times B_K$. Moreover, let $A \subset G$ be a central subgroup. Then the homomorphism $A \times G \longrightarrow G$, $(a, g) \mapsto ag$, induces a map of classifying spaces $B_A \times B_G \longrightarrow B_G$ which is (1) natural, (2) if $G = A$ makes B_A into a commutative topological group, (3) if $G \neq A$ makes B_G into a B_A-space.*

Proof. We construct the inverse map, $\phi: B_H \times B_K \longrightarrow B_{H \times K}$ by

$$\begin{aligned}\{(t_1, \dots, t_n, h_1, \dots, h_n) \times (\tau_1, \dots, \tau_m, k_1, \dots, k_m)\} \mapsto \\ (\lambda_1, \dots, \lambda_{n+m}, \theta_1, \dots, \theta_{n+m})\end{aligned}$$

where the $\lambda_1, \dots, \lambda_{n+m}$ are the terms in the set $(t_1, \dots, t_n, \tau_1, \dots, \tau_m)$ arranged in increasing order, and

$$\theta_i = \begin{cases} (h_s, e_K) & \text{if } \lambda_i = t_s, \\ (e_H, k_v) & \text{if } \lambda_i = \tau_v \ . \end{cases}$$

This identification is ambiguous only when $t_i = \tau_j$ for some i, j. But in this case, since $(e, k_j)(h_i, e) = (h_i, e)(e, k_j)$ the equivalence relations remove the ambiguity. Moreover, when we decompose the product $\sigma^n \times \sigma^m$ into regions where a given (n, m)-shuffle in $\mathcal{S}_{n+m}$ puts the terms $(t_1, \ldots, t_n, \tau_1, \ldots, \tau_m)$ into ascending order, this gives a triangulation of the product where the map is continuous on the interior of each simplex. Since they agree on faces it follows that the map is continuous on $B_H \times B_K \longrightarrow B_{H\times K}$. It is formal to check that this is the inverse map to $B_{p_1} \times B_{p_2}$.

The remainder of the claims in 1.8 follow by expanding the composition $\phi \cdot B_\times : B_A \times B_G \longrightarrow B_{A\times G} \overset{B_\times}{\longrightarrow} B_G$ on points. □

Theorem 1.9 *Let $f: G \longrightarrow G$ be an inner automorphism, $f(g) = kgk^{-1}$ for some $k \in G$, then the induced map $B_f: B_G \longrightarrow B_G$ is homotopic to the identity.*

Proof. We construct the homotopy to the identity explicitly. Let

$$H_t((\tau_1, \ldots, \tau_n, g_1, \ldots, g_n)) = \\ (\tau_1, \ldots, \tau_i, t, \tau_{i+1}, \ldots, \tau_n, f(g_1), \ldots, f(g_i), k, g_{i+1}, \ldots, g_n),$$

where $\tau_i \leq t \leq \tau_{i+1}$. The only points to check are at the boundaries $\tau_i = t$, or $t = \tau_{i+1}$. In the first case note that $kg_i = kg_ik^{-1}k$ so coming from below, i.e. for $t < \tau_i$ gives k, g_i and as t becomes equal to τ_i this is set equal to kg_i via the first relation. Similarly, by the remark above, if t approaches τ_i from above. There is no difference at the upper point. Consequently, H_t is continuous as a function of t. Clearly, $H_0 = \text{id}$, and $H_1 = B_f$. □

When A is a discrete abelian group we have from (1.8) that B_A is an abelian group as well. Moreover, because it is a CW-complex it satisfies the NDR condition, and, because the multiplication $B_\times \cdot \phi$ is CW is cellular, we can iterate the construction obtaining $B_{B_A} = B_A^2$, B_A^3, etc. Note that B_A^n is a $K(A, n)$. Thus we have constructed the $K(A, n)$'s for A any discrete abelian group.

The discussion to this point allows us to make an observation which will be very useful later. The n-fold symmetric product $SP^n(X)$ is the quotient of the n-fold Cartesian product, X^n by the action of the symmetric group, $\mathcal{S}_n$, by permuting coordinates. Thus the points of $SP^n(X)$ can be thought of as unordered n-tuples $\Sigma_1^n x_i$ of points in X. Given any point $x_0 \in X$ there is an inclusion $\subset_{x_0}: X \hookrightarrow SP^n(X)$, $x \mapsto \sum_1^{n-1} x_0 + x$ and we have

Lemma 1.10 *Let A be an Abelian group, then*

$$i_{x_0}: H^*(SP^n(X); A) \longrightarrow H^*(X; A)$$

is onto for X any CW complex and $n \geq 1$.

Proof. We use the equivalence between $H^n(X;A)$ and $[X, K(A,n)]$ and choose B_A^n as our explicit model for $K(A,n)$. Let $\alpha: X \longrightarrow B_A^n$ represent α; then, since B_A^n is path connected for $n \geq 1$, we can suppose that $\alpha(x_0) = e$, the unit of B_A^n. Now, since B_A^n has an abelian multiplication we can extend α to $SP^n(X)$ by $\sum_1^n x_i \mapsto \sum \alpha(x_i)$, so we have factored α through $SP^n(X)$. □

Let $H \subset G$ be any subgroup. The H-orbit space of E_G is given as pt $\times_H E_G$, and the projection $E_G \longrightarrow \text{pt} \times_H E_G$ is a principal H-fibration with contractible total space when H is closed. Consequently pt $\times_H E_G$ is a model for B_H. On the other hand the projection $\text{pt} \times_H E_G \longrightarrow B_G$ is a fibration with fiber G/H. We have shown

Lemma 1.11 *Up to homotopy there is a fibration*

$$G/H \longrightarrow B_H \xrightarrow{B_i} B_G$$

where i is the inclusion $H \subset G$ for any closed subgroup H of G.

While we are on the topic of alternate constructions, here is one which is often very useful. Let G be given as a semi-direct product

$$G = H \times_\alpha K$$

so H is normal in G, K is a subgroup of G and the projection $p: G \longrightarrow G/H$ sends K isomorphically to G/H. Then K acts on H via the inner automorphism action in G. By naturality this induces an action of K on B_H, and we have the associated bundle

$$B_H \times_K E_K \longrightarrow B_K .$$

Theorem 1.12 *The bundle above $B_H \times_K E_K$ is a model for B_G when G is a semi-direct product $H \times_\alpha K$.*

Proof. There is a bundle $E_H \times E_K \longrightarrow B_H \times_K E_K$, with fiber G defined by using the G action defined as follows. For each $g \in G$ write $g = (h,k)$ (this is unique), and set $g(x_H, y_K) = (h \cdot E_k(x_H), k \cdot y_K)$ where the map E_k is the conjugation map discussed above. We easily check that this action is free, the quotient is the space $B_H \times_K E_K$ above, and the projection is a fibering. Since $E_H \times E_K$ is contractible the result follows. □

Remark. More details on these constructions can be found in [Mac], [M3], [M4] and [Ste2].

2. Eilenberg-MacLane Spaces and the Steenrod Algebra $\mathcal{A}(p)$

The goal of this section is to situate the cohomology of finite groups within homotopy theory. The main connection comes from the Steenrod Algebra, which arose from considerations made by Steenrod regarding the symmetric groups, though the operations which generate the Steenrod algebra originally appear in the work of M. Richardson and P.A. Smith, [RS]. The main reference is [SE]; in Chapter IV §7 we describe the construction and its main properties. Some of the results here serve only as an indication of the importance of Steenrod operations in algebraic topology; we hope the interested reader will pursue them further.

Definition 2.1 A *cohomology operation* $\alpha \in \mathcal{O}\{i,j,A,B\}$ for $0 \leq i,j$ and A, B discrete abelian groups, is a natural transformation

$$\alpha\colon\ H^i(X;A) \longrightarrow H^j(X;B)$$

defined on the category of CW-complexes and continuous maps $f\colon X \longrightarrow Y$.

Recall that natural transformation means that the following diagram commutes for all f as above,

$$\begin{array}{ccc} H^i(Y;A) & \xrightarrow{\alpha} & H^j(Y;B) \\ \Big\downarrow f^* & & \Big\downarrow f^* \\ H^i(X;A) & \xrightarrow{\alpha} & H^j(X;B)\,. \end{array}$$

The set $\mathcal{O}\{i,j,A,B\}$ becomes an abelian group when we define $(\alpha+\beta)(x) = \alpha(x) + \beta(x)$, $\forall x \in H^i(X;A)$ and we have

Lemma 2.2 *The group* $\mathcal{O}\{i,j,A,B\} \cong H^j(K(A,i);B)$.

Proof. First, given $x \in H^i(X,A)$ there is a unique homotopy class of maps $X \xrightarrow{x} K(A,i)$ so that the class $\iota \sim \mathrm{id} \in H^i(K(A,i);A) = \mathrm{Hom}_A(A,A)$ maps back to x, $x^*(\iota) = x$. Thus $x^*(\alpha(\iota)) = \alpha(x)$ and α is entirely determined by its value on ι. Conversely, given $\alpha \in H^j(K(A,i);B)$ we can define $\alpha(x) = x^*(\alpha)$. □

Remark. Here if $j > i$ then the operations are identically zero, but the same definitions hold for operations in a generalized cohomology theory, and here there need be no such constraint on the operations.

An operation $\alpha \in \mathcal{O}\{i,j,A,B\}$ need not be a homomorphism. Indeed, we have the factorization of $\alpha(x+y)$ in terms of the composition of maps

$$X \xrightarrow{\Delta} X \times X \xrightarrow{x \times y} B_A^i \times B_A^i \xrightarrow{\times} B_A^i ,$$

since, for the fundmental class, we have $\iota \mapsto \iota \otimes 1 + 1 \otimes \iota \mapsto x + y$. Consequently $\alpha(x+y) = \sum \lambda(x) \cup \lambda'(y) + \text{Ext} - terms$ if $\times^*(\alpha) = \sum \lambda \otimes \lambda' + \text{Ext} - terms$. However, there is a circumstance when an operation must be a homomorphism.

Definition 2.3 A sequence of operations $\alpha_s \in \mathcal{O}\{s, i+s, A, B\}$, $s = 0, 1, \ldots,$ is called *stable* if $\alpha_s(\sigma(x)) = (-1)^i \sigma(\alpha_{s-1}(x))$ where σ is the suspension isomorphism

$$H^*(X; A) \xrightarrow{\sigma} H^{*+1}(\Sigma X; A) ,$$

and $\Sigma X = S^1 \wedge X = S^1 \times X/(e \times X \cup S^1 \times *)$. Such a sequence α_s is said to have degree i, and the set of such operations of degree i is denoted $\mathcal{OS}(i, A, B)$.

A stable operation is always a homomorphism since $\Delta: \Sigma X \longrightarrow \Sigma X \times \Sigma X$ factors through the wedge $\Sigma X \vee \Sigma X \subset \Sigma X \times \Sigma X$ up to homotopy. Moreover, the composition of stable operations is defined in an evident way provided only that the coefficients fit together. In particular the stable operations of type $\mathcal{OS}(*, A, A)$ form an algebra.

Steenrod used methods from group cohomology [SE] to construct operations

$$Sq^i(j) \in \mathcal{O}\{j, j+i, \mathbb{Z}/2, \mathbb{Z}/2\}, \qquad i \geq 0$$

which together form, for each i, a stable operation, denoted Sq^i, as well as operations $\beta(j) \in \mathcal{O}\{j, j+1, \mathbb{Z}/p, \mathbb{Z}/p\}$, $P^i(j) \in \mathcal{O}\{j, j+2i(p-1), \mathbb{Z}/p, \mathbb{Z}/p\}$ for odd primes p. Again the $\beta(j)$ give a stable operation denoted β and called the mod(p) Bockstein. It is the boundary map in the coefficient sequence associated to the exact sequence of coefficients $\mathbb{Z}/p \longrightarrow \mathbb{Z}/p^2 \longrightarrow \mathbb{Z}/p$,

$$\cdots \longrightarrow H^*(X; \mathbb{Z}/p^2) \longrightarrow H^*(X; \mathbb{Z}/p) \xrightarrow{\beta(*)}$$
$$H^{*+1}(X; \mathbb{Z}/p) \longrightarrow H^{*+1}(X; \mathbb{Z}/p^2) \longrightarrow \cdots .$$

Also, the $P^i(j)$ give a stable operation denoted P^i. Composing these operations gives rise to an *algebra* of stable operations, the **Steenrod algebra** $\mathcal{A}(2)$ generated by the Sq^i's for $p = 2$, and the **Steenrod algebra** $\mathcal{A}(p)$ generated by β and the P^i when p is odd.

The construction of these operations is intimately connected with the cohomology of (finite) groups, and we will give the details in Chapter IV, §7, but for now we record the axiomatic descriptions of the $\mathcal{A}(p)$ first given in the book by Steenrod and Epstein [SE].

Axioms for the Steenrod Algebra $\mathcal{A}(2)$

There are elements Sq^i, $i \geq 0$ in $\mathcal{OS}(i, \mathbb{Z}/2, \mathbb{Z}/2)$ which are uniquely specified by the following axioms.
1. $Sq^0 = \mathrm{id}$,
2. If $\dim(x) = n$ then $Sq^n(x) = x^2$,
3. If $i > \dim(x)$ then $Sq^i(x) = 0$,
4. (Cartan formula) $Sq^i(x \cup y) = \sum_0^i Sq^j(x) \cup Sq^{i-j}(y)$.

As a consequence of these axioms one can show that Sq^1 is the Bockstein associated to the exact sequence $\mathbb{Z}/2 \to \mathbb{Z}/4 \to \mathbb{Z}/2$, and [Car] the Adem relations $\mathcal{R}(a,b)$

$$Sq^a Sq^b = \sum_0^{[a/2]} \binom{b-1-j}{a-2j} Sq^{a+b-j} Sq^j$$

for $0 < a < 2b$. Moreover, $\mathcal{A}(2)$ is the graded algebra generated by the Sq^i subject only to the relations $Sq^0 = 1$ and the Adem relations $\mathcal{R}(a,b)$. The first few relations have the form $Sq^1 Sq^2 = Sq^3$, $Sq^1 Sq^1 = 0$, $Sq^2 Sq^2 = Sq^3 Sq^1 = Sq^1 Sq^2 Sq^1$. One can prove that a minimal generating set for $\mathcal{A}(2)$ is $\{1, Sq^1, Sq^2, \ldots, Sq^{2^i}, \ldots\}$. Moreover, a basis for $\mathcal{A}(2)$ as a vector space over $\mathbb{Z}/2$ is the set of monomials $Sq^I = Sq^{i_1} Sq^{i_2} \cdots Sq^{i_r}$, $I = (i_1, i_2, \ldots, i_r)$ with $i_j \geq 2i_{j+1}$ for all $j = 1, \ldots, r-1$. These are called the admissible monomials. The first few are Sq^0, Sq^1, Sq^2, Sq^3, $Sq^{2,1} = Sq^2 Sq^1$, Sq^4, $Sq^{3,1} = Sq^3 Sq^1$.

Axioms for the Steenrod algebra $\mathcal{A}(p)$

There are elements $P^i \in \mathcal{OS}(2(p-1)i, \mathbb{Z}/p, \mathbb{Z}/p)$ for p an odd prime which are uniquely specified by the following axioms.
1. $P^0 = \mathrm{id}$,
2. If $\dim(x) = 2n$ then $P^n(x) = x^p$,
3. If $2i > \dim(x)$ then $P^i(x) = 0$,
4. (Cartan formula) $P^i(x \cup y) = \sum_0^i P^j(x) \cup P^{i-j}(y)$.

Again, as a consequence of these axioms one obtains Adem relations

$$P^a P^b = \sum_0^{[a/p]} (-1)^{a+t} \binom{(p-1)(b-t)-1}{a-pt} P^{a+b-t} P^t$$

if $a < pb$. If $a \leq pb$ then

$$\begin{aligned} P^a \beta P^b &= \sum_0^{[a/p]} (-1)^{a+t} \binom{(p-1)(b-t)}{a-pt} \beta P^{a+b-t} P^t \\ &+ \sum_0^{[(a-1)/p]} (-1)^{a+t-1} \binom{(p-1)(b-t)-1}{a-pt-1} P^{a+b-t} \beta P^t . \end{aligned}$$

As above $\mathcal{A}(p)$ is the graded algebra generated by β and the P^i, subject only to the relations $P^0 = \text{id}$, $\beta^2 = 0$ and the Adem relations above. Again there is a basis for $\mathcal{A}(p)$, consisting of admissible monomials. Let

$$I = (\epsilon_0, i_1, \epsilon_1, i_2, \ldots, i_r, \epsilon_r),$$

with $\epsilon_i = 0$ or 1, then I is admissible if $i_j \geq pi_{j+1} + \epsilon_{j+1}$ for each $j \geq 1$. Corresponding to an admissible monomial is the element of $\mathcal{A}(p)$, $\beta^{\epsilon_0} P^{i_1} \cdots P^{i_r} \beta^{\epsilon_r}$, and these monomials, together with $P^0 = 1$, form a basis for $\mathcal{A}(p)$ over $\mathbb{Z}/p$.

The Cohomology of Eilenberg-MacLane Spaces

We need the following definition in order to give the structure of the cohomology rings of the Eilenberg-MacLane spaces.

Definition 2.4 An *exterior algebra* is a graded algebra of the form

$$\mathbb{F}(e_1, \ldots, e_r, \ldots)$$

subject only to the relations $e_i e_j = -e_j e_i$ for all i, j, and when the characteristic of $\mathbb{F}$ is 2, also the relation $e_j^2 = 0$. Generally the generators are odd dimensional.

The Steenrod operations described above generate $\mathcal{O}\{i, j, \mathbb{Z}/p, \mathbb{Z}/p\}$ when we add in cup products. Precisely, we have the following theorem of H. Cartan and J.P. Serre, [Se1].

Theorem 2.5

1. *Let $p = 2$, and suppose that I is an admissible monomial for $\mathcal{A}(2)$, then the excess of I, $e(I)$, is $i_1 - i_2 - \cdots - i_r$. In these terms $H^*(B^n_{\mathbb{Z}/2}; \mathbb{Z}/2)$ is the polynomial algebra on generators $Sq^I(\iota_n)$ where I runs over all monomials of excess less than n.*
2. *Let p be an odd prime and suppose that I is an admissible monomial for $\mathcal{A}(p)$, then the excess, $e(I)$, is $i_1 - (p-1)(i_2 + \ldots + i_r) - \sum_1^r \epsilon_j$. In these terms $H^*(B^n_{\mathbb{Z}/p}; \mathbb{Z}/p)$ is the tensor product of a polynomial algebra $\mathbb{Z}/p[\cdots, P^I(\iota_n), \cdots]$ when $\dim(P^I(\iota_n))$ is even, and and exterior algebra $E(\cdots, P^I(\iota_n), \cdots)$ when $\dim(P^I(\iota_n))$ is odd, as I runs over all admissible monomials with $e(I) < n$.*

Examples. We have that $H^*(K(\mathbb{Z}/2, 1); \mathbb{F}/2)$ is the polynomial algebra on a single one dimensional generator, $\mathbb{F}_2[\iota]$, while

$$H^*(K(\mathbb{Z}/2, 2); \mathbb{F}_2) = \mathbb{F}_2[\iota_2, Sq^1(\iota_2), Sq^{2,1}(\iota_2), \ldots, Sq^{2^j, 2^{j-1}, \ldots, 2, 1}(\iota_2), \ldots].$$

The Hopf Algebra Structure on $\mathcal{A}(p)$

The Cartan formula defines a homomorphism of algebras $c\colon \mathcal{A}(p) \longrightarrow \mathcal{A}(p) \otimes \mathcal{A}(p)$, $P^i \mapsto \sum_0^i P^j \otimes P^{i-j}$, $\beta \mapsto \beta \otimes 1 + 1 \otimes \beta$. This gives $\mathcal{A}(p)$ the structure of a **cocommutative, coassociative** Hopf algebra. The basic facts in the theory of Hopf algebras are developed in section 2 of Chapter VI, VI.2, and the exposition is independent of the rest of the book. Among the most important elements in a Hopf algebra are the primitives, those θ with $c(\theta) = \theta \otimes 1 + 1 \otimes \theta$. Such elements are characterized by the property that $\theta(a \cup b) = \theta(a) \cup b \pm a \cup \theta(b)$. They correspond to generators of the dual algebra $\mathcal{A}(p)^*$. Milnor analyzed the structure of $\mathcal{A}(p)^*$. His result [Mi] is

Theorem 2.6

1. *There are primitives Q_i in dimension $2^i - 1$ in $\mathcal{A}(2)$, $Q_i = [Sq^{2^{i-1}}, Q_{i-1}]$, and $\mathcal{A}(2)^* = \mathbb{Z}/2[\xi_1, \ldots, \xi_i, \ldots]$ where ξ_i is dual to Q_i.*
2. *Let p be an odd prime. There are primitives Q_i, S_i, in dimensions $2p^i - 1$ and $2(p^i - 1)$ respectively. Moreover, $\mathcal{A}(p)^*$ is the tensor product of a polynomial algebra on generators χ_i dual to the S_i and and exterior algebra on generators τ_i dual to the Q_i.*

For $\mathcal{A}(2)$ the Q_i are characterized by the fact that they are primitive so $Q_i(a \cup b) = Q_i(a) \cup b + a \cup Q_i(b)$, and the fact that $Q_i(e) = e^{2^i}$ when $\dim(e) = 1$. Likewise, the Q_i for p odd are characterized by the same formula for cup products and the fact that $Q_i(e) = \beta(e)^{p^i}$ for $\dim(e) = 1$.

3. Group Cohomology

Let A be an Abelian group, then we define

$$H^*(G; A) \;=\; H^*(B_G; A)$$

and call these groups the cohomology groups of G with (untwisted) coefficients A. If $H \subset G$ is a subgroup, the inclusion $B_H \hookrightarrow B_G$ induces a map in cohomology

$$\left(\mathrm{res}_H^G\right)^* \colon H^*(G; A) \longrightarrow H^*(H; A)$$

called restriction.

Lemma 3.1 *Let $N_G(H)$ be the normalizer of H in G, then there is an action of $N_G(H)/H = W_G(H)$ on $H^*(H; A)$ and $\mathrm{im}\left[\left(\mathrm{res}_H^G\right)^*\right]$ is contained in $H^*(H; A)^{W_G(H)}$.*

Proof. For $g \in N_G(H)$ define a map $\Gamma_g : B_H \longrightarrow B_H$ by

$$(*) \qquad \Gamma_g((t_1, \ldots, t_n, h_1, \ldots, h_n)) = (t_1, \ldots, t_n, gh_1g^{-1}, \ldots, gh_ng^{-1}).$$

Clearly $\Gamma_g \cdot \Gamma_{g'} = \Gamma_{gg'}$ so these maps fit together to give an action map

$$N_G(H) \times B_H \longrightarrow B_H$$

and consequently an action on cohomology rings

$$N_G(H) \times H^*(B_H; A) \longrightarrow H^*(B_H; A).$$

Since $\Gamma_g \simeq \mathrm{id}$ for $g \in H$, (1.9), it follows that $\Gamma_g^* = \mathrm{id}$ if $g \in H$ and the action above factors through

$$\bar{N}_G(H) \times H^*(B_H; A) \longrightarrow H^*(B_H; A).$$

We extend Γ_g to B_G for $g \in N_G(H)$, by the same formula $(*)$, and for each $g \in N_G(H)$ we have a commutative diagram

$$\begin{array}{ccc} B_H & \hookrightarrow & B_G \\ \downarrow{\scriptstyle \Gamma_g} & & \downarrow{\scriptstyle \Gamma_g} \\ B_H & \hookrightarrow & B_G . \end{array}$$

Consequently $\left(\mathrm{res}_H^G\right)^* \cdot \Gamma_g^* = \Gamma_g^* \left(\mathrm{res}_H^G\right)^*$, but Γ_g is obtained from an inner automorphism on B_G, so is homotopic to the identity, and the equation above becomes

$$\left(\mathrm{res}_H^G\right)^* (\alpha) = \Gamma_g^* \left(\mathrm{res}_H^G\right)^* (\alpha)$$

for all $\alpha \in H^*(G; A)$. □

Definition. The group $N_G(H)/H = W_G(H)$ is called the *Weyl group* of H in G.

There is an algebraic reformulation of the definition of group cohomology which allows us to generalize to the case where G acts in some non-trivial way on the coefficients A.

Definition 3.2 Let M be a $\mathbb{Z}(G)$-module, then a resolution of M over G is a long exact sequence of $\mathbb{Z}(G)$-modules and $\mathbb{Z}(G)$-module maps ∂_i,

$$0 \longleftarrow M \xleftarrow{\partial_0} C_0 \xleftarrow{\partial_1} C_1 \xleftarrow{\partial_2} C_2 \longleftarrow \cdots \xleftarrow{\partial_i} C_i \longleftarrow \cdots$$

where each C_i is $\mathbb{Z}(G)$-free.

Remark 3.3 For any $\mathbb{Z}(G)$-module M we can construct a resolution. Indeed, take any set S of $\mathbb{Z}(G)$-generators for M, $m_1, \ldots, m_i, \ldots$ and define a map

$$\partial_0 \colon \coprod_S \mathbb{Z}(G)_{m_i} \longrightarrow M, \quad \partial_0(\sum \theta_i 1_{m_i}) = \sum \theta_i m_i \,.$$

Next, let our new module be the kernel of ∂_0 and apply the same construction above to construct a surjective map $\partial_1' \colon \mathcal{C}_1 \longrightarrow \ker(\partial_0)$. Then, including $\ker(\partial_0) \subset \mathcal{C}_0 = \coprod_S \mathbb{Z}(G)_{m_i}$ defines ∂_1 as the composition of *inclusion* with ∂_1'. Now, we can repeat this construction to obtain a resolution. More generally, we can require the modules in definition 3.2 to be projective (i.e. direct summands in a free module); all the subsequent constructions will work equally well, as the reader can verify.

Example 3.4 An explicit resolution of $\mathbb{Z}$ (with the trivial action) is obtained by taking the cellular chain complex of E_G, since E_G is contractible and G acts freely and cellularly. Explicitly, the cells of E_G have the form $g \times \sigma^n \times (g_1, \ldots, g_n)$ which we write as $g|g_1|g_2|\cdots|g_n|$. Consequently, $\mathcal{C}_n$ can be given as $\coprod \mathbb{Z}(G)|g_1|\cdots|g_n|$ where the $(g_1, \ldots, g_n)$ run over all n-tuples of elements of G with no $g_i = e$. The boundary map ∂ is given by

$$\begin{aligned} \partial(|g_1|\cdots|g_n|) &= g_1|g_2|\cdots|g_n| + \\ &\sum_1^{n-1}(-1)^i|g_1|\cdots|g_ig_{i+1}|\cdots|g_n| + (-1)^n|g_1|\cdots|g_{n-1}| \end{aligned}$$

with the understanding that when $g_ig_{i+1} = e$ that term in the summand is set equal to 0. Thus $\partial(|g_1|g_2|) = g_1|g_2| - |g_1g_2| + |g_1|$, but $\partial(|g_1|g_1^{-1}|) = g_1|g_1^{-1}| + |g_1^{-1}|$.

This resolution is commonly called the *bar construction* and is written $B(\mathbb{Z})$. It is not hard to extend it to a general resolution of a (left) $\mathbb{Z}(G)$-module M provided M is a free $\mathbb{Z}$-module. Set $B(M) = B(\mathbb{Z}) \otimes_{\mathbb{Z}} M$ and define the boundary $\partial(|g_1|\cdots|g_n|m)$ as above, except the last term in the formula above is replaced by

$$(-)^n|g_1|\cdots|g_{n-1}|g_n(m)\,.$$

Proposition 3.5 *Let $\phi \colon M \longrightarrow N$ be a $\mathbb{Z}(G)$-module map, and suppose*

$$0 \longleftarrow M \xleftarrow{\partial_0} \mathcal{C}_0 \xleftarrow{\partial_1} \mathcal{C}_1 \xleftarrow{\partial_2} \mathcal{C}_2 \longleftarrow \cdots$$

$$0 \longleftarrow N \xleftarrow{\partial_0} \mathcal{D}_0 \xleftarrow{\partial_1} \mathcal{D}_1 \xleftarrow{\partial_2} \mathcal{D}_2 \longleftarrow \cdots$$

are two resolutions of M over $\mathbb{Z}(G)$. Then there are $\mathbb{Z}(G)$-maps $\phi_i \colon \mathcal{C}_i \longrightarrow \mathcal{D}_i$, $0 \le i < \infty$, so the diagram

$$(**) \qquad \begin{array}{ccccccccccc} 0 & \longleftarrow & M & \overset{\partial_0}{\longleftarrow} & \mathcal{C}_0 & \overset{\partial_1}{\longleftarrow} & \mathcal{C}_1 & \overset{\partial_2}{\longleftarrow} & \mathcal{C}_2 & \longleftarrow & \cdots \\ & & \downarrow{\scriptstyle\phi} & & \downarrow{\scriptstyle\phi_0} & & \downarrow{\scriptstyle\phi_1} & & \downarrow{\scriptstyle\phi_2} & & \\ 0 & \longleftarrow & N & \overset{\partial_0}{\longleftarrow} & \mathcal{D}_0 & \overset{\partial_1}{\longleftarrow} & \mathcal{D}_1 & \overset{\partial_2}{\longleftarrow} & \mathcal{D}_2 & \longleftarrow & \cdots \end{array}$$

commutes. Moreover, given any other choices ϕ'_0, $\phi'_1, \ldots$ *making* (**) *commute there are maps* $\mu_i: \mathcal{C}_i \longrightarrow \mathcal{D}_{i+1}$, $i = 0, 1, \cdots$ *so that* $\partial_{i+1}\mu_i + \mu_{i-1}\partial_i = \phi'_i - \phi_i$, $0 \le i < \infty$.

Remark. A map $\mu: \mathcal{C}_1 \longrightarrow \mathcal{D}_{i+1}$ with $\partial\mu + \mu\partial = f - g$ is called a chain homotopy.

Proof. To begin the definition of ϕ_0 choose a basis $e_1, \ldots, e_r, \ldots$ for $\mathcal{C}_0$. Since ∂_0 is onto, for each e_i we have $\phi\partial_0(e_i) = \lambda_i$ is in the image of ∂_0 in the second resolution. Say $\lambda_i = \partial_0(f_i)$. Then define $\phi_0(e_i) = f_i$ and extend to a $\mathbb{Z}(G)$-module map by freeness.

Assume that ϕ_j defined so as to satisfy $\partial_j\phi_j = \phi_{j-1}\partial_j$ for $j < k$. Choose a basis $e_1, \ldots, e_r, \ldots$ for $\mathcal{C}_k$ and note that $\partial[\phi_{k-1}(\partial e_r)] = \phi_{k-2}(\partial^2 e_r) = 0$ by assumption. Thus, by the exactness of the second resolution, $\phi_{k-1}(\partial_k(e_r))$ is contained in the image of $\partial_k: \mathcal{D}_k \longrightarrow \mathcal{D}_{k-1}$. Specifically, let $\partial_k(f_r) = \phi_{k-1}(\partial_k e_r)$, define $\phi_k(e_r) = f_r$ and extend to a map ϕ_k by $\mathbb{Z}(G)$ freeness.

To construct the maps μ_k set $\mu_{-1}: M \longrightarrow \mathcal{D}_0$ to be the 0-map, and assume μ_j defined so that $\partial\mu_j + \mu_{j-1}\partial = \phi'_j - \phi_j$ for all $j < k$. Then we proceed substantially as above. Let $e_1, \ldots e_r, \ldots$ be a basis for $\mathcal{C}_k$. We have

$$\partial\mu_{k-1}(\partial e_r) = \partial\left[\mu_{k-1}(\partial e_r)\right] + \mu_{k-2}(\partial^2 e_r) = \phi'_{k-1}(\partial e_r) - \phi_{k-1}(\partial e_r)$$

so

$$\partial(\mu_{k-1}\partial(e_r) - \phi'_k(e_r) + \phi_i(e_r)) = 0$$

and

$$\mu_{k-1}\partial(e_r) - \phi'_i(e_r) + \phi_i(e_r) = \partial_{k+1}(f_r)\ .$$

Consequently, set $\mu_k(e_r) = -f_r$. This completes the inductive step and the proof. □

Let A be any $\mathbb{Z}(G)$-module. We define

$$\mathrm{Ext}^i_{\mathbb{Z}(G)}(M; A)$$

as the ith cohomology group of the cochain complex

$$\mathrm{Hom}_{\mathbb{Z}(G)}(\mathcal{C}_j, A) \;=\; \mathcal{C}^i(M, A)$$

for any resolution of M. From the proposition above, these groups are independent of the particular choice of resolution.

Remark 3.6 $\mathrm{Ext}^i_{\mathbb{Z}(G)}(\mathbb{Z}, A) = H^i(G; A)$ if $\mathbb{Z}$, A are both given as trivial $\mathbb{Z}(G)$-modules, since, as we have seen the cellular chain complex of E_G is a suitable resolution of $\mathbb{Z}$ while $\mathrm{Hom}_{\mathbb{Z}(G)}(\mathcal{C}_*(E_G), A) = \mathrm{Hom}_{\mathbb{Z}}(\mathcal{C}_*(B_G), A)$ if A is a trivial $\mathbb{Z}(G)$-module. Consequently, these Ext groups are true generalizations of the cohomology of G. They are contravariant as functors of G and M, but covariant in A.

Remark 3.7 Even if $\mathbb{Z}(G)$ acts non-trivially on A we often write $H^*(G;A)$ or sometimes $H_t^*(G;A)$ when we want to be explicit about the twisting for $\mathrm{Ext}_{\mathbb{Z}(G)}(\mathbb{Z},A)$. These groups are called the cohomology groups of G with (twisted) coefficients A. Similarly we can define the *homology* groups of G with coefficients in A as $H_*(G,A) = \mathrm{Tor}_*^{\mathbb{Z}G}(\mathbb{Z},A) = H_*(C_*(EG) \otimes_G A)$. However, it turns out that nothing new is obtained using homology instead of cohomology. In fact, later we will see that homology and cohomology can be glued together to form a $\mathbb{Z}$-graded theory, known as Tate cohomology.

Example 3.8 Let $G = \mathbb{Z}/n$, and set $I \subset \mathbb{Z}(G)$ equal to the kernel of the augmentation map $\epsilon\colon \mathbb{Z}(G) \longrightarrow \mathbb{Z}$ defined by $\epsilon(\sum n_i T^i) = \sum n_i$. A $\mathbb{Z}$-basis for I is given by the $n-1$ elements $T-1$, T^2-1, T^3-1, ..., $T^{n-1}-1$. But $T^i - 1 = (T^{i-1} + T^{i-2} + \cdots + 1)(T-1)$, so, as a module over $\mathbb{Z}(G)$, I is generated by $T-1$. It follows that the map $\partial_1\colon \mathbb{Z}(G) \longrightarrow I$ defined by $\partial_1(\sum n_i T^i) = \sum n_i T^i (T-1)$ is surjective. Note that $\partial_1(T^{r-1}) = (T^r - 1) - (T^{r-1} - 1)$ so that the images of 1, T, ..., T^{n-2} are independent over $\mathbb{Z}$. On the other hand $\partial_1(1 + T + T^2 + \cdots + T^{n-1}) = 0$. It follows that the submodule of $\mathbb{Z}(G)$ generated by $\Sigma_G = \sum_{g \in G} g$ generates the kernel of ∂_0, and, moreover, this kernel is a single copy of $\mathbb{Z}$. In particular, since $g\Sigma_G = \Sigma_G$ for all $g \in G$ it follows that the kernel is a copy of the trivial $\mathbb{Z}(G)$ module. But this is the situation we started with. Consequently, we can iterate and a resolution of $\mathbb{Z}$ over $\mathbb{Z}(G)$ is given as

$$\mathbb{Z} \xleftarrow{\epsilon} \mathbb{Z}(G) \xleftarrow{T-1} \mathbb{Z}(G) \xleftarrow{\Sigma_G} \mathbb{Z}(G) \xleftarrow{T-1} \cdots .$$

Passing to Ext groups we have, for example

$$\mathrm{Ext}^*_{\mathbb{Z}(\mathbb{Z}/n)}(\mathbb{Z}, \mathbb{Z}/n) = H^*(\mathbb{Z}/n; \mathbb{Z}/n) = \mathbb{Z}/n$$

for all $* \geq 0$ since $\mathrm{Hom}(\mathbb{Z}(G), \mathbb{Z}/n) = \mathbb{Z}/n$ and $(T-1)^*$ becomes zero here while Σ_G^* becomes multiplication by n. On the other hand, if $A = \mathbb{Z}(\zeta_n)$ with action $T(\theta) = \zeta_n \theta$ where ζ_n is a primitive n^{th} root of unity. Then we find

$$\mathrm{Ext}^i_{\mathbb{Z}(\mathbb{Z}/n)}(\mathbb{Z}, \mathbb{Z}(\zeta_n)) = H_t^i(\mathbb{Z}/n; \mathbb{Z}(\zeta_n)) = \begin{cases} 0 & i \text{ even} \\ \mathbb{Z}(\zeta_n)/(1-\zeta_n) & i \text{ odd.} \end{cases}$$

More generally, for any $\mathbb{Z}(\mathbb{Z}/n)$ module A we have

$$H_t^i(G;A) = \begin{cases} M^G & i = 0 \\ M^G/\Sigma_{\mathbb{Z}/n}(M) & i \text{ even} \\ \ker(\Sigma_{\mathbb{Z}/n})/(T-1)M & i \text{ odd.} \end{cases}$$

We now consider a method to derive resolutions from a free presentation of a group, i.e. an exact sequence

$$N \lhd F(X) \longrightarrow G \longrightarrow 1$$

where $F(X)$ is the free group on a set of generators indexed by the set X. Passing to group rings we have the corresponding exact sequence

$$0 \longrightarrow R \longrightarrow \mathbb{Z}(F(X)) \longrightarrow \mathbb{Z}(G) \longrightarrow 0 .$$

Our first object is to describe the two sided ideal R.

First, we denote by $I(G) \subset \mathbb{Z}(G)$ the augmentation ideal. That is $I(G)$ is the kernel in the augmentation map

$$\epsilon \colon \mathbb{Z}(G) \longrightarrow \mathbb{Z}$$

defined by $\epsilon(\sum n_i g_i) = \sum n_i$. Note that ϵ is a ring homomorphism so $\mathrm{Ker}(\epsilon) = I(G)$ is a 2-sided ideal in $\mathbb{Z}(G)$. As a $\mathbb{Z}$ module, $I(G)$ has a basis consisting of exactly the elements $(g-1)$, as g runs over the elements of G. In the case of the free group $F(X)$ it turns out that $I(F(X))$ is free as a left $\mathbb{Z}(F(X))$ module on generators corresponding to the elements in the set X.

Lemma 3.9 *Let $\{x\} \in I(F(X)) = x-1$. Then*

$$I(F(X)) \cong \coprod_{x \in X} \mathbb{Z}(F(X))\{x\}$$

as a left $\mathbb{Z}(F(X))$ module, where the isomorphism

$$\phi \colon \coprod_{x \in X} \mathbb{Z}(F(X))\{x\} \longrightarrow I(F(X)$$

is defined by $\phi(\sum \alpha_i\{x_i\}) = \sum \alpha_i(x_i - 1)$.

Proof. ϕ is onto. Indeed, if $\alpha_i = x_{i_1}^{\epsilon_1} \cdots x_{i_r}^{\epsilon_r}$ then set $\delta(\alpha_i) = \sum_{j=1}^{r} \beta_j\{x_{i_j}\}$ where $\beta_j = x_{i_1}^{\epsilon_1} \cdots x_{i_{j-1}}^{\epsilon_{j-1}}$ *if* $\epsilon_j = +1$ and $\beta_j = -x_{i_1}^{\epsilon_1} \cdots x_{i_j}^{\epsilon_j}$ *if* $\epsilon_j = -1$. We have directly (e.g. induction on length) $\phi(\delta(\alpha_i)) = \alpha_i - 1$. For example

$$\delta(x_1^{-1}x_2) = x_1^{-1}\{x_2\} - x_1^{-1}\{x_1\}$$

and $\phi\delta(x_1^{-1}x_2) = x_1^{-1}(x_2-1) - x_1^{-1}(x_1-1) = x_1^{-1}x_2 - 1$. Note, in particular that δ has the property $\delta(\alpha_i y) = \alpha_i(\delta y) + \delta(\alpha_i)$. Next, extend δ to

$$\bar{\delta} \colon IF(X) \longrightarrow \sum_{x \in X} \mathbb{Z}(F(X))\{x\}$$

by setting $\bar{\delta}(\alpha_i - 1) = \delta(\alpha_i)$, then $\bar{\delta}(\beta(\alpha_i - 1)) = \delta(\beta\alpha_i) - \delta(\beta) = \beta\delta(\alpha_i)$ so $\bar{\delta}$ is a $\mathbb{Z}(F(X))$ module map, and $\phi\bar{\delta} = \mathrm{id}$.

Now, we check by direct calculation on the $\{x_i\}$ that

$$\bar{\delta} \cdot \phi(\{x\}) = \bar{\delta}(x-1) = \{x\} .$$

Hence, $\bar{\delta} \cdot \phi$ is also the identity and the lemma follows. □

Now we describe the kernel R in the map above.

Lemma 3.10 *If $N \triangleleft F(X)$ is freely generated by $\{y \in Y\}$, then*

$$R = \ker(\pi\colon\ \mathbb{Z}(F(X)) \longrightarrow \mathbb{Z}(F(X)/N))$$

is a free left (right) $\mathbb{Z}(F(X))$ module with generators $\{y-1 \mid y \in Y\}$.

Proof. As before, define

$$\bar{\phi}\colon\ \coprod_{y\in Y} \mathbb{Z}(F(X))\{y\} \longrightarrow R$$

by $\bar{\phi}(\sum \zeta_i\{y_i\}) = \sum \zeta_i(y_i - 1)$. We need to construct an inverse. For each coset $\gamma \in F(X)/N$ choose $a_\gamma \in \gamma$. Then, given $\zeta \in R$ we can write

$$\zeta = \sum m_i a_{\gamma_i} u_i$$

with $u_i \in N$, and this representation is unique. We can rewrite this as

$$\zeta = \sum_{a_j} a_j \left(\sum_{u_i} m_{i,j} u_{i,j} \right)$$

so $\sum_i m_{i,j} = 0$, *and* $\zeta = \sum_{a_j} a_j \sum m_{i,j}(u_{i,j} - 1)$. Thus $\bar{\phi}$ is onto R. Now, suppose there is a relation

$$\sum \zeta_i(y_i - 1) = 0$$

in R, with $\zeta_i \in \mathbb{Z}(F(X))$. Then, as above

$$\sum_j a_{\gamma_j} \sum_{i,k} m_{i,j,k} u_{i,j,k}(y_k - 1) = 0$$

with the $u_{i,j,k} \in N$. But by the first lemma this implies

$$(\sum_i m_{i,j,k} u_{i,j,k}) = 0\,,$$

so

$$\zeta_k = \sum_j a_{\gamma_j} \sum_i m_{i,j,k} u_{i,j,k} = 0$$

and the result follows. □

Lemma 3.11 *If A and B are left ideals of $\mathbb{Z}(F(X))$, freely generated by sets $\{\alpha_i\}$, $\{\beta_j\}$, and A is two sided then AB is freely generated as a left ideal by the elements $\{\alpha_i\beta_j\}$. (Similiarly for right ideals, provided that B is two sided.)*

Proof. AB is generated by all elements $\alpha_i\lambda\beta_j$, $\lambda \in \mathbb{Z}(F(X))$. But $\alpha_i\lambda =$- $\sum \lambda_k\alpha_k$, by assumption, so the set $\{\alpha_i\beta_j\}$ does generate AB. Now, suppose we have

$$\sum_{i,j} \lambda_{i,j}\alpha_i\beta_j = 0 ,$$

then, by the independence of the β_j we must have

$$\sum_i \lambda_{i,j}\alpha_i = 0$$

for each j. But by the independence of the α_i this implies that the $\lambda_{i,j}$ are all 0. The result follows. □

Lemma 3.12 *If A is a left ideal of $\mathbb{Z}(F(X))$ freely generated by a set $\{\alpha_i\}$ then A/RA is freely generated as a $\mathbb{Z}(G)$ module by the cosets $\{\alpha_i + RA\}$.*

Proof. $A = \coprod \mathbb{Z}(F(X))\alpha_i$ so $RA = \coprod R\alpha_i$, and $A/RA \cong \coprod(\mathbb{Z}(F(X))/R)\alpha_i$ but $\mathbb{Z}(F(X))/R = \mathbb{Z}(G)$ from the definitions. The proof is complete. □

We can now state a result due to Gruenberg [Gr1]

Corollary 3.13 *Let G be given, and suppose $G = F(X)/N$ where $F(X)$ is a free group and N is a normal subgroup, then if we set $I = I(\mathbb{Z}(F(X))$ we obtain a free $\mathbb{Z}(G)$ resolution of $\mathbb{Z}$ as*

$$\cdots R^2/R^3 \longrightarrow RI/R^2I \longrightarrow R/R^2 \longrightarrow I/RI \longrightarrow \mathbb{Z}(G) \stackrel{\epsilon}{\longrightarrow} \mathbb{Z} ,$$

all maps being induced by inclusions, and thinking of $\mathbb{Z}(G)$ as identified with $\mathbb{Z}(F(X))/R$.

(Exactness is clear since the map ϵ is just the projection

$$\mathbb{Z}(F(X))/R \longrightarrow \mathbb{Z}(F(X))/I .$$

Freeness over $\mathbb{Z}(G)$ follows from the preceeding lemmas.)

Example 3.14 Let $G = \mathbb{Z}/n$, then

$$1 \longrightarrow n\mathbb{Z} \longrightarrow \mathbb{Z} \longrightarrow G$$

is suitable. So $R = \mathbb{Z}[t,\ t^{-1}](t^n - 1)$, while $I = \mathbb{Z}[t,\ t^{-1}](t-1)$, after identifying $\mathbb{Z}(\mathbb{Z})$ with the polynomial ring $\mathbb{Z}[t]$ with t^{-1} adjoined. Then

$$R^mI = IR^m = \mathbb{Z}[t,t^{-1}](t-1)(t^n-1)^m ,$$

$$R^m = \mathbb{Z}[t,t^{-1}](t^n-1)^m .$$

Moreover,

$$(t^n - 1)^{m+1} = (t^{n-1} + t^{n-2} + \cdots + 1)(t-1)(t^n-1)^m .$$

Hence, the resolution has the form

$$\cdots \longrightarrow \mathbb{Z}(G) \xrightarrow{t-1} \mathbb{Z}(G) \xrightarrow{\sum_t} \mathbb{Z}(G) \xrightarrow{t-1} \mathbb{Z}(G) \xrightarrow{\epsilon} \mathbb{Z}$$

where $\sum_t = t^{n-1} + t^{n-2} + \cdots + 1$, and $G = \mathbb{Z}/n$ with generator t. Hence we recover the usual resolution.

Example 3.15 Let $G = \mathbb{Z}/2 \times \mathbb{Z}/2$, $X = \{s, t\}$, and

$$N \longrightarrow F(X) \longrightarrow G$$

the obvious map. Using standard techniques we calculate that

$$N = F(x_1, \cdots x_5)$$

with $x_1 = s^2$, $x_2 = t^2$, $x_3 = s^{-1}t^2s$, $x_4 = t^{-1}sts$, $x_5 = s^{-1}t^{-1}st$. Then, the associated resolution has the following table of degrees and ranks

degree	1	2	3	4	5	6	$\cdots$
rank	2	5	10	25	50	125	$\cdots$.

The generators are $e = s-1$, $f = t-1$ for I, and $x_1 - 1$, $x_2 - 1$, $x_3 - 1$, $x_4 - 1$, and $x_5 - 1$ for R. The inclusion $R \subset I$ gives

$$\begin{aligned}
&x_1 - 1 \longrightarrow (s+1)e \\
&x_2 - 1 \longrightarrow (t+1)f \\
&x_3 - 1 \longrightarrow s^{-1}((t^2-1)e + (t+1)f) \\
&x_4 - 1 \longrightarrow t^{-1}((st+1)e + (s-1)f) \\
&x_5 - 1 \longrightarrow s^{-1}t^{-1}((s-1)f + (1-t)e)
\end{aligned}$$

and this determines all the differentials in the Gruenberg resolution after we add in the commutation relations

$$\begin{aligned}
&(x_1 - 1)t = t\{x_4(x_5 - 1) + (x_4 - 1)\} \\
&(x_1 - 1)s = s(x_1 - 1) \\
&(x_2 - 1\}t = t(x_2 - 1) \\
&(x_2 - 1)s = s(x_3 - 1) \\
&(x_3 - 1)t = t\{x_5^{-1}x_3(x_5 - 1) + x_5^{-1}(x_3 - 1) - x_5^{-1}(x_5 - 1) \\
&(x_3 - 1)s = s\{x_1^{-1}x_2(x_1 - 1) + x_1^{-1}(x_2 - 1) - x_1^{-1}(x_1 - 1)\} \\
&(x_4 - 1)t = t\{x_2^{-1}x_1x_3(x_5 - 1) + x_2^{-1}x_1(x_3 - 1) \\
&\qquad\qquad + x_2^{-1}(x_1 - 1) - x_2^{-1}(x_2 - 1)\} \\
&(x_4 - 1)s = s\{x_5(x_1 - 1) + (x_5 - 1)\} \\
&(x_5 - 1)t = t\{x_5^{-1}x_3^{-1}(x_2 - 1) - x_5^{-1}x_3^{-1}(x_3 - 1) - x_5^{-1}(x_5 - 1)\} \\
&(x_5 - 1)s = s\{x_1^{-1}(x_4 - 1) - x_1^{-1}(x_1 - 1)\} .
\end{aligned}$$

These are used to write elements of R^nI in terms of elements in R^n, and by a direct iteration can be used to write elements of R^n in terms of basis elements in $R^{n-1}I$. As an exercise, the reader is advised to construct ∂_3 and ∂_4.

Remark 3.16 An alternate resolution of $\mathbb{Z}(\mathbb{Z}/2 \times \mathbb{Z}/2) \xrightarrow{\epsilon} \mathbb{Z}$ is given by tensoring two copies of a resolution of

$$\mathbb{Z}(\mathbb{Z}/2) \xrightarrow{\epsilon} \mathbb{Z}\,.$$

In particular, if we tensor the Gruenberg resolutions constructed above we find that the resulting resolution has ranks as follows

degree	0	1	2	3	4	5	6	$\cdots$	n	$\cdots$
rank	1	2	3	4	5	6	7	$\cdots$	$n+1$	$\cdots$.

Hence, the Gruenberg resolution is far from minimal in general. However, it has the advantage that it is fairly easy to write down and differentials are easy to calculate for it.

If we use the "canonical" free resolution of G, the resulting Gruenberg resolution will turn out to be the Bar construction. In a later section we will discuss minimal resolutions for finite p–groups.

4. Cup Products

In cohomology there is a cup product induced from the diagonal map and the Kunneth formula. In particular, with (untwisted) field coefficients $\mathbb{F}$, this gives a pairing

$$\coprod_{i,j} H^i(G;\mathbb{F}) \otimes_{\mathbb{F}} H^j(G;\mathbb{F}) \longrightarrow H^{i+j}(G;\mathbb{F})$$

making $H^*(G;\mathbb{F})$ into an associative (graded) commutative ring with unit. (Graded commutative means $b \cup a = (-1)^{\dim(a)\dim(b)} a \cup b$.) The pairing is natural with respect to restriction,

$$\left(\mathrm{res}^G_H\right)^* (a \cup b) = \left(\mathrm{res}^G_H\right)^* (a) \cup \left(\mathrm{res}^G_H\right)^* (b)\,.$$

We now extend the notion slightly to define a pairing

$$\left(\mathrm{Ext}^i_{\mathbb{Z}(G)}(\mathbb{Z},\mathbb{F}) = H^i(G;\mathbb{F})\right) \otimes_{\mathbb{F}} \mathrm{Ext}^j_{\mathbb{Z}(G)}(M,\mathbb{F}) \longrightarrow \mathrm{Ext}^{i+j}_{\mathbb{Z}(G)}(M,\mathbb{F})$$

which makes the Ext-groups into graded modules over the ring $H^*(G;\mathbb{F})$.

Lemma 4.1 *There is a chain map in the bar construction for B_G,*

$$\Delta\colon\ (|b_1|\cdots|b_n|m) \mapsto \sum b_1 \ldots b_r|b_{r+1}|\cdots|b_n| \otimes |b_1|\cdots|b_r|(b_{r+1}\cdots b_n m)\,,$$

natural with respect to homomorphisms of groups.

(This is a direct verification.)

The dual map on cochain complexes now gives the desired structures on passing to cohomology. In particular, for $H^*(G;\mathbb{F})$ the pairing becomes

$$\Delta(|b_1|\cdots|b_n|) = \sum |b_{r+1}|\cdots|b_n| \otimes |b_1|\cdots|b_r| \,,$$

and we claim this induces the same product pairing as that induced by the topologically induced $\cup$-product.

This can be seen quite easily. We define a homotopy of the diagonal map

$$\Delta\colon B_G \longrightarrow B_G \times B_G$$

by

$$\begin{aligned} H_t(t_1,\ldots,t_n,g_1,\ldots,g_n) &= \\ &\Big(\big(\underline{(t+1)t_1 - t},\ldots,\underline{(t+1)t_n - t}, g_1,\ldots,g_n\big), \\ &\overline{(t+1)t_1},\ldots,\overline{(t+1)t_n}, g_1,\ldots,g_n)\Big) \end{aligned}$$

where $\overline{\alpha} = \min(\alpha,1)$ and $\underline{\alpha} = \max(0,\alpha)$. Then $H_0 = \Delta$ and

$$\begin{aligned} H_1((t_1,\ldots,t_n,g_1,\ldots,g_n)) &= \\ &((2t_{r+1}-1,\ldots,2t_n-1,g_{r+1},\ldots g_n),(2t_1,\ldots,2t_r,g_1,\ldots,g_r)) \end{aligned}$$

if $t_r \le \frac{1}{2} \le t_{r+1}$.

Corollary 4.2 *Let p be an odd prime, then $H^*(\mathbb{Z}/p;\mathbb{F}_p) = E(v_1) \otimes \mathbb{F}_p[b_2]$, the tensor product of a polynomial algebra on a two dimensional generator and an exterior algebra on a 1-dimensional generator.*

Proof. An embedding of our usual resolution for $\mathbb{Z}$ over $\mathbb{Z}(\mathbb{Z}/p)$ into the bar construction can be constructed quite simply. The one dimensional generator goes to $|T-1|$ (note that $\partial|T-1| = T-1$). The two dimensional generator goes to $|\Sigma_{\mathbb{Z}/p}|T-1|$. The three dimensional generator goes to $|T-1|\Sigma_{\mathbb{Z}/p}|T-1|$, and so on. This embeds the minimal resolution of $\mathbb{Z}/p$ into the bar construction and, in particular, constructs chain representatives there for the generating homology classes. More precisely, the embedding gives rise to the following commutative diagram which shows that it must be a chain equivalence.

$$\begin{array}{ccc}
\vdots & & \vdots \\
\downarrow & & \downarrow \\
\mathcal{C}_2 & \hookrightarrow & B_2(\mathbb{Z}/p) \\
\downarrow & & \downarrow \\
\mathcal{C}_1 & \hookrightarrow & B_1(\mathbb{Z}/p) \\
\downarrow & & \downarrow \\
\mathcal{C}_0 & \hookrightarrow & B_0(\mathbb{Z}/p) \\
\downarrow & & \downarrow \\
\mathbb{F} & = & \mathbb{F}
\end{array}$$

In order to determine the cup product $\alpha \cup \chi$ in cohomology, we choose representative cochains a for α, c for χ and evaluate $\langle a \otimes b, \Delta(w)\rangle$ as w runs over chains representing all the homology classes in $H_{\mathrm{dim}(\alpha)+\mathrm{dim}(\chi)}(B_{\mathbb{Z}/p}; \mathbb{F}_p)$. This determines the image of $\alpha \cup \beta \in \mathrm{Hom}(H_*(B_{\mathbb{Z}/p}; \mathbb{F}_p), \mathbb{F}_p)$, and since $H^*(B_{\mathbb{Z}/p}; \mathbb{F}_p) = \mathrm{Hom}(H_*(B_{\mathbb{Z}/p}; \mathbb{F}_p), \mathbb{F}_p)$ this evaluation determines the entire cup product.

For example choose a representative, a, for the generator of $H^{2i}(B_{\mathbb{Z}/p}; \mathbb{F}_p)$ and c for a generator of $H^{2j}(B_{\mathbb{Z}/p}; \mathbb{F}_p)$. Then

$$\langle a, \underbrace{|T-1|\sigma_{\mathbb{Z}/p}| \cdots |T-1|\sigma_{\mathbb{Z}/p}|}_{i-times}\rangle = 1$$

and similarly for c. Now we apply 4.1 to

$$\Gamma_{i+j} \;=\; \underbrace{|T-1|\sigma_{\mathbb{Z}/p}| \cdots |T-1|\sigma_{\mathbb{Z}/p}|}_{i+j-times}$$

and we see that $\langle a \otimes c, \Delta(\Gamma_{i+j})\rangle = 1$. Similar calculations in the remaining cases complete the proof. □

Corollary 4.3 *When* $p \neq 2$

$$H^*((\mathbb{Z}/p)^n; \mathbb{F}_p) = E(v_1, \ldots, v_n) \otimes \mathbb{F}_p[b_1, \ldots, b_n] ,$$

the tensor product of a polynomial algebra on n two dimensional generators and an exterior algebra on n one dimensional elements.

The Steenrod operations in $H^*((\mathbb{Z}/p)^n; \mathbb{F}_p)$ are completely determined by $P^j(v_i) = 0$, $j > 0$, $\beta(v_i) = b_i$, and $P^1(b_i) = b_i^p$, $P^j(b_i) = \beta(b_i) = 0$, $j > 1$ using the Cartan formula.

The situation when $p = 2$ is somewhat different as the sign -1 no longer affects things since $1 + T = 1 - T$ in this case. The previous calculation in the proof of 4.2 is unchanged but the result is now

Theorem 4.4 *For $p = 2$ we have*

$$H^*((\mathbb{Z}/2)^n; \mathbb{F}_2) = \mathbb{F}_2[e_1, \ldots, e_n] ,$$

a polynomial algebra on n one dimensional generators.

5. Restriction and Transfer

Given a subgroup $H \subset G$ and a resolution of M over $\mathbb{Z}(G)$, by simply forgetting the $\mathbb{Z}(G)$ structure and noting that $\mathbb{Z}(G)$ is free as a $\mathbb{Z}(H)$ module we obtain a resolution of M over $\mathbb{Z}(H)$. The inclusion of complexes clearly corresponds to the restriction map $(\mathrm{res}_H^G)^*$ described in the last section.

Indeed, if A is a $\mathbb{Z}(G)$ module we have the induced map on Hom complexes

$$\phi_{\#}\colon \mathrm{Hom}_{\mathbb{Z}(G)}(\mathcal{C}_i, A) \longrightarrow \mathrm{Hom}_{\mathbb{Z}(H)}(\mathcal{C}_i, A)$$

which gives a map of cochain complexes that defines

$$\left(\mathrm{res}_H^G\right)^* : \mathrm{Ext}^i_{\mathbb{Z}(G)}(M, A) \longrightarrow \mathrm{Ext}^i_{\mathbb{Z}(H)}(M, A)$$

on passing to cohomology. Clearly, if we have $K \subset H \subset G$, inclusions of subgroups, then

$$\left(\mathrm{res}_K^G\right)^* = \left(\mathrm{res}_K^H\right)^* \cdot \left(\mathrm{res}_H^G\right)^* .$$

In case $H \subset G$ has finite index there is a map called the *transfer*, first introduced by Eckmann [E],

$$\mathrm{tr}\colon \mathrm{Ext}^i_{\mathbb{Z}(H)}(M, A) \longrightarrow \mathrm{Ext}^i_{\mathbb{Z}(G)}(M, A)$$

which is defined as follows (on the level of cochains). For

$$\alpha \in C^i_H = \mathrm{Hom}_{\mathbb{Z}(H)}(\mathcal{C}_i, A)$$

set $\mathrm{tr}(\alpha) \in \mathrm{Hom}_{\mathbb{Z}(G)}(\mathcal{C}_i, A) = C^i_G)$ to be the homomorphism given on elements $\lambda \in \mathcal{C}_i$ as

$$\mathrm{tr}(\alpha)(\lambda) = \sum_{i=1}^{[G:H]} g_i \alpha(g_i^{-1}\lambda)$$

where $g_1, \ldots, g_{[G:H]}$ run over a set of left coset representatives for H in G.

In the above sum consider the term $g_i\alpha(g_i^{-1}g\lambda)$. Write $g_i^{-1}g = hg_j^{-1}$ for some other coset representative g_j, so $g_ih = gg_j$, and, since α is H-linear, we have $g_i\alpha(g_i^{-1}g\lambda) = g_ih\alpha(g_j^{-1}\lambda) = gg_j\alpha(g_j^{-1}\lambda)$. Then summing over all the coset representatives it follows that $\mathrm{tr}(\alpha)(g\lambda) = g\mathrm{tr}(\alpha)(\lambda)$ and $\mathrm{tr}(\alpha)$ is, indeed, $\mathbb{Z}(G)$-linear. Moreover, we have

$$\mathrm{tr}(\delta\alpha)(b) = \langle \mathrm{tr}(\delta\alpha), b\rangle = \sum g_i[\alpha(g_i^{-1}\partial b)] = \langle \mathrm{tr}(\alpha), \partial b\rangle = \mathrm{tr}(\alpha)(\partial b)$$

so tr is a map of cochain complexes and induces a map of Ext-groups.

Similarly, if $f: M \to N$ is a $\mathbb{Z}(G)$-module map and $f_\#: C_\#(M) \to C_\#(N)$ any $\mathbb{Z}(G)$-chain map of $\mathbb{Z}(G)$-projective resolutions, then

$$f^\# \mathrm{tr} = \mathrm{tr} f^\#$$

and similarly, if s is a $\mathbb{Z}(G)$-chain homotopy between $f_\#$ and $f'_\#$ then $s^* \mathrm{tr} = \mathrm{tr} s^*$ is a chain homotopy between the transfer maps. Consequently, the transfer is well defined, independent of choice of resolutions or coset representatives.

Clearly, if $K \subset H \subset G$ with $[G: K] < \infty$ we have

$$\mathrm{tr}_H^G \cdot \mathrm{tr}_K^H = \mathrm{tr}_K^G .$$

Lemma 5.1

a. *If $[G: H] < \infty$ then the composite*

$$\mathrm{Ext}_{\mathbb{Z}(G)}(M, A) \xrightarrow{(\mathrm{res}_H^G)^*} \mathrm{Ext}_{\mathbb{Z}(H)}(M, A) \xrightarrow{\mathrm{tr}_G^H} \mathrm{Ext}_{\mathbb{Z}(G)}(M, A)$$

is multiplication by $[G: H]$.

b. *If $H \triangleleft G$ and $[G: H] < \infty$ then the composite*

$$\mathrm{Ext}_{\mathbb{Z}(H)}(M, A) \xrightarrow{\mathrm{tr}_G^H} \mathrm{Ext}_{\mathbb{Z}(G)}(M, A) \xrightarrow{(\mathrm{res}_H^G)^*} \mathrm{Ext}_{\mathbb{Z}(H)}(M, A)$$

is $\alpha \mapsto \sum_1^{[G;H]} g_i^*(\alpha)$ where the g_i run over a set of left coset representatives for H in G.

(b. is immediate from the formulae above. If α is already G-linear, then $\mathrm{tr}(\alpha) = [G: H]\alpha$ already on the level of cochains, so a. also follows directly.)

Corollary 5.2 *Let $p \,|\, |G|$ but assume $[G: H]$ is prime to p (so H contains a p-Sylow subgroup of G if $|G| < \infty$), then*

$$H^*(G; \mathbb{F}_p) \xrightarrow{(\mathrm{res}_H^G)^*} H^*(H; \mathbb{F}_p)$$

is injective if $\mathbb{F}_p$ is the trivial $\mathbb{Z}(G)$ module.

Remark 5.3 More generally $H_t^*(G; M) \xrightarrow{(\mathrm{res}_H^G)^*} H_t^*(H; M)$ is injective when M is any $\mathbb{Z}(G)$-module where multiplication by $[G: H]$, $M \xrightarrow{\times [G:H]} M$ is an isomorphism. From this we obtain the important

Corollary 5.4 *Write $m = |G|$. Then $m\bar{H}_t^*(G, M) = 0$ so, if $\times m: M \to M$ is an isomorphism, it follows that $\bar{H}_t^*(G; M) = 0$.*

The following result is useful in making explicit calculations.

Lemma 5.5 *Let $H \subset G$ have finite index, then $H \times G_1 \subset G \times G_1$ has finite index and, with field coefficients $\mathbb{F}$, $H^*(H \times G_1; \mathbb{F}) = H^*(H; \mathbb{F}) \otimes_{\mathbb{F}} H^*(G_1; \mathbb{F})$, $H^*(G \times G_1; \mathbb{F}) = H^*(G; \mathbb{F}) \otimes_{\mathbb{F}} H^*(G_1; \mathbb{F})$. Moreover, with respect to this splitting* $\mathrm{tr}_{H \times G_1}^{G \times G_1} = \mathrm{tr}_H^G \otimes \mathrm{id}$.

Proof. A resolution of $\mathbb{Z}$ over $\mathbb{Z}(G \times G_1) = \mathbb{Z}(G) \otimes_{\mathbb{Z}} \mathbb{Z}(G_1)$ can be given as $\mathcal{C}(G) \otimes_{\mathbb{Z}} \mathcal{C}(G_1)$ where $\mathcal{C}(G)$ is a resolution of $\mathbb{Z}$ over $\mathbb{Z}(G)$ and $\mathcal{C}(G_1)$ is a resolution of $\mathbb{Z}$ over $\mathbb{Z}(G_1)$. If we select coset representatives for $g_1, \ldots, g_n$ so $G \times G_1 = \sqcup_i g_i H \times G_1$ with each $g_i \in G$, then on the cochain level $\mathrm{tr} = \mathrm{tr} \otimes \mathrm{id}$ and 5.5 follows. □

Transfer and Restriction for Abelian Groups

Let $G = \mathbb{Z}/p^{i+1}$ and $H \subsetneq G$ be the unique subgroup $\mathbb{Z}/p^i \subsetneq \mathbb{Z}/p^{i+1}$. If G is generated by T then H has generator T^p. Now, let $A = \mathbb{Z}(G)$, then, A, as a $\mathbb{Z}(H)$-module is free on p generators, $e_1 = 1$, $e_2 = T$, $e_3 = T^2, \ldots, e_p = T^{p-1}$. An explicit resolution of $\mathbb{Z}$ over A is given as

$$\mathbb{Z} \xleftarrow{\epsilon} A \xleftarrow{T-1} A \xleftarrow{\Sigma_G} A \xleftarrow{T-1} A \longleftarrow \cdots$$

where $\Sigma_G = \sum_{g \in G} g$ for any group G. When we regard this as a resolution of $\mathbb{Z}$ over H the map $T-1$ is given as

$$\begin{aligned} e_1 &\mapsto e_2 - e_1, \\ e_2 &\mapsto e_3 - e_2, \\ &\vdots \\ e_{p-1} &\mapsto e_p - e_{p-1}, \\ e_p &\mapsto T^p e_1 - e_1\,. \end{aligned}$$

Similarly $\Sigma_G e_i = \Sigma_H(e_1 + \cdots + e_p)$. We can map the usual resolution of $\mathbb{Z}$ over $\mathbb{Z}(H)$ into this resolution by

$$\begin{cases} 1_i \mapsto e_1 & \text{when } i \text{ is even}, \\ 1_i \mapsto e_1 + \cdots + e_p & \text{when } i \text{ is odd}. \end{cases}$$

Proposition 5.6 *Cocycles representing the generators in $H^*(H; \mathbb{F}_p)$ are given for the chain complex of $\mathbb{Z}(G)$ above as follows. In even dimensions $b^{2i} \in \mathrm{Hom}_{\mathbb{Z}(H)}(\mathbb{Z}(H), \mathbb{F}_p)$ is defined by $b^{2i}(e_j) = 1$, $1 \leq j \leq p$, while in odd degrees e^{2i+1} is defined by $e^{2i+1}(e_j) = \begin{cases} 1 & j = 1, \\ 0 & j > 1. \end{cases}$ Here, of course, the action on $\mathbb{F}_p$ is trivial.*

Proof. We must check two things, first that the cochains b^{2i} and e^{2i+1} are cocycles, and second that they evaluate non-trivially on chains representing the generators in homology in these dimensions. We have

$$\delta b^{2i}(e_j) = b^{2i}(\partial e_j) = \begin{cases} b^{2i}(e_{j+1} - e_j) = 0 & \text{when } j < p, \\ b^{2i}(T^p e_1 - e_p) = 0 & \text{for } j = p. \end{cases}$$

Similarly

$$\delta e^{2i+1}(e_j) = e^{2i+1}(\partial e_j) = e^{2i+1}(\Sigma_H(e_1 + \cdots + e_p) = e^{2i+1}(\Sigma_H e_1) = |H| e_1 \,,$$

and this is $0 \bmod (p)$. Also, in odd dimensions $e_1 + \cdots + e_p$ represents a generator in homology, while e_1 represents a generator in even dimensions.

□

Corollary 5.7 *Assume $G = \mathbb{Z}/p^i$ with $i > 1$ so that $H \neq \{1\}$. Then we have*

a. $\mathrm{tr}\colon H^j(H; \mathbb{F}_p) \longrightarrow H^j(G; \mathbb{F}_p)$ *is the zero map for j even and the identity when j is odd.*

b. $\mathrm{res}^*\colon H^j(G; \mathbb{F}_p) \longrightarrow H^j(H; \mathbb{F}_p)$ *is zero when j is odd and the identity when j is even.*

Proof. We have

$$\begin{aligned} \mathrm{tr}(b^{2i})(e_1) &= \sum_{j=0}^{p-1} T^j b^{2i}(T^{-j} e_1) = p, \\ \mathrm{tr}(e^{2i+1})(e_1) &= \sum_{j=0}^{p-1} T^j e^{2i+1}(T^{-j} e_1) = 1\,. \end{aligned}$$

To prove (b) note that the inclusions $H \subset G$ induces the chain map of minimal complexes

$$\begin{cases} e_i \longrightarrow (1 + T + \cdots + T^{p-1}) e_1 & \text{in odd degrees,} \\ e_i \longrightarrow e_i & \text{in even degrees.} \end{cases} \quad \square$$

Remark 5.8 The special case $\{1\} \subset \mathbb{Z}/p$ is not covered by (5.7). Here, since

$$H^i(\{1\}; \mathbb{Z}) = \begin{cases} 0 & \text{when } i > 0, \\ \mathbb{Z} & \text{when } i = 0, \end{cases}$$

we see that the transfer is simply multiplication by p in degree 0. In particular, with $\mathbb{F}_p$-coefficients it is identically zero.

Corollary 5.9 *Let $G = (\mathbb{Z}/p)^n$ be an elementary p-group and suppose $H \subsetneq G$ is a proper subgroup. Then* $\mathrm{tr}\colon H^*(H; \mathbb{F}_p) \longrightarrow H^*(G; \mathbb{F}_p)$ *is zero in all degrees.*

Proof. Since the composition of transfers is the transfer of the composition we can assume $H = (\mathbb{Z}/p)^{n-1}$ has index p in G. Moreover, after a change of basis we can assume $H = \{1\} \times (\mathbb{Z}/p)^{n-1} \subset (\mathbb{Z}/p)^n$. Then the result follows from (5.5). □

More generally, we can use (5.5) and (5.7) to give the transfer and restriction explicitly for G any finite abelian group and H any subgroup. As (5.9) is essentially the only case we need in the remainder of this work we leave this calculation to the reader as an (important) exercise.

However, there is a further case where the transfer is very useful. This is when G has an index 2 subgroup H, so we have the extension $H \triangleleft G \longrightarrow \mathbb{Z}/2$. Then the transfer $tr: \mathcal{C}^i_H \longrightarrow \mathcal{C}^i_G$ fits into a short exact sequence

$$0 \longrightarrow \mathcal{C}^i_G \xrightarrow{res^*} \mathcal{C}^i_H \xrightarrow{tr} \mathcal{C}^i_G \longrightarrow 0$$

as long as the action of G on the coefficients A is trivial. Consequently, there is a long exact sequence which turns out to be a special case of the Gysin sequence

5.10
$$\cdots \xrightarrow{\delta} H^i(G;\mathbb{A}) \xrightarrow{res^*} H^i(H;\mathbb{A}) \xrightarrow{tr} H^i(G;\mathbb{A}) \xrightarrow{\delta} H^{i+1}(G;\mathbb{A}) \xrightarrow{res^*} \cdots .$$

When the coefficients $A = \mathbb{F}_2$ the map χ turns out to be $\alpha \mapsto \alpha \cup f$ for all $\alpha \in H^i(G;\mathbb{F}_2)$ where $f \in H^1(G;\mathbb{F}_2)$ is $B^*_\pi(e)$, $e \in H^1(\mathbb{Z}/2;\mathbb{F}/2)$ is the non-zero class, and B_π is the map induced by the projection $G \longrightarrow \mathbb{Z}/2$. This is actually quite easy, just track back using the chain level definition of δ and compare it with the cup product associated to the composition

$$B_G \xrightarrow{\Delta} B_G \times B_G \xrightarrow{B_\pi \times id} B_{\mathbb{Z}/2} \times B_G .$$

As a special case note that the Gysin sequence gives a second proof that $H^*(\mathbb{Z}/2;\mathbb{F}_2) = \mathbb{F}_2[e]$ which is the key step in 4.4.

An Alternate Construction of the Transfer

Geometrically we can view the transfer as follows. First, the map $p: B_H \longrightarrow B_G$ induced by the inclusion $H \hookrightarrow G$ can be thought of as a covering with fiber the set of right cosets of H in G by simply regarding $B_H = * \times_H E_G$ and letting p be the projection onto B_G. Then the chain map $\mathcal{C}_\#(B_G) \longrightarrow \mathcal{C}_\#(B_H)$ associated to the transfer is given as

$$|g_1|\cdots|g_n| \mapsto \sum_{Hg_v} Hg_v|g_1|\cdots|g_n|$$

where the Hg_v run over the right cosets of H in G. That is to say, one takes the sum of all the cells in the inverse image of a cell of B_G.

We can model this more formally be considering $g_1, \ldots, g_n$ as representatives for the right cosets of H in G so $G = \sqcup_{i=1}^{n} Hg_i$, and define a map

$$Tr\colon E_G \longrightarrow (E_G)^n \text{ by } \theta \mapsto (g_1\theta, \ldots, g_n\theta)$$

as θ runs over E_G. (This is the geometric analogue of summing over fibers.) Note that $g\theta \mapsto (g_1 g\theta, \ldots, g_n g\theta)$, and writing

$$g_i g = h_{\sigma_g(i)} g_{\sigma_g(i)}$$

where σ_g is the permutation of cosets associated to g, we obtain a homomorphism of G to the wreath product $G \wr S_n$ which is defined in (IV.1),

$$\lambda\colon G \longrightarrow H \wr S_n, \qquad g \mapsto (h_{\sigma_g^{-1}(1)}, \ldots, h_{\sigma_g^{-1}(n)}, \sigma_g).$$

λ is called the Frobenius map associated to the section $(g_1, \ldots, g_n)$. Thus

$$Tr \cdot g = \lambda(g) Tr,$$

i.e. for all $x \in E_G$, $g \in G$ we have $Tr(xg) = \lambda(g)Tr(x)$. In particular $H \wr S_n$ acts on $(E_G)^n$ and Tr is λ-equivariant. On the other hand $(E_G)^n$ is not $H \wr S_n$ free though it is H^n-free. However, if we take the symmetric product $SP^n(* \times_H E_G)$ then Tr induces a well defined and continuous map

$$\text{tr}\colon B_G \longrightarrow SP^n(* \times_H E_G), \qquad \theta \mapsto \sum_1^n g_i\theta.$$

Using tr we can give a geometric construction of the transfer for $\alpha \in H^m(B_H; A)$ where the A are *untwisted* coefficients.

Let α be represented by

$$\alpha\colon B_H \longrightarrow K(A, m) = B_A^m.$$

From II.1.8 B_A^m is an abelian topological group, so there is a natural extension of α to $SP^n(B_H)$,

5.11 $$\alpha\left(\sum_1^n b_i\right) = \sum_1^i (\alpha(b_i)),$$

and it is a good exercise with chain approximations to verify that $(\alpha \cdot \text{tr})^*(\iota_m) = \text{tr}(\alpha)$ agrees with the previous definition of the transfer.

We can actually carry things a bit further, taking advantage of the Frobenius homomorphism λ.

Let $h\colon E_G \to E_{S_n}$ be any (CW) map which is equivariant with respect to the homomorphism

$$p \cdot \lambda\colon G \longrightarrow H \wr S_n \longrightarrow S_n$$

which gives the permutation action of G on the right cosets of H. Then

$$E_{Tr} \times h\colon E_G \longrightarrow (E_G)^n \times E_{S_n}$$

is λ-equivariant and $(E_G)^n \times E_{S_n}$ is $H \wr S_n$-free and contractible. Consequently $E_{Tr} \times h$ passes to quotients and induces the map

$$\mathrm{tr}_\lambda\colon B_G \longrightarrow (* \times_H E_G)^n \times_{S_n} E_{S_n} \simeq B_{H \wr S_n}\ .$$

tr_λ is, up to homotopy, the map of classifying spaces induced by the homomorphism λ. On points this representation of it has the form

$$x \mapsto \{(\tilde{x}_1, \ldots, \tilde{x}_n, h(x))\}$$

where $\tilde{x}_1, \ldots, \tilde{x}_n$ are the points in the covering $* \times_H E_G \to B_G$ lying over x.

In fact, on the level of homology the map tr_λ is independent of the choice of section $g_1, \ldots, g_n$ used in the definition of λ since we have

Lemma 5.12 *Let $g'_1, \ldots, g'_n$ be a second set of right coset representatives of H in G and λ' the associated Frobenius homomorphism, $\lambda'\colon G \to H \wr S_n$. Then λ' and λ differ by an inner automorphism of $H \wr S_n$.*

Proof. We have $g'_i = \bar{h}_i g_i$, $1 \le i \le n$, so

$$\begin{aligned} g'_i g &= h'_{\sigma_g(i)} g'_{\sigma_g(i)} \\ &= \bar{h}_i g_i g \\ &= \bar{h}_i h_{\sigma_g(i)} \bar{h}^{-1}_{\sigma_g(i)} g'_{\sigma_g(i)} \end{aligned}$$

and conjugation by $(\bar{h}_1, \ldots, \bar{h}_n, 1)$ takes λ to λ'. □

Finally, we note that our second description of the transfer, 5.11, actually factors through the map tr_λ since the map

$$P\colon (* \times_H E_G)^n \times_{S_n} E_{S_n} \longrightarrow SP^n(* \times_H E_G)\ ,$$

$$P(\{x_1, \ldots, x_n, w\}) = \sum_1^n x_i \in SP^n(* \times_H E_G)$$

gives us a factorization of tr as $P \cdot \mathrm{tr}_\lambda$.

This last construction of the transfer, using tr_λ is actually very important in applications since it can be generalized substantially. In fact, given any functorial method of associating cohomology classes $\theta(\alpha) \in H^*(H \wr S_n; A)$ to cohomology classes $\alpha \in H^*(H; A)$ we obtain associated cohomology classes in $H^*(G; A)$. It is this principle, first used by Steenrod, which enables us, in Chapter IV, to construct the Steenrod operations. The principle is also very important in homotopy theory where it provides the basis for both the Kahn-Priddy theorem, [KP], and the Snaith splitting theorem [Sn].

6. The Cartan–Eilenberg Double Coset Formula

We now describe a useful method for computing the restriction map using double cosets which was first developed by Cartan and Eilenberg [CE]. Let G be a finite group and $H, K \subset G$ subgroups. For a given $\mathbb{Z}(G)$-module A, we will consider the composition

$$H^*(K;A) \xrightarrow{\mathrm{tr}_K^G} H^*(G;A) \xrightarrow{(\mathrm{res}_H^G)^*} H^*(H;A)\,.$$

First some notation: $c_x\colon H^*(H;A) \longrightarrow H^*(xHx^{-1};A)$ $(x \in G)$ will denote the isomorphism induced by the homomorphism

$$\mathrm{Hom}_H(C;A) \longrightarrow \mathrm{Hom}_{xHx^{-1}}(C;A)$$

given by $c_x(f)(u) = xf(x^{-1}u)$.

Now take a decomposition of G into double cosets:

$$(*) \qquad G = \coprod_i Hx_iK\,.$$

Remark 6.1 The double coset decomposition can be understood as follows. Given K, $H \subset G$ the left coset decomposition of G over K defines a homomorphism $\phi : G \longrightarrow \mathcal{S}_{[G:K]}$. If we restrict ϕ to H then the image breaks up into separate orbits, or, equivalently, $\phi(H) \subset \mathcal{S}_{k_1} \times \cdots \times \mathcal{S}_{k_r} \subset \mathcal{S}_{[G:K]}$ where r is the number of double cosets and $k_i = |Hg_iK|/|K| = [K : H \cap g_iKg_i^{-1}]$ where Hg_iK is the ith double coset.

Theorem 6.2 *Given the decomposition* $(*)$ *we have*
1. $[G\colon K] = \sum_i [H\colon H \cap x_iKx_i^{-1}]$,
2. $\mathrm{res}_H^G \mathrm{tr}_K^G = \sum_i \mathrm{tr}^H_{H\cap x_iKx_i^{-1}} \cdot \mathrm{res}^{x_iKx_i^{-1}}_{H\cap x_iKx_i^{-1}} \cdot c_{x_i}$.

Proof. Let $W_i = H \cap x_iKx_i^{-1}$ and take a left coset decomposition of H with respect to this subgroup: $H = \coprod_j z_{ji}W_i$. Then we have

$$Hx_i = \coprod_j z_{ji}W_ix_i = \coprod_j z_{ji}\,(Hx_i \cap x_iK)\,.$$

Multiplying by K on the right, we have

$$Hx_iK = \coprod_j z_{ji}(Hx_iK \cap x_iK) = \coprod_j z_{ji}x_iK\,.$$

From $(*)$ we obtain

$$(**) \qquad G = \coprod_{i,j} z_{ji}x_iK,$$

a disjoint union of left cosets. Hence $[G{:}K] = \sum_i [H{:}H \cap x_i K x_i^{-1}]$, proving (1).

For (2), let F_* be a free resolution of $\mathbb{Z}$ over $\mathbb{Z}(G)$, and $\phi \in \mathrm{Hom}_K(F_n, A)$:

$$\begin{aligned}
\mathrm{res}^G_H \cdot \mathrm{tr}^G_K \phi &= \mathrm{res}^G_H \left(\sum_{g \in G/K} g\phi g^{-1} \right) \\
&= \sum_i \left(\sum_j z_{ji} x_i \phi x_i^{-1} z_{ji}^{-1} \right) \quad \text{in } \mathrm{Hom}_H(F_n, A) \\
&= \sum_i \mathrm{tr}^H_{H \cap x_i K x_i^{-1}} \cdot \mathrm{res}^{x_i K x_i^{-1}}_{H \cap x_i K x_i^{-1}} \cdot c_{x_i}(\phi) .
\end{aligned}$$

□

Corollary 6.3 *If $H \triangleleft G$, then for $b \in H^*(H; A)$ we have $\mathrm{res}^G_H \cdot \mathrm{tr}^G_H(b) = \sum xb$, for $x \in G/H$.*

We introduce the notion of stable elements: $a \in H^*(H; A)$ will be called *stable* if $\mathrm{res}^{xHx^{-1}}_{xHx^{-1} \cap H} \cdot c_x(a) = \mathrm{res}^H_{xHx^{-1} \cap H}(a)$ for all $x \in G$. Note that if $H \triangleleft G$, this reduces to $c_x a = a$, i.e. a must be *invariant*. We have:

Proposition 6.4 *If $a \in \mathrm{im}(\mathrm{res}^G_H)$, then a is stable.*

Proof. Let $a = \mathrm{res}^G_H(b)$; note that $c_x(b) = b$ and $c_x(a) = c_x \cdot \mathrm{res}^G_H(b) = \mathrm{res}^G_{xHx^{-1}} \cdot c_x(b) = \mathrm{res}^G_{xHx^{-1}}(b)$ which implies that

$$\begin{aligned}
\mathrm{res}^{xHx^{-1}}_{xHx^{-1}} \cdot c_x(a) &= \mathrm{res}^{xHx^{-1}}_{xHx^{1-} \cap H} \cdot \mathrm{res}^G_{xHx^{-1}}(b) \\
&= \mathrm{res}^G_{xHx^{-1} \cap H}(b) = \mathrm{res}^H_{xHx^{-1} \cap H} \cdot \mathrm{res}^G_H(b) \\
&= \mathrm{res}^H_{xHx^{-1} \cap H}(a) .
\end{aligned}$$

□

Proposition 6.5 *If $a \in H^*(H; A)$ is stable, then $\mathrm{res}^G_H \cdot \mathrm{tr}^G_H(a) = [G{:}H]a$.*

Proof. By theorem 1 we have

$$\begin{aligned}
\mathrm{res}^G_H \cdot \mathrm{tr}^G_H(a) &= \sum_i \mathrm{tr}^H_{x_i H x_i^{-1} \cap H} \cdot \mathrm{res}^{x_i H x_i^{-1}}_{x_i H x_i^{-1} \cap H} c_{x_i}(a) \\
&= \sum_i \mathrm{tr}^H_{x_i H x_i^{-1} \cap H} \cdot \mathrm{res}^H_{x_i H x_i^{-1} \cap H}(a) \\
&= \sum_i [H{:}x_i H x_i^{-1} \cap H] a \\
&= [G{:}H]a .
\end{aligned}$$

□

We now apply these results to the special situation where $H \subseteq G$ contains $\mathrm{Syl}_p(G)$, the p-Sylow subgroup of G.

Theorem 6.6 *Let $G \supseteq H \supseteq \mathrm{Syl}_p(G)$, where $\mathrm{Syl}_p(G)$ is the p-Sylow subgroup of G. Then, for any $\mathbb{Z}(G)$-module A, $\mathrm{tr}_H^G: \bar{H}^*(H;A) \to \bar{H}^*(G;A)_{(p)}$ is surjective,*

$$\mathrm{res}_H^G: H^*(G;A)_{(p)} \to H^*(H;A)_{(p)}$$

is injective, and its image consists of the stable elements in $H^*(H;A)_{(p)}$. In particular, if $H \lhd G$, then G/H acts on $H^*(H;A)$, and

$$H^*(H;A)_{(p)}^{G/H} = \mathrm{im}\left(\mathrm{res}_H^G\right) \cong H^*(G;A)_{(p)} .$$

Proof. Assume $|\mathrm{Syl}_p(G)| = p^s$, $[G;H] = q$; then we may choose l so that $ql \equiv 1 \bmod p^s$. We know from Proposition 3 that the elements of $\mathrm{im}\left(\mathrm{res}_H^G\right)$ are stable. Conversely, assume $a \in H^*(H;A)_{(p)}$ is stable; then

$$l \cdot \mathrm{res}_H^G \mathrm{tr}_H^G(a) = l[G:H]a = lq \cdot a = a$$

(this follows from Proposition 4; note that $p^s a = 0$ if a is positive dimensional) and this implies that $a \in \mathrm{im}\left(\mathrm{res}_H^G\right)$.

Recall that we also have for $b \in H^*(G;A)$

$$\mathrm{tr}_H^G \cdot \mathrm{res}_H^G(b) = [G:H]b ;$$

we deduce our assertions by looking at the p-component of this composition.

In the special case $H \lhd G$, G/H acts on $H^*(H;A)_{(p)}$ and the stable elements are, of course, the invariant ones; hence

$$H^*(G;A)_{(p)} \cong \mathrm{im}\left(\mathrm{res}_H^G\right) = H^*(H;A)_{(p)}^{G/H} . \qquad \square$$

Example 6.7 For the symmetric groups we have

$$\mathcal{S}_{n+1} = \mathcal{S}_n \cup \mathcal{S}_n(n,n+1)\mathcal{S}_n$$

gives the double coset decomposition where $\mathcal{S}_n \subset \mathcal{S}_{n+1}$ is the set of elements leaving $n+1$ fixed. Moreover

$$\mathcal{S}_n \cap (n,n+1)\mathcal{S}_n(n,n+1) = \mathcal{S}_{n-1}$$

where $\mathcal{S}_{n-1}$ leaves both n, $n+1$ fixed, and $c_{(n,n+1)}$ is the identity on $\mathcal{S}_{n-1}$. It follows that

$$H^*(\mathcal{S}_{n+1};\mathbb{F}_p) \xrightarrow{\mathrm{res}_{\mathcal{S}_n}^{\mathcal{S}_{n+1}}} H^*(\mathcal{S}_n;\mathbb{F}_p)$$

is an isomorphism for all p prime to $n+1$.

The stability conditions depend on how H intersects with its conjugates, which in general can be quite involved. We conclude by describing a situation where things simplify. The following result is due to R. Swan, [S1].

Theorem 6.8 *Let G be a finite group such that $\mathrm{Syl}_p(G)$ is abelian; then if A is a trivial G–module*

$$\mathrm{im}\left(\mathrm{res}^G_{\mathrm{Syl}_p(G)}\right) = H^*(\mathrm{Syl}_p(G);A)^{N_G(\mathrm{Syl}_p(G))} .$$

Proof. Define $f_x: x\mathrm{Syl}_p(G)x^{-1} \cap \mathrm{Syl}_p(G) \longrightarrow \mathrm{Syl}_p(G)$ by $f_x(y) = x^{-1}yx$; then it is not hard to see that

$$f_x^*(\alpha) = \mathrm{res}^{x\mathrm{Syl}_p(G)x^{-1}}_{x\mathrm{Syl}_p(G)x^{-1}\cap\mathrm{Syl}_p(G)} \cdot c_x(\alpha), \qquad \alpha \in H^*(\mathrm{Syl}_p(G);A) .$$

Hence α is stable if and only if

$$f_x^*(\alpha) = \mathrm{res}^{\mathrm{Syl}_p(G)}_{x\mathrm{Syl}_p(G)x^{-1}\cap\mathrm{Syl}_p(G)}(\alpha) .$$

Let C be the centralizer of $\mathrm{Syl}_p(G) \cap x\mathrm{Syl}_p(G)x^{-1}$; then there exists $t \in C$ so that $tx\mathrm{Syl}_p(G)x^{-1}t^{-1} = \mathrm{Syl}_p(G)$ (any two Sylow p-subgroups of C are conjugate). For any $y \in \mathrm{Syl}_p(G) \cap x\mathrm{Syl}_p(G)x^{-1}$,

$$f_x(y) = x^{-1}yx = x^{-1}t^{-1}ytx = f_{tx}(y) .$$

We can think of $f_{tx} \in N_G(\mathrm{Syl}_p(G))$, and by the above we obtain a factorizaton

$$\begin{array}{ccc} x\mathrm{Syl}_p(G)x^{-1} \cap \mathrm{Syl}_p(G) & \hookrightarrow & \mathrm{Syl}_p(G) \\ f_x \searrow & & \swarrow f_{tx} \\ & \mathrm{Syl}_p(G) & . \end{array}$$

Hence $f_x^*(\alpha) = \mathrm{res}^{\mathrm{Syl}_p(G)}_{x\mathrm{Syl}_p(g)x^{-1}\cap\mathrm{Syl}_p(G)} \cdot f_{tx}^*(\alpha)$. Assume now that

$$\alpha \in H^*(\mathrm{Syl}_p(G);A)^{N_G(\mathrm{Syl}_p(G))} ,$$

then, since $f_{tx} \in N_G(\mathrm{Syl}_p(G))$, we have $f_x^*(\alpha) = \mathrm{res}^{\mathrm{Syl}_p(G)}_{x\mathrm{Syl}_p(G)x^{-1}\cap\mathrm{Syl}_p(G)}(\alpha)$ and this implies that α is stable which gives the result using (6.6). □

6.9 An Application: $H^*(J_1;\mathbb{F}_2)$

J_1 is the first Janko group, a sporadic simple group of order $8\cdot3\cdot5\cdot7\cdot11\cdot19 = 175,560$ with Sylow 2-subgroup $J_{1(2)} = (\mathbb{Z}/2)^3$. J_1 is the unique simple group which satisfies

1. The Sylow 2-subgroups of G are abelian.
2. G contains an involution t such that $C(t) = \langle t\rangle \times \mathcal{A}_5$.

$N_{J_1}(J_{1(2)})/J_{1(2)}$ is the semi-direct product $\mathbb{Z}/7 \times_T \mathbb{Z}/3$ with generators f of order 3 and g of order 7. Their action can be given in terms of 3×3 matrices as follows:

$$f \mapsto \begin{pmatrix} 1 & 0 & 1 \\ 0 & 0 & 1 \\ 0 & 1 & 1 \end{pmatrix}, \qquad g = \begin{pmatrix} 0 & 0 & 1 \\ 1 & 0 & 0 \\ 0 & 1 & 1 \end{pmatrix}.$$

In III.1.9 we will determine the resulting ring of invariants (or at least its generators, leaving the relations to the reader). This ring will come up again in Chapters VIII and IX. In IX the explicit form of the generators will be very important. The result is

$$\begin{aligned} H^*(J_1;\mathbb{F}_2) &\cong H^*((\mathbb{Z}/2)^3;\mathbb{F}_2)^{\mathbb{Z}/7\times_T\mathbb{Z}/3} \\ &\cong \mathbb{F}_2[x,y,z](\gamma,\mu) \Big/ \begin{matrix} \gamma^2 + y\mu + xz = 0 \\ \mu^2 + x^4 + x^2\mu + y^3 + \gamma z = 0 \end{matrix} \end{aligned}$$

where $\dim(x) = 3$, $\dim(y) = 4$, $\dim(z) = 7$, $\dim(\gamma) = 5$, and $\dim(\mu) = 6$. (This result is due to Chapman [Ch].) The Poincaré series for $H^*(J_1;\mathbb{F}_2)$ is $\dfrac{(1+t^5)(1+t^6)}{(1-t^3)(1-t^4)(1-t^7)}$.

Janko describes J_1 as the subgroup of $\mathrm{GL}_7(\mathbb{F}_{11})$ generated by the following two matrices Y of order 7 and Z of order 5:

$$Y = \begin{pmatrix} 0 & 1 & 0 & 0 & 0 & 0 & 0 \\ 0 & 0 & 1 & 0 & 0 & 0 & 0 \\ 0 & 0 & 0 & 1 & 0 & 0 & 0 \\ 0 & 0 & 0 & 0 & 1 & 0 & 0 \\ 0 & 0 & 0 & 0 & 0 & 1 & 0 \\ 0 & 0 & 0 & 0 & 0 & 0 & 1 \\ 1 & 0 & 0 & 0 & 0 & 0 & 0 \end{pmatrix}$$

$$Z = \begin{pmatrix} -3 & 2 & -1 & -1 & -3 & -1 & -3 \\ -2 & 1 & 1 & 3 & 1 & 3 & 3 \\ -1 & -1 & -3 & -1 & -3 & -3 & 2 \\ -1 & -3 & -1 & -3 & -3 & 2 & -1 \\ -3 & -1 & -3 & -3 & 2 & -1 & -1 \\ 1 & 3 & 3 & -2 & 1 & 1 & 3 \\ 3 & 3 & -2 & 1 & 1 & 3 & 1 \end{pmatrix}.$$

J_1 is an example of a very sparse type of simple group, one with abelian 2-Sylow subgroup. A very nice result in group theory classifying such groups has been proved by Walter, with a simplified proof being given by H. Bender [B].

Theorem 6.10 *If G is a simple group with abelian Sylow 2-subgroups, then $G \cong \mathrm{PSL}_2(\mathbb{F}_q)$, $q \equiv 3, 5 (\mathrm{mod}\ 8)$, $\mathrm{PSL}_2(\mathbb{F}_{2^n})$, $n \geq 2$, J_1, or ${}^2G_2(\mathbb{F}_{3^n})$, n odd $n > 1$.*

Here ${}^2G_2(\mathbb{F}_{3^n})$ is built from the finite groups of Lie type $G_2(\mathbb{F})$ described in (VII.7). The Dynkin diagram for G_2, 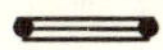 admits a symmetry, reflection, which is not realized as a symmetry of the group except for the fields $\mathbb{F}_{3^n}$. In this case, though, there is a symmetry and ${}^2G_2(\mathbb{F}_{3^n})$ is the subgroup of fixed elements.

The subgroups of $\mathrm{PSL}_2(\mathbb{F}_q)$ occuring above are all isomorphic to $\mathbb{Z}/2 \times \mathbb{Z}/2$. Moreover, since the groups $\mathrm{PSL}_2(\mathbb{F}_q)$ are simple, it follows that the normalizer of $\mathbb{Z}/2 \times \mathbb{Z}/2$ in $\mathrm{PSL}_2(\mathbb{F}_q)$ above must be $\mathcal{A}_4$, and our calculation for $\mathcal{A}_4$ in III.1.3 gives their mod 2 cohomology.

The groups $\mathrm{PSL}_2(\mathbb{F}_{2^n})$ have Sylow 2-subgroups of the form

$$(\mathbb{Z}/2)^n = \left\{ A = \begin{pmatrix} 1 & a \\ 0 & 1 \end{pmatrix} \;\middle|\; a \in \mathbb{F}_{2^n} \right\},$$

with normalizer the semi-direct product of $(\mathbb{Z}/2)^n$ with the subgroup $N = \mathbb{Z}/(2^n - 1) = \mathbb{F}^{\bullet}{}_{2^n}$ embedded in $\mathrm{PSL}_2(\mathbb{F}_{2^n})$ as the diagonal matrices $\xi \mapsto \begin{pmatrix} \xi & 0 \\ 0 & \xi^{-1} \end{pmatrix}$. Since

$$\begin{pmatrix} \xi & 0 \\ 0 & \xi^{-1} \end{pmatrix} \begin{pmatrix} 1 & a \\ 0 & 1 \end{pmatrix} \begin{pmatrix} \xi^{-1} & 0 \\ 0 & \xi \end{pmatrix} = \begin{pmatrix} 1 & \xi^2 a \\ 0 & 1 \end{pmatrix},$$

this action is easily specified, but hard to compute in general, and we have

$$H^*(\mathrm{PSL}_2(\mathbb{F}_{2^n}); \mathbb{F}_2) \cong \mathbb{F}_2[x_1, \ldots, x_n]^{\mathbb{Z}/(2^n-1)}.$$

J_1 was discussed above, and $\mathrm{Syl}_2({}^2G_2(\mathbb{F}_{3^n})) = (\mathbb{Z}/2)^3$ with normalizer isomorphic to that of J_1 hence

$$H^*({}^2G_2(\mathbb{F}_{3^n}); \mathbb{F}_2) = \mathbb{F}_2[x_1, x_2, x_3]^{\mathbb{Z}/7 \times_T \mathbb{Z}/3}.$$

In Chapter III we discuss the techniques for determining these invariant subrings. An important thing to note is that they are closed under the action of the Steenrod algebra $\mathcal{A}(2)$.

7. Tate Cohomology and Applications

We now describe a version of cohomology obtained by using a complete resolution instead of an ordinary one, see (II.3.7) for further remarks. By definition, a *complete resolution* of $\mathbb{Z}$ (the trivial $\mathbb{Z}G$–module) is an acyclic $\mathbb{Z}$–graded complex $\mathcal{F}_*$ of projective $\mathbb{Z}G$–modules which agrees with an ordinary projective resolution of $\mathbb{Z}$ in non–negative dimensions. To construct such a complex we proceed as follows. Let F_* be a free resolution of $\mathbb{Z}$ where $F_0 = \mathbb{Z}G$. Now consider its $\mathbb{Z}$–dual, F^*. This is now a backwards free resolution (a coresolution) of $\mathbb{Z}$, as $\mathbb{Z}G$ is isomorphic to its dual. We now glue the two resolutions where they end (begin) respectively.

$$\begin{array}{ccccccccccc} \cdots & \longleftarrow & C^1 & \longleftarrow & C^0 & & \longleftarrow & & C_0 & \longleftarrow & C_1 & \longleftarrow & \cdots \\ & & & & & \nwarrow & & \swarrow & & & & & \\ & & & & & & \mathbb{Z} & & & & & & \end{array}$$

Let $N = \sum_{g \in G} g$ be the trace (or norm) map from F_0 to F^0, this clearly factors through $\mathbb{Z}$, as the augmentation followed by the inclusion of invariants. Now we define $\mathcal{F}_*$ as the chain complex

$$\mathcal{F}_i = \begin{cases} F_i, & \text{if } i \geq 0 \\ F^{-i-1}, & \text{if } i < 0 \end{cases}$$

endowed with the inherited differentials plus the glueing (norm) map in dimension zero. As for ordinary resolutions, one can show that complete resolutions together with the augmentation map $\mathcal{F}_* \to \mathbb{Z}$ are unique up to chain homotopy, and hence can be used for homological definitions unambiguously.

We can now define

Definition 7.1 For any $\mathbb{Z}G$–module M, the Tate Cohomology of G with coefficients in M is defined for all $i \in \mathbb{Z}$ as

$$\widehat{H}^i(G, M) = H^i(\mathrm{Hom}_G(\mathcal{F}_*, M)) .$$

Denote by I the augmentation ideal in $\mathbb{Z}G$ and by N the norm map as above. Then we can identify Tate Cohomology as follows

$$\widehat{H}^i(G, M) \cong \begin{cases} H^i(G, M), & \text{if } i > 0; \\ H_{-i-1}(G, M), & \text{if } i < -1; \\ M^G/NM, & \text{if } i = 0; \\ \mathrm{Ker}N/IM, & \text{if } i = -1. \end{cases}$$

The above is clear except in dimensions less than -1. For that we use the natural isomorphism $\mathrm{Hom}_G(F^*, M) \cong F_* \otimes_G M$ (derived from the fact that F_* is $\mathbb{Z}G$–free) and the definition for homology (3.7). Note the following simple fact: if $\mathcal{F}$ is a free $\mathbb{Z}G$–module, then $\widehat{H}^*(G, \mathcal{F}) \equiv 0$. This follows from the fact that $\mathcal{F}$ has a trivial projective resolution (itself) and that $\mathcal{F}^G = N\mathcal{F}, \mathrm{Ker}N = I\mathcal{F}$. As a consequence the same must be true for any projective $\mathbb{Z}G$–module. An additional advantage of Tate cohomology is that it allows us to consider homology and cohomology at the same time. It will enjoy the same general properties as ordinary cohomology with respect to short exact coefficient sequences, restriction and transfer, maps induced by group homomorphisms etc.. The following is also a very useful device, known as "dimension–shifting":

Lemma 7.2 *For any finitely generated $\mathbb{Z}G$–module M and integer i, there exists a finitely generated $\mathbb{Z}$–torsion free $\mathbb{Z}G$–module $\Omega^i(M)$ such that for all $n \in \mathbb{Z}$*

$$\widehat{H}^n(G, \Omega^i(M)) \cong \widehat{H}^{n-i}(G, M).$$

Proof. Given any $\mathbb{Z}G$–module M, we may map a free module $\mathcal{F}$ onto it, with kernel K. It follows from the long exact sequence associated to this sequence, that $\widehat{H}^n(G, M) \cong \widehat{H}^{n+1}(G, K)$ for all $n \in \mathbb{Z}$. Now K embeds in a free module as

$$\mathrm{Hom}(\mathbb{Z}, K) \hookrightarrow \mathrm{Hom}(\mathbb{Z}G, K),$$

and the cokernel will be a $\mathbb{Z}$–torsion free module M' with the same cohomology as M, by the same argument. It is then clear that iterating the same type of procedure for this module (known as dimension–shifting), in either direction, we can construct a $\mathbb{Z}$–torsion free module $\Omega^i(M)$, for any $i \in \mathbb{Z}$ with the desired property. □

This allows for considerable flexibility in analyzing a particular cohomology group. Also note that the analogous construction can be done for a module defined over a field of characteristic p. Note that we have said nothing about the uniqueness of the dimension–shift of a module.

To relate cohomology to representation–theoretic properties of a module over a finite p–group, we introduce the notion of a *minimal* projective resolution.

Definition 7.3 Let P be a finite p–group and M a finitely generated $\mathbb{Z}$–torsion free $\mathbb{Z}P$–module. A *minimal* resolution for M over $\mathbb{Z}P$ is a projective resolution $(\mathcal{P}_i, \delta_i)$ of M such that $\mathcal{P}_n$ is a projective module of minimal $\mathbb{Z}$–rank mapping onto $\ker\delta_{n-1}$ for all $n \geq 0$ (for $n = 0$ this means $\mathcal{P}_0$ is a projective of minimal rank mapping onto M).

The above definition can be given for any finite group, but we choose to restrict ourselves to finite p–groups as in that situation there is a simple way of describing a minimal resolution using cohomology. We have

Theorem 7.4 *A $\mathbb{Z}P$–projective resolution $(\mathcal{P}_i, \delta_i)$ of a finitely generated $\mathbb{Z}$–torsion free $\mathbb{Z}P$–module M is minimal if and only if*

$$rk_{\mathbb{Z}}\mathcal{P}_n = |P| \dim\left(H^n(P, (M \otimes \mathbb{F}_p)^*)\right),$$

*where * denotes the usual dual.*

Proof. We start by constructing a projective module $\mathcal{P}_0$ of minimal rank which maps *onto* M. Denote $M_p = M \otimes \mathbb{F}_p$; then observe that the coinvariants $(M_p)_P \neq 0$. This is simply the dual statement to the fact that a finite p–group acting on a finite abelian p–group must have fixed–points. Now the module of coinvariants $M_P = M/IM$ maps onto this, hence it must be non–zero. Assume it is generated by classes $x_1, \ldots, x_r$ in M, where as $(M_p)_P \cong M_P \otimes \mathbb{F}_p$ and M_P has no p'–torsion, $r = \dim (M_p)_P$. Then use these classes to define a map $(\mathbb{Z}P)^r \to M$, denoted by π. By construction, $(\operatorname{coker}\pi)_P = 0$ hence by the argument above, $\operatorname{coker}\pi$ is p' torsion. Choose an integer q prime to p which annihilates it, and let $\mathcal{P}_0 = \pi^{-1}(qM) \subset (\mathbb{Z}P)^r$. Then $\mathcal{P}_0$ is projective because $(\mathbb{Z}P)^r/\mathcal{P}_0$ is p'–torsion, and it clearly maps onto $qM \cong M$. Denote $K_{n+1} = \ker\delta_n$; the coinvariants of any other projective module covering M would have to map onto those of M_p, hence we have shown that the one of minimal rank *has* rank $|P| \dim (M_p)_P = |P| \dim (M_p^*)^P$. Now by iterating this procedure, we can construct a minimal resolution $\mathcal{P}_*$ for M, with $\operatorname{rank}(\mathcal{P}_i) = |P| \dim [(K_i)_p^*]^P$. However, the modules $(K_j)_p^*$ are j-fold dimension–shifts of M_p^*, and by (7.2), $H^i(P, M_p^*) \cong H^1(P, (K_{i-1})_p^*)$. Now from the exact sequence

$$0 \to (K_{i-1})_p^* \to \mathcal{P}_i^* \to (K_i)_p^* \to 0$$

and the fact that the map on the left induces an isomorphism of invariants (dual to the statement about the coinvariants of the original modules, which is true by construction) we deduce from the long exact sequence in cohomology that $H^1(P, (K_{i-1})_p^*) \cong [(K_i)_p^*]^P$. This completes the proof of this result, which is an extension of a theorem due to Swan [S2] which appears in [A1]. □

Now let us consider the partial Euler characteristic of such a resolution. We have

$$\begin{aligned} rk_{\mathbb{Z}}(M) + (-1)^n rk_{\mathbb{Z}}(K_n) &= \sum_{j=0}^{n} (-1)^j rk_{\mathbb{Z}}(\mathcal{P}_j) \\ &= \sum_{j=0}^{n} |P|(-1)^j \dim H^j(P, M_p^*)\,. \end{aligned}$$

As M is $\mathbb{Z}$–torsion free, we have a short exact sequence $0 \to M \to M \to M_p \to 0$. Using the induced long exact sequence in cohomology and using M^* instead of M, we deduce that $\dim H^i(P, M_p^*) = \dim H^i(P, M^*) \otimes \mathbb{F}_p + \dim H^{i+1}(P, M^*) \otimes \mathbb{F}_p$. Substituting in the equation above yields:

$$\begin{aligned} rk_{\mathbb{Z}}(K_n) &= |P|[\dim H^{n+1}(P, M^*) \otimes \mathbb{F}_p + (-1)^n rk_{\mathbb{Z}}(M^*)^P] \\ &\quad + (-1)^{n+1} rk_{\mathbb{Z}}(M^*)\,. \end{aligned}$$

We observe that if M (or equivalently M^*) is not projective, then both sides in this equation must be positive. In addition the dual of M can be used to prove such a formula for all of Tate Cohomology, and in particular we obtain

$$0 < |P| \dim \widehat{H}^n(P, M) \otimes \mathbb{F}_p + (-1)^{n-1}[|P| rk_{\mathbb{Z}}(M^P) - rk_{\mathbb{Z}}(M)]\,.$$

Denote by $\gamma_P(M) = |P| rk_{\mathbb{Z}}(M^P) - rk_{\mathbb{Z}}(M)$. Clearly if $\gamma_P(M) = 0$, $\widehat{H}^n(P, M)$-$\neq 0$ for all $n \in \mathbb{Z}$. If $\gamma_P(M) < 0$, then all the odd cohomology is non-zero, and finally if $\gamma_P(M) > 0$, then all the even cohomology must be non–zero. As by (7.2) every module is cohomologous to a torsion–free one, we have proved

Proposition 7.5 *Let P be a finite group and M a finitely generated non-projective $\mathbb{Z}P$–module. Then either $\widehat{H}^i(P,M) \neq 0$ for all even values of i, or it is non-zero for all odd values of i.*

We remark that the invariant $\gamma_P(M)$ depends only on the *rational* representation type of the module M. A simple corollary is that for a torsion module over a p–group either *all* its Tate cohomology groups are *non–trivial* or they are *all trivial.* Also note the case when $M = \mathbb{Z}$; then $\gamma_P(M) = |G| - 1 > 0$, and we deduce that the even dimensional Tate cohomology of a finite p–group is non–trivial in every degree. This result was originally proved by Kuo [Ku].

A $\mathbb{Z}G$–module is said to be cohomologically trivial if $\widehat{H}^i(H,M) = 0$ for all subgroups $H \subseteq G$. We can now state the main application of Tate cohomology in this context, an extension of a result due to Nakayama and Rim ([N], [R]) recently proved by A. Adem [A1]:

Theorem 7.6 *Let M be a finitely generated torsion–free $\mathbb{Z}G$–module, where G is a finite group. Then the following three statements are equivalent:*

1. *M is cohomologically trivial.*
2. *M is $\mathbb{Z}G$–projective.*
3. *$\widehat{H}^i(\mathrm{Syl}_p(G), M) = 0$ for two values of i of different parity, $\forall$ primes p dividing $|G|$.*

Proof. To prove the theorem note that by the preceding result, all three statements are equivalent to having M projective as a module restricted to all the $\mathrm{Syl}_p(G)$. We will now show that this is equivalent to projectivity over the whole group. To do this we must show that under any of the three conditions on M, any exact sequence $0 \to K \to F \to M \to 0$ with F free will split. Apply $\mathrm{Hom}(*, K)$ to it; this is the same as requiring that in the sequence

$$0 \longrightarrow \mathrm{Hom}_G(M,K) \longrightarrow \mathrm{Hom}_G(F,K) \longrightarrow \mathrm{Hom}_G(K,K) \longrightarrow \mathrm{Ext}^1_{\mathbb{Z}G}(\mathbb{Z}, \mathrm{Hom}(M,K)) \longrightarrow \dots$$

the identity map from K to itself should go to zero under the connecting homomorphism. However, from the fact that M is projective over every p–Sylow subgroup of G it follows that so is $\mathrm{Hom}(M,K)$, and hence $\mathrm{Ext}^1_{\mathbb{Z}P}(\mathbb{Z}, \mathrm{Hom}(M,K)) = 0$ for all $P = \mathrm{Syl}_p(G)$. Using restriction–transfer we conclude that $\mathrm{Ext}^1_{\mathbb{Z}G}(\mathbb{Z}, \mathrm{Hom}(M,K)) = 0$ and the proof of the theorem is complete. □

Now, suppose G is a finite p-group and A is a G module with $|A| = p^n$ for some $n < \infty$. From the above result we have

Corollary 7.7 *If G, A satisfy the assumptions above and $H^1(G;A) = 0$ then $H^1(N;A) = 0$ for any subgroup $N \subset G$.*

Proof. Choose M cohomologous to A but torsion–free. Then clearly $\gamma_G(M) = 0$, hence either all the cohomology is non–zero or M is cohomologically trivial. □

Similarly, we obtain a result due to Gaschütz [G].

Corollary 7.8 *If G, A satisfy the assumptions above and $H^1(G;A) = 0$ then $H^i(G;A) = 0$ for all $i \geq 1$.*

Example 7.9 Let $G = \mathbb{Z}/p^n$, $A = \mathbb{Z}/p^v$ and suppose the action of G on A is given by the rule

$$t(a) = a^{1+p^{v-n}} ,$$

and more generally $(\alpha + \beta)(a) = a^{\lambda(\alpha)+\lambda(\beta)}$ with $-1(a) = a^{-1}$. Then

$$(t-1)(a) \quad = \quad a^{p^{v-n}}$$

and

$$(1 + t + \cdots + t^{p^n - 1})(a) \quad = \quad a^{p^n} ,$$

consequently, in order to compute $H^*(G;A)$ we use the exact sequence

$$0 \longrightarrow A \xrightarrow{p^{m-n}} A \xrightarrow{p^n} A \xrightarrow{p^{v-n}} A \longrightarrow \cdots$$

and $H^0(G;A) = \mathbb{Z}/p^n$, while $H^i(H;A) = 0$ for all $i \geq 1$.

This is essentially the only thing that can happen for $\mathbb{Z}/p^n$ actions on cyclic p-groups when p is odd. However, when $p = 2$ there are two exotic actions of $\mathbb{Z}/2$ on $\mathbb{Z}/2^v$. First there is the action $t(a) = a^{-1}$ closely related to the construction of the dihedral group, and second, there is the action

$$t(a) \quad = \quad a^{2^{v-1}-1}$$

which is used in the construction of the semi-dihedral group as a semi-direct product. The dihedral action does not have trivial $H^1(\mathbb{Z}/2; \mathbb{Z}/2^v)$, but the corresponding group for the semi-dihedral action is trivial.

In the next section these results will be used to study the group of outer automorphisms of a p–group.

8. The First Cohomology Group and Out(G)

In this section we study the relation between the first cohomology group of G and the group Out(G) of outer automorphisms of G. The connection is particularly striking in the case where G is a p-group, where we will be able to prove a theorem of Gaschütz which shows that Out(G) is non-trivial and $|\mathrm{Out}(G)|$ is divisible by p.

Definition 8.1 Let $A \triangleleft G$, then $Aut(G,\ A,\ G/A) \subset \mathrm{Aut}(G)$ is the subgroup of elements τ in Aut(G) which fix A pointwise, and so that the induced automorphism

$$\bar{\tau}\colon G/A \longrightarrow G/A$$

is also the identity.

Remark 8.2 If τ, $\tau' \in \mathrm{Aut}(G, A, G/A)$ and A is abelian, then $\tau\tau' = \tau'\tau$ since we can write

$$\tau(g) \;=\; \lambda(g)g, \quad \lambda(g) \in A\,,$$

and

$$\tau'(g) \;=\; \lambda'(g)g, \quad \lambda'(g) \in A\,,$$

while

$$\tau(a) \;=\; \tau'(a) \;=\; a, \quad \textit{so}\ \lambda(a) \;=\; \lambda'(a) \;=\; 1, \ \textit{all}\ a \in A\,.$$

But this implies $\tau(\tau'(g)) = \tau(\lambda'(g)g) = \lambda'(g)\lambda(g)g = \tau'(\tau(g))$, and the remark follows.

The connection between $H^1(G/A; A))$ and Out(G) is made explicit in the next result.

Theorem 8.3 *Suppose $A \triangleleft G$, and A is abelian, then there is a well defined action (conjugation) of G/A on A, and we have*

$$H^1(G/A; A) = \mathrm{Aut}(G, A, G/A)/(A \cap \mathrm{Inn}(G))\,.$$

(Note that if $a \in A$ and $e \in G/A$ then $(aea^{-1})e^{-1} = a(ea^{-1}e^{-1}) \in A$ so a induces the identity on G/A and A, hence $A \cap \mathrm{Inn}(G)$ is contained in Aut(G, A, G/A), so the group above makes sense.)

Proof. We verify this using the bar resolution. Let $\phi\colon B_1(G/A) \longrightarrow A$ be a given G invariant cochain, so

$$\phi(\sum n_{i,j}e_{i,j}|e_j|) \;=\; \prod e_{i,j}(\phi(|e_j|)^{n_{i,j}})\,.$$

We have

$$\delta\phi(|e_i|e_j|) = \phi(e_i|e_j| - |e_ie_j| + |e_i|) = e_i(\phi(|e_j|) \cdot \phi(|e_i|) \cdot \phi(|e_ie_j|)^{-1} .$$

Hence ϕ is a cocycle if and only if

$$\phi(|e_ie_j|) = e_i(\phi(|e_j|) \cdot \phi(|e_i|), \textit{ all } (e_i,\ e_j) \in (G/A - \{1\})^2 . \qquad \square$$

Proposition 8.4 *Let $\lambda: G/A \longrightarrow A$ be a map with $\lambda(1) = 1$, then the map*

$$\tau: G \longrightarrow G$$

defined by

$$\tau(g) = (\lambda\pi(g)) \cdot g$$

is an automorphism of G if and only if $\bar{\lambda}$: $B_1(G/A) \longrightarrow A$ defined by the formula

$$\bar{\lambda}(\sum n_i g_i|e|) = \prod(g_i(\lambda(e)))^{n_i}$$

is a cocycle.

Proof. If $\tau(gg') = \tau(g)\tau(g')$ we must have

$$\lambda(\pi(gg'))gg' = \lambda\pi(g) \cdot g \cdot \lambda\pi(g') \cdot g' = \lambda\pi(g) \cdot g[\lambda\pi(g')]g \cdot g'$$

for all $g,\ \pi \in G$. But this is exactly the condition that $\bar{\lambda}$ be a cocycle. The converse is now clear as well.

Thus we can identify the set of 1-cocycles with $\mathrm{Aut}(G,\ A,\ G/A)$. Now, consider a coboundary. We have

$$B_0(G/A) = \mathbb{Z}(G/A) ,$$

and hence $\mathrm{Hom}_{\mathbb{Z}(G/A)}(B_0(G/A),\ A) \cong A$. Under this identification we find

$$\delta(a)(|e|) = (a)(e-1) = e(a) \cdot a^{-1} ,$$

and the associated automorphism is given by

$$g \longrightarrow \pi(g)(a)a^{-1}g = gag^{-1}a^{-1}g = a^{-1}ga$$

since $gag^{-1} \in A$. Thus the coboundaries are precisely the inner automorphisms of G obtained from the elements of A. The result follows. $\square$

Corollary 8.5 *If A is a maximal abelian normal subgroup of G then the induced map*

$$H^1(G/A; A) \longrightarrow \mathrm{Out}(G)$$

is an injection.

Proof. If $\alpha \in G$, $\alpha a \alpha^{-1} = a$ for all $a \in A$, and $\{\alpha\}b\{\alpha^{-1}\} = b$ for all $b \in G/A$, then $\{\alpha\} \in C(G/A)$ and additionally

$$\{\alpha\} \in \mathrm{Ker}(\tau\colon\ G/A \longrightarrow \mathrm{Aut}(A))\ ,$$

(here τ is the homomorphism induced by the conjugation action of G/A on A). But this implies that the subgroup $\langle A,\ \alpha\rangle$ is abelian and normal in G, so A is not maximal unless $\alpha \in A$. □

Remark 8.6 The above proof is equivalent to the observation that if A is normal and abelian, then it is maximal only if $\tau\colon C(G/A) \longrightarrow \mathrm{Aut}(A)$ is an injection.

From now on we will assume that G is a finite p-group for a fixed prime p. To begin our study of Out(G) we consider two special cases.

Proposition 8.7 *Let G be a finite p-group, and suppose that every normal abelian subgroup $A \subset G$ is cyclic. Suppose, moreover, that there exists an A as above with $H^1(G/A; A) = 0$, then p must be 2, and G must be a semi-dihedral group*

$$G \cong \mathbb{Z}/2^n \times_{(2^{n-1}-1)} \mathbb{Z}/2\ .$$

Proof. $H^1(G/A; A) = 0$ implies $H^2(G/A; A) = 0$ so $G = A \times_\tau G/A$, for some homomorphism

$$\tau\colon\ G/A \longrightarrow \mathrm{Aut}(A)\ .$$

But $A = \mathbb{Z}/p^m$ by assumption so

$$\mathrm{Aut}(A) \ = \ \begin{cases} \mathbb{Z}/(p-1) \times \mathbb{Z}/p^{m-1} & p \text{ odd} \\ \mathbb{Z}/2 \times \mathbb{Z}/2^{m-2} & p = 2 \end{cases}\ .$$

Suppose $V = \mathrm{Ker}(\tau)$, and $V \neq \{1\}$, then $A \times C(V) \triangleleft G$ is a normal abelian subgroup of G which is not cyclic, a result that violates our assumptions. Thus τ must be an injection, and

$$G \ = \ \mathbb{Z}/p^m \times_{1+p^{m-v}} \mathbb{Z}/p^v\ ,$$

or in case $p = 2$, there is also the possibility that G is a dihedral or semi-dihedral group, or finally that

$$G \ = \ \mathbb{Z}/2^m \times (\mathbb{Z}/2 \times \mathbb{Z}/2^w)$$

with $v > 0$. In the first or last case, if b generates the $\mathbb{Z}/p^m$, and a generates the $\mathbb{Z}/p^v$, then the subgroup

$$\langle b^{p^{m-1}},\ c^{p^{v-1}}\rangle$$

is abelian and normal. So these cases can't occur.

On the other hand, if G is a dihedral group, we can calculate easily that $H^1(G/A, A) \neq 0$ for any normal abelian subgroup A, but if G is semi-dihedral then we have

$$H^1(\mathbb{Z}/2;\mathbb{Z}/2^m) = 0,$$

and the proposition follows. □

Proposition 8.8 *Let G be a finite p-group and suppose A is a maximal normal abelian subgroup of G. Suppose A is not cyclic but $H^1(G/A;A) = 0$, so $G \cong A \times_\tau G/A$. Suppose, finally, that there is a maximal subgroup $G/A \subset N \subset G$ ($|G{:}\,N| = p$), and the center $C(N)$ is not contained in $C(G)$, then G must be the dihedral group D_8 and $A = \mathbb{Z}/2 \times \mathbb{Z}/2$.*

Proof. Note that since A is maximal, our previous argument shows that

$$\tau\colon G/A \longrightarrow \mathrm{Aut}(A)$$

is injective. Hence $C(G) \subset A$. Also, $N \cap A = B$ has index p in A. Now pick $r \in C(N)$ so that r does not belong to B but $r^p \in B$. (Such an r exists by our assumptions.) Then r is certainly not contained in A so $\langle A,\ r\rangle$ is a degree p extension of A.

On the other hand, since $H^1(G;A) = 0$, the results of the last section imply that

$$H^1(\langle A,\ r\rangle/A;A) = H^1(\mathbb{Z}/p;A) = 0$$

so

$$\langle A,\ r\rangle \cong A \times_\tau \mathbb{Z}/p\,.$$

But r centralizes B, so in the exact sequence

$$1 \longrightarrow B \longrightarrow A \longrightarrow \mathbb{Z}/p \longrightarrow 1$$

r acts trivially on both B and $\mathbb{Z}/p$. This induces a long exact sequence in cohomology

$$\cdots \longrightarrow H^v(\mathbb{Z}/p;B) \longrightarrow H^v(\mathbb{Z}/p;A) \longrightarrow H^v(\mathbb{Z}/p;\mathbb{Z}/p) \longrightarrow H^{v+1}(\mathbb{Z}/p;B) \longrightarrow \cdots.$$

By assumption, the terms with coefficient group A are 0 for $v > 0$. Thus, the coboundary map is an isomorphism. On the other hand

$$H^v(\mathbb{Z}/p;\mathbb{Z}/p) = \mathbb{Z}/p \quad \textit{all } v > 0\,,$$

while, since $\mathbb{Z}/p$ acts trivially on B,

$$H^v(\mathbb{Z}/p;B) = (\mathbb{Z}/p)^{\mathrm{rank}(B)}\,.$$

Hence, $\mathrm{rank}(B)$ must be 1, and we have that

$$B = \mathbb{Z}/p^w\,.$$

By assumption A is not cyclic so

$$A = B \times \mathbb{Z}/p = \mathbb{Z}/p^w \times \mathbb{Z}/p\,,$$

with generators s, u respectively. We have $rsr^{-1} = s$, so we must have $rur^{-1} = us^m$, and since $H^2(\mathbb{Z}/p; A) = 0$ (from Gaschütz lemma in the previous section) we must have

$$u \prod_{i=1}^{p-1} r^i u r^{-i} = s^k$$

with $g.c.d.(k,p) = 1$. But k can be rewritten as

$$k = m(1 + 2 + \cdots + (p-1)) = m \cdot p \cdot (p-1)/2 \,.$$

Hence, $p = 2$. Additionally, since $r^2 u r^2 = u = us^{2m}$ we have that $s^2 = 1$ and

$$A = \mathbb{Z}/2 \times \mathbb{Z}/2 \,.$$

The fact that $G = D_8$ now follows easily. □

With these special cases out of the way, here now is the proof of the non-triviality of Out(G) (due to Gaschütz [G]) promised at the beginning of this section.

Theorem 8.9 *Suppose G is a finite, non-abelian p-group with $|G| > p$, then $|\mathrm{Out}(G)| = pq$ with $q \geq 1$.*

Proof. If $A \subset G$ is maximal, normal, abelian and

$$H^1(G/A; A) \neq 0 \,,$$

then we are done. With A as above, if there is an $N \lhd G$ with $G/N = \mathbb{Z}/p$ and $C(N) \subset C(G)$ then consider the homomorphism

$$\phi \colon G \xrightarrow{\pi} G/N \xrightarrow{\rho} C(N) \subset G$$

where $\rho \colon \mathbb{Z}/p \to C(N)$ is some injection. We have $\mathrm{Ker}(\phi) = N$ and so ϕ^2 is trivial. Define

$$\mu \colon G \longrightarrow G$$

by $\mu(g) = g \cdot \phi(g)$. Clearly, μ is a homomorphism, but since $\mu^2 = \mathrm{id}$, μ is actually an automorphism.

Suppose μ were an inner automorphism. Then there is an $h \in G$ and $\mu(g) = hgh^{-1}$ for all $g \in G$. Since $\mu(n) = n$, for all $n \in N$ it follows that $h \in C_G(N)$. If h does not belong to N then $C_G(N) \cdot N = G$ and $\langle h, C(N) \rangle = C_G(N)$ so $h \in C(G)$, and μ is the identity. If $h \in N$ then $h \in C(N) \subset C(G)$ and once more μ is the identity. Hence μ must be an outer automorphism.

But all remaining cases are classified by the previous two propositions, and it is easy to show the theorem is true for D_8 and the semi-dihedral groups. Indeed, for the semi-dihedral groups this is true because we have

$$D_{2^n} = \mathbb{Z}/2^{n-1} \times_{-1} \mathbb{Z}/2 \subset \mathbb{Z}/2^n \times_{-1+2^{n-1}} \mathbb{Z}/2 = SD_{2^{n+1}} ,$$

and $C(D_{2^n}) = C(SD_{2^{n+1}})$. Thus, the construction above gives an outer automorphism of $SD_{2^{n+1}}$. For

$$D_8 = \{x, y \mid x^2 = y^2 = (xy)^4 = 1\} ,$$

the map interchanging x and y is a suitable outer automorphism. The theorem follows. □

Remark. 8.9 also holds if p is odd and G is abelian since in this case G either has a direct summand $\mathbb{Z}/p^i$ with $i \geq 2$ and hence $|\mathrm{Out}|$ is divisible by $(p-1)p^{i-1}$, or is an elementary p-group with $\mathrm{Out}(G) = GL_n(\mathbb{F}_p)$. When $p = 2$ and G is abelian 8.9 continues to hold provided $|G| > 4$.

Chapter III. Modular Invariant Theory

0. Introduction

In this chapter we discuss the role of classical invariant theory in determining and analyzing the cohomology of finite groups. Typically, one has a subgroup of the form $H = (\mathbb{Z}/p)^n \subset G$ and we note that

$$\text{im}(\text{res}^*: H^*(G; \mathbb{F}_p) \longrightarrow H^*(H; \mathbb{F}_p))$$

is contained in the ring of invariants under the action of $N_G(H)/H$ on $H^*(H; \mathbb{F}_p)$, (II.3.1). In some cases, see e.g. (II.6.8), it is possible to describe the entire cohomology ring of G in this way, but more often they contribute important but incomplete portions which are assembled using restriction maps to give the most important pieces of $H^*(G; \mathbb{F}_p)$, see (IV.5).

In the first section we give some of the basic techniques for determining rings of modular invariants. In §2 we study the Dickson algebras, the rings $\mathbb{F}_p[x_1, \ldots, x_n]^{GL_n(\mathbb{F}_p)}$. These algebras play a basic role in describing the cohomology rings of the symmetric groups in VI. In §3 we apply some of the results of §2 to prove a very important theorem of Serre. Then in §4 we determine the groups $H^*((\mathbb{Z}/p)^n; \mathbb{F}_p)^{S_n}$ which are very useful in understanding the cohomology of the groups of Lie type, VII. Finally, in §5 we prove the Cardenas-Kuhn theorem, one of the most important and effective tools in the theory, and one which underlies most of the calculational results in the remainder of the book.

This is by no means an exhaustive exposition of invariant theory, which is a vast subject with many ramifications into algebra and combinatorics. A good reference for further discussions is [St].

1. General Invariants

We begin by considering the group $\mathcal{A}_4$, the alternating group on 4 letters. It can be written as a normal extension

$$1 \longrightarrow K \overset{\triangleleft}{\longrightarrow} \mathcal{A}_4 \longrightarrow \mathbb{Z}/3 \longrightarrow 1$$

where $K \cong \mathbb{Z}/2 \times \mathbb{Z}/2$ is the Klein group with generators x, y and the generator T of $\mathbb{Z}/3$ acts by $T(x) = y$, $T(y) = x + y$. Passing to cohomology it follows from (II.3.1) and (II.4) that

$$\mathrm{im}(H^*(\mathcal{A}_4; \mathbb{F}_2)) \hookrightarrow H^*(K; \mathbb{F}_2) \;=\; \mathbb{F}_2[a,b]$$

(where a is dual to x and b is dual to y) is contained in the subring of invariants

$$\mathbb{F}_2[a,b]^{\mathbb{Z}/3} .$$

The action is given by $T(a) = b$, $T(b) = a + b$ on the generators, and it extends multiplicatively to the entire polynomial algebra. Thus $T(a^2b) = (T(a))^2T(b) = b^2(a+b) = b^2a + b^3$.

How can we obtain the ring of invariants $\mathbb{F}_2[a,b]^{\mathbb{Z}/3}$? The simplest method, dating from the 19th century, is as follows. If we consider the effect of field extension

$$\mathbb{F}_2[a,b] \;\Rightarrow\; \mathbb{F}_4 \otimes_{\mathbb{F}_2} \mathbb{F}_2[a,b] \;=\; \mathbb{F}_4[a,b] ,$$

with the same action of $\mathbb{Z}/3$, we claim first

$$(*) \qquad \mathbb{F}_4[a,b]^{\mathbb{Z}/3} \;=\; \mathbb{F}_4 \otimes_{\mathbb{F}_2} \left(\mathbb{F}_2[a,b]^{\mathbb{Z}/3}\right) .$$

Indeed, the action of $\mathbb{Z}/3$ extends to a (graded) module structure over the group ring $\mathbb{F}_2(\mathbb{Z}/3)$ by setting $(\sum \lambda_i T^i)w \;=\; \sum \lambda_i(T^i(w))$. But, since 2 is prime to the order of $\mathbb{Z}/3$, the group ring is semi-simple by Maschke's Theorem and thus splits as a direct sum

$$\mathbb{F}_2(\mathbb{Z}/3) \;=\; \mathbb{F}_2 \oplus I \text{ where } I \cong \mathbb{F}_4 .$$

Here the action of $\mathbb{Z}/3$ on the $\mathbb{F}_2$ summand is trivial, and the action on I is by multiplication by the units $\mathbb{Z}/3 \cong \mathbb{F}_4^\bullet$. (Explicitly, the idempotents which give the splitting are $(1 + T + T^2)$ for the $\mathbb{F}_2$ summand and $(T + T^2)$ for the $\mathbb{F}_4$ summand.) Consequently, in each dimension we have a splitting

$$\mathbb{F}_2[a,b]_i \;=\; I\mathbb{F}_2[a,b]_i \oplus \mathbb{F}_2[a,b]_i^{\mathbb{Z}/3} .$$

This splitting is preserved on tensoring up, i.e. tensoring over $\mathbb{F}_2$ with a larger field.

Lemma 1.1 *Let $\mathbb{F}$ be any field of characteristic prime to the order of the finite group G, and let $\mathbb{K}/\mathbb{F}$ be a splitting field for G, i.e. a field over which the group ring decomposes as a direct sum $\sum M_i(\mathbb{K})$ where $M_i(\mathbb{K})$ is the $i \times i$ matrix ring over $\mathbb{K}$. Then, there is one and only one $\mathbb{K}$ summand in the sum above for which the action map $G \times \mathbb{K} \longrightarrow \mathbb{K}$ is trivial.*

Proof. Under these circumstances the regular representation, i.e. the representation via left multiplication on the group ring, has the decomposition

$\sum n_i V_i$ where the V_i run over the irreducible representations and n_i is the dimension of V_i. Consequently, the trivial representation can only appear once, and the result follows. □

In particular, applied to the situation above, we see that $\mathbb{F}_4$ is a splitting field for $\mathbb{Z}/3$ and $\mathbb{F}_4 \otimes_{\mathbb{F}_2} I = \mathbb{F}_4 \oplus \mathbb{F}_4$ where the first representation is $T \mapsto \zeta_3$, a primitive third root of unity, while the second representation is $T \mapsto \zeta_3^2$. The claim, $(*)$, follows immediately from this.

More generally the argument above shows

Theorem 1.2 *Let G be a finite group, and suppose the order of G is prime to the characteristic of a field $\mathbb{F} = \mathbb{F}_{p^r}$, then, given an action of G on the n dimensional vector space $\mathbb{F}\langle x_1, \ldots, x_n\rangle$ with basis $x_1, \ldots, x_n$, it extends to an action on the polynomial ring $\mathbb{F}[x_1, \ldots, x_n]$ and, for any extension field $\mathbb{K}/\mathbb{F}$ we have*

$$(\mathbb{K}[x_1, \ldots, x_n])^G = \mathbb{K} \otimes_{\mathbb{F}} \left(\mathbb{F}[x_1, \ldots, x_n]^G\right) .$$

On the other hand, after tensoring up, we can split the vector space into the irreducible representations afforded in $\mathbb{K}\langle x_1, \ldots, x_n\rangle$, and that usually makes the calculation of the invariants in degree i much easier.

In counting the number of invariants we often find the *Poincaré series* convenient. Here the Poincaré series of an invariant subalgebra $(\mathcal{R}^*)^G \subseteq \mathcal{R}^*$ of a graded $\mathbb{F}$–algebra is defined as

$$p_G(\mathcal{R}^*)(t) = \sum_{i=0}^{\infty} \dim_{\mathbb{F}} (\mathcal{R}^i)^G t^i .$$

Returning to our example, we have $\mathbb{F}_4\langle a, b\rangle = \mathbb{F}_4 \oplus \mathbb{F}_4$, where the action on the first summand is multiplication by ζ_3, while the action on the second is multiplication by ζ_3^2. Indeed, a specific choice for the generator f_1 for the first summand is $a + \zeta_3 b$, while $f_2 = a + \zeta_3^2 b$. Consequently, $T f_1^i f_2^j = \zeta_3^{i+2j} f_1^i f_2^j$, and the invariants now have a basis of the form

$$f_1^i f_2^j \text{ with } i + 2j \equiv 0 \bmod 3 .$$

This equation implies $j \equiv i \bmod 3$, so the general solution is $f_1 f_2 f_1^{3l} f_2^{3m}$, $f_1^2 f_2^2 f_1^{3l} f_2^{3m}$, or $f_1^{3l} f_2^{3m}$. It follows that as a ring

$$\mathbb{F}_4[a, b]^{\mathbb{Z}/3} \cong \mathbb{F}_4[f_1^3, f_2^3](f_1 f_2)$$

with the single (cubic) relation $(f_1 f_2)^3 = (f_1^3)(f_2^3)$. In particular this ring has Poincaré series $\dfrac{1 + x^2 + x^3}{(1 - x^3)^2}$, and so must the ring of invariants $\mathbb{F}_2[a, b]^{\mathbb{Z}/3}$. However, we cannot assert that the relation is that holding above. In fact, calculating, we have

$$\begin{aligned} f_1 f_2 &= (a+\zeta_3 b)(a+\zeta_3^2 b) = a^2+ab+b^2 \\ f_1^3 &= a^3+\zeta_2 a^2 b+\zeta_3^2 ab^2+b^3 \\ f_2^3 &= a^3+\zeta_3^2 a^2 b+\zeta_3 ab^2+b^3. \end{aligned}$$

Thus, generators in degree three over $\mathbb{F}_2$ are $A = f_1^3 + f_2^3 = a^2b + ab^2$, and $B = \zeta_3 f_1^3 + \zeta_3^2 f_2^3 = a^3 + a^2b + b^3$. Reversing the calculation $f_1^3 = \zeta_3^2 A + B$, $f_2^3 = \zeta_3 A + B$. It follows that for these generators the invariant algebra becomes

$$\mathbb{F}_2[a,b]^{\mathbb{Z}/3} = \mathbb{F}_2[A,B](C)/(C^3 + A^2 + B^2 + AB)\,,$$

since $A^2 + B^2 + AB$ is the expansion of $f_1^3 f_2^3$. Applying II.6.8 we obtain

Theorem 1.3 $H^*(\mathcal{A}_4; \mathbb{F}_2) \cong \mathbb{F}_2[a,b]^{\mathbb{Z}/3} \cong \mathbb{F}_2[A,B](C)/(C^3+A^2+B^2+AB)$ *with deg*$(A) =$ *deg*$(B) = 3$, *and deg*$(C) = 2$.

Using II.5.1 we can in fact prove

Theorem 1.4 *Let K be normal in G and suppose $[G{:}\,K]$ is prime to p, then*

$$\mathrm{im}\,(\mathrm{res}^*(H^*(G;\mathbb{F}_p) \longrightarrow H^*(K;\mathbb{F}_p))$$

is exactly

$$\left(\sum_{t\in G/K} t\right) H^*(K;\mathbb{F}_p) = H^*(K;\mathbb{F}_p)^{G/K}\,.$$

We now discuss some techniques which enable us to do expicit calculations in special cases. Here is an extension of (1.4) which is very useful in building up invariant subrings for G from the knowledge of the invariants for certain subgroups.

Lemma 1.5 *Let p be a prime, G a finite group, and $G = \sqcup g_i H$ be a coset decomposition with respect to H where $Syl_p(G) \subset H$. Let V be any $\mathbb{F}_p(G)$-module and V^H the $\mathbb{F}_p(H)$-invariants. Then $\sum g_i$ maps V^H to itself and has image exactly V^G.*

Proof. Let g be an element of G. Then $g\sum g_i = \sum g_{\varphi_g(i)} h_i(g)$ using the notation preceeding (II.5.10). In particular, for $v \in V^H$, $g\sum g_i(v) = \sum g_i(v)$, so $\sum g_i(v) \in V^G \subset V^H$. Moreover $\sum g_i(\sum g_i(v)) = |G{:}\,H| \sum g_i(v)$, and since $|G{:}\,H|$ is prime to p it follows that $\sum g_i$ restricted to V^G is just multiplication by a unit. □

This process in many cases can be used with computers to obtain explicit rings of invariants. However, it does require techniques for handling the invariants of p-groups in order to be effective. This is usually more difficult.

Sometimes it is possible to filter V via invariant submodules $V_1 \subset V_2 \subset \cdots \subset V_k = V$ where we already know the invariants on the quotients V_i/V_{i-1} and piece together the structure of the invariants on V from that of the quotients. The key technique here involves the consideration of a single extension

$$V_1 \subset V \xrightarrow{\pi} V/V_1 .$$

This gives rise to the long exact sequence in cohomology

$$H^0(G; V_1) \longrightarrow H^0(G; V) \xrightarrow{\pi^*} H^0(G; V/V_1) \xrightarrow{\delta} H^1(G; V_1) \longrightarrow \cdots$$

which is useful because $H^0(G; A) = A^G$. From this point of view the most important step in the determination of V^G is determining the connecting homomorphism δ. Remember that this is determined by lifting the element in $C^0(V/V_1)$ which represents the cohomology class to $C^0(V)$, taking the coboundary there, and then lifting this coboundary back to $C^1(V_1)$. Here are some examples of how this process works.

Lemma 1.6

a. *Let $\mathbb{Z}/2$ act on $\mathbb{F}_2[x_1, x_2]$ by interchanging the generators then*

$$\mathbb{F}_2[x_1, x_2]^{\mathbb{Z}/2} = \mathbb{F}_2[x_1 + x_2, x_1 x_2] .$$

b. *Let $\mathbb{Z}/2$ act on $\mathbb{F}_2[x_1, \ldots, x_4]$ by interchanging the first and second generators and the third and fourth generators. Then*

$$\mathbb{F}_2[x_1, x_2, x_3, x_4]^{\mathbb{Z}/2} = \mathbb{F}_1[x_1 + x_2, x_3 + x_4, x_1 x_2, x_3 x_4](1, M_2)$$

where $M_2 = x_1 x_3 + x_2 x_4$ satisfies the relation

$$M_2^2 + (x_1 + x_2)(x_3 + x_4)M_2 + (x_1 + x)^2 x_3 x_4 + (x_3 + x_4)^2 x_1 x_2 = 0 .$$

Proof. Let $A = \mathbb{F}_2[x_1 + x_2, x_1 x_2] \subset \mathbb{F}_2[x_1, x_2]$. Then we can write $\mathbb{F}_2[x_1, x_2]$ as the direct sum $A \oplus A x_1$. In particular there is an exact sequence of $\mathbb{F}_2(\mathbb{Z}/2)$ modules

$$A \longrightarrow \mathbb{F}_2[x_1, x_2] \longrightarrow A x_1,$$

and $\mathbb{F}_2[x_1, x_2]$ can be written as an extension of trivial $\mathbb{F}_2(\mathbb{Z}/2)$ modules. Passing to cohomology

$$H^*(\mathbb{Z}/2; A) \cong H^*(\mathbb{Z}/2; \mathbb{F}_2) \otimes A = \coprod e^i A ,$$

where e^i is the non-zero element in $H^i(\mathbb{Z}/2; \mathbb{F}_2) = \mathbb{F}_2$. This decomposition is natural with respect to δ in the sense that $\delta(e^i \otimes a) = e^i \cup \delta(1)a$ and $\delta(1) = e^1 \otimes (x_1 + x_2)$ since $\delta(x_1) = (1 + T)x_1 = (x_1 + x_2) \in H^1(\mathbb{Z}/2; \mathbb{F}_2) \otimes A$. Thus δ is an injection and (a.) follows.

To prove (b.) let $B \subset \mathbb{F}_2[x_1, \ldots, x_4]$ be given as $A \otimes A = \mathbb{F}_2[x_1 + x_2, x_1x_2, x_3 + x_4, x_3x_4]$. Consequently the polynomial ring decomposes as $B \oplus Bx_1 \oplus Bx_3 \oplus Bx_1x_3$. Let $V_1 = B$, $V_2 = V_1 \oplus Bx_1 \oplus Bx_3$. Then in the exact sequence the quotient V_2/V_1 is just two copies of B. δ is determined by $\delta(x_1) = e^1 \otimes (x_1 + x_2)$, $\delta(x_3) = e^1 \otimes (x_3 + x_4)$. Consequently, δ has kernel $B\{(x_1 + x_2)x_3 + (x_3 + x_4)x_1\}$ which lifts back to BM_2.

It remains to determine the extension by Bx_1x_3. Using the determination of δ we can equally determine $H^1(\mathbb{Z}/2; V_2)$, as $e^1 \otimes (BM_2 \oplus (B \oplus B)/im\delta)$ and $\delta(x_1x_3) = e^1 \otimes M_2$. This completes the proof. □

Corollary 1.7 *Let the dihedral group $D_8 = Syl_2(S_4)$ act on $\mathbb{F}_2[x_1, \ldots, x_4]$ by permuting coordinates. (It has generators $(1,2)$, $(1,3,2,4)$ and contains the subgroup $\langle(1,2),(3,4)\rangle$.) Then, setting $\sigma_1 = x_1+x_2+x_3+x_4$, $\sigma_4 = x_1x_2x_3x_4$ we have*

$$\mathbb{F}_2[x_1, \ldots x_4]^{D_8} = \mathbb{F}_2[\sigma_1, (x_1 + x_2)(x_3 + x_4), x_1x_2 + x_3x_4, \sigma_4](1, M_3)$$

where $M_3 = (x_1 + x_2)x_3x_4 + (x_3 + x_4)x_1x_2$.

Proof. The invariants under $(\mathbb{Z}/2)^2 = \langle(1,2),(3,4)\rangle$ are clearly $A \otimes A = \mathbb{F}_2[x_1 + x_2, x_1x_2, x_3 + x_4, x_3x_4]$, and the remaining generator for D_8 can be taken as $(1,3)(2,4)$. But this acts on the previous invariant subring by exchanging the generators in pairs. Thus the result follows from (1.6). □

Corollary 1.8 *Let $K \subset D_8$ be the Klein group $\langle(1,2)(3,4),(1,3)(2,4)\rangle$, then*

$$\mathbb{F}_2[x_1, \ldots, x_4]^K = \mathbb{F}_2[\sigma_1, (x_1 + x_2)(x_3 + x_4), x_1x_2 + x_3x_4, \sigma_4](1, M_2, M_3, M_2M_3)\,.$$

Proof. Use the subgroup $\langle(1,2)(3,4)\rangle$ and (1.6) to determine the invariants under this subgroup. Now apply $(1,3)(2,4)$. Note that it leaves M_2 invariant, and exchanges the polynomial generators in pairs. The proof is now direct. □

Remark. These last two calculations, although treated here primarily as examples, are actually very important in numerous calculations of $H^*(G; \mathbb{F}_2)$ for explicit G which occur in the study of the sporadic groups of rank three and four some of which are discussed in Chapter VIII.

Example. Using the results above one can prove that

$$\mathbb{F}_2[x_1, \ldots, x_4]^{\mathcal{A}_4} = \mathbb{F}_2[\sigma_1, \sigma_2, \sigma_3, \sigma_4](1, b_6)$$

where $\sigma_2 = x_1x_2 + x_1x_3 + x_1x_4 + x_2x_3 + x_2x + 4 + x_3x_4$, $\sigma_3 = x_1x_2x_3 + x_1x_2x_4 + x_1x_3x_4 + x_2x_3x_4$ and b_6 can be given as the sum $\sum x_ix_j^2x_k^3$ where (i,j,k) runs over the images of $g(1,2,3)$ for $g \in \mathcal{A}_4$.

Example 1.9 The cohomology of J_1. In II.6.9 we noted that $H^*(J_1;\mathbb{F}_2) = \mathbb{F}_2[x_1,x_2,x_3]^{\mathbb{Z}/7\times_T\mathbb{Z}/3}$ where the group $\mathbb{Z}/7\times_T\mathbb{Z}/3$ has the matrix generators f,g described in II.6.9. Using the techniques here we determine this ring of invariants in two steps as follows, first determining $\mathbb{F}_2[x_1,x_2,x_3]^{\mathbb{Z}/7}$, and then using the projection operator $1+f+f^2$ on the resulting invariants to obtain the answer.

To obtain the $\mathbb{Z}/7$ invariants tensor with $\mathbb{F}_8$ and change bases so $\mathbb{F}_8\otimes_{\mathbb{F}_2}\mathbb{F}_2[x_1,x_2,x_3] = \mathbb{F}_8[y_1,y_2,y_3]$ with $g(y_1)=\zeta_7 y_1$, $g(y_2)=\zeta_7^2 y_2$, $g(y_3)=\zeta_7^4 y_3$ since the eigenvalues of $g = \begin{pmatrix} 0 & 0 & 1\\ 1 & 0 & 0\\ 0 & 1 & 1\end{pmatrix}$ are ζ_7, ζ_7^2 and ζ_7^4. Note that $\mathcal{A} = \mathbb{F}_8[y_1^8,y_2^8,y_3^8] \subset \mathbb{F}_8[y_1,y_2,y_3]^{\mathbb{Z}/7}$ and the set of cosets of $\mathcal{A}$ in the full polynomial ring can be identified with the set $\mathbb{F}_8\{y_1^iy_2^jy_3^k \mid 0\le i,j,k\le 7\}$, and a monomial is invariant if and only if $i+2j+4k\equiv 0 \bmod (7)$. Thus the invariants are identified with a hyperplane in $\mathbb{F}_7^3$ and, including 1 there are exactly 49 monomial generators. Note that with respect to this new basis the action of f is $f(y_1)=y_2$, $f(y_2)=y_3$, $f(y_3)=y_1$ so most of these g invariant monomial generators are free under the action of f. Explicitly there are 42 free monomials in the basis of the form v, $f(v)$, and $f^2(v)$, where the v run over the set $y_1y_2^3$, $y_1^2y_3^3$, $y_1y_3^5$, $y_1^2y_2^4y_3$, $y_1^2y_2^6$, $y_1y_2^4y_3^2$, $y_1^4y_2^5$, $y_1y_2^6y_3^2$, $y_1^4y_3^6$, $y_1^2y_2^3y_3^5$, $y_1^2y_2^5y_3^4$, $y_1y_2^5y_3^6$, $y_1^3y_2^4y_3^6$, and $y_1^3y_2^6y_3^5$ together with the six fixed generators $(y_1y_2y_3)^i$, $1\le i\le 6$. Thus, for each of the free generators the projection operator gives a copy of $\mathcal{A}$, while each fixed generator gives a term $\mathcal{A}^{\mathbb{Z}/3}(y_1y_2y_3)^i$.

To find $\mathcal{A}^{\mathbb{Z}/3}$ tensor with $\mathbb{F}_4$ obtaining $\mathbb{F}_{64}[y_1^7,y_2^7,y_3^7]$ which has a new basis z_1, z_2, z_3 with $f(z_1)=z_1$, $f(z_2)=\zeta_3 z_2$, $f(z_3)=\zeta_3^2 z_3$, and the invariant subring is $\mathbb{F}_{64}[z_1, z_2z_3, z_2^3](1, z_3^3)$. Thus we have determined the Poincaré series for $H^*(J_1;\mathbb{F}_2)$ as

$$
\begin{aligned}
&\frac{(1-x^{21})(1+x^{21})}{(1-x^3)(1-x^7)(1-x^{14})(1-x^{21})}\\
&+\frac{x^4+x^5+x^6+x^7+2x^8+2x^9+2x^{10}+x^{11}+x^{12}+x^{13}+x^{14}}{(1-x^7)^3}\\
&=\frac{1+x^{21}}{(1-x^3)(1-x^7)(1-x^{14})}+\frac{x^4(1+x^4)}{(1-x)(1-x^7)^2}\\
&=\frac{1+x^4(1+x+x^2)(1+x^4)(1+x^7)+x^{21}}{(1-x^3)(1-x^7)(1-x^{14})}
\end{aligned}
$$

but the numerator in this last expression can be written

$$
\frac{(1+x^5)(1+x^6)(1-x^{14})}{(1+x^2)(1-x^2)}
$$

so the entire Poincaré series simplifies to

$$\frac{(1+x^5)(1+x^6)}{(1-x^3)(1-x^4)(1-x^7)} = \frac{(1+x^3)(1+x^5)(1+x^6)}{(1-x^4)(1-x^6)(1-x^7)}$$

as asserted in II.6.9.

The rest of the determination of the invariant subring in the original polynomial algebra $\mathbb{F}_2[x_1, x_2, x_3]$ is direct and depends only on getting explicit generators. In the next section we will determine the ring of invariants under the entire $GL_3(\mathbb{F}_2)$ as the Dickson algebra $\mathbb{F}_2[d_4, d_6, d_7]$ where the d_i are given explicitly in Example 2.3. Here, note that the eigenvectors of g, y_1, y_2, and y_3, discussed above are explicitly given as $(\zeta_7 x_1 + x_2 + \zeta_7^2 x_3)$, $(\zeta_7^2 x_1 + x_2 + \zeta_7^4 x_3)$, $(\zeta_7^4 x_1 + x_2 + \zeta_7 x_3)$ respectively so that the invariant element in dimension 3, $x = y_1 y_2 y_3$ can be written

$$x = x_1^3 + x_2^3 + x_3^3 + x_1 x_2^2 + x_1^2 x_3 + x_2^2 x_3, \tag{1.10}$$

after expanding out. From the axioms for the Steenrod algebra in (II.2) we have

$$\gamma = Sq^2(x) = x_1^5 + x_2^5 + x_3^5 + x_1 x_2^4 + x_1^4 x_3 + x_2^4 x_3$$

since the ring of invariants must be closed under the action of $\mathcal{A}(2)$. Finally, we can check that $x^2 \neq d_6$ so we have all the generators in II.6.9 now given explicitly where $y = d_4$, $\mu = d_6$, and $z = d_7$.

2. The Dickson Algebra

In this section we study the ring of invariants $\mathbb{F}_p[x_1, \ldots, x_n]^{\mathrm{GL}_n(\mathbb{F}_p)}$. These results will be remarkably useful in the sequel. We need the matrices

$$V_n(x_1, \ldots, x_n) = \begin{pmatrix} x_1 & x_1^p & \cdots & x_1^{p^{n-1}} \\ x_2 & x_2^p & \cdots & x_2^{p^{n-1}} \\ \vdots & \vdots & \ddots & \vdots \\ x_n & x_n^p & \cdots & x_n^{p^{n-1}} \end{pmatrix}.$$

Let $D_n(x_1, \ldots, x_n) = det(V_n)$, so $D_n(x_1, \ldots, x_n)$ is a homogeneous polynomial of degree $\left(\frac{p^n - 1}{p-1}\right)$. Note that $D_n \neq 0$ since, for example, the term $x_1 x_2^p x_3^{p^2} \cdots x_n^{p^{n-1}}$ occurs once and once only in its expansion.

Let f_1, f_2, …, f_n be some other basis for the $\mathbb{F}_p$-vector space spanned by $x_1, \ldots, x_n$, so $f_i = \sum a_{ij} x_j$, $1 \leq i \leq n$. Let A be the $n \times n$ matrix with entries a_{ij} so we have $AV_n(x_1, \ldots, x_n) = V(f_1, \ldots, f_n)$ and $D_n(f_1, \ldots, f_n) \pm det(A) D_n(x_1, \ldots, x_n)$. Thus, $D_n(x_1, \ldots, x_n)$ is an $\mathrm{SL}_n(\mathbb{F}_p)$ invariant but not a $\mathrm{GL}_n(\mathbb{F}_p)$ invariant. However the term $D_n(x_1, \ldots, x_n)^{p-1}$ is, in fact, a $\mathrm{GL}_n(\mathbb{F}_p)$ invariant.

Examples 2.1 For $n = 2$, $p = 2$, $D_2(x_1, x_2) = x_1x_2(x_2 + x_1)$. For $n = 2$, $p = 3$, $D_2(x_1, x_2) = x_1x_2^3 - x_1^3x_2 = x_1x_2(x_2 - x_1)(x_2 - 2x_1)$. More generally, for $n = 2$ and arbitrary p we have

$$D_2(x_1, x_2) \;=\; x_1x_2^p - x_1^px_2 \;=\; x_1x_2\prod_{\theta=1}^{p-1}(x_2 - \theta x_1)\,.$$

Now consider the $(n \times n)$ minors of $V_{n+1}(x_1, \ldots, x_{n+1})$ with respect to the last row, and let

$$D_{n,i}(x_1, \ldots, x_n) \;=\; det\begin{pmatrix} x_1 & x_1^p & \cdots & \hat{x}_1^{p^{i-1}} & \cdots & x_1^{p^n} \\ x_2 & x_2^p & \cdots & \hat{x}_2^{p^{i-1}} & \cdots & x_2^{p^n} \\ \vdots & \vdots & \ddots & \vdots & \ddots & \vdots \\ x_n & x_n^p & \cdots & \hat{x}_n^{p^{i-1}} & \cdots & x_n^{p^n} \end{pmatrix}.$$

In particular

$$\begin{aligned} D_{n,n+1}(x_1, \ldots, x_n) &= D_n(x_1, \ldots, x_n) \\ D_{n,1} &= D_n(x_1^p, \ldots, x_n^p) \\ &= (D_n(x_1, \ldots, x_n))^p\,. \end{aligned}$$

Clearly $D_{n,i}(f_1, \ldots, f_n) = det(G)D_{n,i}(x_1, \ldots, x_n)$ for $f_1, \ldots, f_n$ and G as above.

Expanding $D_{n+1}(x_1, \ldots, x_n, x)$ with respect to the bottom row we have

$$D_{n+1}(x_1, \ldots, x_n, x) \;=\; D_{n,n+1}x^{p^n} - D_{n,n}x^{p^{n-1}} + \cdots + (-)^nD_{n,1}x$$

or $D_{n+1}(x_1, \ldots, x_n, x)$ is equal to

(*)

$$D_n(x_1, \ldots, x_n)\left[x^{p^n} - d_1x^{p^{n-1}} + \cdots + (-)^{n-1}d_{n-1}x^p + (-)^n(D_n)^{p-1}x\right]$$

where $d_j = D_{n,n+1-j}/D_n$. Here, note that the element d_j is contained in the quotient field $\mathcal{Q}(\mathbb{F}_p[x_1, \ldots, x_n])$, and that it is a $\mathrm{GL}_n(\mathbb{F}_p)$ invariant. But more is true.

Lemma 2.2

a. $D_{n+1}(x_1, \ldots, x_{n+1}) = D_n(x_1, \ldots, x_n)\prod(x_{n+1} - f)$ *where f runs over all vectors in the n-dimensional vector space over $\mathbb{F}_p$ spanned by $x_1, \ldots x_n$.*

b. $d_j \in \mathbb{F}_p[x_1, \ldots, x_n]$, $1 \le j \le n-1$.

Proof. Regard (*) as an expansion of $D_{n+1}(x_1, \ldots, x_n, x)$ over the polynomial ring in x with coefficients in the quotient field $\mathcal{Q}(\mathbb{F}_p[x_1, \ldots, x_n])$. Now consider the polynomial

$$p_n(x) = x^{p^n} - d_1 x^{p^{n-1}} + \cdots + (-)^{n+1} D_n^{p-1} x\,.$$

Clearly, if we set $x = f \in \langle x_1, \ldots, x_n \rangle$, then the last row of

$$V_{n+1}(x_1, \ldots, x_n, f)$$

is a linear combination of the first n rows so $D_{n+1}(x_1, x_2, \ldots, x_n, f) = 0$. Since $D_n(x_1, \ldots, x_n) \neq 0$ it follows from $(*)$ that $p_n(f) = 0$, and there is a factorization over the quotient field

$$(**) \qquad p_n(x) = \prod_{f \in \langle x_1, \ldots, x_n \rangle} (x - f)\,.$$

Since the roots f all belong to $\mathbb{F}_p[x_1, \ldots, x_n]$ it follows that the coefficients of $p_n(x)$ also are contained in $\mathbb{F}_p[x_1, \ldots, x_n]$ and the lemma follows. □

Examples 2.3 When $p = 2$, $n = 2$, we have

$$\begin{aligned} d_1 &= x_1^2 + x_1 x_2 + x_2^2 \\ d_2 &= D_2 = x_1^2 x_2 + x_1 x_2^2\,. \end{aligned}$$

When $n = 3$ we have

$$\begin{aligned} d_1 &= x_1^4 + x_1^2(x_2^2 + x_2 x_3 + x_3^2) + x_1(x_2^2 x_3 + x_2 x_3^2) + x_2^4 + x_2^2 x_3^2 + x_3^4 \\ d_2 &= x_1^4(x_2^2 + x_2 x_3 + x_3^2) + x_1^2(x_2^4 + x_2^2 x_3^2 + x_3^4) + x_1 x_2^4 x_3 + \\ &\quad x_1 x_2 x_3^4 + x_2^4 x_3^2 + x_2^2 x_3^4 \\ d_3 &= D_3 = x_1^4 x_2^2 x_3 + x_1^4 x_2 x_3^2 + x_1^2 x_2^4 x_3 + x_1^2 x_2 x_3^4 + x_1 x_2^4 x_3^2 + x_1 x_2^2 x_3^4\,. \end{aligned}$$

The following theorem of L.E. Dickson is basic.

Theorem 2.4 *The ring of invariants $\mathbb{F}_p[x_1, \ldots, x_n]^{\mathrm{GL}_n(\mathbb{F}_p)}$ is a polynomial algebra, and is generated by the elements d_1, d_2, …, $d_n = D_n^{p-1}$ constructed above, i.e.*

$$\mathbb{F}_p[x_1, \ldots, x_n]^{\mathrm{GL}_n(\mathbb{F}_p)} = \mathbb{F}_p[d_1, \ldots, d_{n-1}, D_n^{p-1}]\,.$$

Proof. (Following C. Wilkerson [W]): First note that $\mathbb{F}_p[x_1, \ldots x_n]$ is integral over the ring spanned by $d_1, \ldots, d_{n-1}, D_n^{p-1}$ which we will temporarily denote $\mathcal{A}_n$. This is so since the integral elements are closed under sum and product, and from $(**)$ contain $x_1, x_2, \ldots, x_n$. Consider next the quotient fields $\mathcal{Q}(\mathcal{A}_n)$ and $\mathcal{Q}(\mathbb{F}_p[x_1, \ldots, x_n])$.

Lemma 2.5 *$\mathcal{Q}(\mathbb{F}_p[x_1, \ldots, x_n])$ is Galois over $\mathcal{Q}(\mathcal{A}_n)$ with Galois group* $\mathrm{GL}_n(\mathbb{F}_p)$.

Proof 2.5 First $\mathcal{Q}(\mathbb{F}_p[x_1,\ldots,x_n])$ is generated by $x_1,\ldots,x_n$ over $\mathcal{Q}(\mathcal{A}_n)$. But these and the elements $f_i = \sum_1^n a_i x_i$ are precisely the roots of $p_n(x)$ and are all contained in $\mathcal{Q}(\mathbb{F}_p[x_1,\ldots,x_n])$. Thus $\mathcal{Q}(\mathbb{F}_p[x_1,\ldots x_n])$ is the splitting field of $p_n(x)$ over the field $\mathcal{Q}(\mathcal{A}_n)$. Moreover, since $p_n(x)$ has no repeated roots it is separable and it follows that the extension is Galois.

We now determine the Galois group G. Since $\mathbb{F}_p \subset \mathcal{A}_n$ it follows that for $g \in G$ we have $g(\sum n_i x_i) = \sum n_i g(x_i)$, $n_i \in \mathbb{F}_p$. Moreover, since g acts to permute the roots of $p_n(x)$, and, indeed is determined by its action on these roots, it follows that $g(\sum n_i x_i) = \sum n_i(\sum g_{ij} x_j)$ so $g \in \mathrm{GL}_n(\mathbb{F}_p)$. Conversely, given $g \in \mathrm{GL}_n(\mathbb{F}_p)$ it fixes $\mathcal{A}_n$ and hence $\mathcal{Q}(\mathcal{A}_n)$, meanwhile acting as an automorphism in $\mathcal{Q}(\mathbb{F}[x_1,\ldots,x_n])$ by construction, so $g \in G$. The lemma follows. □

We return to the proof of Dickson's theorem. Consider again the integral extension $\mathbb{F}_p[x_1,\ldots,x_n]$ over $\mathcal{A}_n$. Since $\mathcal{Q}(\mathbb{F}_p[x_1,\ldots,x_n])$ is a finitely generated extension of $\mathcal{Q}(\mathcal{A}_n)$, it follows that $\mathbb{F}_p[x_1,\ldots,x_n]$ is a finitely generated extension of the integral closure of $\mathcal{A}_n$ in $\mathcal{Q}(\mathcal{A}_n)$. In particular the integral closures of $\mathcal{A}_n$ and $\mathbb{F}_p[x_1,\ldots,x_n]$ both have the same transcendence degree n. But $\mathcal{A}_n$ is generated by precisely n elements over $\mathbb{F}_p$. It follows that these elements are polynomially independent, and so $\mathcal{A}_n = \mathbb{F}_p[d_1,\ldots,d_n = D_n^{p-1}]$.

On the other hand a polynomial ring in n variables over a field $\mathbb{F}$ is integrally closed in its quotient field (see e.g. [La]). It follows, also that $\mathbb{F}_p[x_1,\ldots,x_n]$ is integrally closed in its quotient field. Moreover

$$\mathbb{F}_p[x_1,\ldots,x_n] \cap \mathcal{Q}(\mathcal{A}_n)$$

is the ring of $\mathrm{GL}_n(\mathbb{F}_p)$ invariants in $\mathbb{F}_p[x_1,\ldots,x_n]$.

But $\mathbb{F}_p[x_1,\ldots,x_n]$ is integral over $\mathcal{A}_n$, so the intersection above is integral over $\mathcal{A}_n$ in its quotient field, and, since $\mathcal{A}_n$ is integrally closed, therefore equals $\mathcal{A}_n$. The proof is complete. □

Remark 2.6 The degree of $d_j(x_1,\ldots,x_n)$ as a (homogeneous) polynomial in $x_1,\ldots,x_n$ is $p^n - p^{n-j} = p^{n-j}(p^j - 1)$. In particular, in the range of dimensions $< p^n$ the p-adic expansion of any product $d_1^{i_1} d_2^{i_2} \cdots d_{n-1}^{i_{n-1}} D_n^{(p-1)j}$ has the form $s_1 p^{n-1} + s_2 p^{n-2} + \cdots s_{n-1} p + s_n$ with $p - 1 \geq s_1 \geq s_2 \geq s_3 \cdots \geq s_n \geq 0$, and for any such p-adic expansion there is a unique product as above having this degree. Thus, in degrees less than p^n the invariants only occur when the p-adic expansion has the form above, and in each such dimension the space of invariants is one dimensional.

Example 2.7 Consider the regular embedding

$$(\mathrm{reg})\colon\ (\mathbb{Z}/p)^n \hookrightarrow \mathcal{S}_{p^n}$$

which takes $g \in (\mathbb{Z}/p)^n$ to the permutation of the p^n points of $(\mathbb{Z}/p)^n$ induced by $h \mapsto g + h$. The normalizer of $(\mathrm{reg})(\mathbb{Z}/p)^n$ is the semi-direct product

$(\mathbb{Z}/p)^n \times_\alpha \mathrm{Aut}((\mathbb{Z}/p)^n)$. (See (VI.1) for details.) Indeed, let $\beta \in \mathrm{Aut}((\mathbb{Z}/p)^n)$. Then $\beta(h+g) = \beta(h) + \beta(g)$ so the permutation which we again call β, $h \mapsto \beta(h)$ satisfies $\beta \cdot p_g(h) = \beta(g+h) = \beta(g) + \beta(h) = p_{\beta(g)}(\beta(h))$ and $\beta \cdot p_g \beta^{-1}(h) = p_{\beta(g)}(h)$. Similarly for the converse.

But $\mathrm{Aut}((\mathbb{Z}/p)^n) = \mathrm{GL}_n(\mathbb{Z}/p)$! Consequently, when we pass to cohomology we have

$$\begin{aligned} H^*(\mathcal{S}_{p^n};\mathbb{F}_p) \xrightarrow{(\mathrm{reg})^*} & H^*((\mathbb{Z}/p)^n;\mathbb{F}_p)^{\mathrm{GL}_n(\mathbb{F}_p)} \\ &= (E(v_1,\dots,v_n)\otimes\mathbb{F}_p[b_1,\dots,b_n])^{\mathrm{GL}_n(\mathbb{F}_p)} \end{aligned}$$

by (II.4.3). Clearly, the polynomial elements in this image are contained in the Dickson algebra $\mathbb{F}_p[b_1,\dots,b_n]^{\mathrm{GL}_n(\mathbb{F}_p)}$ which we have just analyzed. In the next section we will show that $\mathrm{im}(reg)^* \cap \mathbb{F}_p[b_1,\dots,b_n]$ is precisely equal to $\mathbb{F}_p[b_1,\dots,b_n]^{\mathrm{GL}_n(\mathbb{F}_p)}$.

Previously we proved that a complete set of invariants for the action of $\mathrm{GL}_n(\mathbb{F}_p)$ on $\mathbb{F}_p[x_1,\dots,x_n]$ can be obtained using determinants (Dickson's Theorem). We will now outline the generalization of this result (due to Mui [Mu]) to the calculation of the invariants of the $\mathrm{GL}_n(\mathbb{F}_p)$ action on $\mathcal{A} = \mathbb{F}_p[x_1,\dots,x_n]\otimes E(e_1,\dots,e_n)$, an algebra isomorphic to the mod p cohomology of $(\mathbb{Z}/p)^n$, p an odd prime. We will therefore assume that the x_i are in degree 2 and the e_i in degree 1.

We begin by constructing some invariants for this action.

Definition 2.8

a. In $\mathbb{Z}[x_1,\dots,x_n]\otimes E_{\mathbb{Z}}(e_1,\dots,e_n)$, let $\overline{D}_{n,s_1,\dots,s_k}$ be defined as

$$\frac{1}{k!}\mathrm{Det}\begin{pmatrix} e_1 & \cdots & e_1 & x_1 & \cdots & \hat{x}_1^{p^{s_1}} & \cdots & \hat{x}_1^{p^{s_k}} & \cdots & x_1^{p^{n-1}} \\ e_2 & \cdots & e_2 & x_2 & \cdots & \hat{x}_2^{p^{s_1}} & \cdots & \hat{x}_2^{p^{s_k}} & \cdots & x_2^{p^{n-1}} \\ \vdots & \ddots & \vdots & \vdots & \ddots & \vdots & \ddots & \vdots & \ddots & \vdots \\ e_n & \cdots & e_n & x_n & \cdots & \hat{x}_n^{p^{s_1}} & \cdots & \hat{x}_n^{p^{s_k}} & \cdots & x_n^{p^{n-1}} \end{pmatrix}$$

where $0 \le s_1 < s_2 < \cdots < s_k \le n-1$ and the matrix for the preceeding determinant is filled out with k columns of the form $\begin{pmatrix} e_1 \\ e_2 \\ \vdots \\ e_n \end{pmatrix}$ to have n rows and columns.

b. Let $D_{n,s_1,\dots,s_k} \in \mathcal{A}$ be the mod p reduction of $\overline{D}_{n,s_1,\dots,s_k}$.

Note in particular that

$$D_{n,0,1,\dots,n-1} = e_1 e_2 \dots e_n\ .$$

Also, if $A \in \mathrm{GL}_n(\mathbb{F}_p)$, then $(AD)_{n,s_1,\dots,s_k} = \mathrm{Det}(A)\, D_{n,s_1,\dots,s_k}$.

Using our previous notation, we see that the elements

$$\mathcal{M}_{n,s_1,\ldots,s_k} = D_{n,s_1,\ldots,s_k} D_n^{p-2}$$

for $0 \leq s_1 < s_2 < \ldots < s_k \leq n-1$ will be $\mathrm{GL}_n(\mathbb{F}_p)$ invariants. We can now state Mui's generalization of Theorem 2.4:

Theorem 2.9

$$(\mathbb{F}_p[x_1,\ldots,x_n]\otimes E(e_1,\ldots,e_n))^{\mathrm{GL}_n(\mathbb{F}_p)} \cong \mathbb{F}_p[d_1,\ldots,d_n] \oplus \bigoplus \mathbb{F}_p[d_1,\ldots,d_n]\mathcal{M}_{n,s_1,\ldots,s_k}$$

where a double summation is taken over $k = 1,\ldots,n$ and $0 \leq s_1 < \ldots < s_k \leq n-1$. Furthermore the generators satisfy the relations

$$\mathcal{M}_{n,s}^2 = 0,$$
$$\mathcal{M}_{n,s_1} \cdots \mathcal{M}_{n,s_k} = (-1)^{k(k-1)/2}\mathcal{M}_{n,s_1,\ldots,s_k} d_n^{k-1}.$$

This extension of Dickson's result is proved by inductively showing that any invariant decomposes as above, using the fact that the ring of invariants is finitely generated over the (polynomial) Dickson algebra. The case $n = 2$ was originally proved by H. Cardenas, [Card]. We refer the reader to [Mu] for full details for the general case.

3. A Theorem of Serre

The following theorem of Serre, [Se2], is a key result in the study of the general structure of the cohomology ring of a finite p-group. It shows that for any non-elementary p-group G, the subring of the cohomology ring $H^*(G;\mathbb{F}_p)$ generated by the Bocksteins of the elements in $H^1(G;\mathbb{F}_p) \cong \mathrm{Hom}(G;\mathbb{F}_p)$ has transcendence degree strictly less than the dimension of $H^1(G;\mathbb{F}_p)$ as a vector space over the field $\mathbb{F}_p$. It should be compared with the corresponding situation for an elementary p-group (p odd) $(\mathbb{Z}/p)^r$ with cohomology ring

$$H^*((\mathbb{Z}/p)^r;\mathbb{F}_p) \cong E[e_1,\ldots,e_r] \otimes \mathbb{F}_p[b_1,\cdots b_r]$$

where b_i is the mod(p) Bockstein of e_i $i = 1,2,\ldots r$.

Theorem 3.1 *Let G be a p-group, not elementary abelian (i.e. $G \ncong (\mathbb{Z}/p)^k$). Then there is an $n \geq 1$ and elements $v_1,\ldots,v_n \in H^1(G;\mathbb{F}_p)$ such that $\beta(v_1)\cup \cdots \cup \beta(v_n) = 0$ where β is the mod p Bockstein.*

Proof. We first want to construct a central extension. Since

$$H_1(G;\mathbb{Z}) = \frac{G}{[G,G]}$$

it follows that $\pi: G \longrightarrow H_1(G;\mathbb{Z}/p) = H_1(G;\mathbb{Z}) \otimes_{\mathbb{Z}} \mathbb{Z}/p$ is an epimorphism. Let $G' = \ker(\pi)$, and now do this once more, that is to say we let G'' be the kernel of $\pi': G' \longrightarrow H_1(G';\mathbb{Z}/p)$. Notice that $G/G' = H_1(G;\mathbb{Z}/p) \cong (\mathbb{Z}/p)^r$ and similarly $G'/G'' \cong (\mathbb{Z}/p)^s$.

Both G' and G'' are characteristic subgroups of G. That is, they are taken to themselves by every automorphism of the p-group G, since every automorphism of G induces one of $H_*(G;\mathbb{Z}/p)$, and consequently, takes $\ker(\pi)$ to itself. In particular G'' is normal in G. Hencc we have the extension sequence

$$(*) \qquad 0 \longrightarrow \begin{matrix} G'/G'' \\ \| \\ (\mathbb{Z}/p)^s \end{matrix} \longrightarrow G/G'' \longrightarrow \begin{matrix} G/G' \\ \| \\ (\mathbb{Z}/p)^r \end{matrix} \longrightarrow 0\,.$$

Let the data for this extension be given by

$$\sigma\colon\ G/G' \longrightarrow \mathrm{Aut}(G'/G'') = \mathrm{Aut}((\mathbb{Z}/p)^s)\,.$$

Since G'/G'' is abelian its group of inner automorphisms is trivial and consequently σ is a homomorphism. Note also that $\mathrm{Aut}((\mathbb{Z}/p)^s) = \mathrm{GL}_s(\mathbb{F}_p)$, the general linear group over the finite field $\mathbb{F}_p$. In particular, since G/G' is a p-group, the image of σ is contained in a p-Sylow subgroup of $\mathrm{GL}_s(\mathbb{F}_p)$.

The order of $\mathrm{GL}_s(\mathbb{F}_p)$ is $(p^s-1)(p^s-p)(p^s-p^2)\cdots(p^s-p^{s-1}) = p^{\frac{s(s-1)}{2}}w$ with w prime to p. (This is well known, and easily checked by simply looking at the possible images of the basis vectors $e_1, e_2, \ldots, e_s$.) Moreover, a p-Sylow subgroup can be given as the upper triangular matrices with 1's along the diagonal. Moreover, we can assume $\mathrm{im}(\sigma)$ is contained in this subgroup.

Let $V \subset (\mathbb{Z}/p)^s = G'/G''$ be the vector subspace spanned by $e_2, \ldots, e_s$ and form the quotient $\frac{G'/G''}{V} = \mathbb{Z}/p$. In particular we can write the quotient $\bar{G} = G/(G''\langle e_2, \ldots, e_s\rangle)$ as a central extension $\mathbb{Z}/p \longrightarrow \bar{G} \longrightarrow G/G'$ and we have the commutative diagram

$$(3.2) \qquad \begin{array}{ccccccccc} 0 & \longrightarrow & \mathbb{Z}/p & \longrightarrow & \bar{G} & \xrightarrow{\pi'} & G/G' & \longrightarrow & 0 \\ & & \uparrow & & \uparrow & & \| & & \\ 0 & \longrightarrow & (\mathbb{Z}/p)^s & \longrightarrow & G/G'' & \longrightarrow & G/G' & \longrightarrow & 0\,. \end{array}$$

Since $(*)$ is a *totally* non-trivial extension we have that (3.2) is a non-trivial central extension. Let $\alpha \in H^2(G/G')$ be the class of the extension (3.2). Since the extension is non-trivial $\alpha \neq 0$. Thus $\alpha = \sum \lambda_i b_i + \sum \tau_{ij} e_i e_j$ where at least one of the coefficients in not 0. Consequently we have that $\pi'^*: H^1(G/G';\mathbb{F}_p) \longrightarrow H^1(\bar{G};\mathbb{F}_p)$ is an isomorphism as is thus the projection induced map $H^1(\bar{G};\mathbb{F}_p) \longrightarrow H^1(G;\mathbb{F}_p)$.

Note also that $\alpha \in H^2(G/G';\mathbb{F}_p)$ is in the kernel of

$$\pi^*\colon H^2(G/G';\mathbb{F}_p) \longrightarrow H^2(\bar{G};\mathbb{F}_p)\,.$$

There are now two cases:

1. One or more of the λ_i in the expression for α above are non-zero. In this case we have

$$\begin{aligned}\alpha^p &= \sum \lambda_i b_i^p + \sum \tau_{ij}(e_ie_j)^p \\ &= \sum \lambda_i b_i^p \\ &= \Big(\sum \lambda_i b_i\Big)^p \\ &= [\beta(\sum \lambda_i e_i)]^p\end{aligned}$$

since $\beta(e_i) = b_i$. Thus, the theorem is true in this case by setting $v_i = \pi^*(\sum \lambda_i e_i)$.

2. $\alpha = \sum \tau_{ij} e_i e_j$. We may assume $\tau_{12} \neq 0$. Now form $\tau_{12}^{-1}\alpha \cup e_3e_4\cdots e_n = e_1e_2e_3\cdots e_n$. Let $\mathcal{A}(p)$ be the mod p Steenrod algebra and $Q_i \in \mathcal{A}(p)$ be the ith odd dimensional Milnor primitive element which acts by $Q_i(e) = b^{p^i}$. We have

$$\pi^*(Q_{n-1}Q_{n-2}\cdots Q_2Q_1\beta(e_1e_2\cdots e_n)) = \\ \pi^*(Q_{n-1}\cdots Q_2Q_1\beta(\tau_{12}^{-1}\alpha \cup e_3\cdots e_n)) = 0\,.$$

But calculating explicitly we have

$$Q_{n-1}\cdots Q_2Q_1\beta(e_1\cdots e_n) = S(b_1b_2^p\cdots b_n^{p^{n-1}})$$

where, in this case S denotes the symmetric polynomial given by summing over all permutations of $1,\ldots,n$, and then multiplying by the sign of the permutation. In particular we have

$$(3.3)\qquad Q_{n-1}\cdots Q_2Q_1\beta(e_1\cdots e_n) = det\begin{pmatrix} b_1 & b_2 & \cdots & b_n \\ b_1^p & b_2^p & \cdots & b_n^p \\ \vdots & \vdots & \ddots & \vdots \\ b_1^{p^{n-1}} & b_2^{p^{n-1}} & \cdots & b_n^{p^{n-1}} \end{pmatrix}.$$

This is the polynomial $D_n(b_1,\ldots,b_n)$ considered in our discussion of the Dickson algebra, and is (up to sign) the product of all the linear summands $(b_{i_1} + a_{i_1+1}b_{i_1+1} + \cdots a_nb_n)$, where the leading coefficient is always 1. But each of these linear terms is the mod p Bockstein of the corresponding term in the e_i's. Thus this case, too, is finished and the proof is complete.

□

4. The Invariants in $H^*((\mathbb{Z}/p)^n; \mathbb{F}_p)$ Under the Action of $\mathcal{S}_n$

In this section we calculate some invariants which will be used in an essential way to calculate the cohomology of the general linear groups away from their characteristic in Chapter VII.

Let $\mathbb{F}$ be an arbitrary field having any characteristic except 2. In the polynomial ring in n variables over $\mathbb{F}$, $\mathbb{F}[x_1, \ldots, x_n]$, recall that the ith symmetric monomial is the sum

$$\sigma_i = \sum_{j_1 < j_2 < \cdots < j_i} x_{j_1} x_{j_2} \cdots x_{j_i} .$$

These σ_i are invariant under the action of the symmetric group $\mathcal{S}_n$ on $\mathbb{F}[x_1, \ldots, x_n]$ by permuting the x_j, and in fact, one of the fundamental theorems of the theory of equations states that

$$\mathbb{F}[x_1, \ldots, x_n]^{\mathcal{S}_n} = \mathbb{F}[\sigma_1, \ldots, \sigma_n] .$$

Now we consider the discriminant $D_n = \prod_{1 \leq i < j \leq n}(x_j - x_i)$. Clearly,

$$D_n^2 \in \mathbb{F}[\sigma_1, \ldots, \sigma_n]$$

so D_n is algebraic and even integral over $\mathbb{F}[\sigma_1, \ldots, \sigma_n]$. Moreover, the rational function field $\mathbb{F}(x_1, \ldots, x_n)$ is Galois over $\mathbb{F}(\sigma_1, \ldots, \sigma_n)$ with Galois group

$$\mathrm{Gal}\,(\mathbb{F}(x_1, \ldots, x_n)/\mathbb{F}(\sigma_1, \ldots, \sigma_n)) = \mathcal{S}_n .$$

D_n is invariant under the alternating group $\mathcal{A}_n$ and since the index of $\mathcal{A}_n$ in $\mathcal{S}_n$ is 2 it follows that $\mathbb{F}(\sigma_1, \ldots, \sigma_n)(D_n) = \mathbb{F}(x_1, \ldots, x_n)^{\mathcal{A}_n}$ provided that the characteristic of $\mathbb{F}$ is not 2 which is one of our current assumptions.

Proposition 4.1 *The integral closure of* $\mathbb{F}[\sigma_1, \ldots, \sigma_n]$ *in the quotient field* $\mathbb{F}(\sigma_1, \ldots, \sigma_n)(D_n)$ *is exactly*

$$\mathbb{F}[\sigma_1, \ldots, \sigma_n] + \mathbb{F}[\sigma_1, \ldots, \sigma_n] D_n .$$

That is to say, an element $\alpha \in \mathbb{F}(\sigma_1, \ldots, \sigma_n)(D_n)$ *is integral over* $\mathbb{F}[\sigma_1, \ldots, \sigma_n]$ *if and only if we can write* $\alpha = A + BD_n$ *with* A *and* B *contained in* $\mathbb{F}[\sigma_1, \ldots, \sigma_n]$.

Proof. Consider the field extension $\mathbb{F}(\sigma_1, \ldots, \sigma_n)(D_n)$ over $\mathbb{F}(\sigma_1, \ldots, \sigma_n)$. The Galois group of this extension is $\mathbb{Z}/2$ and if T is a generator, then $T(D_n) = -D_n$, so for $\alpha = A + BD_n$ we have that α is a root of

$$x^2 - 2Ax + (A^2 - B^2 D_n^2)$$

and, for $B \neq 0$ this is the minimal polynomial for α. Consequently, α is integral if and only if (1) $2A \in \mathbb{F}[\sigma_1, \ldots, \sigma_n]$ and (2), $A^2 - B^2D_n^2 \in \mathbb{F}[\sigma_1, \ldots, \sigma_n]$. Hence we need to show that if

$$B = C/\lambda, \qquad \text{with } C, \lambda \in \mathbb{F}[\sigma_1, \ldots, \sigma_n], \quad \text{while } B^2D_n^2 \in \mathbb{F}[\sigma_1, \ldots, \sigma_n]$$

then λ must divide C and B is already contained in $\mathbb{F}[\sigma_1, \ldots, \sigma_n]$. To do this we look at

$$\left(\frac{C}{\lambda}\right) D_n \in \mathbb{F}(x_1, \ldots, x_n)\,.$$

Since the integral closure of $\mathbb{F}[\sigma_1, \ldots, \sigma_n]$ in $\mathbb{F}(x_1, \ldots, x_n)$ is $\mathbb{F}[x_1, \ldots, x_n]$ it follows that

$$\left(\frac{C}{\lambda}\right) D_n \in \mathbb{F}[x_1, \ldots, x_n]$$

and we have

$$CD_n = \lambda f$$

for some non-zero $f \in \mathbb{F}[x_1, \ldots, x_n]$.

It is standard that $\mathbb{F}[x_1, \ldots, x_n]$ is a unique factorization domain, see e.g. [Hu], pg.164. Thus we have unique decompositions into irreducibles (up to multiplication by a non-zero constant coming from $\mathbb{F}$), $D_n = \prod_{1 \leq i < j \leq n}(x_j - x_i)$, $C = \prod c_i$, $f = \prod f_j$, $\lambda = \prod \lambda_k$, and

$$\prod c_i \prod (x_j - x_i) = \prod \lambda_k \prod f_j$$

so it follows by the uniqueness of decomposition into irreducibles in a UFD that each λ_k is either equal to a c_i or an $(x_j - x_i)$.

If all the λ_k are c_i's then we are done. So assume $\lambda_k = (x_j - x_i)$. Since λ is invariant under the action of $\mathcal{S}_n$ it follows that $(x_r - x_s)$ is also a divisor of λ for any $r > s$ and thus D_n divides λ.

In this case let $\lambda' = \lambda/D_n$. Clearly $\lambda' = \prod \lambda'_k$ with λ'_k irreducible, and each λ'_k is a factor of C so λ' divides C. Also

$$CD_n/\lambda = C/\lambda' = C'\,,$$

and clearly for $g \in \mathcal{S}_n$ we have $g(C') = (-1)^g C'$. Now, let us assume by induction that the result is true for $n-1$. (Its truth for $n = 2$ is clear.) Then expand

$$C' = A_r x_n^r + A_{r-1}x_n^{r-1} + \cdots + A_0$$

with each $A_j \in \mathbb{F}[x_1, \ldots, x_{n-1}]$. We have for each $g \in \mathcal{S}_{n-1} \subset \mathcal{S}_n$, $(g(x_n) = x_n)$, that $g(A_j) = (-1)^g A_i$, so, by our inductive assumption, since each A_j is integral over $\mathbb{F}[\bar{\sigma}_1, \ldots, \bar{\sigma}_n]$ where $\bar{\sigma}_i$ denotes the ith symmetric polynomial in $x_1, \ldots, x_{n-1}$, we have

$$A_j = L_i(\bar{\sigma}_1, \ldots, \bar{\sigma}_{n-1})D_{n-1}$$

and D_{n-1} divides C'. From this, using the Galois action it follows that D_n divides C' and $C' = C''D_n$ with C'' invariant under the action of $\mathcal{S}_n$. Also, C'' is integral over $\mathbb{F}[\sigma_1, \ldots, \sigma_n]$ so, since $\mathbb{F}[\sigma_1, \ldots, \sigma_n]$ is integrally closed in its quotient field $\mathbb{F}(\sigma_1, \ldots, \sigma_n)$ it follows that $C'' \in \mathbb{F}[\sigma_1, \ldots, \sigma_n]$. The proof of the proposition is complete. □

Theorem 4.2 *Let $\mathcal{A} = \mathbb{F}[x_1, \ldots, x_n] \otimes E(e_1, \ldots, e_n)$ be a tensor product of a polynomial algebra on even dimensional generators and an exterior algebra on odd dimensional generators where $\mathbb{F}$ is a field of characteristic not equal to two. Let $\mathcal{S}_n$ act by permutation on the x_j's and the e_i's, then*

$$\mathcal{A}^{\mathcal{S}_n} = \mathbb{F}[\sigma_1, \ldots, \sigma_n] \otimes E(f_1, f_3, \ldots, f_{2n-1})$$

where

$$f_{2i-1} = \sum_{j=1}^{n} \left(\frac{\partial}{\partial x_j} \sigma_i \right) e_j \, .$$

Example. $f_1 = \sum e_i$ while $f_3 = \sum_i (\sum_{j \neq i} x_j) e_i$.

Proof of 4.2 Let $\alpha \in \mathcal{A}^{\mathcal{S}_n}$. Then

$$\alpha = \sum_I A_I e_{i_1} \cdots e_{i_r}$$

where $I = (i_1, \ldots, i_r)$ with $1 \leq i_1 < \cdots < i_r \leq n$ and $A_I \in \mathbb{F}[x_1, \ldots, x_n]$. Moreover, since α is invariant under $\mathcal{S}_n$ it follows that, given $(i_1, \ldots, i_r)$ as above, there is a $g \in \mathcal{S}_n$ with $g(t) = i_t$ for $1 \leq t \leq r$, and $g(A_{(1,\ldots,r)}) = A_{(i_1,\ldots,i_r)}$. Thus α is completely determined by A_0, A_1, $A_{(1,2)}$, and so on until $A_{(1,2,\ldots,n)}$. Also, each $A_{(1,\ldots,r)}$ satisfies $g(A_{(1,\ldots,r)}) = (-1)^g A_{(1,\ldots,r)}$ for any $g \in \mathcal{S}_r \subset \mathcal{S}_n$ where $\mathcal{S}_r$ fixes the last $n-r$ coordinates, while $g(A_{(1,\ldots,r)}) = A_{(1,\ldots,r)}$ if $g \in \mathcal{S}'_{n-r}$ where $\mathcal{S}'_t$ fixes the first $n-t$ coordinates.

It follows from (4.1) that

$$A_{(1,\ldots,r)} = B_{(1,\ldots,r)} D_r$$

where $B_{(1,\ldots,r)}$ is invariant under $\mathcal{S}_r \times \mathcal{S}'_{n-r}$ and hence belongs to

$$\mathbb{F}[\sigma_1, \ldots, \sigma_r] \otimes \mathbb{F}[\sigma'_1, \ldots, \sigma'_{n-r}]$$

where $\sigma'_i = \sigma_i(x_{r+1}, \ldots, x_n)$. And, of course, if $A_{(1,\ldots,r)}$ has the form above then there is a unique invariant under $\mathcal{S}_n$ of the form $A_{(1,\ldots,r)} e_1 \cdots e_r + \cdots$. The Poincaré series for such invariants has the form

$$\begin{aligned} &\frac{x^{2r(r-1)/2} x^r}{(1-x^2)\cdots(1-x^{2r})(1-x^2)\cdots(1-x^{2n-2r})} \\ &\quad = \frac{1}{(1-x^2)\cdots(1-x^{2n})} \left[x^{r^2} \cdot \frac{(1-x^{2r+2})\cdots(1-x^{2n})}{(1-x^2)\cdots(1-x^{2n-2r})} \right] \\ &\quad = \frac{1}{(1-x^2)\cdots(1-x^{2n})} T_r(n) \, . \end{aligned}$$

Next, we claim that there is an identity

$$\sum_{r=0}^{n} T_r(n) = (1+x)(1+x^3)\cdots(1+x^{2n-1}).$$

We verify this for $n = 2$ where the sum is

$$\begin{aligned} 1 + x\frac{(1-x^4)}{(1-x^2)} + x^4 &= 1 + x(1+x^2) + x^4 \\ &= 1 + x + x^3 + x^4 \\ &= (1+x)(1+x^3). \end{aligned}$$

Now, the proof of the claim proceeds by induction. To obtain the inductive step note that

$$T_r(n+1) - T_r(n) = \frac{x^{2n+2-2r}(1-x^{2r})}{(1-x^{2n+2-2r})} T_r(n) = x^{2n+1} T_{r-1}(n).$$

The claim follows, and the Poincaré series for the ring of invariants is

$$\frac{(1+x)(1+x^3)\cdots(1+x^{2n-1})}{(1-x^2)(1-x^4)\cdots(1-x^{2n})}$$

which is the Poincaré series of the asserted ring in the Theorem.

Of course, this is not yet a complete proof. Certainly, the elements in the ring of the theorem are invariants, but it is not yet clear that this ring actually injects into the invariants – there might be some relations. However, to prove that the ring actually injects it suffices to show that $f_1 f_3 \cdots f_{2n-1} \neq 0$. This in turn can be done by induction. Write

$$f_{2m-1}^{n} = f_{2m-1}^{n-1} + f_{2m-1}^{n-1} x_n + \bar{\sigma}_{m-1} e_n$$

where the superscript on the f's denotes that it involves only the first $n-1$ of the variables or all of them depending on whether it is $n-1$ or n. Using the formula above and multiplying out we find

$$f_1 f_3 \cdots f_{2n-1} = (f_1^{n-1} f_3^{n-1} \cdots f_{2n-3}^{n-1})(\bar{\sigma}_{n-1} + \bar{\sigma}_{n-2} x_n + \cdots + x_n^{n-1}) e_n$$

which is non-zero by our inductive assumption. Now the proof is complete. □

5. The Cardenas-Kuhn Theorem

As we have seen, when $L \subset G$ there is an action of $N_G(L)$ on $H^*(L;\mathbb{F}_p)$ where the action of $L \triangleleft N_G(L)$ is the identity and

$$\mathrm{im}\,(\mathrm{res}^*\colon\ H^*(G;\mathbb{F}_p) \longrightarrow H^*(L;\mathbb{F}_p))$$

is contained in the subring of invariants under this action. Since the action of L is trivial the action factors through an action of the *Weyl group* of L in G,

$$W_G(L) \;=\; N_G(L)/L\,,$$

and we can write

$$\mathrm{im}\,(\mathrm{res}^*\colon\ H^*(G;\mathbb{F}_p) \longrightarrow H^*(L;\mathbb{F}_p)) \subseteq H^*(L;\mathbb{F}_p)^{W_G(L)}\,.$$

Let K be a third subgroup so that $L \subset K \subset G$. Since $W_K(L) \subseteq W_G(L)$ we have an inclusion of rings

$$\mathrm{res}^*\colon\ H^*(L;\mathbb{F}_p)^{W_G(L)} \subseteq H^*(L;\mathbb{F}_p)^{W_K(L)}$$

and corresponding to the transfer in group cohomology we have a map going in the opposite direction,

$$\tau\colon\ H^*(L;\mathbb{F}_p)^{W_K(L)} \longrightarrow H^*(L;\mathbb{F}_p)^{W_G(L)}$$

obtained by taking a left coset decomposition of $W_G(L)$ with respect to $W_K(L)$, $W_G(L) = \sqcup_i g_i W_K(L)$, and setting $\tau(\alpha) = \sum_i g_i(\alpha)$. We have evidently that

$$\mathrm{im}(\tau) \subseteq H^*(L;\mathbb{F}_p)^{W_G(L)}$$

and $\tau \cdot \mathrm{res}^*$ is multiplication by the index of $W_K(L)$ in $W_G(L)$, that is to say, by $|W_G(L){:}\, W_K(L)|$. Also, τ is an $H^*(L;\mathbb{F}_p)^{W_G(L)}$-module map, since $\tau(v\alpha) = v\tau(\alpha)$ for any $v \in H^*(L;\mathbb{F}_p)^{W_G(L)}$, though τ is not a ring map in general. In the case when $W_K(L)$ contains a p-Sylow subgroup of $W_G(L)$ the map τ is surjective and the discussion above shows

Proposition 5.1 *Suppose $L \subsetneq K \subsetneq G$ is a sequence of subgroups and $W_K(L)$ contains a p-Sylow subgroup of $W_G(L)$, then τ splits $H^*(L;\mathbb{F}_p)^{W_K(G)}$ as a direct sum $H^*(L;\mathbb{F}_p)^{W_G(L)} \oplus \mathrm{Ker}(\tau)$ of $H^*(L;\mathbb{F}_p)^{W_G(L)}$-modules.*

There are certain important cases in which we can be more precise about

$$\mathrm{im}\,(\mathrm{res}^*\colon\ H^*(G;\mathbb{F}_p) \longrightarrow H^*(L;\mathbb{F}_p))\,.$$

Definition 5.2 $L \subsetneq K \subsetneq G$ is a *closed system*, also called a weakly closed system, if every subgroup of K which is conjugate to L in G is already conjugate to L in K.

For closed systems we have

Lemma 5.3 *Suppose $L \subsetneq K \subsetneq G$ is a closed system with L an elementary p-group. Then there is a commutative diagram*

$$\begin{array}{ccc} H^*(K;\mathbb{F}_p) & \xrightarrow{\text{tr}^*} & H^*(G;\mathbb{F}_p) \\ \Big\downarrow \text{res}^* & & \Big\downarrow \text{res}^* \\ H^*(L;\mathbb{F}_p)^{W_K(L)} & \xrightarrow{\tau} & H^*(L;\mathbb{F}_p)^{W_G(L)} . \end{array}$$

Proof. Consider the double coset decomposition $G = \sqcup_i Lg_iK$. If $g^{-1}Lg \subset K$ then $g^{-1}Lg$ is already conjugate to L in K, say $g^{-1}Lg = k^{-1}Lk$ for an appropriate $k \in K$. Consequently we can assume $g^{-1}Lg = L$. In the remaining cases $g^{-1}Lg \cap K$ is a proper subgroup of $g^{-1}Lg$ as g runs over the g_i above in the double coset decomposition.

Sublemma 5.4 *Suppose Lg_iK and Lg_jK both satisfy $g^{-1}Lg = L$, then the double cosets are equal if and only if the left cosets $\phi(g_i)W_K(L)$ and $\phi(g_j)W_K(L)$ are equal in $W_G(L)$, where $\phi\colon N_G(L) \longrightarrow W_G(L)$ is the projection.*

Proof of 5.4 We can regard the double cosets LgK as corresponding to the orbits of the left cosets of K in G under the action of L. From this point of view the coset gK is fixed if and only if $g^{-1}Lg \subset K$, and hence, under our closure assumption and choices of the g_i, if and only if $g_i \in N_G(L)$. Moreover the total variation of g_i is $Lg_iN_K(L) = g_iN_K(L)$. □

We now return to the proof of the lemma. The double coset formula gives that $\text{res}^* \cdot \text{tr}^*\colon H^*(K;\mathbb{F}_p) \longrightarrow H^*(G;\mathbb{F}_p) \longrightarrow H^*(L;\mathbb{F}_p)$ can be written as the sum of the compositions

$$H^*(K;\mathbb{F}_p) \xrightarrow{\text{res}^*} H^*(g_i^{-1}Lg_i \cap K;\mathbb{F}_p)$$
$$\xrightarrow{g_i^*} H^*(L \cap g_iKg_i^{-1};\mathbb{F}_p) \xrightarrow{(\text{tr})^*} H^*(L;\mathbb{F}_p) .$$

Since L is p-elementary and $\text{tr}^* \equiv 0$ for any $V \subsetneq L$ we need only sum over the $g_i \in N_G(L)$, and the theorem follows. □

From this we obtain the Cardenas-Kuhn theorem,

Theorem 5.5 *Let $L \subsetneq K \subsetneq G$ be a closed system with L a p-elementary subgroup of G. Suppose also that $|W_G(L)\colon W_K(L)|$ is prime to p, i.e. $W_K(L)$ contains a p-Sylow subgroup of $W_G(L)$, then the image of*

$$\text{res}^*\colon H^*(G;\mathbb{F}_p) \longrightarrow H^*(L;\mathbb{F}_p)$$

is exactly equal to the intersection

$$H^*(L;\mathbb{F}_p)^{W_G(L)} \cap \mathrm{im}\,(\mathrm{res}^*\colon H^*(K;\mathbb{F}_p)\longrightarrow H^*(L;\mathbb{F}_p))\ .$$

Example 5.6 The 2-Sylow subgroup of the sporadic simple group M_{12} is given as the semi-direct product $(\mathbb{Z}/4\times\mathbb{Z}/4)\times_T(\mathbb{Z}/2\times\mathbb{Z}/2)$, where if a, b are the generators of the two copies of $\mathbb{Z}/4$ and c, d generate the two copies of $\mathbb{Z}/2$ then $ca^ib^jc = a^{-i}b^{-j}$, $dad = a$, $dbd = ab^{-1}$. The subgroup $V = (\mathbb{Z}/2)^3 = \langle a^2, b^2, c\rangle$ is normal in this group with $W_{\mathrm{Syl}_2(M_{12})}(V) = D_8$, and we will see in Chapter VIII that the image of restriction $H^*(\mathrm{Syl}_2(M_{12});\mathbb{F}_2)\xrightarrow{\mathrm{res}^*}H^*(V;\mathbb{F}_2)$ is exactly $H^*(V;\mathbb{F}_2)^{D_8}$. Moreover, it is true that $V\subset\mathrm{Syl}_2(M_{12})\subset M_{12}$ is a closed system. Thus the Cardenas-Kuhn theorem gives us the image in this group of $H^*(M_{12};\mathbb{F}_2)$.

This group $\mathrm{Syl}_2(M_{12})$ occurs in several other contexts as well. It turns out that there is a unique non-split extension

$$(\mathbb{Z}/2)^3\xrightarrow{\triangleleft}E\longrightarrow\mathrm{GL}_3(\mathbb{F}_2)$$

which has $\mathrm{Syl}_2(M_{12})$ as its 2-Sylow subgroup. (Here the action of $\mathrm{GL}_3(\mathbb{F}_2)$ on $(\mathbb{Z}/2)^3 = \mathbb{F}_2^3$ is the usual one.) Consequently, $W_E(V) = \mathrm{GL}_3(\mathbb{F}_2)$ contains $W_{\mathrm{Syl}_2(M_{12})}(V)$ as a 2-Sylow subgroup and

$$\mathrm{im}\,(\mathrm{res}^*\colon\ H^*(E;\mathbb{F}_2)\longrightarrow H^*(V;\mathbb{F}_2))$$

is exactly $\mathbb{F}_2[e_1,e_2,e_3]^{\mathrm{GL}_3(\mathbb{F}_2)} = \mathbb{F}_2[d_4,d_6,d_7]$.

This group E is contained in the exceptional (continuous) Lie group G_2 as a maximal (finite) subgroup, and the composition

$$H^*(B_{G_2};\mathbb{F}_2)\xrightarrow{\mathrm{res}^*}H^*(E;\mathbb{F}_2)\xrightarrow{\mathrm{res}^*}H^*(V)^{\mathrm{GL}_3(\mathbb{F}_2)}$$

is an isomorphism in cohomology.

Remark. A brief comment on the history of the Cardenas-Kuhn theorem might be in order. The original idea for this decompostion occurs in H. Cardenas' thesis [Card], where it was exploited only for the groups $\mathcal{S}_{p^2}$. The senior author of this book in 1965, and H. Mui in approximately 1974 both realized that this could be extended to give the critical step in studying the ring structure of the groups $H^*(\mathcal{S}_n;\mathbb{F}_p)$, but both formulations were restricted to the situation occurring in the symmetric groups so their work should properly be regarded as subsumed in Cardenas'.

However, N. Kuhn gave the very useful general formulation of Cardenas' idea that we proved above in [K] and since Kuhn's formulation is applicable in a very large number of cases including most of the groups of Lie type and some of the sporadic groups we felt that calling this basic calculational result the Cardenas-Kuhn theorem was appropriate.

6. Discussion of Related Topics and Further Results

There are some further special cases where calculations have been made or where there are unexpected connections between the topics in this chapter and other areas of mathematics.

The Dickson Algebras and Topology

As we point out in (VIII.5), the cohomology of the classifying space of the fourteen dimensional compact Lie group G_2, $H^*(BG_2; \mathbb{F}_2) = \mathbb{F}_2[d_4, d_6, d_7]$ is a copy of the Dickson algebra $\mathbb{F}_2[x_1, x_2, x_3]^{GL_3(\mathbb{F}_2)}$.

It is known that the Dickson algebra $\mathbb{F}_2[x_1, \dots, x_n]^{GL_n(\mathbb{F}_2)}$ for $n \geq 5$ cannot be realized as the mod (2) cohomology ring of any topological space, and there were claims in the literature that neither could the Dickson algebra for $n = 4$. But recently W. Dwyer and C. Wilkerson, [DW], have actually constructed such a space, V_4. Its loop space ΩV_4 has finite mod (2) cohomology, $H^*(\Omega V_4; \mathbb{F}_2) = \mathbb{F}_2[x_7]/(x_7^4) \otimes E(x_{11}, x_{13})$, but is not the mod(2) homotopy type of any Lie group. It is conjectured that there is a connection between the space V_4 and the Conway sporadic group Co_3, in much the same way that BG_2 is related to M_{12} in [M2] (as described in (IX.3)) but the exact connection, if any, awaits a cohomological analysis of the Conway sporadic group Co_3.

There is also a connection between the J_1 and G_2, due to F. Cohen, which we discuss in VIII.5, IX.3, which shows that in an appropriate sense, at the prime 2, B_{J_1} can be regarded as the total space of a fibration with fiber G_2 and base B_{G_2}.

The Ring of Invariants for $Sp_{2n}(\mathbb{F}_2)$

Recently P. Kropholler, motivated by conversations with J.F. Adams, studied the invariants in $\mathbb{F}_2[x_1, \dots, x_{2n}]$ for the groups of Lie type $Sp_{2n}(\mathbb{F}_2)$, (see (VII.3) for a discussion of the finite Chevalley groups of symplectic type). He found the following structure for the ring of invariants

$$\begin{aligned}\mathbb{F}_2[x_1, \dots, x_{2n}]^{Sp_{2n}(\mathbb{F}_2)} &= \\ &\mathbb{F}_2[\bar{w}_3, \gamma_5, \dots, \gamma_{2^{n-1}+1}, d_{2^{2n-1}}, \dots, d_{2^{2n}-2^{2n-2}}](1, \gamma_{2^n+1})\end{aligned}$$

where $\bar{w}_3 = \sum_{i \neq j} x_i^2 x_j$, $Sq^2(\bar{w}_3) = \gamma_5$, and for $i \geq 2$ we have $Sq^{2^i}(\gamma_{2^i+1}) = \gamma_{2^{i+1}+1}$. As an example there is a classic isomorphism $Sp_4(\mathbb{F}_2) = \mathcal{S}_6$, and

$$\mathbb{F}_2[x_1, x_2, x_3, x_4]^{\mathcal{S}_6} = \mathbb{F}_2[\bar{w}_3, \gamma_5, d_8, d_{12}](1, \gamma_9),$$

where $Sq^2(\bar{w}_3) = \gamma_5 = \sum x_i^4 x_j$, $Sq^4(\gamma_5) = \gamma_9 = \sum x_i^8 x_j$, and $Sq^4(d_8) = d_{12}$.

The Invariants for Subgroups of $GL_4(\mathbb{F}_2)$

There are other special isomorphisms like that of $\mathcal{S}_6$ with $Sp_4(\mathbb{F}_2)$ between some of the smaller finite Chevalley groups in different families. For example (VII.3.8), $SL_2(\mathbb{F}_4) = Sp_2(\mathbb{F}_4) \cong A_5$. This and many others are consequences of the classical isomorphism $GL_4(\mathbb{F}_2) \cong A_8$. (See (VI.6.6) for an explicit construction of the matrices which realize this isomorphism.) There is a second A_5 in A_8. This A_5 acts on $\mathbb{F}_2[x_1, \ldots, x_4]$ by tensoring over $\mathbb{F}_2$ from the action of $\mathcal{S}_5$ on $(\mathbb{Z})^4$ where $(\mathbb{Z})^4$ consists of those elements $(n_1, \ldots, n_5) \in (\mathbb{Z})^5$ with $\sum n_i = 0$, and $\mathcal{S}_5$ acts by permuting coordinates. These two A_5's are both contained in A_6 and, in fact, are the two non-conjugate copies of A_5 there.

These groups occur as parts of maximal subgroups in many of the sporadic groups. For example the semi-direct product $(\mathbb{Z}/2)^4 \times_T A_6$ is maximal in the Mathieu group M_{22}, while $(\mathbb{Z}/2)^4 \times_T A_7$ is maximal in M_{23} and is the normalizer of both maximal 2-elementary subgroups $(\mathbb{Z}/2)^4$ in the group M^cL. Likewise, A_5 is important in the Janko groups J_2 and J_3 where one has a maximal group of the form $\mathbb{Z}/2 \to G \to (\mathbb{Z}/2)^4 \times_T A_5$.

Using the techniques sketched in (III.1) we have been determining the rings of invariants for these actions on $\mathbb{F}_2[x_1, \ldots, x_4]$. In [AM1] the invariants for the first A_5 were determined. They have Poincaré series with denominator $(1 - x^5)^2(1 - x^{12})^2$ and numerator

$$\begin{aligned} 1 + 2x^3 + 3x^6 + x^8 + 6x^9 + 2x^{11} + 9x^{12} + x^{14} + 10x^{15} \\ + x^{16} + 9x^{18} + 2x^{19} + 6x^{21} + x^{22} + 3x^{24} + 2x^{27} + x^{30} . \end{aligned}$$

The second A_5 has invariant subring of the form

$$\mathbb{F}_2[\sigma_2, \sigma_3, \sigma_4, \sigma_5](1, b_{10}).$$

The invariants under A_6 are a degree two integral extension of the invariants for $Sp_4(\mathbb{F}_2)$,

$$\mathbb{F}_2[x_1, \ldots, x_4]^{A_6} = \mathbb{F}_2[\overline{w}_3, \gamma_5, d_8, d_{12}](1, \gamma_9, b_{15}, \gamma_9 b_{15}) ,$$

and finally

$$\mathbb{F}_2[x_1, \ldots, x_4]^{A_7} = \mathbb{F}_2[d_8, d_{12}, d_{14}, d_{15}](1, b_{18}, b_{21}, b_{22}, b_{23}, b_{24}, d_{27}, b_{45}) .$$

Further details can be found in [AM2].

Chapter IV. Spectral Sequences and Detection Theorems

0. Introduction

In this chapter we introduce the main computational techniques for determining the cohomology of finite groups. We start with the various forms of the Lyndon-Hochschild-Serre spectral sequence, first from a geometric point of view in §1 and then from a purely algebraic point of view, following work of A. Liulevicius and C.T.C. Wall, in §2.

The work in §1 is particularly well adapted to wreath products, and we determine the cohomology ring of $G \wr \mathbb{Z}/p$ in terms purely of $H^*(G; \mathbb{F}_p)$. Then, in §2 we determine the rings $H^*(G; \mathbb{F}_2)$ where G is the dihedral group D_{2^n} or the generalized quaternion group Q_{2^n}.

Section 3 concentrates on chain level techniques. In particular we use these techniques to get some information on cup products and the ring structure in $H_*(G; \mathbb{F}_p)$ induced from the map $B_G \times B_G \longrightarrow B_G$ defined in (II.1.8).

Section 4 considers detection theorems for the cohomology of wreath products and we prove some theorems of D. Quillen which are very important in understanding the cohomology of symmetric groups and finite groups of Lie type.

Then §5 proves most of the general structure theorems for the ring $H^*(G; \mathbb{F}_p)$, Evens' finite generation theorem, and many of Quillen's results on the role of the p-elementary subgroups of G in determining $H^*(G; \mathbb{F}_p)$. In particular we describe Quillen's proof, [Q1], of the Atiyah-Evens-Swan conjecture on the Krull dimension of $H^*(G; \mathbb{F}_p)$.

The groups of Krull dimension at most one at all primes are the periodic groups and they are studied and their cohomology rings determined in §6. We also review the results of Suzuki-Zassenhaus which completely classify such groups there.

Finally, in §7 we use the ideas and techniques developed in this chapter to sketch a modification of Steenrod's construction of the Steenrod Squares and p-power operations.

These results summarize most of the techniques available for calculations in group cohomology until the more recent work of Peter Webb which we discuss in Chapter V.

1. The Lyndon–Hochschild–Serre Spectral Sequence: Geometric Approach

Let $H \subset G$ be a normal subgroup, and consider the induced surjection

$$B_p\colon B_G \longrightarrow B_{G/H}\ .$$

Note that $B_p^{-1}(*)$ consists of those points of the form $\{t_1, \ldots, t_r, g_1, \ldots, g_r\}$ with each $g_i \in H$ and this is just $B_H \subset B_G$. Every continuous map $f\colon X \longrightarrow Y$ can be turned into a (Serre) fibration by replacing X by the space of paths in the mapping cylinder of f which start at X, $E^{M(f)}_{X,M(f)}$. (Projecting a path to its initial point gives the equivalence to X and projecting to its endpoint gives the map to $Y \simeq M(f)$.) The fiber over $y \in M(f)$ is the set of paths which end at y, $E^{M(f)}_{X,y}$. Consequently there is a natural inclusion of B_H into the homotopy fiber of the map B_p, the space of paths $E^{B_{G/H}}_{B_G\ *}$, in the mapping cylinder of B_p by sending (b,t) to (b,t) in the mapping cylinder for $b \in B_H$.

Lemma 1.1 *The inclusion $B_H \subset E^{B_{G/H}}_{B_G\ *}$ sending $\theta \in B_G$ to $\gamma(t) = (t, \theta)$ in the mapping cylinder is a homotopy equivalence.*

Proof. First, from the homotopy exact sequence of the fibration

$$\cdots \longrightarrow \pi_{i+1}(B_{G/H}) \longrightarrow \pi_i(E^{B_{G/H}}_{B_G\ *}) \longrightarrow \pi_i(B_G) \longrightarrow \pi_i(B_{G/H}) \longrightarrow \cdots$$

we see that $E^{B_{G/H}}_{B_G\ *}$ has the homotopy type of B_H, since, by a theorem of Milnor, the path space $E^{B_{G/H}}_{B_G\ *}$ has the homotopy type of a CW-complex. Next, from the factorization of the inclusion $B_H \subset B_G$ through $E^{B_{G/H}}_{B_G\ *}$ constructed above

$$\begin{array}{ccc} B_H & \hookrightarrow & B_G \\ \Big\downarrow \lambda & & \Big\downarrow = \\ E^{B_{G/H}}_{B_G\ *} & \longrightarrow & B_G \end{array}$$

we see that λ induces an isomorphism in homotopy. Consequently, by Whitehead's theorem, it is a homotopy equivalence. □

Corollary 1.2 *There is a spectral sequence converging to $H^*(G; A)$ for untwisted coefficients A with $E_2^{i,j}$ term $H^i(G/H; H^j_t(H; A))$.*

This is just the Serre spectral sequence for the fibering above. [CE], [Brown] and [Sp] provide good descriptions of spectral sequences for readers interested in more background material. The spectral sequence above is called the Lyndon–Hochschild-Serre spectral sequence ([HS], [L]). Sometimes it can actually be analyzed using homotopy theory or geometric constructions. More often, in order to study the differentials we need an algebraic reformulation which will be given in the next section.

Here, though, we will consider a special case where geometric methods prove the spectral sequence collapses and $E_2 = E_\infty$ for a small, but very important class of groups.

Wreath Products

Let $\mathbb{Z}/p$ act on $\underbrace{G \times G \times \cdots \times G}_{p-times}$ by cyclic shifting, i.e.

$$T(g_1, g_2, \ldots, g_p) = (g_p, g_1, g_2, \ldots, g_{p-1})$$

where T is a selected generator for $\mathbb{Z}/p$. Then, with respect to this twisting, define the wreath product $G \wr \mathbb{Z}/p$ as the semi-direct product $G^p \times_T \mathbb{Z}/p$. A classifying space for $G \wr \mathbb{Z}/p$ can be explicitly given as

$$(B_G)^p \times_{\mathbb{Z}/p} E_{\mathbb{Z}/p}$$

where $\mathbb{Z}/p$ also acts on $(B_G)^p$ by shifting coordinates cyclically. In this case, since $(B_G)^p = B_{G^p}$, the homotopy fibering above becomes an actual fibering

$$(1.3) \qquad (B_G)^p \longrightarrow (B_G)^p \times_{\mathbb{Z}/p} E_{\mathbb{Z}/p} \longrightarrow B_{\mathbb{Z}/p}$$

where the normal subgroup H is $(G)^p$.

We now consider the Lyndon–Hochschild–Serre spectral sequence in this case, with $A = \mathbb{F}_p$. First we study the E_2-term, so we study the action of $\mathbb{Z}/p$ on

$$\underbrace{H^*(B_G; \mathbb{F}_p) \otimes \cdots \otimes H^*(B_G; \mathbb{F}_p)}_{p-times} = H^*((B_G)^p; \mathbb{F}_p)\ .$$

Lemma 1.4 *Let Λ be a graded $\mathbb{F}_p$-vector space, $\Lambda = \coprod_i \Lambda_i$. Then as a $\mathbb{F}_p(\mathbb{Z}/p)$-module the p-fold tensor product $\Lambda^p = \Lambda \otimes \cdots \otimes \Lambda$ where the action of $\mathbb{Z}/p$ is by cyclic shifting, is a direct sum of free modules with trivial modules, the trivial ones generated by elements of the form $(\lambda_i \otimes \lambda_i \otimes \cdots \otimes \lambda_i)$ as the λ_i run over a $\mathbb{F}_p$-basis for Λ.*

Proof. A $\mathbb{F}_p$ basis for $\Lambda \otimes \cdots \otimes \Lambda$ is given by the elements $\lambda_{i_1} \otimes \cdots \otimes \lambda_{i_p}$ as the λ_{i_j} run over a $\mathbb{Z}/p$-basis for Λ. Moreover, the action of $\mathbb{Z}/p$ preserves (possibly up to sign) this basis. Suppose, then, that $T^i(\lambda_{i_1} \otimes \cdots \otimes \lambda_{i_p}) = \lambda_{i_1} \otimes \cdots \otimes \lambda_{i_p}$. Since p is prime, T^i generates $\mathbb{Z}/p$ unless $i \equiv 0 \bmod (p)$, and we can assume T fixes this element as well. But this implies $\lambda_{i_1} = \lambda_{i_2} = \cdots = \lambda_{i_p}$. Consequently, except in this case, the T^i on a basis element give a total of p distinct basis elements, and consequently a copy of the free module $\mathbb{F}_p(\mathbb{Z}/p)$. $\square$

This lemma makes the calculation of the E_2-term routine for $G \wr \mathbb{Z}/p$. However, it remains to study the differentials. In order to do this we enlarge the domain of discussion. The fibration (1.3) above is a special case of a fibering associated to an arbitrary space X,

$$X^p \longrightarrow X^p \times_{\mathbb{Z}/p} E_{\mathbb{Z}/p} \longrightarrow B_{\mathbb{Z}/p} \tag{1.5}$$

where, as above, the action on X^p is by cyclically shifting coordinates. In this context the lemma above gives directly

Corollary 1.6 *Let λ_{i_j} be a $\mathbb{F}_p$-basis for $H^*(X;\mathbb{F}_p)$, then, for the fibering (1.5) above, the E_2-term of the Serre spectral sequence has the form $E_2^{0,j} = H^j(X^p;\mathbb{F}_p)^{\mathbb{Z}/p}$, while, for $i > 0$, $E_2^{i,j}$ equals 0 unless $j = pr$, and then $E_2^{i,pr} \cong H^r(X;\mathbb{F}_p)$ with explicit generators $(\lambda_{i_j} \otimes \cdots \otimes \lambda_{i_j}) \cup \theta_i$ where θ_i is a generator for $H^i(B_{\mathbb{Z}/p};\mathbb{F}_p) = E_2^{i,0}$.*

The next result is of basic importance in topology as well as in studying the cohomology of groups.

Theorem 1.7 *Let X be a CW-complex. Then the Serre spectral sequence for the fibering (1.5) above collapses, i.e. for that fibering $E_2^{i,j} = E_\infty^{i,j}$ for all i, j.*

Proof. We start by observing that any element in $E_2^{0,j}$ corresponding to an invariant for a free $\mathbb{F}_p(\mathbb{Z}/p)$ module is an infinite cycle. The map

$$X^p \simeq X^p \times E_{\mathbb{Z}/p} \longrightarrow X^p \times_{\mathbb{Z}/p} E_{\mathbb{Z}/p}$$

is a $\mathbb{Z}/p$-covering. Consequently we have the transfer map

$$\mathrm{tr}\colon\ H^*(X^p;\mathbb{F}_p) \longrightarrow H^*(X^p \times_{\mathbb{Z}/p} E_{\mathbb{Z}/p};\mathbb{F}_p)$$

and $(\mathrm{res} \cdot \mathrm{tr})^* = (1 + T + \cdots + T^{p-1})\colon H^*(X^p;\mathbb{F}_p) \longrightarrow H^*(X^p;\mathbb{F}_p)$. But the image of Σ_T is exactly the subset of $H^*(X^p;\mathbb{F}_p)^{\mathbb{Z}/p}$ associated to the free $\mathbb{F}_p(\mathbb{Z}/p)$-modules. Thus, these classes must survive to E_∞.

It remains to consider the classes

$$\underbrace{\lambda \otimes \cdots \otimes \lambda}_{p}, \quad \lambda \in H^n(X;\mathbb{F}_p)\,.$$

Since X is a CW-complex there is a map $(\lambda)\colon X \longrightarrow K(\mathbb{Z}/p, n)$ so that $(\lambda)^*(\iota_n) = \lambda$ where $K(\mathbb{Z}/p, n)$ is an Eilenberg-MacLane space and $\iota_n \in H^n(K(\mathbb{Z}/p, n);\mathbb{F}_p)$ is the fundamental class. Consequently we have maps of fibrations

$$\begin{array}{ccc} X^p & \xrightarrow{(\lambda)^p} & (K(\mathbb{Z}/p,n))^p \\ \downarrow & & \downarrow \\ X^p \times_{\mathbb{Z}/p} E_{\mathbb{Z}/p} & \xrightarrow{(\lambda)^p \times \mathrm{id}} & K(\mathbb{Z}/p,n)^p \times_{\mathbb{Z}/p} E_{\mathbb{Z}/p} \\ \downarrow & & \downarrow \\ B_{\mathbb{Z}/p} & \xrightarrow{\mathrm{id}} & B_{\mathbb{Z}/p} \end{array} \tag{1.8}$$

and passing to Serre spectral sequences, at E_2, the class $(\iota_n)^p$ maps back to $(\lambda)^p$ in $E_2^{0,np}$. Consequently, if $(\iota_n)^p$ is an infinite cycle, then $(\lambda)^p$ must be also.

We now turn our attention to the right hand fibering in (1.8). Note that $K(\mathbb{Z}/p,n)$ is $(n-1)$-connected, so, below dimension np all the terms on the fiber (the vertical line $E_2^{0,*}$) come from invariants of $\mathbb{F}_p(\mathbb{Z}/p)$-free modules, and we have $E_2^{i,j} \equiv 0$ for $0 < j < np$ if $i \geq 1$. Consequently, if $(\iota_n)^p$ has any non-trivial differential it must be d_{np+1}, the transgression, taking $E^{0,np}$ to $E^{np+1,0}$.

The terms on the line $E_\infty^{*,0}$ are identified with the image of $H^*(B_{\mathbb{Z}/p};\mathbb{F}_p)$ in the cohomology of the total space of the fibration, and the fibration (1.8) has a section, $y \mapsto (x,x,\ldots,x) \times_{\mathbb{Z}/p} \tilde{y}$ where $\tilde{y}$ projects to y in $B_{\mathbb{Z}/p}$ and $(x,x,\ldots,x)$ is some point on the p-fold diagonal in X^p. Consequently, the cohomology of the base injects into that of the total space, and it is not possible that any term along the base line $(E^{*,0})$ can be in the image of any differential.

It follows that $(\iota_n)^p$ is, indeed, an infinite cycle, and the proof is complete. □

Remark 1.9 The argument above using (1.8) shows more generally that the class $\underbrace{\lambda \otimes \cdots \otimes \lambda}_{n}$ survives to E_∞ for $H^*(\mathbb{Z}/p \wr \mathcal{S}_n;\mathbb{F}_p)$, $n < \infty$.

Thus we have a complete calculation of $H^*(\mathbb{Z}/p \wr \mathbb{Z}/p;\mathbb{F}_p)$ and more generally,

$$H^*(\underbrace{\mathbb{Z}/p\wr\mathbb{Z}/p\wr\cdots\wr\mathbb{Z}/p}_{i\ times};\mathbb{F}_p)$$

for any i. Later we will study these groups much more carefully, since, for example, $\underbrace{\mathbb{Z}/p\wr\cdots\wr\mathbb{Z}/p}_{i\ times}$ is the p-Sylow subgroup of the symmetric group $\mathcal{S}_{p^i}$.
We now provide an explicit calculation using this result.

Example. The dihedral group of order 8, denoted D_8, can be identified with $\mathbb{Z}_2 \wr \mathbb{Z}_2$. First we note that

$$(\mathbb{F}_2[x_1,y_1])^{\mathbb{Z}_2} = \mathbb{F}_2[\sigma_1,\sigma_2]$$

where the $\mathbb{Z}_2$ acts by exchanging generators, and the σ_i are the usual symmetric classes. Observe that $\sigma_1 = x_1 + y_1$ is a trace class and hence will multiply trivially with elements from the cohomology of the base, generated by a 1–dimensional polynomial class e. Therefore we obtain

$$H^*(D_8) \cong \mathbb{F}_2[e, \sigma_1, \sigma_2]/e\sigma_1 .$$

The action of the Steenrod Algebra is determined by $Sq^1(\sigma_2) = (\sigma_1 + e)\sigma_2$, a fact which will be verified in (2.7).

Remark 1.10 This construction and 1.7 provide the basis for the construction of the Steenrod operations. We will give details in IV.7. Also, a different proof of 1.7, working at the chain level, is implicit in [SE]. In some ways that proof is more general than the one given here, but this one provides us with better control of cup products.

Central Extensions

Consider a central extension $\mathbb{Z}/p \overset{\triangleleft}{\longrightarrow} E \overset{\pi}{\longrightarrow} G$ and the associated fibration of classifying spaces

$$B_{\mathbb{Z}/p} \overset{j}{\longrightarrow} B_E \overset{B_\pi}{\longrightarrow} B_G. \tag{1.11}$$

Lemma 1.12 *The fibration (1.11) is a principal fibration with classifying map $B_k: B_G \to B(B_{\mathbb{Z}/p}) = K(\mathbb{Z}/p, 2)$. In particular, there is a unique cohomology class $k \in H^2(G; \mathbb{F}_p)$ which generates the kernel of*

$$B_\pi^*:\ H^2(B_G; \mathbb{F}_p) \longrightarrow H^2(B_E; \mathbb{F}_p) .$$

Moreover, k can be identified with the class of the extension described in I.6.8.

Proof. Since $\mathbb{Z}/p$ is central in E the multiplication map $\mathbb{Z}/p \times E \to E$ is a homomorphism of groups. Consequently, it induces a map of classifying spaces $B_{\mathbb{Z}/p} \times B_E \to B_E$ which defines the fiber preserving action of $B_{\mathbb{Z}/p}$ on B_E and gives B_π the structure of a principal fibration. Then the Steenrod classification theorem, II.1, constructs the classifying map. Since $B(B_{\mathbb{Z}/p})$ is a $K(\mathbb{Z}/p, 2)$ we have that the set of homotopy classes of maps $f: X \to K(\mathbb{Z}/p, 2)$ is identified with $H^2(X; \mathbb{F}_p)$ for X a CW complex, and the identification is explicit $f \leftrightarrow f^*(\iota)$ for the specific element $\iota \in H^2(K(\mathbb{Z}/p, 2); \mathbb{F}_p)$ corresponding to the identity homomorphism, id: $\mathbb{Z}/p \to \mathbb{Z}/p$.

Finally, the identification of k with the element corresponding to the class of I.6.8 follows by an explicit chain level calculation. We omit the details of this step but in summary one proceeds as follows. First a choice of splitting $E = G \times \mathbb{Z}/p$ is made so that $\mathbb{Z}/p$ acts from the right. This gives an explicit choice for the two dimensional cocycle representing the extension. Then one checks that the map on the two skeleton, $B_{G,2}$, of B_G to $B(B_{\mathbb{Z}/p})$ defined by

$|g_1|g_2| \mapsto ||l(g_2)^{-1}l(g_2)^{-1}l(g_1g_2)|| \in B(B_{\mathbb{Z}/p})$ lifts to give a map of principal fibrations $B_{\mathbb{Z}/p} \to \{B_\pi^{-1}(B_{G,2}) \to B_{G,2}$ to the fibration over the two skeleton of $B(B_{\mathbb{Z}/p})$. After this one extends the map by using the principal action and extending over lifts of the individual cells of B_G. □

This next result is very useful theoretically, and provides a nice application of our result on the cohomology of wreath products and the Frobenius map discussed in (II.5).

Lemma 1.13 *Let $\mathbb{Z}/p \to E \to G$ be a central extension, then there is a finite n so that $E_n = E_\infty$ in the spectral sequence of the fibration (1.11).*

Proof. The Frobenius map associated to the inclusion $\mathbb{Z}/p \hookrightarrow E$ gives a homomorphism $\varphi: E \to \mathbb{Z}/p \wr S_{|G|}$, and, from (1.9), $\varphi^*(\underbrace{b \otimes \cdots \otimes b}_{|G|})$ is non-zero and restricts to $b^{|G|}$ in $H^*(\mathbb{Z}/p; \mathbb{F}_p)$ since $\mathbb{Z}/p$ is central in E. Thus the class $b^{|G|}$ in $E_2^{0,2|G|}$ is an infinite cycle in the spectral sequence. Now write

$$E_2 = \mathbb{F}_p[b^{|G|}] \otimes \left(E_2^{*,0} \oplus \cdots \oplus E_2^{*,2|G|-1}\right).$$

An easy induction shows that $E_{2|G|} = E_\infty$. □

There is actually quite a bit more that can be said about the structure of the differentials in (1.11) using the fact that (1.11) is an induced fibration. The Lyndon–Hochschild–Serre spectral sequence for (1.11) has E_2-term $H^*(G; \mathbb{F}_p) \otimes H^*(\mathbb{Z}/p; \mathbb{F}_p)$ and there is a map from the Serre spectral sequence of the fibration

$$B_{\mathbb{Z}/p} \longrightarrow E(B_{\mathbb{Z}/p}) \longrightarrow B(B_{\mathbb{Z}/p})$$

with E_2-term $H^*(K(\mathbb{Z}/p, 2); \mathbb{F}_p) \otimes H^*(\mathbb{Z}/p; \mathbb{F}_p)$ to the spectral sequence of the extension due to the naturality of the Serre spectral sequence and the commutativity of the diagram of fibrations

$$\begin{array}{ccc}
B_{\mathbb{Z}/p} & \xrightarrow{=} & B_{\mathbb{Z}/p} \\
\downarrow j & & \downarrow j \\
B_E & \xrightarrow{E(k)} & E(B_{\mathbb{Z}/p}) \\
\downarrow B_\pi & & \downarrow \\
B_G & \xrightarrow{k} & B(B_{\mathbb{Z}/p}).
\end{array}$$

In particular the map is the identity on $H^*(\mathbb{Z}/p; \mathbb{F}_p)$ and is the map

$$k^*\colon H^*(K(\mathbb{Z}/p,2);\mathbb{F}_p)\longrightarrow H^*(G;\mathbb{F}_p)\,.$$

The map also commutes with differentials as well as cup products due to naturality, and consequently the differentials in the spectral sequence for $K(\mathbb{Z}/p,2)$ imply differentials in the Lyndon–Hochschild–Serre spectral sequence of the central extension.

In the spectral sequence for $K(\mathbb{Z}/p,2)$ the differentials are completely determined by the Kudo and Serre transgression theorems. We describe these results now. First the classes on the fiber have differentials as follows: $d_2(e_1) = \iota_2$, then $d_3(b) = \beta(\iota_2)$. After that the differentials are given by $d_{2p^i+1}(b^{p^i}) = P^{p^{i-1}}P^{p^{i-2}}\cdots P^1\beta(\iota_2)$. These give all the differentials when $p=2$, but when p is odd there are also the differentials coming from the Kudo transgression theorem. They are given as follows: $d_{2p-1}(b^{p-1}\beta(\iota_2) = \beta P^1\beta(\iota_2)$, and generally, $d_{2p^i(p-1)+1}(b^{p^i(p-1)}d_{2p^i+1}b^{p^i}) = \beta d_{p^{i+1}+1}(b^{p^{i+1}})$.

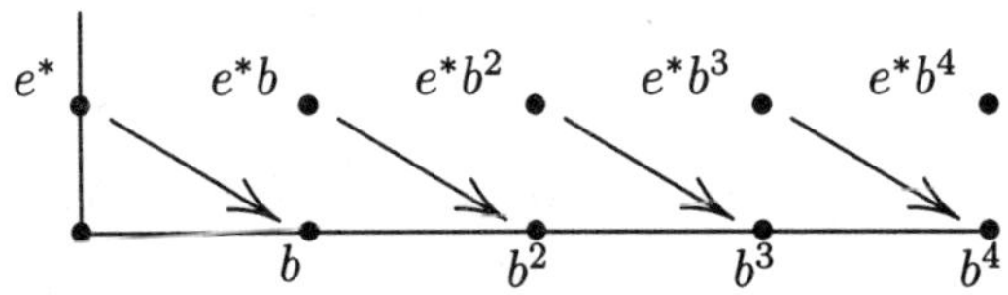

The form of the first differential.

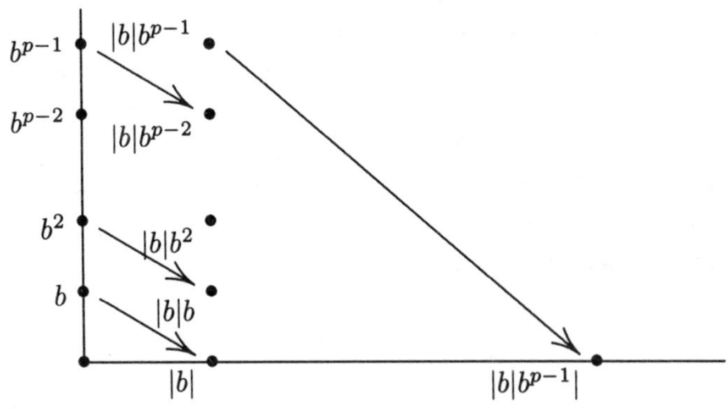

The form of the higher differentials and the Kudo differential.

A Lemma of Quillen-Venkov

Consider an extension of the form $G\overset{\triangleleft}{\longrightarrow}E\overset{\pi}{\longrightarrow}\mathbb{Z}/p$. Let $u\in H^2(E;\mathbb{F}_p)$ be $\pi^*(b)$ where $b\in H^2(\mathbb{Z}/p;\mathbb{F}_p)$ is the Bockstein of the fundamental class. Then we have the following useful result of [QV].

Lemma 1.14 *If w is contained in the kernel of the restriction map*

$$\text{res}^*\colon\ H^*(E;\mathbb{F}_p)\longrightarrow H^*(G;\mathbb{F}_p)$$

then w^2 is contained in the ideal $(u)\subset H^*(E;\mathbb{F}_p)$.

Proof. The Lyndon-Hochschild-Serre spectral sequence of the extension has E_2-term $H^*(\mathbb{Z}/p;H^*(G;\mathbb{F}_p))$, and viewing the spectral sequence as in the illustrations above we can prove by induction that $\cup b\colon E_r^{i,j}\longrightarrow E_r^{i+2,j}$ is surjective for $i\geq 0$ and injective for $i\geq r-1$. (The failure of injectivity in the range $i<r-1$ comes since the $E_r^{i,j}$ terms are identically 0 for $i<0$.) In particular, writing $E_\infty^{i,n-i}=F^i(H^n(E;\mathbb{F}_p))/F^{i+1}(H^n(E;\mathbb{F}_p))$, and using the finiteness of the filtration we see that $F^2(H^n(E;\mathbb{F}_p))$ can be identified with the ideal (u). On the other hand, since w is in the kernel of res^*, it lies in $F^1(H^*(E;\mathbb{F}_p))$, so w^2 is contained in $F^2(H^*(E;\mathbb{F}_p))$ and (1.14) follows. □

2. Change of Rings and the Lyndon–Hochschild–Serre Spectral Sequence

Let $H\overset{\triangleleft}{\longrightarrow}G\longrightarrow G/H$ and let

$$\cdots\longrightarrow\mathcal{C}_n\longrightarrow\cdots\longrightarrow\mathcal{C}_1\longrightarrow\mathcal{C}_0\longrightarrow\mathbb{Z}$$

be a resolution of $\mathbb{Z}$ over $\mathbb{Z}(H)$. Then we have

Lemma 2.1
1. *$\mathbb{Z}(G)\otimes_{\mathbb{Z}(H)}\mathbb{Z}=\mathbb{Z}(G/H)$ as a left $\mathbb{Z}(G)$-module*
2. *$\cdots\mathbb{Z}(G)\otimes_{\mathbb{Z}(H)}\mathcal{C}_n\longrightarrow\cdots\longrightarrow\mathbb{Z}(G)\otimes_{\mathbb{Z}(H)}\mathcal{C}_0\longrightarrow\mathbb{Z}(G/H)$ is a $\mathbb{Z}(G)$ resolution of $\mathbb{Z}(G/H)$.*

Proof. Let $H\cup g_2H\cup g_3H\cup\cdots\cup g_mH\cup\cdots$ be a coset decomposition of G. Then $\mathbb{Z}(G)\otimes_{\mathbb{Z}(H)}\mathbb{Z}$ as a (free) $\mathbb{Z}$-module has a basis $\langle 1\rangle$, $\langle g_2\rangle$, $\langle g_3\rangle$, $\cdots$, $\langle g_m\rangle$, and the action of G on these basis elements is given by $g\langle g_i\rangle=g_rh\langle g_i\rangle=\langle g_rhg_i\rangle=\langle g_rg_i\rangle$ when we write $g=g_rh$ in the decomposition above. This gives the first claim. Now the second is clear. □

Corollary 2.2 *Let A be a $\mathbb{Z}(G)$-module regarded as a $\mathbb{Z}(H)$-module via the inclusion $\mathbb{Z}(H)\subset\mathbb{Z}(G)$, then there is an isomorphism*

$$\text{Ext}^i_{\mathbb{Z}(G)}(\mathbb{Z}(G/H),A)\cong\text{Ext}^i_{\mathbb{Z}(H)}(\mathbb{Z},A)\ .$$

Proof. $\text{Hom}_{\mathbb{Z}(G)}(\mathbb{Z}(G),A)=\text{Hom}_{\mathbb{Z}(H)}(\mathbb{Z}(H),A)=A$. Thus the map

$$\mathbb{Z}(G)\otimes_{\mathbb{Z}(H)}\colon\ \mathcal{C}\longrightarrow\mathbb{Z}(G)\otimes_{\mathbb{Z}(H)}\mathcal{C}$$

on passing to cochain complexes yields an isomorphism

$$\mathrm{Hom}_{\mathbb{Z}(H)}(\mathcal{C}, A) \longrightarrow \mathrm{Hom}_{\mathbb{Z}(G)}(\mathbb{Z}(G) \otimes_{\mathbb{Z}(H)} \mathcal{C}, A)$$

and the result follows. □

Example 2.3 Let D_{2n} be the dihedral group

$$D_{2n} = \{\tau, T \mid \tau^n = T^2 = 1, T\tau T = \tau^{-1}\} .$$

Let $H = \langle \tau \rangle = \mathbb{Z}/n$. There is then an isomorphism

$$\mathrm{Ext}^i_{\mathbb{Z}(D_{2n})}(\mathbb{Z}(T), A) \cong \mathrm{Ext}^i_{\mathbb{Z}(\mathbb{Z}/n)}(\mathbb{Z}, A) = H^i_t(\mathbb{Z}/n; A) .$$

The equation above is a typical example of the *change of rings principle*, i.e. the specification of Ext terms for a module over A as Ext terms for a simpler module over B where $B \subset A$ is a subring. This change of rings principle allows us to give an algebraic construction of the Lyndon–Hochschild-Serre spectral sequence. The idea is to first resolve $\mathbb{Z}$ over $\mathbb{Z}(G/H)$, then resolve each term of the resolution over $\mathbb{Z}(G)$, lift the boundary map in the resolution of $\mathbb{Z}$ and then make a guess that the boundary in the resulting doubly graded complex is of the form $\partial + (-1)^\epsilon d_1$ where d_1 is the lifted boundary and ∂ is the boundary in the resolutions of the terms in the original resolution of $\mathbb{Z}$, and then start adding correction terms to make the resulting guess into an actual differential. Thus, to begin we obtain a diagram

$$\begin{array}{ccccccccc}
 & & \vdots & & \vdots & & \vdots & & \ddots \\
 & & \downarrow \partial & & \downarrow \partial & & \downarrow \partial & & \cdot \\
 & & \mathcal{C}_{01} & & \mathcal{C}_{11} & & \mathcal{C}_{21} & & \cdot \\
 & & \downarrow \partial & & \downarrow \partial & & \downarrow \partial & & \cdot \\
 & & \mathcal{C}_{00} & & \mathcal{C}_{10} & & \mathcal{C}_{20} & & \cdot \\
 & & \downarrow \epsilon_0 & & \downarrow \epsilon_1 & & \downarrow \epsilon_2 & & \cdot \\
\mathbb{Z} & \overset{\epsilon}{\longleftarrow} & B_0 & \overset{d_0}{\longleftarrow} & B_1 & \overset{d_1}{\longleftarrow} & B_2 & \overset{d_2}{\longleftarrow} &
\end{array}$$

where the vertical sequences consist of $\mathbb{Z}(G)$ free resolutions of the B_i's. Next, as indicated, we lift the d_i's to obtain chain maps between the vertical columns, so the diagram above now takes the form

$$
\begin{array}{ccccccccc}
 & & \vdots & & \vdots & & \vdots & & \ddots \\
 & & \downarrow \partial & & \downarrow \partial & & \downarrow \partial & & \cdot \\
 & & \mathcal{C}_{01} & \xleftarrow{d_{10}} & \mathcal{C}_{11} & \xleftarrow{d_{11}} & \mathcal{C}_{21} & \xleftarrow{d_{12}} & \\
 & & \downarrow \partial & & \downarrow \partial & & \downarrow \partial & & \cdot \\
 & & \mathcal{C}_{00} & \xleftarrow{d_{10}} & \mathcal{C}_{10} & \xleftarrow{d_{11}} & \mathcal{C}_{20} & \xleftarrow{d_{12}} & \\
 & & \downarrow \epsilon_0 & & \downarrow \epsilon_1 & & \downarrow \epsilon_2 & & \cdot \\
\mathbb{Z} & \xleftarrow{\epsilon} & B_0 & \xleftarrow{d_0} & B_1 & \xleftarrow{d_1} & B_2 & \xleftarrow{d_2} & .
\end{array}
$$

If $G = H \times G/H$ is a Cartesian product the diagram above could be achieved by tensoring a resolution of $\mathbb{Z}$ over H with one of $\mathbb{Z}$ over G/H so we would have explicitly $\mathcal{C}_{ij} = \mathcal{R}_j \otimes B_i$ where $\mathcal{R}_j$ is the jth term in a resolution of $\mathbb{Z}$ over H, and

$$d = \partial_H \otimes 1 + (-1)^j 1 \otimes \partial_B$$

would be the total differential. This motivates attempting to use the same formula with the diagram above. The difficulty is that, in general

$$
\begin{aligned}
d^2 &= (\partial + (-1)^j d_{1j})^2 \\
&= \partial^2 + (-1)^j \partial d_{1j} + (-1)^{j-1} d_{1,j-1}\partial + (-1) d_{1,j-1} d_{1j} \\
&= (-1) d_{1,j-1} d_{1j}
\end{aligned}
$$

is not necessarily zero. However, $\epsilon d_{1,j-1} d_{1,j} = d_{j-1} d_j \epsilon = 0$, so the *chain map* $d_{1,j-1} d_j$ is homotopic to zero, and we can construct a family of maps $d_{2j}: \mathcal{C}_{ji} \to \mathcal{C}_{j-2,i+1}$ so that $\partial d_{2j} + d_{2j}\partial = d_{1,j-1} d_{1j}$. We add these in, making a new (prospective) differential $d = \partial + (-1)^j d_{1j} + d_{2j}$. Next we square this operator to find the deviation from being a differential, correct that deviation and iterate the procedure. We obtain a result due to A. Liulevicius, [Li], and C.T.C. Wall, [Wa],

Theorem 2.4 *In the situation above there exist a series of $\mathbb{Z}(G)$-linear maps $d_{rs}: \mathcal{C}_{sj} \to \mathcal{C}_{s-r,j+r-1}$, so that if we set $\mathcal{D}_m = \coprod_0^{t=m} \mathcal{C}_{t,m-t}$, $d = \partial + (-1)^s d_{1s} + \sum_{l=2}^m d_{l*}$, then $d^2 = 0$ and the resulting complex is a resolution of $\mathbb{Z}$ over $\mathbb{Z}(G)$. Specifically, each d_{r*} can be chosen so that*

(1) $d_0 = \partial$ *is the vertical differential,*

(2) $d_i \epsilon_i = \epsilon_{i-1} d_{1i}$,

(3) $\sum_{i=0}^k d_{i,*} d_{k-i,*} = 0$ *for each* k;

conversely, any map with properties (1), (2), (3) is a differential which makes the complex above acyclic.

Proof. We show first that any d with properties (1), (2), and (3) above makes the total complex acyclic. Filter $\mathcal{C}_{ij}$ by $F^p\mathcal{C} = \sum_{i\le p}\mathcal{C}_{ij}$. The differential d preserves filtration, and the associated spectral sequence converges to $H_*(\mathcal{C})$. The differential in E^0 is precisely $d_0 = \partial$. Hence $E^1 = \mathcal{B}$, the resolution of $\mathbb{Z}$ by $\mathbb{Z}(G/H)$ free modules with which we started. Moreover, (2) implies that d^1 is exactly the differential d_*, the differential in $\mathcal{B}$. Since $\mathcal{B}$ is a resolution of $\mathbb{Z}$ it is acyclic, and $E^2 = E^\infty = E^\infty_{00} = \mathbb{Z}$, hence $\mathcal{C}$ is acyclic.

To complete the proof we must construct the maps d_{r*} for $r \ge 3$. To begin we set $d_{rs} = 0$ if $s < r$. Now, suppose that d_{rs} has been defined on all $\mathcal{C}_{ij}$ with $i + j < v$. Let $f = -\sum_{i=1}^r d_i d_{r-i}$. We claim there is a map d_r so that $d_0 d_r = f$. To prove this, it suffices to show that $d_0 f = 0$ and $\epsilon_* f = 0$, but this is direct

$$\begin{aligned} d_0 f &= -\sum_{i=1}^{r} d_0 d_i d_{r-i} = \sum_{i=1}^{r}\sum_{j=1}^{i} d_j d_{i-j} d_{r-i} \\ &= \sum_{j=1}^{r} d_j \sum_{i=j}^{r-j} d_{i-j} d_{r-i} = 0. \end{aligned}$$

which completes the proof. □

We now illustrate the explicit resolution techniques above, by constructing resolutions for the dihedral groups D_{2n} and the quaternion group $\mathcal{Q}_8$.

The Dihedral Group D_{2n}

We will apply the Lyndon–Hochschild–Serre spectral sequence to the situation obtained by using the index two subgroup $\mathbb{Z}/n \triangleleft D_{2n}$ with quotient $\mathbb{Z}/2$. The associated resolution of $\mathbb{Z}$ over $\mathbb{Z}(\mathbb{Z}/2)$ has differentials which are alternately $T-1$ and $T+1$. Consequently, to begin we need to lift these maps to chain maps of resolutions of $\mathbb{Z}$ over $\mathbb{Z}(\mathbb{Z}/n)$.

Proposition 2.5 *Let $D_{2n} = \{T, \tau \mid T^2 = \tau^n = 1, T\tau T = \tau^{-1}\}$. Let $\Sigma_\tau = 1+\tau+\tau^2+\cdots+\tau^{n-1}$, and $\mathcal{C} = \mathbb{Z}(D_{2n})$, then the following is a commutative diagram of chain maps*

$$\begin{array}{cccccccccc}
\mathcal{C} & \xleftarrow{\tau-1} & \mathcal{C} & \xleftarrow{\Sigma_\tau} & \mathcal{C} & \xleftarrow{\tau-1} & \mathcal{C} & \xleftarrow{\Sigma_\tau} & \mathcal{C} & \longleftarrow \\
\downarrow{\scriptstyle T-1} & & \downarrow{\scriptstyle -(\tau T+1)} & & \downarrow{\scriptstyle -(T+1)} & & \downarrow{\scriptstyle \tau T-1} & & \downarrow{\scriptstyle T-1} & \\
\mathcal{C} & \xleftarrow{\tau-1} & \mathcal{C} & \xleftarrow{\Sigma_\tau} & \mathcal{C} & \xleftarrow{\tau-1} & \mathcal{C} & \xleftarrow{\Sigma_\tau} & \mathcal{C} & \longleftarrow .
\end{array}$$

Proof. Note that $\tau T\tau = \tau\tau^{-1}T = T$ so

1. $-(\tau T+1)(\tau-1) = -T-\tau+\tau T+1$, $(\tau-1)(T-1) = \tau T - T - \tau + 1$.

2. $-\Sigma_\tau(\tau T+1) = -\Sigma_\tau(T+1) = -(T+1)\Sigma_\tau$.
3. $(\tau T-1)(\tau-1) = T-\tau-\tau T+1$
   $-(\tau-1)(T+1) = -\tau T-\tau+T+1$.
4. $(T-1)\Sigma_\tau = \Sigma_\tau(T-1) = \Sigma_\tau(\tau T-1)$. □

Remark 2.6 The first non-trivial dihedral group $D_8 = \mathbb{Z}/2 \wr \mathbb{Z}/2$. Hence, 1.6 gives the structure of $H^*(D_8;\mathbb{F}_2)$. In particular, the Lyndon–Hochschild–Serre spectral sequence collapses in this case. But for the general case we still need to make an explicit calculation. We turn to this now.

From the proposition we obtain the following diagram for D_{2n} which we have turned on its side for convenience.

$$
\begin{array}{cccccccccc}
\downarrow & & \downarrow & & \downarrow & & \downarrow & & \downarrow & \\
B_3 & \xleftarrow{\epsilon_3} & C & \xleftarrow{\tau-1} & C & \xleftarrow{\Sigma_\tau} & C & \xleftarrow{\tau-1} & C & \xleftarrow{\Sigma_\tau} \\
\downarrow & T-1 & \downarrow & T-1 & \downarrow & -(\tau T+1) & \downarrow & -(T+1) & \downarrow & \tau T-1 \\
B_2 & \xleftarrow{\epsilon_2} & C & \xleftarrow{\tau-1} & C & \xleftarrow{\Sigma_\tau} & C & \xleftarrow{\tau-1} & C & \xleftarrow{\Sigma_\tau} \\
\downarrow & T+1 & \downarrow & T+1 & \downarrow & -(\tau T-1) & \downarrow & -(T-1) & \downarrow & \tau T+1 \\
B_1 & \xleftarrow{\epsilon_1} & C & \xleftarrow{\tau-1} & C & \xleftarrow{\Sigma_\tau} & C & \xleftarrow{\tau-1} & C & \xleftarrow{\Sigma_\tau} \\
\downarrow & T-1 & \downarrow & T-1 & \downarrow & -(\tau T+1) & \downarrow & -(T+1) & \downarrow & \tau T-1 \\
B_0 & \xleftarrow{\epsilon_0} & C & \xleftarrow{\tau-1} & C & \xleftarrow{\Sigma_\tau} & C & \xleftarrow{\tau-1} & C & \xleftarrow{\Sigma_\tau} \\
\downarrow \epsilon_0 & & & & & & & & & \\
\mathbb{Z} & & & & & & & & &
\end{array}
$$

Note that $(\tau T-1)(\tau T+1) = (\tau T)^2-1 = 0 = (\tau T+1)(\tau T-1)$ so $d_2 \equiv 0$ and the construction of the higher differentials stops. It follows that alternating the signs on the columns above defines the total differential for the complex above as $\partial_\tau + (-1)^j d_1$. For example, if n is even and we take our coefficients as $\mathbb{F}_2$ all the differentials are zero and $H^i(D_{4m};\mathbb{F}_2) = (\mathbb{Z}/2)^{i+1}$. It follows, in particular, that this resolution is *minimal* in the sense that, in each dimension, the number of free summands cannot be reduced.

We now apply the above result using induction and the central extensions $\mathbb{Z}/2 \xrightarrow{\triangleleft} D_{2^n} \xrightarrow{\pi} D_{2^{n-1}}$ to determine the rings $H^*(D_{2^n};\mathbb{F}_2)$ for all $n \geq 3$. Here, to be explicit we use the presentations for D_{2^n} given as

$$D_{2^n} \; = \; \{x, y \mid x^2 = y^2 = (xy)^{2^{n-1}} = 1\}\,.$$

$D_{2^n}/[D_{2^n}, D_{2^n}] = \mathbb{Z}/2 \times \mathbb{Z}/2$ and $H^1(D_{2^n};\mathbb{F}_2) = \mathrm{Hom}(D_{2^n}/[D_{2^n}, D_{2^n}], \mathbb{F}_2) = (\mathbb{Z}/2)^2$. We can choose explicit generators here as $\bar{x}$ dual to x, and $\bar{y}$ dual to

y. In what follows there will be no possibility of confusion so we will drop the bars and write these generators as x and y, respectively.

The k-invariant for D_8 is xy. To see this note that it is natural with respect to restriction. So we have the diagram

$$\begin{array}{ccc}
\mathbb{Z}/2 & \overset{=}{\hookrightarrow} & \mathbb{Z}/2 \\
\Big\downarrow \vartriangleleft & & \Big\downarrow \vartriangleleft \\
(\mathbb{Z}/2)^2 & \hookrightarrow & D_8 \\
\Big\downarrow \pi & & \Big\downarrow \pi \\
\langle x\rangle & \hookrightarrow & \langle x, y\rangle
\end{array}$$

with a similar diagram over the inclusion of $\langle y\rangle$. Thus the restrictions of k to $H^2(\langle x\rangle;\mathbb{F}_2)$ and to $H^2(\langle y\rangle,\mathbb{F}_2)$ are both zero. Consequently, since the k-invariant is non-trivial it can only be xy.

It follows that $xy = 0$ in $H^*(D_8;\mathbb{F}_2)$. More generally we have

Theorem 2.7 *For the dihedral group D_{2^n} we have*

$$H^*(D_{2^n};\mathbb{F}_2) = \mathbb{F}_2[x, y, w]/(xy = 0)$$

where x and y are one dimensional and w is two dimensional. Moreover, x and y are in the image from $H^*(D_8;\mathbb{F}_2)$ under the projection $D_{2^n}\longrightarrow D_8$, and w is uniquely specified by the condition that $\text{res}^*(w) = 0$ *in $H^2(\langle x\rangle;\mathbb{F}_2)$ and $H^2(\langle y\rangle;\mathbb{F}_2)$. Finally, $Sq^1(w) = (x + y)w$.*

Proof. We use induction and the central extension $D_{2^n}\longrightarrow D_{2^{n-1}}$. The key step is to determine the k-invariant.

There are two subgroups $(\mathbb{Z}/2)^2$ contained in D_{2^n}: $G_I(n) = \langle x, (xy)^{2^{n-2}}\rangle$, and $G_{II}(n) = \langle y, (xy)^{2^{n-2}}\rangle$. It is clear that $\pi^{-1}(G_i(n-1)) = D_8(i)$ in D_{2^n} for $i = I$ or II. Thus the restriction of the k invariant to $H^2(G_i(n-1);\mathbb{F}_2)$ is a k-invariant for D_8, but on restricting all the way to $\langle x\rangle$ or $\langle y\rangle$, the k-invariant must go to zero since, if not then the resulting extension over $\langle x\rangle$ would be $\mathbb{Z}/4$, and similarly for $\langle y\rangle$. This shows that k cannot be x^2, y^2 or $x^2 + y^2$. Hence w must be a term in the k-invariant and, since w restricts to 0 in both $H^2(\langle x\rangle;\mathbb{F}_2)$ and $H^2(\langle y\rangle;\mathbb{F}_2)$ it follows that w is the k-invariant.

We now apply the spectral sequence of the central extension

$$\mathbb{Z}/2 \vartriangleleft D_{2^n} \overset{\pi}{\longrightarrow} D_{2^{n-1}} .$$

We have $d_2(e) = w$, so $d_2(e^{2j+1}\theta) = e^{2j}w\theta$, and $E_3 = \mathbb{F}_2[x,y]/(xy = 0) \otimes \mathbb{F}_2[e^2]$. But for this E_3-term there are exactly $n+1$ elements in each dimension n, so, from the chain level determination of $H_*(D_{2n};\mathbb{F}_2)$ in (2.6), we see that $E_3 = E_\infty$. This shows that $H^*(D_{2^n};\mathbb{F}_2) = \mathbb{F}_2[x,y]/(xy = 0) \otimes \mathbb{F}_2[w]$ as desired. Of course, w is not uniquely determined yet, but it is when we make use of the additional requirement that the restrictions of w to the groups $H^2(\langle x\rangle;\mathbb{F}_2)$ and $H^2(\langle y\rangle;\mathbb{F}_2)$ are both zero. (This argument also shows that the k-invariant for the group D_8 must be xy.)

It remains to show that $Sq^1(w) = (x+y)w$. This is again checked by using the restrictions to the two subgroups $G_i(n)$. For example, restricting to $G_I(n)$ we see that $w \mapsto xa + a^2$ or xa where a is dual to $(xy)^{2^{n-2}}$. In either case $Sq^1(w) \mapsto x^2a + xa^2$ which is not zero. However, $H^3(D_{2^n};\mathbb{F}_2) \doteq \langle x^3, y^3, xw, yw\rangle$ so $Sq^1(w) = \lambda x^3 + \tau x\mathrm{res}^*(w)$ and the only way this is possible is if $\mathrm{res}^*(w) = xa + a^2$ and the image of $Sq^1(w)$ is the image of xw. A similar argument holds for the restriction to $H^3(G_{II}(n);\mathbb{F}_2)$ and $Sq^1(w) = (x+y)w$ as asserted. □

The Quaternion Group $\mathcal{Q}_8$

Now we consider the quaternion group $\mathcal{Q}_8$. As in the case of the dihedral groups we first need to lift the boundary maps in the B sequence to chain maps. We have

Proposition 2.8 *Let $\mathcal{Q}_8$ be the quaternion group $\{\tau, T \mid \tau^2 = T^2 = (\tau T)^2\}$, let $C = \mathbb{Z}(\mathcal{Q}_8)$, then the following diagram commutes*

$$
\begin{array}{ccccccc}
 & \xleftarrow{\tau-1} & & \xleftarrow{\Sigma_\tau} & & \longleftarrow \\
\Big\downarrow{\scriptstyle T-1} & & \Big\downarrow{\scriptstyle -(T\tau^3+1)} & & \Big\downarrow{\scriptstyle -(T+1)} & \\
C & \xleftarrow{\tau-1} & C & \xleftarrow{\Sigma_\tau} & C & \longleftarrow \\
\Big\downarrow{\scriptstyle T+1} & & \Big\downarrow{\scriptstyle -T\tau^3+1} & & \Big\downarrow{\scriptstyle -(T-1)} & \\
C & \xleftarrow{\tau-1} & C & \xleftarrow{\Sigma_\tau} & C & \longleftarrow \\
\Big\downarrow{\scriptstyle T-1} & & \Big\downarrow{\scriptstyle -(T\tau^3+1)} & & \Big\downarrow{\scriptstyle -(T+1)} & \\
C & \xleftarrow{\tau-1} & C & \xleftarrow{\Sigma_\tau} & C & \longleftarrow .
\end{array}
$$

Proof. Note that $T(\tau-1) = (\tau^{-1}-1)T = T(\tau^3-1) = T(\tau^2+\tau+1)(\tau-1)$. Thus $(\tau-1)(T-1) = (T(\tau^2+\tau+1)-1)(\tau-1)$. But, since $\Sigma_\tau(\tau-1) = 0$ we are free to modify the map by subtracting $T\Sigma_\tau$, and we obtain

$$(\tau-1)(T-1) = -(T\tau^3+1)(\tau-1)\,.$$

Next, note that $\Sigma_\tau \tau = \Sigma_\tau$, $\Sigma_\tau T = T\Sigma_\tau$. Consequently, $-\Sigma_\tau(T\tau^3+1) = -(T+1)\Sigma_\tau$. The arguments for the next column are similar. □

Now, the situation changes, since $d_1^2 = T^2 - 1 = \tau^2 - 1$ at all points. Thus, d_2 is non-trivial. In fact, taking account of the fact that d_1 is given by the vertical arrows in the propsition above, except we change the signs on every odd column, we calculate directly that

$$d_2 = \begin{cases} -(\tau + 1) & C_{i,2j} \longrightarrow C_{i-2,2j+1},\ i \geq 2 \\ 0 & \text{otherwise}. \end{cases}$$

Next we need to check d_3. Note that $d_1 d_2 + d_2 d_1 = -T\Sigma_\theta$ when we are on an even row. Consequently, $d_3 \colon C_{i,0} \longrightarrow C_{i-3,2}$ is multiplication by T if $i \geq 3$. To obtain d_3 on the next row, we must also take account of the term $d_3 d_0$, so $d_3 \colon C_{i,1} \longrightarrow C_{i-3,3}$ is multiplication by $-T\tau^3$ for $i \geq 3$. For the next row we see that d_3 becomes zero. Then the process above repeats. This analysis shows more or less formally that we have

Theorem 2.9 *In the Lyndon-Hochschild-Serre spectral sequence for $\mathcal{Q}_8$ where $H = \langle \tau \rangle$, and with $\mathbb{F}_2$-coefficients, we have $E_2^{i,j} = E_3^{i,j} = \mathbb{F}_2$ for $i \geq 0$, $j \geq 0$. However, $d_3 \neq 0$. Indeed, $d_3 \colon E_3^{i,j+2} \longrightarrow E_3^{i+3,j}$ is an isomorphism for $j = 0, 1$. Consequently,*

$$E_4^{i,j} = E_\infty^{i,j} = \begin{cases} \mathbb{F}_2 & i = 0,\ 1,\ 2,\ j = 4s,\ 4s+1 \\ 0 & \textit{otherwise}. \end{cases}$$

In particular $H^*(\mathcal{Q}_8; \mathbb{F}_2)$ is periodic with period 4, and

$$\begin{aligned} H^0(\mathcal{Q}_8; \mathbb{F}_2) &= H^3(\mathcal{Q}_8; \mathbb{F}_2) = \mathbb{F}_2, \\ H^1(\mathcal{Q}_8; \mathbb{F}_2) &= H^2(\mathcal{Q}_8; \mathbb{F}_2) = (\mathbb{F}_2)^2. \end{aligned}$$

(That the differentials are as stated is clear. Then, note that the E_4-term only has classes in the bidegrees stated, and, since the higher differentials go from (i, j) to at least $(i + 4, j - 3)$ there can be no further differentials.)

We now determine the rings $H^*(\mathcal{Q}_{2^n}; \mathbb{F}_2)$ for all $n \geq 3$. We will use the central extension $\mathbb{Z}/2 \xrightarrow{\triangleleft} \mathcal{Q}_{2^n} \xrightarrow{\pi} D_{2^{n-1}}$, and we start with $\mathcal{Q}_8$ (see also V.1.9).

Lemma 2.10 *In the central extension*

$$\mathbb{Z}/2 \xrightarrow{\triangleleft} \mathcal{Q}_8 \xrightarrow{\pi} (\mathbb{Z}/2)^2$$

the k-invariant in $H^2((\mathbb{Z}/2)^2; \mathbb{F}_2)$ is $x^2 + y^2 + xy$. In particular $H^*(\mathcal{Q}_8; \mathbb{F}_2) = \mathbb{F}_2[e_4](1, x, y, x^2, y^2, x^2y = xy^2)$, and $x^3 = y^3 = 0$.

Proof. The three classes x^2, y^2 and $x^2 + y^2$ all give copies of $(\mathbb{Z}/4) \times \mathbb{Z}/2$ as the resulting extensions while the three classes xy, $x^2 + xy$, and $xy + y^2$ all give D_8. The only remaining non-trivial class is $x^2 + y^2 + xy$, which must, therefore be the k-invariant for $\mathcal{Q}_8$.

Now we consider the spectral sequence associated to the central extension with E_2-term $H^*((\mathbb{Z}/2)^2; \mathbb{F}_2) \otimes H^*(\mathbb{Z}/2; \mathbb{F}_2) = \mathbb{F}_2[x, y, e]$. We know that

$d_2(e) = x^2+xy+y^2$ determines the d_2-differential and $E_3 \cong \mathbb{F}_2[x,y]/(x^2+y^2+xy)\otimes\mathbb{F}_2[e^2]$. Then the next differential $d_3(e^2) = Sq^1(x^2+y^2+xy) = x^2y+xy^2$, and by comparing with (2.9), there are no further differentials. □

The generalization to $\mathcal{Q}_{2^n}$ is similar, but the ring structure changes.

Lemma 2.11 *In the central extension*

$$\mathbb{Z}/2 \xrightarrow{\triangleleft} \mathcal{Q}_{2^n} \xrightarrow{\pi} D_{2^{n-1}}, \qquad n \geq 4$$

the k-invariant is $x^2+y^2+w \in H^2(D_{2^{n-1}};\mathbb{F}_2)$. Consequently, $H^*(\mathcal{Q}_{2^n};\mathbb{F}_2) = \mathbb{F}_2[x,y,e_4]/(xy=0, x^3=y^3)$ for $n \geq 4$.

Proof. Let $j\colon \mathbb{Z}/2^{n-2} \hookrightarrow D_{2^{n-1}}$ be the inclusion $T \mapsto xy$. Then we have the commutative diagram of extensions

$$\begin{array}{ccc} \mathbb{Z}/2 & \overset{=}{\hookrightarrow} & \mathbb{Z}/2 \\ \downarrow & & \downarrow \\ \mathbb{Z}/2^{n-1} & \hookrightarrow & \mathcal{Q}_{2^n} \\ \downarrow & & \downarrow \\ \mathbb{Z}/2^{n-2} & \overset{j}{\hookrightarrow} & D_{2^{n-1}} \end{array}$$

so that $j^*(k) \neq 0$ in $H^2(\mathbb{Z}/2^{n-2};\mathbb{F}_2)$, but since $n \geq 4$ both x^2 and y^2 map to zero, so the coefficient of w in the k-invariant must be 1. A similar diagram holds for $\langle x\rangle \subset D_{2^{n-1}}$ and $\langle y\rangle$ which both give $\mathbb{Z}/4$ as the extension. Consequently the k invariant restricts non-trivially here as well, but since w restricts to zero in both, it follows that the k-invariant is as asserted.

We now apply the spectral sequence of the central extension. The details are virtually identical to those above so are omitted. □

Remark 2.12 The $\mathcal{Q}_{2^n}$ all have periodic cohomology with period four. Indeed, periodic (and minimal) resolutions are given explicitly in [CE], page 253. Such groups have been extensively studied both by group theorists and by topologists. We will discuss them more completely in IV.6.

Remark 2.13 The groups $\mathcal{Q}_8$ and D_8 are examples of extraspecial 2-groups. These groups are built up as central products of copies of $\mathcal{Q}_8$, D_8, and possibly a copy of $\mathbb{Z}/4$. More precisely, suppose that G and H are finite 2-groups having the property that there is a unique copy of $\mathbb{Z}/2$ in the center of each, then $G * H$ is the quotient $G \times H/\Delta(\mathbb{Z}/2)$. We have the evident associative and commutative properties of $*$, $(G*H)*K \cong G*(H*K)$, $G*H \cong H*G$. Then the extraspecial 2-groups are inductively defined, starting with $G = \{1\}$ as the $*$ sums $G*H$ where G is extraspecial and H is either D_8, $\mathcal{Q}_8$ or $\mathbb{Z}/4$ as long as the resulting $G*H$ has a unique $\mathbb{Z}/2$ in its center. One can verify the relation $D_8 * D_8 \cong \mathcal{Q}_8 * \mathcal{Q}_8$ and also $\mathbb{Z}/4 * \mathbb{Z}/4 = \mathbb{Z}/4 \times \mathbb{Z}/2$ which is not extraspecial. Hence it is easy to write down all of these groups.

Each extraspecial 2-group is given as a central extension of the form $\mathbb{Z}/2 \triangleleft G \to (\mathbb{Z}/2)^n$ and the K-invariant in $H^2((\mathbb{Z}/2)^n;\mathbb{F}_2)$ is a symmetric non-singular form. In [Q4] Quillen shows that the associated Lyndon-Hochschild-Serre spectral sequence with $\mathbb{F}_2$ coefficients for any extraspecial 2-group has the property that $E_i^{r,s} = E_i^{r,0} \otimes E_i^{0,s}$ for each i and $E_i^{0,*} = \mathbb{F}_2[e^{2^j}]$ for an appropriate j.

This is a property which actually seems to happen quite infrequently for central extensions of 2-groups. Normally, in the higher terms of the spectral sequence one expects indecomposables to appear "in the middle", i.e. in $E_i^{r,s}$ with neither r nor s equal to 0.

3. Chain Approximations in Acyclic Complexes

In general one prefers to not work on the chain level, but sometimes it is unavoidable. For this reason we now discuss a basic technique for constructing chain maps.

The situation is this: we are given a free complex over $\mathcal{R}(G)$ where $\mathcal{R} = \mathbb{Z}$ or a field $\mathbb{Q}$, $\mathbb{F}_q$, etc..

$$0 \leftarrow \mathcal{R} \xleftarrow{\epsilon} \mathcal{C}_0 \leftarrow \mathcal{C}_1 \leftarrow \mathcal{C}_2 \leftarrow \cdots$$

which is acyclic. This means there is an $\mathcal{R}$-linear map $\phi: \mathcal{R} \to \mathcal{C}_0$, which is, first, a chain map, and second, a chain homotopy equivalence (over $\mathcal{R}$). (This is just a fancy way of saying that $H_*(\mathcal{C}_*) = H_0(\mathcal{C}_*) = \mathcal{R}$.) In particular, the composite map $\epsilon \cdot \phi: \mathcal{R} \to \mathcal{R}$ is the identity. Note that ϵ is $\mathcal{R}(G)$-linear, where $\mathcal{R}$ is given the trivial $\mathcal{R}(G)$ action. However, ϕ cannot be, in general, $\mathcal{R}(G)$-linear.

As a consequence of the chain equivalence ϕ, there is an *$\mathcal{R}$-linear contracting homotopy* $s: \mathcal{C}_i \to \mathcal{C}_{i+1}$, $i \geq 0$ so that

$$\partial s + s\partial = 1 - \phi\epsilon .$$

Examples 3.1

1. Let $G = \mathbb{Z}$ with generator T, then a resolution of $\mathbb{Z}$ over $\mathbb{Z}(G)$ is given as

$$\mathbb{Z} \xleftarrow{\epsilon} \mathbb{Z}(\mathbb{Z})[e_0] \xleftarrow{T-1} \mathbb{Z}(\mathbb{Z})[e_1] ,$$

$$s(e_0) = 0, \text{ but } sT^i e_0 = \begin{cases} (T^{i-1} + T^{i-2} + \cdots + 1)e_1 & i > 0 \\ -(T^{-i} + T^{-i+1} + \cdots + T^{-1})e_1 & i < 0 . \end{cases}$$

2. Let $G = \mathbb{Z}/2$, then in the standard resolution of $\mathbb{Z}$ over $\mathbb{Z}(\mathbb{Z}/2)$,

$$\mathbb{Z} \xleftarrow{\epsilon} \mathbb{Z}(\mathbb{Z}/2)[e_0] \xleftarrow{T-1} \mathbb{Z}(\mathbb{Z}/2)[e_1] \xleftarrow{T+1} \mathbb{Z}(\mathbb{Z}/2)[e_2] \xleftarrow{T-1} \cdots$$

we set $s(e_i) = 0$, $s(Te_i) = e_{i+1}$.

3. Let $G = \mathbb{Z}/p$ where p is an odd prime, then a resolution of $\mathbb{Z}$ over $\mathbb{Z}(G)$ is given by

$$\mathbb{Z} \xleftarrow{\epsilon} \mathbb{Z}(\mathbb{Z}/p)[e_0] \xleftarrow{T-1} \mathbb{Z}(\mathbb{Z}/p)[e_1] \xleftarrow{\Sigma_T} \mathbb{Z}(\mathbb{Z}/p)[e_2] \longleftarrow \cdots .$$

A contracting homotopy for the resolution above is given by the formulae

$$sT^j e_{2i+1} = \begin{cases} 0 & j \neq p-1 \\ e_{2i+2} & j = p-1, \end{cases}$$

$$sT^j e_{2i} = \begin{cases} 0 & j = 0 \\ (T^{j-1} + T^{j-2} + \cdots + 1)e_{2i+1} & j \neq 0, \end{cases}$$

for all $i \geq 0$.

In practice, we often have to work with a chain complex

$$\mathcal{D} = \mathcal{C}_* \otimes_{\mathcal{R}} \mathcal{E}_* ,$$

where $\mathcal{C}_*$ is free over $\mathcal{R}(G)$ while $\mathcal{E}$ is free over $\mathcal{R}(H)$. Since the tensor product of two acyclic complexes, free over $\mathcal{R}$, is free over $\mathcal{R}$ and acyclic, the above complex becomes a free resolution of $\mathcal{R}$ over $\mathcal{R}(G) \otimes_{\mathcal{R}} \mathcal{R}(H) = \mathcal{R}(G \times H)$. Then, given s_G, a contracting homotopy for $\mathcal{C}_*$, and s_H, a contracting homotopy for $\mathcal{E}_*$, we have

Lemma 3.2 *In the situation above, a contracting homotopy for the complex $\mathcal{C}_* \otimes_{\mathcal{R}} \mathcal{E}_*$ is given by the formula $s^{\otimes} = s_G \otimes 1 + \phi\epsilon \otimes s_H$.*

Proof. This is a direct calculation. We have

$$\begin{aligned} \partial^{\otimes} s^{\otimes} + s^{\otimes} \partial^{\otimes} &= (\partial \otimes 1 + (-1)^{sgn} \otimes \partial)(s_G \otimes 1 + \phi\epsilon \otimes s_H) \\ &\quad + (s_G \otimes 1 + \phi\epsilon \otimes s_H)(\partial \otimes 1 + (-1)^{sgn} \otimes \partial) \\ &= (\partial s_G + S_G \partial) \otimes 1 + \phi\epsilon \otimes (\partial s_H + s_H \partial) \\ &\quad + (-1)^{sgn} \otimes \partial(s_G \otimes 1) + s_G \otimes 1((-1)^{sgn} \otimes \partial), \end{aligned}$$

but these last two terms cancel out, and expanding out the first two we obtain

$$(1 - \phi\epsilon) \otimes 1 + \phi\epsilon \otimes (1 - \phi\epsilon) = 1 \otimes 1 - \phi\epsilon \otimes \phi\epsilon ,$$

which is the result. □

Example 3.3 Let $\mathcal{V}_*$ be a free resolution of $\mathbb{Z}$ over the ring $\mathcal{R}(G)$. Then, given p, and the usual resolution $\mathcal{W}_*$ of $\mathbb{Z}$ over $\mathbb{Z}(\mathbb{Z}/p)$, we can construct a free action of the wreath product $G \wr \mathbb{Z}/p$ on the chain complex

$$\underbrace{\mathcal{V}_* \otimes \cdots \otimes \mathcal{V}_*}_{p \text{ times}} \otimes \mathcal{W}_*$$

by setting

$$T(v_1 \otimes \cdots \otimes v_p \otimes \theta) = (-1)^{\dim(v_p)\sum_1^{p-1}\dim(v_i)}(v_p \otimes v_1 \otimes \cdots \otimes v_{p-1} \otimes T\theta),$$

while $(g_1, \cdots, g_p)$ acts as $(g_1 v_1 \otimes \cdots g_p v_p \otimes \theta)$. Then, given an $\mathcal{R}$-linear contracting homotopy s_G for $\mathcal{V}_*$, and the contracting homotopy in (3) of the examples above, the formula above gives a contracting homotopy, $s^{\otimes}$, for chain complex above, and it can be used to study the wreath product.

Now suppose that

$$(3.4) \qquad \mathcal{M}_0 \longleftarrow \mathcal{M}_1 \longleftarrow \mathcal{M}_2 \longleftarrow \cdots$$

is a second $\mathcal{R}(G)$ free complex, but we do not assume it is acyclic. Next, suppose that $\lambda\colon (H_0(\mathcal{M}_*) = \mathcal{M}_0/\mathrm{im}\partial(\mathcal{M}_1)) \longrightarrow \mathcal{R}$ is a given $\mathcal{R}(G)$-linear map. Then we can construct an *explicit* chain map $\hat{\lambda}\colon \mathcal{M}_* \longrightarrow \mathcal{C}_*$ so that $\epsilon\cdot\hat{\lambda}_0\colon H_0(\mathcal{M}_*) \longrightarrow \mathcal{R}$ is the original map λ, as follows. First, on $\mathcal{M}_0$, choose a $\mathcal{R}(G)$-basis, $e_1, \ldots, e_i, \ldots$, and define $\hat{\lambda}_)(e_i) = \phi \cdot \lambda(e_i)$. Then extend to $\mathcal{M}_0$ by $\mathcal{R}(G)$-freeness. Next, iteratively, define $\hat{\lambda}_i$ on $\mathcal{M}_i$ by choosing a basis $e_1, \ldots, e_j, \ldots$, defining $\hat{\lambda}_i(e_j) = s\hat{\lambda}_{i-1}(\partial(e_j))$, and, as before, extend to $\mathcal{M}_i$ by $\mathcal{R}(G)$-freeness. Since

$$\begin{aligned} \partial\hat{\lambda}_i(e_r) &= \partial s\hat{\lambda}_{i-1}(\partial(e_r)) \\ &= (1 - s\partial)\hat{\lambda}(\partial(e_r)) \\ &= \hat{\lambda}(\partial(e_r)) \end{aligned}$$

it follows that this operation constructs a chain map extending $\hat{\lambda}_0$.

Actually, there is no reason to assume that $\mathcal{M}_*$ is free over $\mathcal{R}(G)$. It is sufficient to assume that it is free over a ring $\mathcal{B}$, and there is a (unitary) ring homomorphism $\tau\colon \mathcal{B} \longrightarrow \mathcal{R}(G)$. Then we can use the same formula above to construct a map $\hat{\lambda}_*\colon \mathcal{M}_* \longrightarrow \mathcal{C}_*$ which satisfies the condition

$$(3.5) \qquad \hat{\lambda}_i(b \cdot m) = \tau(b) \cdot \hat{\lambda}_i(m),$$

provided, only that we start with a map $\lambda\colon H_0(\mathcal{M}_*) \longrightarrow \mathcal{R}$ which satisfies (3.5) above.

Example 3.6 Chain approximation to the diagonal. A map

$$\Delta\colon\ \mathbb{Z}(G) \longrightarrow \mathbb{Z}(G) \otimes \mathbb{Z}(G) = \mathbb{Z}(G \times G)$$

is given by $\Delta(g) = g \times g$ for each $g \in G$. Consequently, the identity map $\mathbb{Z} \longrightarrow \mathbb{Z}$ extends to a map of resolutions $\mathcal{C}(\mathbb{Z}(G))_* \xrightarrow{\Delta_*} \mathcal{C}(\mathbb{Z}(G \times G))_*$ which we can construct explicitly from the formula above. For example, suppose $G = \mathbb{Z}/2$. Then a resolution of $\mathbb{Z}$ over $\mathbb{Z}(\mathbb{Z}/2)$ is given as above, and a resolution of $\mathbb{Z}$ over $\mathbb{Z}(\mathbb{Z}/2 \times \mathbb{Z}/2)$ can be given as $\mathcal{C}(\mathbb{Z}(\mathbb{Z}/2))_* \otimes_{\mathbb{Z}} \mathcal{C}(\mathbb{Z}(\mathbb{Z}/2))$! Since

$$\begin{aligned} \mathcal{C}(\mathbb{Z}(\mathbb{Z}/2))_0 \otimes_{\mathbb{Z}} \mathcal{C}(\mathbb{Z}(\mathbb{Z}/2))_0 &= \mathbb{Z}(\mathbb{Z}/2)[e_0] \otimes_{\mathbb{Z}} \mathbb{Z}(\mathbb{Z}/2)[e_0] \\ &= \mathbb{Z}(\mathbb{Z}/2 \times \mathbb{Z}/2)[e_0 \otimes e_0], \end{aligned}$$

(more generally the resolution above in degree i is free over $\mathbb{Z}(\mathbb{Z}/2 \times \mathbb{Z}/2)$ on generators $e_r \otimes e_{i-r}$), we can assume that that map in degree 0 takes e_0 to $e_0 \otimes e_0$. Using the map extension process we now obtain

$$\begin{aligned} \hat{\lambda}(e_1) &= s\hat{\lambda}(\partial(e_1)) \\ &= s(T \times T - 1)(e_0 \otimes e_0) \\ &= e_1 \otimes Te_0 + e_0 \otimes e_1 \,. \end{aligned}$$

Next,

$$\begin{aligned} \hat{\lambda}(e_2) &= s\hat{\lambda}(\partial(e_2)) \\ &= s(T \times T + 1)(e_1 \otimes Te_0 + e_0 \otimes e_1) \\ &= e_2 \otimes e_0 + e_1 \otimes Te_1 + e_0 \otimes e_2 \,. \end{aligned}$$

Proceeding in this way, it is direct to see that the explicit chain map is given by

$$\hat{\lambda}(e_i) = \sum_{j=0}^{i} e_j \otimes T^j e_{i-j} \,.$$

Example 3.7 Multiplication for a commutative group. The multiplication map $u\colon \mathbb{Z}/2 \times \mathbb{Z}/2 \longrightarrow \mathbb{Z}/2$, $(a,b) \mapsto ab$, is a homomorphism. It therefore induces a homomorphism

$$\mathbb{Z}(\mathbb{Z}/2 \times \mathbb{Z}/2) \longrightarrow \mathbb{Z}(\mathbb{Z}/2)$$

and consequently a *chain map*

$$\mu\colon \ \mathcal{C}_*(\mathbb{Z}(\mathbb{Z}/2)) \otimes_{\mathbb{Z}} \mathcal{C}_*(\mathbb{Z}(\mathbb{Z}/2)) \longrightarrow \mathcal{C}_*(\mathbb{Z}(\mathbb{Z}/2)) \,.$$

An explicit μ can be given as follows:

Theorem 3.8 *The chain map μ constructed using the contracting homotopy above is associative, graded commutative ($\mu(a \otimes b) = (-1)^{\dim(a)\dim(b)}\mu(b \otimes a)$), and*

$$\begin{aligned} \mu(e_{2i} \otimes e_{2j}) &= \binom{i+j}{i} e_{2(i+j)} \\ \mu(e_1 \otimes e_{2i}) &= e_{2i+1} \\ \mu(e_1 \otimes e_{2i+1}) &= 0 \,. \end{aligned}$$

Proof. The construction of the chain map starts as follows

$$\begin{aligned}
\mu(e_0 \otimes e_0) &= e_0 \\
\mu(e_1 \otimes e_0) &= s(T-1)e_0 \\
&= e_1, \\
\mu(e_0 \otimes e_1) &= s(T-1)e_0 \ = \ e_1, \\
\mu(e_1 \otimes e_1) &= s\{\mu((T-1)e_0 \otimes e_1 - e_1 \otimes (T-1)e_0)\} \\
&= s(0) \ = \ 0\,.
\end{aligned}$$

Now, assume the formula of the theorem is true for $e_i \otimes e_j$ with $i+j \leq n$. Then in dimension $n+1$ we have

1. If $n+1$ is even, then

$$\begin{aligned}
e_{2w} \otimes e_{2l} &\mapsto s\mu(Te_{2w-1} \otimes e_{2l} + e_{2w} \otimes Te_{2l-1} \\
&= s\left[\binom{w+l-1}{w-1} + \binom{w+l-1}{w}\right] Te_n \\
&= \binom{w+l}{w} e_{n+1}\,.
\end{aligned}$$

while

$$e_{2w+1} \otimes e_{2l+1} \mapsto s\left[\binom{w+l}{w} - \binom{w+l}{w}\right] Te_n \ = \ 0\,.$$

2. When $n+1$ is odd, then

$$\begin{aligned}
e_{2w} \otimes e_{2l+1} &\mapsto s\mu(Te_{2w-1} \otimes e_{2l+1} + e_{2w} \otimes Te_{2l}) \\
&= s\binom{w+l}{w} Te_{2(w+l)} \\
&= \binom{w+l}{w} e_{2(w+l)+1}\,.
\end{aligned}$$

Moreover, the assumed commutativity of the formula in degree n implies the same in degree $n+1$. □

Remark 3.9 Passing to homology, the map above gives a map

$$H_*(\mathbb{Z}/p;\mathbb{F}_p) \otimes H_*(\mathbb{Z}/p;\mathbb{F}_p) \xrightarrow{\mu} H_*(\mathbb{Z}/p;\mathbb{F}_p)\,.$$

This is called the Pontrjagin product. For the case when $p=2$ the result above shows that $H_*(\mathbb{Z}/2;\mathbb{F}_2) = E(e_1, e_2, e_4, \ldots, e_{2^i}, \ldots)$, is an exterior algebra. More generally, the same arguments can be extended to the case of p odd to show that $H_*(\mathbb{Z}/p;\mathbb{F}_p) = E(e_1) \otimes \Gamma_p[\gamma_2]$, where $\Gamma_p[\gamma_2]$ is the divided power algebra on the two dimensional generator γ_2. (The divided power algebra has generators γ_{2i} in each even dimension $2i$ and multiplication given by the rule $\gamma_{2i} \cdot \gamma_{2j} = \binom{i+j}{i}\gamma_{2(i+j)}$.)

Perhaps the most important example for applications is the following.

Example 3.10 Let $G = \mathbb{Z} \times \mathbb{Z}/p$, $H = \mathbb{Z} \wr \mathbb{Z}/p$, and let $h: G \longrightarrow H$ be the homomorphism

$$\Delta^p \times \text{id}: \ (g, \lambda) \mapsto (\underbrace{(g, g, \cdots, g)}_{p \ times}, \lambda) .$$

Then we can construct an explicit chain approximation

$$\mu(h): \ C(\mathbb{Z}(\mathbb{Z})) \otimes_{\mathbb{Z}} C(\mathbb{Z}(\mathbb{Z}/p)) \longrightarrow C(\mathbb{Z}(\mathbb{Z}))^p \otimes C(\mathbb{Z}(\mathbb{Z}/p))$$

by these techniques using $s^{\otimes}$ in the complex for the wreath product.

To begin we may assume $C(\mathbb{Z}(\mathbb{Z}))$ is the resolution discussed above

$$\mathbb{Z} \xleftarrow{\epsilon} \mathbb{Z}(\mathbb{Z})[e_0] \xleftarrow{T-1} \mathbb{Z}(\mathbb{Z})[e_1] ,$$

$\mu(h)(e_1 \otimes e_0) = \left(\sum 1^j \otimes e_1 \otimes T^{p-j-1}\right) \otimes e_0$, $\mu(h)(1 \otimes e_j) = 1^p \otimes e_j$. The particular class which is of most interest in the sequel is the image $\mu(h)(e_1 \otimes e_{p-1})$. When $p = 2$ we obtain, explicitly

$$\begin{aligned} \mu(h)(e_1 \otimes e_1) &= s^{\otimes}((T \times T - 1 \times 1)e_0 \otimes e_0 \otimes e_1 \\ &\quad - (\tau - 1)(e_1 \otimes T + 1 \otimes e_1) \otimes e_0) \\ &= (e_1 \otimes Te_0 + e_0 \otimes e_1) \otimes e_1 - s^{\otimes} T \otimes e_1 \otimes \tau e_0 \\ &= (e_1 \otimes Te_0 + e_0 \otimes e_1) \otimes e_1 - e_1 \otimes e_1 \otimes \tau e_0 . \end{aligned}$$

The most important class here turns out to be the last term, $-(e_1 \otimes e_1) \otimes \tau e_0$. More generally, for odd p, we have

Theorem 3.11 *In the chain approximation constructed above for $\mathbb{Z} \times \mathbb{Z}/p$ with p odd, we have that, after reducing mod the action of $\mathbb{Z} \wr \mathbb{Z}/p$,*

$$\mu(h)(e_1 \otimes e_{p-1}) = \pm \left(\frac{p-1}{2}\right)! \underbrace{(e_1 \otimes \cdots \otimes e_1)}_{p \ times} \otimes e_0 + \sum_{i>0} \theta_{i,p} \otimes e_i .$$

Proof. The proof is a close study of the way in which the iterate operator $s^{\otimes}(\tau - 1)s^{\otimes}\Sigma_\tau$ acts. It turns out that the only term at any stage, which can reach the last class under these operations has the form of a tensor product of e_1's and T's in the first p positions – *no ones* – tensored with $\tau^l e_0$ in the rightmost position. Moreover, the effect of looking at $s^{\otimes}(\tau)s^{\otimes}\Sigma_\tau$ shows that it produces classes with two *adjacent* e_1's in the first two positions. After iterating this operator $(p-1)/2$ times we achieve terms of the form

$$\lambda(l, j) e_1^{2j} \otimes T \otimes e_1^{p-2j-1} \otimes \tau^l e_0 .$$

For each j, the sum $\sum_l \lambda(l,j)$ is $\left(\frac{(p-1)}{2}-1\right)!$ corresponding to the order in which the pairs of adjacent $e_1 \otimes e_1$'s appear, except for the leftmost, which we know just appeared. Then applying the final $s^{\otimes}\Sigma_\tau$ we obtain a sum

$$\sum_{l,j} \lambda(l,j) \underbrace{e_1 \otimes \cdots \otimes e_1}_{p\ times} \otimes \tau^{l+2j} e_0 .$$

(The signs are all positive since we are always permuting an even number of e_1's.) But now, summing the $\lambda(l,j)$ over both l, j, gives $(\frac{p-1}{2})!$ as asserted. □

We will see later that this formula is the essential step in defining and proving the basic properties for the Steenrod p^{th} power operations. It is also the basic step in proving a key lemma which is crucial in our study of the cohomology of the symmetric groups and many of the groups of Lie type. We turn to some of this in the next section.

4. Groups With Cohomology Detected by Abelian Subgroups

We begin with a topological application of the last result in the previous section. The following result is due to N. Steenrod, and is found in [SE].

Theorem 4.1 *Let X be a CW-complex, and consider the map*

$$\Delta^p \times \mathrm{id}\colon\ X \times B_{\mathbb{Z}/p} \longrightarrow X^p \times_{\mathbb{Z}/p} E_{\mathbb{Z}/p} .$$

Then, for all $\alpha \in H^*(X;\mathbb{F}_p)$, we have

$$(\Delta^p \times \mathrm{id})^*(\gamma_p(\alpha)) = \pm \left(\frac{p-1}{2}\right)^{\dim(\alpha)} \alpha \times b^{(\frac{p-1}{2})\dim(\alpha)} + \sum_{j>0} \alpha_j \otimes \theta_j$$

where b is a generator for $H^2(\mathbb{Z}/p;\mathbb{F}_p)$, and $\dim(\alpha_j) = \dim(\alpha) + j$.

Proof. For $X = S^1$ this is a restatement of the last result in the previous section. Suppose the result is true for a generator, e^n, of $H^n(S^n;\mathbb{F}_p) = \mathbb{F}_p$. We wish to show it for a generator of $H^{n+1}(S^{n+1};\mathbb{F}_p)$. To do this consider $S^1 \times S^n$ and the projection $S^1 \times S^n \longrightarrow S^{n+1}$ which induces an isomorphism of cohomology groups in dimension $n+1$. Consequently, if the result is true for $S^1 \times S^n$ it is true for S^{n+1}. But the following diagram commutes

$$\begin{array}{ccc}
S^1 \times S^n \times B_{\mathbb{Z}/p} & \xrightarrow{\text{shuff}\cdot\Delta} & (S^1 \times B_{\mathbb{Z}/p}) \times (S^n \times B_{\mathbb{Z}/p}) \\
\big\downarrow {\scriptstyle \Delta^p \times \text{id}} & & \big\downarrow {\scriptstyle (\Delta^p \times \text{id})^2} \\
(S^1 \times S^n)^p \times_{\mathbb{Z}/p} E_{\mathbb{Z}/p} & \xrightarrow{p} & (S^1)^p \times_{\mathbb{Z}/p} E_{\mathbb{Z}/p} \times (S^n)^p \times_{\mathbb{Z}/p} E_{\mathbb{Z}/p}.
\end{array}$$

From the naturality of the construction of $\gamma_p(\alpha)$ we can assume $p^*(\gamma_p(e^1) \otimes \gamma_p(e^n)) = \gamma_p(e^1 \otimes e^n)$, and from this, the inductive assumption and the commutativity of the diagram above, the result holds for S^n.

Next, consider the map $e^n \colon S^n \to K(\mathbb{Z}/p, n)$ where $(e^n)^*(\iota_n) = e^n$. The following diagram commutes

$$\begin{array}{ccc}
S^n \times B_{\mathbb{Z}/p} & \xrightarrow{e^n \times \text{id}} & K(\mathbb{Z}/p, n) \times B_{\mathbb{Z}/p} \\
\big\downarrow {\scriptstyle \Delta^p \times \text{id}} & & \big\downarrow {\scriptstyle \Delta^p \times \text{id}} \\
(S^n)^p \times_{\mathbb{Z}/p} E_{\mathbb{Z}/p} & \xrightarrow{(e^n)^p \times \text{id}} & K(\mathbb{Z}/p, n)^p \times_{\mathbb{Z}/p} E_{\mathbb{Z}/p}
\end{array}$$

and $(e^n)^* \colon H^n(K(\mathbb{Z}/p, n); \mathbb{F}_p) \to H^n(S^n; \mathbb{F}_p)$ is an isomorphism. Consequently, the truth of the result for e^n implies it for ι_n. But since ι_n is universal, the theorem follows. □

This result has very strong consequences for the cohomology of finite groups.

Definition 4.2 Let G be a finite group. We say that $H^*(G; \mathbb{F}_p)$ is *detected by abelian subgroups* if there is a family of abelian subgroups $H_i \subset G$ so that

$$\coprod_i \left(\text{res}^G_{H_i}\right)^* \colon\ H^*(G; \mathbb{F}_p) \longrightarrow \coprod_i H^*(H_i; \mathbb{F}_p)$$

is an injection.

Note that if G_1, G_2 both have mod p cohomology detected by abelian subgroups then the subgroups $H_i(G_1) \times H_j(G_2)$ detect $H^*(G_1 \times G_2; \mathbb{F}_p)$, so the family of such groups is closed under products. Moreover, if G is such that a subgroup H contains a Sylow p-subgroup of G and H has mod p cohomology detected by abelian subgroups, then the same is true for G, since $\left(\text{res}^G_H\right)^* \colon H^*(G; \mathbb{F}_p) \hookrightarrow H^*(H; \mathbb{F}_p)$ must be an injection.

The following result, due to D. Quillen, [Q6], but anticipated in large measure by M. Nakaoka, [Na], will show that the set of such groups is really rather large.

Theorem 4.3 *Let $H^*(G; \mathbb{F}_p)$ be detected by abelian subgroups, then the same property holds for $H^*(G \wr \mathbb{Z}/p; \mathbb{F}_p)$.*

Proof. Let $\{H_i\}$ detect $H^*(G;\mathbb{F}_p)$. Then, our previous calculation of $H^*(G \wr \mathbb{Z}/p;\mathbb{F}_p)$ shows that the collection of subgroups $\{H_i \wr \mathbb{Z}/p\}$ detect $H^*(G \wr \mathbb{Z}/p;\mathbb{F}_p)$. Thus, we are reduced to proving the result for an abelian group G. The theorem will, thus, follow directly from the following result.

Lemma 4.4 *Let G be abelian, then G^p and $G \times \mathbb{Z}/p$ embedded as a subgroup of $G \wr \mathbb{Z}/p$ via the inclusion $\Delta^p \times \mathrm{id}$, detect $H^*(G \wr \mathbb{Z}/p;\mathbb{F}_p)$.*

Proof 4.4 We have already seen that

$$H^*(G\wr\mathbb{Z}/p;\mathbb{F}_p) = H^*(G^p;\mathbb{F}_p)^{\mathbb{Z}/p} + \{\gamma(\tau) \cup \theta_i\}$$

where θ_i comes from $H^i(\mathbb{Z}/p;\mathbb{Z}/p)$, and $\gamma(\tau)$ restricts to $\underbrace{\tau \otimes \cdots \otimes \tau}_{p\ times}$ in $H^*(G^p;\mathbb{F}_p)$. Consequently, it is only necessary to check the result on the elements $\gamma(\tau) \cup \theta_i$. In the group $\Delta^p \times \mathrm{id}(G \times \mathbb{Z}/p)$ this element has image $(\Delta^p \times \mathrm{id})^*(\gamma(\tau)) \cup (1 \otimes \theta_i)$. Thus, the lemma follows from Steenrod's theorem above since it implies that under the assumptions above

$$(\Delta^p \times \mathrm{id})^*(\gamma(\tau)) = \pm\gamma \otimes b^{\dim(\gamma)\frac{p-1}{2}} + \sum_{j>0} D_j(\gamma) \otimes \Lambda_j$$

where $\dim(D_j(\gamma)) = \dim(\gamma) + j$ for p an odd prime, and it is equal to $\gamma \otimes (e_1)^{\dim(\gamma)} + \sum_{j>0} D_j(\gamma) \otimes e_1^{\dim(\gamma)-j}$ for $p = 2$. □

This finishes the proof of the theorem. □

In Chapter VI one of the crucial facts which makes the determination of the cohomology of the symmetric groups possible will be proved: that $\mathrm{Syl}_p(\mathcal{S}_n)$ is a product of wreath products of the form $\underbrace{\mathbb{Z}/p\wr\mathbb{Z}/p\wr\cdots\wr\mathbb{Z}/p}_{j\ times}$ with $p^j < n$. Thus the cohomology of $\mathcal{S}_n$ is detected by its elementary abelian p-subgroups for every n.

But even more is true. Let $G = \mathrm{GL}_n(\mathbb{F}_q)$ for some finite field $\mathbb{F}_q$. We will see in chapter VII that for p odd the Sylow p-subgroup of G is a product of wreath products having the form $\mathbb{Z}/p^l \wr \underbrace{\mathbb{Z}/p\wr\cdots\wr\mathbb{Z}/p}_{j\ times}$ for appropriate l, j, depending on q and n as long as p does not divide q. Moreover, a similar result is true for many other groups of Lie type. Consequently, (4.3) is a very powerful tool in analyzing the cohomology of many of the most important finite groups.

5. Structure Theorems for the Ring $H^*(G;\mathbb{F}_p)$

There are a number of general things which can be said about the structure of the ring $H^*(G;\mathbb{F}_p)$ when G is a finite group. Among them is the result of L. Evens [Ev1] that it is Noetherian, in particular finitely generated, with only a finite number of relations. Also, Quillen, in his landmark paper [Q1], proved a series of beautiful theorems relating the structure of $H^*(G;\mathbb{F}_p)$ to the set of conjugacy classes of elementary p-groups in G. He showed

1. that the Krull dimension, (the number of generators of the largest polynomial subring contained in $H^*(G;\mathbb{F}_p)$, an alternate definition is given in 5.4), of $H^*(G;\mathbb{F}_p)$ is equal to the largest rank of any elementary p-subgroup of G,
2. that the intersection of the kernels of the restriction maps

$$H^*(G;\mathbb{F}_p) \to H^*(A;\mathbb{F}_p)$$

   as A runs over the set of p-elementary subgroups of G is contained in the Radical of $H^*(G;\mathbb{F}_p)$, and
3. that the set of minimal prime ideals of $H^*(G;\mathbb{F}_p)$ is in one to one correspondence with the set of conjugacy classes of maximal p-elementary subgroups of G.

In this section we prove Evens' theorem and the first two of Quillen's results. (Note: Venkov [V] obtained an independent proof of the finite generation of group cohomology; this is described in [Q1], pp. 554-555.)

Evens–Venkov Finite Generation Theorem

Recall the definition of Noetherian ring: a ring is (left or right) *Noetherian* if and only if, for each ascending sequence of (left or right) ideals $I_1 \subset I_2 \subset \cdots$, there is an n so that $I_j = I_{j+1}$ if $j \geq n$. We use the same definition for the *graded* rings which occur in cohomology, and a similar definition for Noetherian modules. In particular quotients and ideals are Noetherian for a Noetherian ring. Also, a finitely generated exterior algebra over a Noetherian ring is Noetherian, as is a polynomial ring on a finite number of generators over a (graded) commutative Noetherian ring. (This last is just Hilbert's theorem.)

Lemma 5.1 *Let E be a finite p-group, then $H^*(E;\mathbb{F}_p)$ is Noetherian.*

Proof. It is clear that $H^*(\mathbb{Z}/p;\mathbb{F}_p)$ is Noetherian. Now we proceed by induction. We can give E as a central extension $\mathbb{Z}/p \overset{\triangleleft}{\longrightarrow} E \overset{\pi}{\longrightarrow} G$. Let us assume that $H^*(G;\mathbb{F}_p)$ is Noetherian. Then the E_2-term in the Lyndon-Hochschild-Serre spectral sequence of the extension is Noetherian as is E_r for any finite $r \geq 2$. By (1.13) the spectral sequence collapses at level $|G|$, and an easy induction over the filtration gives the result for E. □

Using (5.1) we can prove the Evens–Venkov result. For further details and generalizations see for example the discussions in the books [Ev2] and [Be].

Theorem 5.2 *Let G be a finite group, then $H^*(G;\mathbb{F}_p)$ is Noetherian for any p dividing the order of G.*

Proof. Consider the inclusion $\mathrm{Syl}_p(G) \hookrightarrow G$. Suppose that $I_1 \subset I_2 \subset \cdots \subset I_r \cdots$ is any infinite sequence of increasing ideals in $H^*(G;\mathbb{F}_p)$. Associated to this is the sequence of ideals in $H^*(\mathrm{Syl}_p(G);\mathbb{F}_p)$:

$$res^*(I_1) \cup H^*(\mathrm{Syl}_p(G);\mathbb{F}_p) \subset res^*(I_2) \cup H^*(\mathrm{Syl}_p(G);\mathbb{F}_p) \subset \cdots$$

which are all equal from some n on by (5.1). However, we have $tr^*(res^*(a) \cup b) = a \cup tr^*(b)$, so $tr^*\left(res^*(I_r) \cup H^*(\mathrm{Syl}_p(G);\mathbb{F}_p)\right) = I_r$, and (5.2) follows. □

The Quillen-Venkov Theorem

As we mentioned in the introduction to this section Quillen [Q1] proved that the intersections of the kernels of restriction to $H^*(A;\mathbb{F}_p)$ is contained in the Radical of $H^*(G;\mathbb{F}_p)$ for any finite group G. His original proof involved an analysis along lines similar to, but harder than, the considerations in the next chapter. However, a much more elementary proof was given in [QV] and we present that now.

Theorem 5.3 *Let G be a finite group. If $w \in H^*(G;\mathbb{F}_p)$ restricts to zero on every elementary abelian p-subgroup of G, then w is nilpotent.*

Proof. We argue by induction on the order of G. So assume the theorem is true for all groups of order less than $|G|$. Consequently, a power of w restricts to zero on every *subgroup* of G. In particular, if G is not a p-group then this power of w restricts to zero in $H^*(\mathrm{Syl}_p(G);\mathbb{F}_p)$ and is thus zero.

Consequently it suffices to prove (5.3) for finite p-groups. For any index p-subgroup $H \subset G$ we know that H is normal and we have an extension $H \overset{\triangleleft}{\longrightarrow} G \overset{\pi}{\longrightarrow} \mathbb{Z}/p$. By (1.14) w^{2n} is in the ideal $(u(H))$ where $u = \beta(\pi^*(e))$. But a sufficiently high power of w is thus divisible by every element $\beta(v) \in H^2(G;\mathbb{F}_p)$ and (III.3.1) now implies that this power of w is, in fact, zero. (5.3) follows. □

The Krull Dimension of $H^*(G;\mathbb{F}_p)$

We now sharpen (5.3) somewhat to determine the asymptotic growth rate of group cohomology. Atiyah, Evens, and Swan conjectured that it is exactly the p-rank of G, and Quillen proved this in [Q1], see for example [Ev2, pg. 103].

Definition 5.4 Let $\{V_n\}_{n\in\mathbb{N}}$ be a graded vector space with $\dim(V_n) < \infty$ for all $n \in \mathbb{N}$. We define the *growth rate* of V_* as

$$\gamma(V_*) = \min\left\{ s \in \mathbb{N} \,\middle|\, \lim_{n\to\infty} \frac{\dim(V_n)}{n^s} = 0 \right\}$$

We want to calculate $\gamma(H^*(G;\mathbb{F}_p)$ which coincides with the Krull dimension of the associated graded commutative ring. From (5.3) the largest polynomial algebra contained in $H^*(G;\mathbb{F}_p)$ has degree at most the p-rank of G. On the other hand, Evens' finite generation theorem shows that as a module over the polynomial parts of $H^*(G;\mathbb{F}_p)$ it is finitely generated.

This shows that the Krull dimension is at most the p-rank of G. To establish the reverse inequality it suffices to show that $H^*(G;\mathbb{F}_p)$ contains a polynomial subalgebra of degree p-rank G. To this end, consider the regular representation $G \subset \mathcal{S}_{|G|}$. Restricting to a maximal p-elementary subgroup $A \subset G$ it is easy to see that the image is $[G{:}\,A]$ copies of the regular representation of A. We have constructed the Dickson algebra $H^*(A;\mathbb{F}_p)^{GL(A)}$ in (III.2), and (VI.1), more particularly (VI.1.2), (VI.1.3), (VI.1.7) show that the elements $d_i^{[G:A]}$ are present in the image of restriction from $H^*(G;\mathbb{F}_p)$. But these classes are transcendentally independent and the result follows.

In [Q1] Quillen left the following as a problem: given a finitely generated $\mathbb{F}_pG$–module M, find a formula for the asymptotic growth rate of $H^*(G,M)$. His ideas can in fact be adapted to this situation, and this has led to the development of a whole series of cohomological techniques for representation theory, often referred to as complexity theory. We refer the reader to the recent book of L.Evens, [Ev2], where this subject is dealt with in full detail.

We would like to finish this section by giving a more conceptual approach to the previous results, which was Quillen's original point of view (see [Q1]). A well-known result in the complex representation theory of finite groups is that characters are determined on hyperelementary subgroups by restriction (Brauer's Theorem). For mod p cohomology it is clear that p-elementary groups should play a somewhat similar role. More precisely, let $A_p(G)$ denote the family of all non–trivial p–elementary abelian subgroups in G; clearly this is closed under conjugation and intersection. More formally speaking, we may think of $A_p(G)$ as a category whose objects are the subgroups $(\mathbb{Z}/p)^n \subset G$, with morphisms induced by inclusion and conjugation.

Definition 5.5 Let $A \in A_p(G)$ and denote $H_A^* = H^*(A,\mathbb{F}_p)$. Then we define

$$\lim_{A\in A_p(G)} H_A^*$$

as the sequences $(x_A) \in \prod_{A\in A_p(G)} H_A^*$ such that $\mathrm{res}^A_{A'}(x_A) = x_{A'}$ if $A' \subset A$ and $c_g^*(x_A) = x_{A'}$ if $A' = g^{-1}Ag$.

Using the restrictions, we can construct a map

$$\phi\colon H^*(G,\mathbb{F}_p) \longrightarrow \lim_{A\in A_p(G)} H_A^*\,.$$

We can now state Quillen's main theorem on this topic which strengthens (5.3) by also giving information about the image of restriction to the cohomology of the p-elementary subgroups of G.

Theorem 5.6 *The map ϕ has nilpotent kernel and cokernel.*

A map satisfying the above is known as an F–isomorphism. This result is the analogue in cohomology of theorems such as that of Brauer in representation theory. The situation is evidently far more complicated in cohomology.

6. The Classification and Cohomology Rings of Periodic Groups

The results of (IV.5) on the Krull dimension imply that it is a good invariant for organizing groups. In this section we study the groups G which have Krull dimension one at all primes p which divide $|G|$.

Definition 6.1 A finite group G is *periodic* and of *period* $n > 0$ if and only $H^i(G;\mathbb{Z}) \cong H^{i+n}(G;\mathbb{Z})$ for all $i \geq 1$ where the G action on $\mathbb{Z}$ is trivial.

We have already seen that if G is cyclic then it is periodic and if G is the quaternion group Q_{2^n}, (IV.2.10- IV.2.12) show that it is periodic as well. Periodic groups come up naturally in topology when one considers finite groups acting on spheres.

Lemma 6.2 *If G acts freely and orientation preserving on a sphere S^{n-1} then G must be periodic of period n.*

Proof. If G is a finite group and acts freely on X there is a standard fibration which is used to study the quotient,

$$X \longrightarrow X/G \xrightarrow{\varphi} B_G$$

where φ classifies the principal G fibering $X\to X/G$. In our case, consider the spectral sequence of the fibration

$$S^n \longrightarrow S^n/G \longrightarrow B_G\,.$$

The E_2-term is $H^*(G; H^*(S^n; \mathbb{Z}))$ and the assumption that the action preserves orientation implies that the action of G on $H^*(S^n; \mathbb{Z})$ is trivial. Consequently this E_2-term has the form of two rows $E_2^{i,0} \cong E_2^{i,n-1} \cong H^i(G; \mathbb{Z})$, while all the other rows are identically zero. It follows that there is only one differential, $d_n: E_2^{i,n-1} \xrightarrow{d_n} E_2^{i+n,0}$. On the other hand, the total space of the fibering is $(n-1)$-dimensional, so $H^j(S^n/G; \mathbb{Z}) = 0$ for $j > n-1$ and $E_\infty^{l,m} = 0$ for $l + m \geq n$. From this $E_2^{0,j} \cong H^j(S^{n-1}/G; \mathbb{Z})$ for $j < n-1$, and otherwise d_n must be an isomorphism for each i. □

Since S^3 is a continuous group any subgroup acts freely, so the generalized quaternion groups, the binary tetrahedral group T, the binary octahedral group O, and the binary icosahedral group all act freely on S^3 and hence are periodic. (We discuss them more precisely after 6.10.) But there also exist periodic groups which do not act freely on any sphere.

Lemma 6.3 *$\mathcal{S}_3$ is periodic with period four.*

Proof. $H^*(\mathcal{S}_3; \mathbb{F}_3) = H^*(\mathbb{Z}/3; \mathbb{F}_3)^{\mathbb{Z}/2} = \mathbb{F}_3[b^2] \otimes E(eb)$ while $H^*(\mathcal{S}_3; \mathbb{F}_2) = H^*(\mathbb{Z}/2; \mathbb{F}_2) = \mathbb{F}_2[e]$. Passing to integral cohomology we have

$$H^i(\mathcal{S}_3; \mathbb{Z}) = \begin{cases} \mathbb{Z} & i = 0, \\ \mathbb{Z}/2 & i \cong 2 \bmod (4), \\ \mathbb{Z}/6 & i \cong 0 \bmod (4), \\ 0 & i \text{ odd.} \end{cases}$$

and the result follows. □

Milnor, [Mi2], proved that $\mathcal{S}_3$ cannot act freely on any sphere. In fact he proved that a necessary condition for a periodic group G to act freely on a sphere is that any element of order 2 in G must be central.

Remark. The proof of (6.3) illustrates the important fact that G is periodic only if its $\mathbb{F}_p$ cohomology is periodic for each p and then the period is the least common multiple of its p-periods. This basic result follows directly from (6.4) and (6.6) below which determine the possible p-Sylow subgroups of G.

Lemma 6.4 *G is periodic only if the only p-elementary subgroups of G are cyclic, which is true only if all the abelian subgroups of G are cyclic.*

Proof. Indeed, if $(\mathbb{Z}/p)^n \subset G$ for $n \geq 2$, then $\mathbb{F}_p[a, b] \subset H^*(G; \mathbb{F}_p)$ by the Krull dimension results of (IV.5), for appropriate a, b. But in this case $\dim(H^i(G; \mathbb{F}_p))$ is unbounded. On the other hand from the coefficient sequence

$$0 \to \mathbb{Z} \xrightarrow{\cdot p} \mathbb{Z} \to \mathbb{F}_p \to 0$$

we get the long exact sequence of cohomology groups

$$\xrightarrow{\delta} H^i(G;\mathbb{Z}) \xrightarrow{\cdot p} H^i(G;\mathbb{Z}) \longrightarrow H^i(G;\mathbb{F}_p) \xrightarrow{\delta} H^{i+1}(G;\mathbb{Z}) \longrightarrow \cdots ,$$

and it follows that $\dim(H^i(G;\mathbb{Z}) \otimes \mathbb{F}_p)$ is also unbounded, so G cannot be periodic. □

Remark. Using a spectral sequence argument with $\mathbb{F}_p$ coefficients in 6.2, we infer that $\mathbb{Z}/p \times \mathbb{Z}/p$ cannot act freely on a sphere, and hence if a finite group acts freely on a sphere (preserving orientation or not) it must be periodic (see V.0.7).

Definition 6.5 A $\mathcal{P}$-group G is any finite group G which satisfies the condition that *every abelian subgroup of G is cyclic.*

The $\mathcal{P}$-groups have been completely classified by Suzuki and Zassenhaus. It turns out, using the classification theorem that every $\mathcal{P}$-group is periodic. We will review the classification shortly. But for now we give further details on the structure of $\mathcal{P}$-groups.

Corollary 6.6 *If G is a $\mathcal{P}$-group then* $\mathrm{Syl}_p(G)$ *is cyclic for p odd and either cyclic or generalized quaternion for $p = 2$.*

Proof. We wish to show that if H is any finite p-group where the only abelian subgroups are cyclic, then H is $\mathbb{Z}/p^r$ if p is odd. This is surely true if $|H| = p$, so assume the truth of the assertion for $|H| < p^r$. Since H is a p-group it contains a normal subgroup, N, of index p, and by assumption $N = \mathbb{Z}/p^{r-1}$. Hence H is an extension of the form

$$\mathbb{Z}/p^{n-1} \xrightarrow{\triangleleft} H \longrightarrow \mathbb{Z}/p \,.$$

If $r = 2$ the action of $\mathbb{Z}/p$ on N is trivial and H is either $\mathbb{Z}/p \times \mathbb{Z}/p$ or $\mathbb{Z}/p^2$. The first case is impossible, so the second case holds. Otherwise $r > 2$. Then there are two possible actions of $\mathbb{Z}/p$ on $\mathbb{Z}/p^{r-1}$, multiplication by $1 + p^{r-2}$, or the trivial action. In the first case $H^2(\mathbb{Z}/p; \mathbb{Z}/p^{r-1}) = 0$, so the extension is the semi-direct product, which contains $(\mathbb{Z}/p)^2$ as a subgroup, so there remains only the second case. Here $H^2(\mathbb{Z}/p; \mathbb{Z}/p^r) = \mathbb{Z}/p$ with representatives $(p-1)$ copies of $\mathbb{Z}/p^r$ and $\mathbb{Z}/p^{r-1} \times \mathbb{Z}/p$. Thus the case p odd is proved.

When $p = 2$ we once more assume that the result is true for $|H| < 2^r$, where we know the only possibility for $|H| = 4$ is $\mathbb{Z}/4$. There are four actions of $\mathbb{Z}/2$ on $\mathbb{Z}/2^r$, multiplication by -1, $1 + 2^{r-2}$, $1 - 2^{r-2}$ and the trivial action. We have $H^2(\mathbb{Z}/2; \mathbb{Z}/2^{r-1}) = \mathbb{Z}/2$ for the first and the fourth, but $H^2(\mathbb{Z}/2; \mathbb{Z}/2^{r-1}) = 0$ for the second and third. Thus we are reduced to the first and fourth cases. In the first case the extensions are the dihedral group and the generalized quaternion group. In the fourth case they are the cyclic group and $\mathbb{Z}/2 \times \mathbb{Z}/2^{r-1}$.

When the subgroup of index two is $\mathcal{Q}_{2^{r-1}}$ we are in the situation of $H^2(\mathbb{Z}/2;\mathbb{Z}/2) = \mathbb{Z}/2$. Here the two extensions are the semi-direct product and the generalized quaternion group of order 2^r. □

Corollary 6.7 *Let G be periodic of period n.*

1. *For each p dividing $|G|$ there is a unique subgroup $\mathbb{Z}/p \subset \mathrm{Syl}_p(G)$ and $\mathbb{Z}/p \subset \mathrm{Syl}_2(G)$ is weakly closed in G. Consequently*

$$\mathrm{im}\,(\mathrm{res}^*\colon H^*(G;\mathbb{F}_p)\longrightarrow H^*(\mathbb{Z}/p;\mathbb{F}_p))$$

   is exactly equal to $H^*(\mathbb{Z}/p;\mathbb{F}_p)^{W_G(\mathbb{Z}/p)}$ for p odd.
2. *For $\mathrm{Syl}_2(G) = \mathbb{Z}/2^r$, the restriction image is $\mathbb{F}_2[e^2]$, the polynomial algebra on a two dimensional generator.*
3. *When $\mathrm{Syl}_2(G) = \mathcal{Q}_{2^n}$ we have*

$$\mathrm{im}\,(\mathrm{res}^*\colon H^*(G;\mathbb{F}_2)\longrightarrow H^*(\mathbb{Z}/2;\mathbb{F}_2))$$

   is $\mathbb{F}_2[e^4]$, the polynomial algebra on one generator in dimension four.

Proof. The fact of weak closure is clear. Moreover, we have seen in (IV.2.10), (IV.2.12), that the images of the restriction maps are respectively $\mathbb{F}_p[b_2]$ for p odd or $p = 2$ and $G = \mathbb{Z}/2^r$, while the image of restriction is $\mathbb{F}_2[e^4]$ for $G = \mathcal{Q}_{2^n}$. Of course, in these last two cases the Weyl group is trivial. □

Corollary 6.8 *Let G be periodic, then $H^*(G;\mathbb{F}_p) = \mathbb{F}_p[b^i] \otimes E(e_{2i-1})$ for i odd, where i divides $(p-1)$. Also, if $\mathrm{Syl}_2(G) = \mathbb{Z}/2^r$ then $H^*(\mathrm{Syl}_2(G);\mathbb{Z}) = H^*(G;\mathbb{Z})_2$ and $H^*(G;\mathbb{F}_2) = E(e) \otimes \mathbb{F}_2[b_2]$.*

Proof. The integral cohomology of $\mathbb{Z}/p^r$ is given as $\mathbb{Z}/p^r$ in each even dimension and is zero in odd dimensions. Consequently, since $H^*(G;\mathbb{Z})_p$ is a direct summand of $H^*(\mathrm{Syl}_p(G);\mathbb{Z})$ it follows that $H^*(G;\mathbb{Z})_p = \mathbb{Z}/p^r$ in each dimension $2im$ and is zero otherwise. Now apply the coefficient exact sequence in the proof of (6.5). The argument for 2 is similar. □

The situation which holds when $\mathrm{Syl}_2(G) = \mathcal{Q}_{2^r}$ is more complex. In order to understand it we review the classification of $\mathcal{P}$-groups as determined by Suzuki and Zassenhaus.

The Classification of Periodic Groups

To begin we need $\mathrm{Aut}(\mathcal{Q}_8)$.

Lemma 6.9 $\mathrm{Aut}(\mathcal{Q}_8) \cong S_4$, *the symmetric group on 4 letters.*

Proof. $\mathcal{Q}_8/[\mathcal{Q}_8, \mathcal{Q}_8] = \mathbb{Z}/2 \times \mathbb{Z}/2$. Hence, there is a homomorphism

$$e\colon \operatorname{Aut}(\mathcal{Q}_8) \longrightarrow \mathrm{GL}_2(\mathbb{F}_2) = \mathcal{S}_3 = \operatorname{Aut}(\mathbb{Z}/2 \times \mathbb{Z}/2).$$

Moreover, e is onto since $\{x \leftrightarrow y\}$, $\{x \mapsto xy,\ y \mapsto x\}$ both induce automorphisms of $\mathcal{Q}_8$, and their images generate $\mathcal{S}_3$. The kernel of e is the set of automorphisms

$$f(x) = xy^{2a}, \quad f(y) = y^{1+2b}$$

and the inner automorphisms ($\cong \mathbb{Z}/2 \times \mathbb{Z}/2$) give all such possibilities. Moreover, it is direct to check that the extension

$$\mathbb{Z}/2 \times \mathbb{Z}/2 \longrightarrow \operatorname{Aut}(\mathcal{Q}_8) \longrightarrow \mathcal{S}_3$$

identifies $\operatorname{Aut}(\mathcal{Q}_8)$ with the group of affine transformations $\operatorname{Aff}(\mathbb{F}_2) \cong \mathcal{S}_4$. □

The Groups T_i and O_i^*

The action of $\mathbb{Z}/3$ on $\mathcal{Q}_8$ gives rise to a semi-direct product, $\mathcal{Q}_8 \times_\alpha \mathbb{Z}/3 = T$. T is the binary tetrahedral group. It is also isomorphic to $\mathrm{SL}_2(\mathbb{F}_3)$. Indeed, we construct an embedding by first embedding $\mathcal{Q}_8 \subset \mathrm{SL}_2(\mathbb{F}_3)$ by setting

$$x \mapsto \begin{pmatrix} -1 & -1 \\ -1 & 1 \end{pmatrix}, \quad y \mapsto \begin{pmatrix} 0 & 1 \\ -1 & 0 \end{pmatrix}.$$

Next extend to T, by sending the element t of order 3 to $\begin{pmatrix} 1 & 1 \\ 0 & 1 \end{pmatrix}$. A presentation of $\mathrm{SL}_2(\mathbb{F}_3)$ can be given as

$$\mathrm{SL}_2(\mathbb{F}_3) = \{\gamma, y \mid \gamma^3 = y^2 = (y\gamma)^3\}$$

where $\gamma = -t$.

The subgroup $\mathcal{S}_3 \subset \operatorname{Aut}(\mathcal{Q}_8)$ has two distinct extensions by $\mathcal{Q}_8$ since $H^2(\mathcal{S}_3; \mathbb{Z}/2) = \mathbb{Z}/2$. The first is the semi-direct product, but the second O^*, is the binary octahedral group. It is obtained from T by extending T by $\mathbb{Z}/2$. Hence, we have the extension diagram

$$1 \longrightarrow T \longrightarrow O^* \longrightarrow \mathbb{Z}/2 \longrightarrow 1.$$

Specifically, let the new generator be z, and set

$$z(x) = xy^{-1}, \ z(y) = y^{-1}, \ x(t) = t^{-1}, \ z^2 = x^2 = -I.$$

Note that

$$(zx)^2 = zxz^{-1}z^2x = xyx = y.$$

Hence, $z(zx)z^{-1} = zxy^{-1} = (zx)^7 = (zx)^{-1}$, and the subgroup generated by x, y, z is isomorphic to $\mathcal{Q}_{16}$.

More generally we set

$$T_i = \mathcal{Q}_8 \times_\alpha \mathbb{Z}/3^i$$

with $\mathbb{Z}/3^i \longrightarrow \mathbb{Z}/3 \subset \mathrm{Aut}(\mathcal{Q}_8)$ giving the extension. Each T_i also admits an extension in the evident way by $\mathbb{Z}/2$ generalizing the extension above,

$$1 \longrightarrow T_i \longrightarrow O_i^* \longrightarrow \mathbb{Z}/2 \longrightarrow 1 .$$

These groups are clearly all $\mathcal{P}$-groups.

The Groups $\mathrm{SL}_2(\mathbb{F}_p)$

Theorem 6.10 $\mathrm{SL}_2(\mathbb{F}_p)$ *is a $\mathcal{P}$-group for every prime p.*

Proof. $|\mathrm{SL}_2(\mathbb{F}_p)| = p(p^2-1) = p(p-1)(p+1)$. A subgroup isomorphic to $\mathbb{Z}/p = \left\{\begin{pmatrix} 1 & t \\ 0 & 1 \end{pmatrix}\right\}$. A subgroup isomorphic to $\mathbb{Z}/(p-1) = \{\begin{pmatrix} \alpha & 0 \\ 0 & \alpha^{-1} \end{pmatrix}\}$. The group $\mathbb{F}^{\cdot}_{p^2}$ embeds in $\mathrm{GL}_2(\mathbb{F}_p)$ when we think of the vector space $(\mathbb{F}_p)^2$ as the field $\mathbb{F}_{p^2}$, so that $\mathbb{F}_p$-linear isomorphisms are obtained by multiplying by the elements of $\mathbb{F}^{\cdot}_{p^2}$. However, the determinant of $\{\alpha\}$ (multiplication by $\alpha \in \mathbb{F}^{\cdot}_{p^2}$) is $N(\alpha) = \alpha^{1+p}$. Hence, the subgroup $\mathrm{Ker}N \cap \mathrm{SL}_2(\mathbb{F}_p)$ which is contained in $\mathrm{SL}_2(\mathbb{F}_p)$ is isomorphic to $\mathrm{Ker}(N) = \mathbb{Z}/(p+1)$. Since $p-1$, p, and $p+1$ are coprime except for the factor 2 which is common to $p-1, p+1$, we are done except for checking the structure of the 2-Sylow subgroup.

There are two cases. First, assume that $p \equiv 1 \bmod (4)$, then the subgroup

$$\left\{\begin{pmatrix} 0 & 1 \\ -1 & 0 \end{pmatrix}, \begin{pmatrix} \alpha & 0 \\ 0 & \alpha^{-1} \end{pmatrix}\right\}$$

has order $2(p-1)$, and is clearly a quaternion group. Second, if $p \equiv 3 \bmod (4)$, then, if $W \subset \mathbb{Z}/(p+1)$ is the 2-Sylow subgroup, the action of the Galois automorphism g is via inversion. However, $\det(g) = -1$. To correct this use $\omega \in \mathbb{F}_{p^2}$ for which $w^{1+p} = -1$. Then $(g\omega)^2 = -1$, and $\det(g\omega) = -\det(\omega) = 1$. Thus the subgroup generated by $g\omega, \mathrm{Ker}(N)$ is again a quaternion group and the result follows. □

Remark. $\mathrm{SL}_2(\mathbb{F}_5)$ is the binary icosahedral group. It has a presentation

$$\mathrm{SL}_2(F_5) = \{r, s \mid r^2 = s^3 = (rs)^5\} ,$$

see for example [Co, p. 2]. For the remaining odd primes H. Behr and J. Mennicke, [BM], give presentations as

$$\mathrm{SL}_2(\mathbb{F}_p) = \{A, B \mid A^p = 1,\ (AB)^3 = B^2,\ B^4 = (A^2BA^{\frac{1}{2}(p+1)}B)^3 = 1\} .$$

Indeed, we can map this group to $\mathrm{SL}_2(\mathbb{F}_p)$ by

$$A \mapsto \begin{pmatrix} 1 & 1 \\ 0 & 1 \end{pmatrix}, B \mapsto \begin{pmatrix} 0 & 1 \\ -1 & 0 \end{pmatrix} .$$

The Groups $TL_2(\mathbb{F}_p)$

With the demonstration that the $\mathrm{SL}_2(\mathbb{F}_p)$ are periodic and the introduction of the generalized binary tetrahedral and octahedral groups T_i^* and O_i^* we have almost completed the list of periodic groups. There remains one more family which we construct now.

$\mathrm{Out}(\mathrm{SL}_2(\mathbb{F}_2)) = \mathbb{Z}/2$ with generator, γ, acting as conjugation by the matrix $\begin{pmatrix} \omega & 0 \\ 0 & 1 \end{pmatrix}$ where ω is a non-square of $\mathbb{F}_p^{\cdot}$. Consequently we have the group

$$U_{2,p} = \mathrm{SL}_2(\mathbb{F}_p) \times_\alpha \mathbb{Z}/(p-1)$$

where $\alpha(x^i)$ is the automorphism conjugation by $\begin{pmatrix} \omega^i & 0 \\ 0 & 1 \end{pmatrix}$. Let $L \subset U_{2,p}$ be the subgroup generated by $\begin{pmatrix} \omega^{-1} & 0 \\ 0 & \omega \end{pmatrix} \cdot \gamma^2$.

Lemma 6.11 *L is central (in particular normal) in $U_{2,p}$.*

Proof. If $\tau \in \mathrm{SL}_2(\mathbb{F}_p)$ then γ^2 acts on τ as does $\begin{pmatrix} \omega^2 & 0 \\ 0 & 1 \end{pmatrix}$. But

$$\begin{pmatrix} \omega^{-1} & 0 \\ 0 & \omega \end{pmatrix} = \begin{pmatrix} \omega & 0 \\ 0 & \omega \end{pmatrix} \begin{pmatrix} \omega^2 & 0 \\ 0 & 1 \end{pmatrix},$$

hence the two actions cancel out. Also, γ commutes with $\begin{pmatrix} \omega^{-1} & 0 \\ 0 & \omega \end{pmatrix}$, so the action is trivial on $\mathbb{Z}/(p-1)$ as well, and the lemma follows. □

Definition 6.12 We write $TL_2(\mathbb{F}_p)$ for the quotient group $U_{2,p}/L$.

$TL_2(\mathbb{F}_p)$ is given as an extension

$$1 \longrightarrow \mathrm{SL}_2(\mathbb{F}_p) \overset{\triangleleft}{\longrightarrow} TL_2(\mathbb{F}_p) \longrightarrow \mathbb{Z}/2 \longrightarrow 1$$

with extension data $\gamma^2 = \begin{pmatrix} \omega & 0 \\ 0 & \omega^{-1} \end{pmatrix}$, γ acting on $\mathrm{SL}_2(\mathbb{F}_p)$ via conjugation by $\begin{pmatrix} \omega & 0 \\ 0 & 1 \end{pmatrix}$.

Remark 6.13 There is a second subgroup $L' \subset U_{2,p}$ defined as

$$\left\langle - \begin{pmatrix} \omega^{-1} & 0 \\ 0 & \omega \end{pmatrix} \cdot \gamma^2 \right\rangle.$$

When $p \equiv 1 \bmod (4)$ the two quotients are isomorphic, but for $p \equiv 3 \bmod (4)$ we could choose $\omega = -1$, in which case $L' = \langle \gamma^2 \rangle$, and the resulting quotient is not a $\mathcal{P}$-group. Moreover, since $H^2(\mathbb{Z}/2;\mathbb{Z}/2) = \mathbb{Z}/2$, these are the only two extensions of $\mathrm{SL}_2(\mathbb{F}_p)$ by $\mathrm{Out}(\mathrm{SL}_2(\mathbb{F}_p))$.

Lemma 6.14 *$TL_2(\mathbb{F}_p)$ is a $\mathcal{P}$-group.*

Proof. As usual there are two cases.

a. $p \equiv 1 \bmod 4$. Then a 2-Sylow subgroup of $\mathrm{SL}_2(\mathbb{F}_p)$ has the form

$$\mathrm{SL}_2(\mathbb{F}_p)_2 = \left\langle \begin{pmatrix} 0 & 1 \\ -1 & 0 \end{pmatrix}, \begin{pmatrix} \omega & 0 \\ 0 & \omega^{-1} \end{pmatrix} \right\rangle .$$

Now

$$\begin{aligned}\begin{pmatrix} \omega & 0 \\ 0 & 1 \end{pmatrix}\begin{pmatrix} 0 & 1 \\ -1 & 0 \end{pmatrix}\begin{pmatrix} \omega^{-1} & 0 \\ 0 & 1 \end{pmatrix} &= \begin{pmatrix} 0 & \omega \\ -\omega^{-1} & 0 \end{pmatrix} \\ &= \begin{pmatrix} \omega & 0 \\ 0 & \omega^{-1} \end{pmatrix}\begin{pmatrix} 0 & 1 \\ -1 & 0 \end{pmatrix} \\ &\in \mathrm{SL}_2(\mathbb{F}_p)_2,\end{aligned}$$

and

$$\begin{pmatrix} 0 & 1 \\ -1 & 0 \end{pmatrix}\gamma\begin{pmatrix} 0 & -1 \\ 1 & 0 \end{pmatrix} = \gamma^{-1} .$$

Hence, the 2-Sylow subgroup of $TL_2(\mathbb{F}_p)$ is a quaternion group.

b. $p \equiv 3 \bmod (4)$. Here $\begin{pmatrix} -1 & 0 \\ 0 & 1 \end{pmatrix}$ gives the Galois action on $\mathbb{F}^{\cdot}_{p^2} \subset \mathrm{GL}_2(\mathbb{F}_p)$, and we can assume that γ acts in this way. Let $v \in \mathbb{F}^{\cdot}_{p^2}$ satisfy $v^{p+1} = -1$, then

$$\left\langle \begin{pmatrix} -1 & 0 \\ 0 & 1 \end{pmatrix} \cdot v,\ v^2 \right\rangle$$

generates a 2-Sylow subgroup of $\mathrm{SL}_2(\mathbb{F}_p)$. We have $\gamma v \gamma^{-1} = v^p = -v^{-1}$ so $(\gamma \begin{pmatrix} -1 & 0 \\ 0 & 1 \end{pmatrix} v)\gamma^{-1} = -(\begin{pmatrix} -1 & 0 \\ 0 & 1 \end{pmatrix} v)v^{-2}$, and

$$\begin{aligned}(\gamma\begin{pmatrix} -1 & 0 \\ 0 & 1 \end{pmatrix} v)^2 &= (\gamma\begin{pmatrix} -1 & 0 \\ 0 & 1 \end{pmatrix} v\gamma^{-1})(\gamma^2\begin{pmatrix} -1 & 0 \\ 0 & 1 \end{pmatrix} v\gamma^{-2})\gamma^2 \\ &= -\begin{pmatrix} -1 & 0 \\ 0 & 1 \end{pmatrix} vv^{-2}(-\begin{pmatrix} -1 & 0 \\ 0 & 1 \end{pmatrix} v)v^2 = -v^2.\end{aligned}$$

Also

$$\begin{aligned}\gamma\gamma\begin{pmatrix} -1 & 0 \\ 0 & 1 \end{pmatrix} v\gamma^{-1} &= \gamma\begin{pmatrix} -1 & 0 \\ 0 & 1 \end{pmatrix}\{-v^{-1}\} \\ &= -(\gamma\begin{pmatrix} -1 & 0 \\ 0 & 1 \end{pmatrix} v)v^{-2} = [\gamma\begin{pmatrix} -1 & 0 \\ 0 & 1 \end{pmatrix} v]^{-1}.\end{aligned}$$

It follows that this group is again a quaternion group and the result follows. □

The Suzuki-Zassenhaus Classification Theorem

In summary we have the Suzuki-Zassenhaus theorem (See [Wo], [DM])

Theorem 6.15 *A complete table of $\mathcal{P}$-groups is given by*

Family	*Definition*	*Conditions*
I	$\mathbb{Z}/a \times_\alpha \mathbb{Z}/b$	$(a,b)=1$
II	$\mathbb{Z}/a \times_\beta (\mathbb{Z}/b \times Q_{2^i})$	$(a,b)=(ab,2)=1$
III	$\mathbb{Z}/a \times_\gamma (\mathbb{Z}/b \times T_i)$	$(a,b)=(ab,6)=1$
IV	$\mathbb{Z}/a \times_\tau (\mathbb{Z}/b \times O_i^*)$	$(a,b)=(ab,6)=1$
V	$(\mathbb{Z}/a \times_\alpha \mathbb{Z}/b) \times \mathrm{SL}_2(\mathbb{F}_p)$	$(a,b)=(ab,p(p^2-1))=1$
VI	$\mathbb{Z}/a \times_\mu (\mathbb{Z}/b \times TL_2(\mathbb{F}_p))$	$(a,b)=(ab,p(p^2-1))=1$

Remark 6.16 All the groups G in the list except those in family I for which α is injective on the 2-Sylow subgroup of $\mathbb{Z}/b$ satisfy the Milnor condition that there is at most one element of order 2 in G, and it is central.

The Mod(2) Cohomology of the Periodic Groups

From the classification result above the only cases which need to be considered now are the cases IV, V, VI. We will first determine the situation in case V and then an easy spectral sequence argument will give the remaining two cases. To begin we need a lemma.

Lemma 6.17 *Let $Q_8 \subset \mathrm{SL}_2(\mathbb{F}_p)$ where p is an odd prime. Then the normalizer of Q_8 in $\mathrm{SL}_2(\mathbb{F}_p)$ contains an elmenent T with $T^3 = 1$ which acts as the non-trivial outer automorphism of Q_8 of order 3.*

Proof. Since p is odd it follows that the group ring $\mathbb{F}_p(Q_8)$ is semi-simple. On the other hand Q_8 has four distinct 1-dimensional representations, $(x \mapsto \pm 1, y \mapsto \pm 1)$. It also has a non-commutative representation $\varphi\colon Q_8 \longrightarrow M_2(\mathbb{F}_p)$, defined as

$$\begin{cases} x \mapsto \begin{pmatrix} a & b \\ b & -a \end{pmatrix} & a^2 + b^2 = -1 \text{ in } \mathbb{F}_p, \\ y \mapsto \begin{pmatrix} 0 & 1 \\ -1 & 0 \end{pmatrix} & \end{cases}$$

for some choice of the pair (a, b). This representation must be irreducible, and it follows that

$$\mathbb{F}_p(Q_8) = \mathbb{F}_p \oplus \mathbb{F}_p \oplus \mathbb{F}_p \oplus \mathbb{F}_p \oplus M_2(\mathbb{F}_p)\,.$$

Thus there is only one non-commutative representation, up to conjugation, of $\mathcal{Q}_8 \subset \mathrm{GL}_2(\mathbb{F}_p)$. Hence, if we set $\Psi(x) = \varphi(y)$, $\Psi(y) = \varphi(xy)$, there is $T \in \mathrm{GL}_2(\mathbb{F}_p)$ with $T\varphi(x)T^{-1} = \Psi(x)$, $T\varphi(y)T^{-1} = \Psi(y)$. It follows that T^3 is contained in the center of $M_2(\mathbb{F}_p)$ so T satisfies $T^3 - bI = 0$. But this equation cannot be irreducible over $\mathbb{F}_p$ since otherwise $\mathbb{F}_pI + \mathbb{F}_pT + \mathbb{F}_pT^2$ would be a field of degree 3 over $\mathbb{F}_p$ contained in $M_2(\mathbb{F}_p)$ which is impossible. It follows that there is an $a \in \mathbb{F}_p$ with $a^3 = b$, so $a^{-1}T \in \mathrm{SL}_2(\mathbb{F}_p)$ satisfies the asserted conditions. □

Corollary 6.18 $H^*(\mathrm{SL}_2(\mathbb{F}_p); \mathbb{F}_2) = \mathbb{F}_2[e_4] \otimes E(b_3)$, *a polynomial algebra on a four dimensional generator tensored with an exterior algebra on a three dimensional generator.*

Proof. If $\mathrm{Syl}_2(\mathrm{SL}_2(\mathbb{F}_p)) = \mathcal{Q}_8$ we have $H^*(\mathrm{SL}_2(\mathbb{F}_p); \mathbb{F}_2) \subset H^*(\mathcal{Q}_8; \mathbb{F}_2)^T$, but this invariant ring is $\mathbb{F}_2[e_4] \otimes E(b_3)$. On the other hand by (6.7.3), the cohomology ring cannot be smaller than this.

Let $\mathrm{Syl}_2(\mathrm{SL}_2(\mathbb{F}_p)) = \mathcal{Q}_{2^n}$ with $n > 3$. Then, if x and y generate this copy of $\mathcal{Q}_{2^n}$ and $x^{2^{n-2}} = y^2, y^4 = 1$ there are two copies of $\mathcal{Q}_8$,

$$\mathcal{Q}_{8,1} = \langle x^{2^{n-3}}, y\rangle, \text{ and } \mathcal{Q}_{8,2} = \langle xy, y\rangle .$$

It is direct that the two restriction maps

$$\mathrm{res}_1^* \oplus \mathrm{res}_2^* \colon H^*(\mathcal{Q}_{2^n}; \mathbb{F}_2) \longrightarrow H^*(\mathcal{Q}_{8,1}; \mathbb{F}_2) \oplus H^*(\mathcal{Q}_{8,2}; \mathbb{F}_2)$$

together are injective, so $\mathcal{Q}_{8,1}$ and $\mathcal{Q}_{8,2}$ detect mod(2) cohomology. Now the remainder of the argument is clear. □

It remains to discuss the groups $H^*(O_i^*; \mathbb{F}_2)$ and $H^*(TL_2(\mathbb{F}_p); \mathbb{F}_2)$. In both cases, corresponding to the extension data

$$\begin{array}{ccccc} O_i & \xrightarrow{\triangleleft} & O_i^* & \longrightarrow & \mathbb{Z}/2 \\ \mathrm{SL}_2(\mathbb{F}_p) & \xrightarrow{\triangleleft} & TL_2(\mathbb{F}_p) & \longrightarrow & \mathbb{Z}/2 \end{array}$$

we have a spectral sequence with E_2-term $H^*(\mathbb{Z}/2; \mathbb{F}_2) \otimes E(b_3) \otimes \mathbb{F}_2[e_4]$. In each case there must be a differential on b_3 since we know that $e_4 \in H^4(O_i; \mathbb{F}_2)$ or $e_4 \in H^4(\mathrm{SL}_2(\mathbb{F}_p); \mathbb{F}_2)$ must be in the image of restriction, and we know that the Krull dimension is one for the extension group. But the only possible differential on b_3 is $\delta_4(b_3) = e^4$. Then $E_3 = \mathbb{F}_2[e_4](1, e, e^2, e^3)$, and there can be no further differentials. Thus we have determined the mod(2) cohomology of all the periodic groups.

7. The Definition and Properties of Steenrod Squares

We use the notation $\Gamma_p(X)$ for the space $X^p \times_{\mathbb{Z}/p} E_{\mathbb{Z}/p}$ studied in §1. Recall that $\Gamma_p(X)$ is given as the total space of a fibering

$$X^p \xrightarrow{j} \Gamma_p(X) \xrightarrow{\pi} B_{\mathbb{Z}/p} .$$

The space was originally introduced by Steenrod and used in [SE] to explicitly construct the Steenrod P power operations. We review that construction here.

There is the map of (4.1),

$$\Delta^p \times \mathrm{id}\colon\ X \times B_{\mathbb{Z}/p} \longrightarrow \Gamma_p(X)$$

and these constructions are natural with respect to maps $f\colon X \longrightarrow Y$ as is illustrated, for example in (1.8), so we have the commutative diagram

$$\begin{array}{ccc} X \times B_{\mathbb{Z}/p} & \xrightarrow{\Delta^p \times \mathrm{id}} & \Gamma_p(X) \\ \Big\downarrow{\scriptstyle f\times \mathrm{id}} & & \Big\downarrow{\scriptstyle \Gamma(f)} \\ Y \times B_{\mathbb{Z}/p} & \xrightarrow{\Delta^p \times \mathrm{id}} & \Gamma_p(Y) . \end{array}$$

Theorem 7.1 *Let X be a CW complex and $\alpha \in H^n(X, \mathbb{F}_p)$, with $n \geq 1$, then there is a unique class $\Gamma(\alpha) \in H^{np}(\Gamma_p(X), \mathbb{F}_p)$ so that*
1. $(j^*)^*\Gamma(\alpha) = \underbrace{\alpha \otimes \alpha \cdots \otimes \alpha}_{p\ times}$ *in* $H^*(X^p; \mathbb{F}_p)$
2. *if* $f\colon X \longrightarrow Y$ *is a continuous map* $\Gamma(f)^*(\Gamma(\alpha)) = \Gamma(f^*(\alpha))$
3. $\Gamma(\alpha \cup \beta) = \Gamma(\alpha) \cup \Gamma(\beta)$.

Proof. The existence of $\Gamma(\alpha)$ is the argument in the proof of (1.7). Diagram (1.8) shows how to reduce the construction of $\Gamma(\alpha)$ to the construction of $\Gamma(\iota)$ where $\iota \in H^n(K(\mathbb{Z}/p, n); \mathbb{F}_p)$ is the fundamental class. But uniqueness is not demonstrated there.

To see uniqueness, note that $K(\mathbb{Z}/p, n)$ is connected for $n \geq 1$. Hence the choice of basepoint in $K(\mathbb{Z}/p, n)$ is immaterial and we can consider the reduced construction with fiber the smash product $F_p(K(\mathbb{Z}/p, n) = \underbrace{K(\mathbb{Z}/p, n) \wedge \cdots \wedge K(\mathbb{Z}/p, n)}_{p\ times}$. Here, the fiber is $(np-1)$-connected with

$$H^{np}(F_p(K(\mathbb{Z}/p, n); \mathbb{F}_p) = \mathbb{F}_p^{\otimes p} \cong \mathbb{F}_p .$$

Moreover, the basepoint $* \in F_p(K(\mathbb{Z}/p, n))$ gives a lifting, L, of $B_{\mathbb{Z}/p}$ to $F_p(K(\mathbb{Z}/p, n)) \times_{\mathbb{Z}/p} E_{\mathbb{Z}/p}$ so we define $\bar{\Gamma}(\iota)$ as the unique class in

$$H^{np}(F_p(K(\mathbb{Z}/p, n) \times_{\mathbb{Z}/p} E_{\mathbb{Z}/p}; \mathbb{F}_p)$$

which is in the kernel of L^* and restricts to $\underbrace{\iota \otimes \cdots \otimes \iota}_{p\ times}$ on the fiber. Then, by naturality, $\Gamma(\alpha)$ is given as the pullback under the composition

$$\Gamma(X) \xrightarrow{\Gamma(\alpha)} \Gamma(K(\mathbb{Z}/p, n)) \xrightarrow{\pi} F_p(K(\mathbb{Z}/p, n)) \times_{\mathbb{Z}/p} E_{\mathbb{Z}/p}$$

of $\bar{\Gamma}(\iota)$. This gives the first two statements. To see the third consider the universal construction for the cup-product of an n dimensional class α and and m dimensional class β,

$$K(\mathbb{Z}/p, n) \wedge K(\mathbb{Z}/p, m) \xrightarrow{\iota_n \otimes \iota_m} K(\mathbb{Z}/p, n+m) .$$

We now pass to the Γ_p-constructions and build the obvious diagram. Define

$$e\colon \Gamma_p(K(\mathbb{Z}/p, n) \times K(\mathbb{Z}/p, m)) \longrightarrow \Gamma_p(K(\mathbb{Z}/p, n)) \times \Gamma_p(K(\mathbb{Z}/p, m))$$

by $e\{(x_1, y_1), \ldots, (x_p, y_p), v\} = (\{(x_1, \ldots, x_p), v\}, \{(y_1, \ldots, y_p), v\})$. Then the diagram

$$\begin{array}{ccc}
K(\mathbb{Z}/p, p(n+m)) & \xleftarrow{\Gamma(\iota)} & \Gamma_p(K(\mathbb{Z}/2, n+m)) \\
\uparrow{\scriptstyle \iota \otimes \iota} & & \uparrow{\scriptstyle \Gamma(\iota \otimes \iota)} \\
K(\mathbb{Z}/p, np) \wedge K(\mathbb{Z}/p, mp) & & \Gamma_p(K(\mathbb{Z}/p, n) \wedge K(\mathbb{Z}/p, m)) \\
\nwarrow \gamma \times \gamma & & \swarrow e \\
& \Gamma_p(K(\mathbb{Z}/p, n)) \times \Gamma_p K((\mathbb{Z}/p, m)) &
\end{array}$$

commutes up to homotopy. From this the third statement follows. □

The Squaring Operations

We now show how to construct the Steenrod squaring operations using the spaces $\Gamma_2(X)$.

Definition 7.2 Let X be a CW-complex and suppose $\alpha \in H^n(X, \mathbb{F}_2)$. Then

$$(\Delta^p \times \mathrm{id})^*(\Gamma(\alpha)) = \sum Sq^i(\alpha) \otimes e^{n-1}$$

defines the *Steenrod squares* of α.

Of course we must show that the classes defined in this way satisfy the axioms of (II.2). In particular $Sq^n(\alpha) = \alpha^2$ follows from the commutative diagram

$$
\begin{array}{ccc}
X & \xrightarrow{\Delta} & X \times X \\
\Big\downarrow {\scriptstyle \mathrm{id}\times\mathrm{pt}} & & \Big\downarrow {\scriptstyle j} \\
X \times B_{\mathbb{Z}/2} & \xrightarrow{\Delta\times\mathrm{id}} & \Gamma_2(X),
\end{array}
$$

since $(\mathrm{id} \times \mathrm{pt})^* \sum \theta_{n+i} \otimes e^{n-i} = \theta_{2n}$ but $\Delta^* j^*(\Gamma(\alpha)) = \Delta^*(\alpha \otimes \alpha) = \alpha^2$.

Since $K(\mathbb{Z}/2, n)$ is $(n-1)$-connected it follows that $Sq^i(\iota_n) = 0$ for $i < 0$, so, by naturality, $Sq^i(\alpha) = 0$ for $i < 0$ as well.

Next we verify the Cartan formula,

Theorem 7.3

$$Sq^i(\alpha \cup \beta) = \sum_{r=o}^{i} Sq^i(\alpha) \cup Sq^{i-r}(\beta).$$

Proof. Since $\Gamma(\alpha) \cup \Gamma(\beta) = \Gamma(\alpha \cup \beta)$ 7.2 gives

$$
\begin{aligned}
(7.4) \qquad \Big(\sum Sq^i(\alpha) \otimes e^{n-i}\Big) \cup \Big(\sum Sq^r(\beta) \otimes e^{m-r}\Big) \\
= \sum Sq^t(\alpha \cup \beta) \otimes e^{n+m-t}
\end{aligned}
$$

and since $e^r \cup e^q = e^{p+q}$, 7.3 follows on expanding the left-hand side of (7.4). □

Lemma 7.5 *Let $\sigma: \tilde{H}^n(X, \mathbb{F}_2) \to H^{n+1}(\Sigma X, \mathbb{F}_2)$ be the suspension isomorphism, then*

1. $Sq^i(\sigma(x)) = \sigma(Sq^i(x))$
2. $Sq^0(x) = (x)$.

Proof. By (3.11) and naturality $\Delta^*(\Gamma(e^1)) = e^1 \otimes e^1 + 1 \otimes (e^1)^2$ for $e^1 \in H^1(RP^\infty; \mathbb{F}_2)$. Thus, since $RP^\infty = K(\mathbb{Z}/2, 1)$, $Sq^0(\iota_1) = \iota_1$ as desired. Now take the suspension $\sigma: \Sigma K(\mathbb{Z}/2, n) \to K(\mathbb{Z}/2, n+1)$ so $\sigma^*(\iota_{n+1}) = \sigma(\iota_n)$. Notice that σ^* is an isomorphism in dimension $n+1$. Hence if we knew the first statement then the second assertion would follow for the universal model and hence for all X.

To show (1) note $\Sigma X = S^1 \wedge X$ is a factor space of $S^1 \times X$. Hence $\Gamma(\sigma(\alpha)) = \Gamma(e) \cup \Gamma(\alpha)$ and

$$
\begin{aligned}
\Delta^* \Gamma(\sigma(\alpha)) &= e^1 \otimes e \cup \Sigma Sq^i(\alpha) \otimes e^{n-i} \\
&= \Sigma \sigma Sq^i(\alpha) \otimes e^{n+1-i}.
\end{aligned}
$$

□

Corollary 7.6 *The Sq^i maps are homomorphisms, that is to say*

$$Sq^i(\alpha + \beta) = Sq^i(\alpha) + Sq^i(\beta).$$

Proof. $\Gamma(\alpha + \beta) = \Gamma(\alpha) + \Gamma(\beta) + E$ where E is in the image of the transfer $(1 + T)$. But $\Delta \times \mathrm{id}(1 + T) \cong 0$ so additivity follows. □

The P-power Operations for p Odd

There are essential differences here from the case of the squares. One starts, as before, with formal operations $D^i(\alpha)$ given by

Definition. Let X be a CW complex and $\alpha \in H^n(X; \mathbb{F}_p)$ then we set

$$(\Delta^p \times \mathrm{id})^*(\Gamma(\alpha)) = \sum D^i(\alpha) \otimes \Theta_{n-i}$$

where $\Theta_{n-i} = b^{(n(p-1)-i)/2}$ if i is even and is $eb^{(n(p-1)-i-1)/2}$ when i is odd.

Then one shows that these D^i operations commute with suspension up to a non-zero coefficient determined in (3.11) as above and that they satisfy the Cartan formula, and are linear. However, to show that most of them vanish, and to relate them to the P-power operations and mod (p) Bocksteins requires the extension of the construction from $X^p \times_{\mathbb{Z}/p} E_{\mathbb{Z}/p}$ to $X^p \times_{S_p} E_{S_p}$. It was the necessity of this extension which was one of the main reasons for Steenrod's interest in the cohomology of the symmetric groups. Finally, the remaining D_i's are still not quite the operations $P^i(\alpha)$ or $\beta P^i(\alpha)$, but have the form $\mu_{i,n} P^i(\alpha)$ and $\nu_{i,n} \beta P^i(\alpha)$ where $\mu_{i,n}$ and $\nu_{i,n}$ are non-zero constants made necessary again by the coefficient in (3.11).

We omit the details here as they are readily available in [SE] and our major interest in the remainder of the book is with the situation at the prime 2.

Chapter V. G-Complexes and Equivariant Cohomology

0. Introduction to Cohomological Methods

Let G be a finite group and X a space on which G acts. In this chapter we will describe a cohomological analysis of X which involves $H^*(G;\mathbb{F}_p)$ in a fundamental way. First developed by Borel and then by Quillen, this approach is the natural generalization of classical Smith Theory. After reviewing the basic constructions and a few examples, we will apply these techniques to certain complexes defined from subgroups of a group G, first introduced by K. Brown and D. Quillen. Using results due to Brown and P. Webb we will show how these complexes provide a systematic method for approaching the cohomology of simple groups which will be discussed later on. One of the aims of this chapter is to expose the reader to the important part played by group cohomology in the theory of finite transformation groups. By no means is this a complete account; in addition it requires a different background than the previous chapters have. We recommend the texts [AP], [Bre1] as excellent sources for those wanting to learn more about group actions.

For technical reasons we assume that the G-spaces which appear are CW-complexes with a compatible cellular G-action (G-CW complexes). Note that any compact manifold with a G-action can be given the structure of a finite G-CW complex. We denote the cellular complex of X by $C_*(X;\mathcal{R})$ ($\mathcal{R}$ the coefficient ring). We have an elementary lemma to begin

Lemma 0.1 *Each chain group $C_*(X;\mathcal{R})$ is isomorphic to a direct sum of permutation modules, i.e.*

$$C_i(X;\mathcal{R}) \cong \bigoplus_{\sigma_i \in X^{(i)}/G} \mathcal{R}_{\sigma_i} \otimes_{G_{\sigma_i}} \mathcal{R}G$$

where $\mathcal{R}_\sigma$ is the sign representation on $\mathcal{R}$ defined by the orientation on the cell σ.

Proof. This follows immediately from the decomposition of X into cellular orbits, keeping track of whether elements $g \in G$ preserve or reverse the orientation of the cells. □

Note the case in which the G-action is free (i.e. $G_x = \{1\}$ for all $x \in X$); then $C_*(X;\mathcal{R})$ will be a chain complex of **free** $\mathcal{R}G$-modules. Recall that EG denotes a contractible, free G-CW complex which is always infinite dimensional for G finite. We define

Definition 0.2 The *Borel construction* on a G-space X is the orbit space

$$X \times_G EG = (X \times EG)/G$$

where G acts diagonally on $X \times EG$.

For $X = *$, $X \times_G EG = BG$; the natural projection $X \times EG \longrightarrow EG$ induces a map

$$\text{0.3} \qquad \begin{array}{ccc} X & \hookrightarrow & X \times_G EG \\ & & \downarrow \\ & & BG\,. \end{array}$$

Because the action on EG is free, the vertical map is a fibration with fiber X. Hence, associated to (0.3) there is a spectral sequence with E_2-term

$$\text{0.4} \qquad E_2^{p,q} = H^p(BG;\mathcal{H}^q(X;\mathcal{R})) \Rightarrow H^{p+q}(X \times_G EG;\mathcal{R})$$

Here $\mathcal{H}^*(X;\mathbb{R})$ is a twisted coefficient system as $G = \pi_1(BG)$ may act non-trivially on it. The term $E_2^{p,q}$ may be identified with the group cohomology with coefficients, $H^p(G; H^q(X;\mathcal{R}))$.

Remark. This spectral sequence is originally due to J. Leray, H. Cartan, and R. Lyndon in various forms provided that G is discrete and the action is sufficiently reasonable. Many of the applications before Borel's work are discussed in chapters XV and XVI of [CE].

Definition 0.5 The *G-equivariant cohomology* of X is

$$H_G^*(X;\mathcal{R}) = H^*(X \times_G EG;\mathcal{R})\,.$$

Note that if X is a *free* G-complex, the map $X \times_G EG \longrightarrow X/G$ is also a fibration with contractible fiber, hence $X \times_G EG \simeq X/G$. Also note that the arguments given in Chapter V §5 can be adapted to show that if $H^*(X;\mathcal{R})$ is a finitely generated $\mathcal{R}$–module, then $H_G^*(X;\mathcal{R})$ is a finitely generated $\mathcal{R}$–algebra.

The analysis of the spectral sequence (0.4) was first undertaken by A. Borel. It yields numerous restrictions on finite group actions on familiar spaces such as spheres, projective spaces, etc., (see [Bo1]). We give two simple applications of these techniques to illustrate their utility.

Example 0.6 Let X be a finite complex with the mod p homology of a point, and assume that a finite p-group P acts on X. We will prove the following celebrated result due to P. Smith

Theorem. *Under the conditions above, $X^P \neq \emptyset$ and in fact X^P has the mod p cohomology of a point.*

Choose $\mathbb{Z}/p \cong C \subseteq P$ central; then $X^P = (X^C)^{P/C}$, and so by induction on $|P|$ it suffices to prove the result for $P \cong \mathbb{Z}/p$. In this case (0.4) becomes

$$E_2^{p,q} \cong \begin{cases} H^p(P;\mathbb{F}_p) & q = 0 \\ 0 & q > 0 \end{cases} \Rightarrow H^{p+q}(X \times_P EP;\mathbb{F}_p)\,.$$

Hence it collapses and so

$$H^{p+q}(X \times_P EP;\mathbb{F}_p) \cong H^{p+q}(P;\mathbb{F}_p)\,.$$

On the other hand, the inclusion $X^P \hookrightarrow X$ induces a map

$$j_P^*\colon H^*(X \times_P EP;\mathbb{F}_p) \longrightarrow H^*(X^P \times BP) \cong H^*(X^P) \otimes H^*(BP)\,.$$

In sufficiently high dimensions j_P^* will be an isomorphism because P acts freely off X^P. X is finite and so $H^*(X \times_P EP, X^P \times BP) = 0$ for $* \gg 0$. So for large $*$ we have $H^*(X^P) \otimes H^*(BP) \cong H^*(BP)$, which implies $X^P \sim_p *$.

This theorem also admits a very interesting extension due to T. Chang and T. Skjelbred, [CS] which is not needed in the sequel but is well worth pointing out.

Theorem. *Let $G = \mathbb{Z}/p$ and $K = \mathbb{F}_p$ or $G = S^1$ and $K = \mathbb{Q}$. Suppose that G acts on the compact Poincaré duality space X of formal dimension n. Then each connected component of the fixed set satisfies Poincaré duality over K and, if $G \neq \mathbb{Z}/2$, has formal dimension congruent to n* mod (2).

Example 0.7 Now let us assume G is a finite group acting freely on $X = S^n$ (the n-sphere). In this framework we can also recall the classical result due to P. Smith proved in Chapter IV:

Theorem. *If a finite group G acts freely on a sphere S^n, then all its abelian subgroups are cyclic.*

Remark. Recently this result has been extended to products of spheres of the same dimension, namely, if $(\mathbb{Z}/p)^r$ (p odd, or if $p = 2$ then $n \neq 1,3,7$) acts freely on $(S^n)^k$, then $r \leq k$. (See the papers by G. Carlsson [Ca] and Adem-Browder, [AB].)

At this point it is worthwhile to point out that all of the preceding constructions can be derived algebraically from $C_*(X)$. Let $\mathcal{F}_*$ be a $\mathbb{Z}G$-free resolution of the trivial G-module $\mathbb{Z}$. Consider the double co-complex

$$\mathrm{Hom}_G(\mathcal{F}_*, C^*(X;\mathcal{R})) .$$

Then it is not hard to show that

$$H^*_G(X;\mathcal{R}) \cong H^*(\mathrm{Hom}_G(\mathcal{F}_*, C^*(X;\mathcal{R}))) .$$

Note that we endow the double complex with the usual mixed differential:

$$\partial\partial \ \mathrm{Hom}_G(\mathcal{F}_r, C^s(X)) \longrightarrow \mathrm{Hom}_G(\mathcal{F}_{r+1}, C^s(X)) \oplus \mathrm{Hom}_G(\mathcal{F}_r, C^{s+1}(X))$$

which is given by

$$\partial(f) = f \cdot \partial_{\mathcal{F}}^{r+1} + (-1)^{r+s}\partial_C \cdot f .$$

For completeness we recall the Cartan-Eilenberg terminology

Definition 0.8 The *G-hypercohomology* of a G-cochain complex C^* is defined as

$$H^*(G;C^*) = H^*(\mathrm{Hom}_G(\mathcal{F}_*, C^*))$$

where $\mathcal{F}_*$ is a free (projective) resolution of the trivial $\mathbb{Z}G$-module $\mathbb{Z}$.

Hence we may say that the equivariant cohomology of a G-CW complex is isomorphic to the hypercohomology of its cellular cochain complex. This has certain technical advantages, above all if a specific cellular decomposition is available. In addition we may filter the double complex in (0.8) using either of two filtrations associated to the "vertical" and "horizontal" directions respectively. This yields two spectral sequence which applied to $C^*(X;\mathcal{R})$ become

$$\left.\begin{array}{l}(I)\ E_2^{p,q} = H^p(G; H^q(X;\mathcal{R})) \\ (II)\ E_1^{p,q} = H^q(G; C^p(X,\mathcal{R}))\end{array}\right\} \Rightarrow H_G^{p+q}(X;\mathcal{R}) .$$

The spectral sequence (I) is canonically isomorphic to (0.4). As for (II), this can be identified with the E_1-term of the Leray spectral sequence associated to

$$\begin{array}{c} X \times_G EG \\ \big\downarrow \\ X/G . \end{array}$$

In the general situation the E_1-term is not so easy to handle, but the E_2-term can be identified with

$$E_2^{p,q} = H^p(X/G; \mathcal{H}_G^q) \Rightarrow H_G^{p+q}(X)$$

where $\mathcal{H}_G^*$ is the sheaf on X/G associated to the presheaf $V \mapsto H_G^*(\pi^{-1}V)$, $\pi: X \times_G EG \to X/G$. In the case of a finite G-CW complex (assuming constant isotropy on the cells) the d_1-differential is just the map induced by the coboundary operator on $C^*(X)$ and $E_1^{p,q} = H^q(G, C^p(X))$.

Recall that in Chapter II we defined Tate cohomology of $\mathbb{Z}G$-modules; instead of an ordinary projective resolution we used a complete resolution i.e. an acyclic $\mathbb{Z}$-graded complex of projective $\mathbb{Z}G$-modules which in non-negative degrees coincides with an ordinary projective resolution of $\mathbb{Z}$. Taking such a complete resolution $\mathcal{F}_*$, we can define Tate hypercohomology groups

$$\hat{H}^*(G; C^*) \cong H^*(\mathrm{Hom}(\mathcal{F}_*; C^*)) .$$

In case $C^* = C^*(X)$, we obtain the equivariant Tate cohomology of X, first introduced by R.G. Swan [S3],

$$\hat{H}^*(G; C^*(X; \mathbb{Z})) = \hat{H}_G^*(X) . \tag{0.9}$$

The main technical advantage is the disappearance of the orbit space for free actions:

Theorem 0.10 *If X is a finite dimensional free G-CW complex then*

$$\hat{H}_G^*(X) \equiv 0 .$$

Proof. There is a spectral sequence analogous to (II), say

$$(\widehat{II}) \quad \hat{E}_2^{p,q} = \hat{H}^q(G, C^p(X)) \Rightarrow \hat{H}_G^{p+q}(X) .$$

This converges because X is finite dimensional. Now each $C^p(X)$ is G-free, hence G-acyclic, so $\hat{E}_2^{p,q} \equiv 0$ and the result follows. □

Note how the analogous spectral sequence $(\hat{I})$ will not involve the orbit space, hence strengthening cohomological arguments.

1. Restrictions on Group Actions

In this section we will outline some instances of how the machinery described in §0 can be applied to transformation groups. As this is not our main topic of interest, we urge the reader to consult the original sources for the foundational results [Bo1], [Q1].

The first result to take note of is the localization theorem. This result shows that for certain groups the equivariant cohomology contains substantial information about the fixed point set after inverting certain cohomology classes, and hence this makes the entire spectral sequence approach quite powerful. More precisely we have the localization theorem of Borel and Quillen

Theorem 1.1 *Let $G = (\mathbb{Z}/p)^r$ (p prime) act on a finite complex X. Then the inclusion of the fixed point set $X^G \hookrightarrow X$ induces an isomorphism*

$$H^*(X \times_G EG; \mathbb{F}_p)[e_A^{-1}] \xrightarrow{\cong} H^*(X^G \times BG; \mathbb{F}_p)[e_A^{-1}]$$

where $e_A \in H^{2p^r-2}(G; \mathbb{F}_p)$ is the product of the Bocksteins of the non–zero elements in $H^1(G, \mathbb{F}_p) \cong \mathrm{Hom}(G, \mathbb{F}_p)$, and $[e_A^{-1}]$ denotes localization by the multiplicative system of powers of e_A.

Note that here we are considering equivariant cohomology as a graded $H^*(G; \mathbb{F}_p)$-module.)

The next theorem, again due to D. Quillen, is an important structural result for equivariant cohomology, which can be proved using finite generation arguments as in IV.5 (see [Q1]).

Theorem 1.2 *Let X be a finite dimensional G-complex and denote*

$$p_G(X)(t) = \sum_{i=0}^{\infty} \dim(H^i(X \times_G EG; \mathbb{F}_p))t^i \ .$$

Then $p_G(X)(t)$ is a rational function of the form $p(t)/\prod_{i=1}^{n}(1-t^{2n})$ and the order of the pole at $t = 1$ is equal to

$$\max\{n | (\mathbb{Z}/p)^n \subseteq G \textit{ fixes a point } x \in X\} \ .$$

This is the version of the result for G-spaces which we discussed previously for the case $X = \mathrm{pt}$. This result was the starting point for the idea of introducing varieties associated to cohomology rings. As we have seen, this may be considered as a special case of $H^*(G; M)$ where M is the cellular complex associated to a G-space. This theorem however, has its natural extension to any coefficient module M. In other words, the asymptotic growth rate of $H^*(G; M)$ (known as the complexity of M) is determined on the p-elementary abelian subgroups of G. The proof of this (due to Alperin and Evens [AE]) is an algebraic formulation of Quillen's result, and has important applications in modular representation theory.

The above results are not too interesting in the case of free actions. For this situation Tate cohomology comes in handy because we obtain a spectral sequence which abuts to a zero term. First a

Definition 1.3

i. Let A be a finite abelian group. We define its *exponent*, $exp(A) = \min\{n \in \mathbb{N} | n \cdot A = 0\}$

ii. Given $x \in A$ we define

$$exp(x) = exp(\langle x \rangle) .$$

Using restriction and transfer it is elementary to see that $|G| \cdot \hat{H}^*_G(X) \equiv 0$ for any finite dimensional G-complex X; hence Tate cohomology provides an interesting sequence of $\mathbb{Z}$-graded torsion modules. For free actions we have a theorem of W. Browder, [Brow],

Theorem 1.4 *Let X be a free, connected G-CW complex. Then*

$$|G| \left| \prod_{i=1}^{\infty} exp(\hat{H}^{-i-1}(G; H^i(X;\mathbb{Z}))) \right. .$$

Proof. The proof we give is from [A2]. Consider the spectral sequence

$$\hat{E}_2^{p,q} = \hat{H}^p(G; H^q(X;\mathbb{Z})) \Rightarrow \hat{H}_G^{p+q}(X) \equiv 0 .$$

Look at the $E_r^{0,0}$-terms; we have exact sequences

$$E_{r+1}^{-r-1,r} \longrightarrow E_{r+1}^{0,0} \longrightarrow E_{r+2}^{0,0} \longrightarrow 0$$

for $r = 1, 2, \ldots, \dim(X)$. From this we obtain

$$exp(E_{r+1}^{0,0})/exp(E_{r+2}^{0,0}) \left| exp(E_{r+1}^{-r-1,r}) \right. .$$

Multiplying all these out we obtain

$$exp(\hat{H}^0(G;\mathbb{Z})) = |G| \left| \prod exp(E_{r+1}^{-r-1,r}) \right| \prod_{i=1}^{\infty} exp(\hat{H}^{-r-1}(G; H^r(X;\mathbb{Z})))$$

completing the proof. □

Remark. Tensoring $C^*(X;\mathbb{Z})$ with a torsion-free $\mathbb{Z}G$-module M and applying the same proof yields

$$exp(\hat{H}^0(G;M)) \left| \prod_{i=1}^{\infty} exp(\hat{H}^{-r-1}(G; H^r(X) \otimes M)) \right. .$$

This result has a few interesting consequences which we now briefly describe.

Corollary 1.5 *If $(\mathbb{Z}/p)^r$ acts freely on a connected complex X with homologically trivial action, then at least r of the cohomology groups $\tilde{H}^*(X;\mathbb{Z}_{(p)})$ must be non-zero.*

Proof. Under the above hypotheses, if $G = (\mathbb{Z}/p)^r$ then

$$p \cdot \hat{H}^*(G; H^*(X; \mathbb{Z}_{(p)})) = 0 .$$

(Kunneth formula.) □

In particular this shows that $(\mathbb{Z}/p)^{r+1}$ cannot act freely, homologically trivially on $(S^n)^r$.

Corollary 1.6 *For a finite group G, let $n(G) = \min\{n|G$ acts feeely, homologically trivially on an n-dimensional connected complex $\}$. Then $n(G) \geq \max_{p||G|}\{p\text{-rank}(G)\} + 1$.*

Corollary 1.7 *Let M be a finitely generated torsion free $\mathbb{Z}G$-module, G a finite group. Then there exists an integer N (depending only on G) such that*

$$\bigoplus_{i=1}^{N} \hat{H}^{*+i}(G; M) \neq 0$$

for all $* \in \mathbb{Z}$, or else $\hat{H}^*(G; M) \equiv 0$.

Proof. Take a faithful unitary representation

$$\phi\colon\ G \hookrightarrow U(n) .$$

Then G will act freely and cohomologically trivially on $U(n)$, which has the homology type of $S^1 \times S^3 \times \cdots \times S^{2n-1}$. From the remark after (1.4) we see that

$$exp(\hat{H}^0(G; M)) \,\Big|\, \prod_{r=1}^{n^2} exp(\hat{H}^{-r-1}(G; M)) \ .$$

Using M^* instead of M and identifying $\hat{H}^i(G; M) \cong \hat{H}^{-i}(G; M^*)$ (Tate duality) yields

$$exp(\hat{H}^0(G; M)) \,\Big|\, \prod_{r=1}^{n^2} exp(\hat{H}^{r+1}(G; M)) \ .$$

By dimension shifting this can be generalized to

$$exp(\hat{H}^k(G; M)) \,\Big|\, \prod_{r=1}^{n^2} exp(\hat{H}^{r+k+1}(G; M)) \qquad \text{for all } k \in \mathbb{Z} .$$

Suppose now that for some $k \in \mathbb{Z}$, $\hat{H}^{r+k+1}(G; M) = 0$ for $r = 0, 1, \ldots, n^2$. Then $\hat{H}^k(G; M) = 0$ and consequently $\hat{H}^*(G; M) = 0$ for $* \leq k + n^2 + 1$. On the other hand, using the dual again yields

$$exp(\hat{H}^{-k}(G;M)) \Bigg| \prod_{r=1}^{n^2} exp(\hat{H}^{-r-k-1}(G;M)) \qquad \text{for all } k \in \mathbb{Z}\,.$$

Dimension shifting yields

$$exp(\hat{H}^{k+n^2+2}(G;M)) \Bigg| \prod_{r=1}^{n^2} exp(\hat{H}^{k+r}(G;M))\ .$$

As before we get $\hat{H}^*(G;M) = 0$ for $* \geq k + n^2 + 1$ and we conclude that $\hat{H}^*(G;M) \equiv 0$. □

Remark A different proof of this result appeared in [BCR] using purely algebraic methods. Notice that the restrictions on M are not important as any finitely generated $\mathbb{Z}G$-module is cohomologous to a torsion free one.

Recall (see [MS]) that

$$H^*(BU(n), \mathbb{Z}) \cong \mathbb{Z}[x_1, x_2, \ldots, x_n]$$

where each x_i is 2i–dimensional. Now given a representation $\phi: G \hookrightarrow U(n)$ as before, the polynomial generators $x_i \in H^{2i}(BU(n);\mathbb{Z})$ can be pulled back under ϕ to define

$$\phi^*(x_{2i}) = c_i(\phi) \in H^{2i}(G;\mathbb{Z}) \qquad i = 1, \ldots, n$$

the ith Chern class of ϕ. Under this map, the cohomology of G is a finitely generated $H^*(BU(n);\mathbb{Z})$–module. These classes naturally carry some torsion; the following quantifies this.

Theorem 1.8 *Let G be a finite group and $\phi: G \hookrightarrow U(n)$ a faithful unitary representation of G with Chern clases $c_1(\phi), \ldots, c_n(\phi)$; then*

$$|G| \Bigg| \prod_{i=1}^{n} exp(c_i(\phi))\ .$$

Proof. Consider the Borel construction on $U(n)$ and its associated spectral sequence

$$E_2^{p,q} = H^p(G; H^q(U(n))) \Rightarrow H^{p+q}(U(n)/G)\,.$$

Now $H^*(U(n);\mathbb{Z}) \cong \Lambda_{\mathbb{Z}}(x_1, \ldots, x_{2n-1})$ and by construction these classes transgress down to $c_i(\phi) \in H^{2i}(G;\mathbb{Z})$ for $i = 1, \ldots, n$. This implies that $[exp(c_i(\phi))] \cdot x_{2i-1} \in E_2^{0,2i-1}$ is a permanent cocycle for $i = 1, \ldots, n$, which implies that $[\prod exp(c_i(\phi))] \cdot \mu_{U(n)} \in \text{im}(i^*)$ where $\mu_{U(n)}$ is the top cohomology class on the fiber and $i: U(n) \to U(n)/G$ is the projection. However, i is a covering map of oriented manifolds, hence it has degree $|G|$ on the top class from which the result follows. □

Example 1.9 We now compute the cohomology of $\mathcal{Q}_8$ using the action on S^3 as a subgroup. The spectral sequence (I) with integral coefficients for this action degenerates into the long exact sequence (for $i \geq 0$)

$$\ldots \to H^{i-4}(\mathcal{Q}_8) \to H^i(\mathcal{Q}_8) \to H^i(S^3/\mathcal{Q}_8) \to H^{i-3}(\mathcal{Q}_8) \to \ldots .$$

From this we deduce that $H^2(S^3/\mathcal{Q}_8) \cong H^2(\mathcal{Q}_8) \cong \mathbb{Z}/2 \times \mathbb{Z}/2$, and as the action is free, $H^4(\mathcal{Q}_8) \cong \mathbb{Z}/8$. Hence we have that

$$H^*(\mathcal{Q}_8, \mathbb{Z}) \cong \mathbb{Z}[\alpha_2, \beta_2, \gamma_4]/\mathcal{R}$$

where $\mathcal{R}$ is the set of relations $2\alpha, 2\beta, 8\gamma, \alpha\beta, \alpha^2, \beta^2$.

A similar analysis yields

$$H^*(\mathcal{Q}_8, \mathbb{F}_2) \cong \mathbb{F}_2[x_1, y_1, z_4]/\mathcal{R}'$$

and the relations in this case are $x^2 + xy + y^2, x^2y + xy^2$. These are obtained easily using the additive structure (computed in Ch.IV) and symmetry considerations.

2. General Properties of Posets Associated to Finite Groups

For the remainder of this chapter we will specialize to certain G-complexes which are defined from the lattice of subgroups in G. The point of this is that their equivariant structure is an important device for tackling the p-local structure of G and its mod p cohomology.

As before let G be a finite group and consider the collection $S_p(G)$ which is the set of all finite p-subgroups of G which are non-trivial. Under inclusion this becomes a partially ordered set (poset for short) endowed with a natural G-action induced by conjugation. In the usual way we can associate a G-simplicial complex to it denoted by $|S_p(G)|$. We recall how this is constructed: its vertices are the elements of $S_p(G)$ and its simplices are the non-empty finite chains in $S_p(G)$. Note that the isotropy subgroups of the vertices are the normalizers of the corresponding p-subgroups of G. Similarly we denote by $A_p(G)$ the poset of p-elementary abelian subgroups of G which are not trivial, and its associated space by $|A_p(G)$.

These two functors

$$\{\text{Finite groups}\} \xrightarrow{|A_p()|, |S_p()|} \{G\text{-simplicial sets}\}$$

were first introduced by K. Brown, [Brown], and D. Quillen, [Q5]. The objective was to construct a natural complex which distilled the p-local structure of the group G as well as to provide analogues of Tits buildings for general finite groups.

Given a map $f: X \longrightarrow Y$ of posets and $y \in Y$ we define

$$\begin{aligned} f|_y &= \{x \in X | f(x) \leq y\} \\ {}_y|f &= \{x \in X | f(x) \geq y\}. \end{aligned}$$

The following result is important for determining the homotopy type of posets. Let $f: X \longrightarrow Y$ as above:

Proposition 2.1 *Assume $|f|_y|$ is contractible for all $y \in Y$ (respectively $|_y|f|$ is contractible for all $y \in Y$). Then $|f|$ is a homotopy equivalence.*

Proof. We sketch the proof in the simply connected case; fundamental groups must be dealt with using twisted coefficients. For more on this, see [Q]. Consider the Leray spectral sequence associated to $|X| \xrightarrow{f} |Y|$:

$$E^2_{p,q} = H_p(|Y|; \mathcal{H}_q) \Rightarrow H_{p+q}(|X|; \mathbb{Z})$$

where $\mathcal{H}_q$ is the sheaf which comes from the pre-sheaf associating to any open subset $\mathcal{U} \subset |Y|$ the group

$$H_q(|f^{-1}|(\mathcal{U}; \mathbb{Z})) .$$

As f is a map of posets $|f|$ is a map of simplicial complexes and the sheaf can be computed combinatorially. In fact we have that $\mathcal{H}_q$ can be identified with the local coefficient system $y \mapsto H_q(|f_y|)$. By hypothesis we have that $f|_y$ is contractible for all $y \in Y$ and hence the E^2-term becomes

$$E^2_{p,q} = \begin{cases} H_p(|Y|; \mathbb{Z}) & q = 0 \\ 0 & \text{otherwise} \end{cases}$$

and it follows that f induces a homology isomorphism of simply connected CW-complexes. By Whitehead's theorem this implies that $|f|$ is a homotopy equivalence. In the case ${}_y|f$ contractible the above result is proved using X^{op}, Y^{op} instead. □

A subset K of a poset X is said to be closed if $x' \leq x \in K$ implies $x' \in K$. Let Z be a closed subset of a product of posets $X \times Y$; and $p_1: Z \longrightarrow X$, $p_2: Z \longrightarrow Y$ the projections. We can identify the fiber of p_1 over $x \in X$ with the subposet

$$Z_x = \{y \in Y | (x, y) \in Z\};$$

similarly for p_2 we have

$$Z_y = \{x \in X | (x, y) \in Z\} .$$

Then we have

Lemma 2.2 *If Z_x is contractible for each $x \in X$ then $p_1: Z \longrightarrow X$ is a homotopy equivalence.*

Proof. ${}_x|p_1 = \{(x', y) \in Z | x' \geq x\}$; define $\phi: Z_x \longrightarrow {}_x|p_1$ and $\nu: {}_x|p_1 \longrightarrow Z_x$ by $\phi(y) = (x, y)$, $\nu(x', y) = y$. Then $\nu \cdot \phi(y) = y$, $\phi \cdot \nu(x', y) \leq (x', y)$.

If two maps of posets $f, g: W_1 \longrightarrow W_2$ satisfy $f(w) \leq g(w)$ for all $w \in W_1$ then it is easy to verify that $|f|$ and $|g|$ are homotopic (indeed, f and g determine a map $\{0 < 1\} \times W_1 \longrightarrow W_2$ and $|\{0 < 1\}|$ is a 1-simplex).

Hence ν and ϕ are homotopy inverses and so ${}_x|p_1$ is contractible for all $x \in X$ and it follows that $|Z| \simeq_{p_1} |X|$. □

Corollary 2.3 *If Z_x, Z_y are contractible for each $x \in X$ and $y \in Y$, then $|X|$ and $|Y|$ are homotopy equivalent.*

Proof. By 2.2 $|X| \simeq |Z| \simeq |Y|$. □

We can now analyze the homotopy theoretic properties of the poset spaces $S_p(G)$ and $A_p(G)$. To begin we have

Lemma 2.4 *If P is a non-trivial p-group then $A_p(P)$ is contractible.*

Proof. Let $B \subseteq P$ be the subgroup of the center of P consisting of the identity and all the elements of order precisely p; then $B > 1$ since $P > 1$. Hence, in $A_p(G)$ we have $G \leq AB \geq A$ for all $A \in A_p(P)$, so $|\mathrm{id}_{A_p(P)}| \simeq ct$ and $A_p(P) \simeq *$. □

We use this to prove

Proposition 2.5 *The inclusion $i: A_p(G) \hookrightarrow S_p(G)$ is a homotopy equivalence.*

Proof. Let $P \in S_p(G)$, then $i|_P = A_p(P)$, a contractible poset, and hence $|A_p(G)| \simeq |S_p(G)|$. □

Next we have

Proposition 2.6 *If G has a non-trivial normal p-subgroup then $A_p(G)$ is contractible.*

Proof. We show $S_p(G)$ is contractible. If $1 < K \triangleleft G$, then for any $P \in S_p(G)$ we have $P \leq PK \geq K$ in $S_p(G)$, hence $|S_p(G)| \simeq *$. □

Now if K is a non-trivial normal p-subgroup of G then the subgroup E of elements of order 1 or p in the center of K is characteristic in K, hence E is a normal p-torus in G. In this case 2.6 can be reformulated as the statement that if $|A_p(G)|^G \neq \emptyset$ then $|A_p(G)|$ is contractible. We note that the converse of this is still an open problem due to D. Quillen:

Conjecture. *If $A_p(G)$ is contractible then G has a non-trivial normal p-subgroup.*

The cases $rk_p(G) = 1, 2$ have been settled affirmatively since then $A_p(G)$ is a finite set of points and a graph (respectively) in which case contractibility always implies a G-fixed point.

Now suppose that G is the finite group of rational points of a semisimple algebraic group defined over a finite field $\mathbb{K}$ of characteristic p. Then one may associate a "building" Π to G in the sense of Tits [T]. It is a simplicial complex of dimension $m-1$ with a G-action where m is the rank of the underlying algebraic group over $\mathbb{K}$. Let $\Delta \subset \Pi$ be a simplex; the correspondence

$$\Delta \mapsto \text{stabilizer of } \Delta$$

estabishes a contravariant isomorphism between the poset of simplices in Π and parabolic subgroups in G. $A_p(G)$ is closely related to this complex.

Theorem 2.7 *The poset $A_p(G)$ is homotopy equivalent to the building Π. Hence $A_p(G)$ has the homotopy type of a bouquet of $(m-1)$-dimensional spheres.*

Proof. We provide the proof for the simple case $G = \mathrm{SL}_n(\mathbb{K})$. Let X be the poset of simplices in Π and define $Z \subset X \times A_p(G)$ as

$$Z = \{(x, A) | x \in X^A\} = \{(x, A) | A \subseteq G_x\}.$$

For $G = \mathrm{SL}_n(\mathbb{K})$, Π can be identified with the simplicial complex $|T(\mathbb{K}^n)|$ where $T(\mathbb{K}^n)$ denotes the poset of proper subspaces of $\mathbb{K}^n$. Note that $m = n-1$, differs from $rk_p(G)$. Now it is clear that an element of G_x leaves **each point** of x fixed, hence $x' \leq x$ implies $G_{x'} \supset G_x$ and so Z is closed. Hence we need only show that $Z_x = A_p(G_x)$ and $Z_A = X^A$ are contractible.

Consider a simplex x; it is a flag

$$0 < V_1 < \cdots < V_r < \mathbb{K}^n, \qquad r \geq 1$$

with stablizer G_x. The elements of G_x which induce the identity on each quotient of the flag form a non-trivial normal p-subgroup of G_x, hence $|A_p(G_x)| \simeq *$. Now let Π^H be the simplicial complex associated to the poset $T(\mathbb{K}^n)^H$ of proper H-invariant subspaces of $\mathbb{K}^n$. If H is a p-subgroup of G, $W^H > 0$ for $W \in T(\mathbb{K}^n)^H$, hence this poset is contractible; $W \geq W^H \leq (\mathbb{F}^n)^H$. Thus the X^A are contractible and we conclude $|A_p(G)| \simeq \Pi$. □

We now introduce the following notation for a G-complex X: the *singular set* of the action is, by definition, the subcomplex

$$\mathcal{S}_G(X) = \{x \in X | G_x \neq \{1\}\} .$$

We have a key lemma,

Lemma 2.8 *Let $H \subseteq G$ be a non-trivial p-subgroup. Then $|A_p(G)|^H$ is contractible.*

Proof. Let A be a non-trivial p-torus in G normalized by a p-group $H \subseteq G$; then $A^H \neq \{1\}$. Denote by E the p-torus of central elements of order dividing p in H. We have

$$A^H \leq A, \qquad A^H \leq A^H E, \qquad E \leq A^H E$$

whence the result follows. □

We can now prove the following proposition which will be vital for our later study of the cohomology of simple groups.

Theorem 2.9 *Let $P = \mathrm{Syl}_p(G)$; then $\mathcal{S}_P(|A_p(G)|)$ is contractible.*

Proof. Replace $|A_p(G)|$ by its barycentric subdivision $|X|$, where X is the poset of simplices in $|A_p(G)|$. Then $\mathcal{S}_P(|X|)$ is the subcomplex

$$\bigcup_{H \leq P} |X|^H = \bigcup_{H} |X^H| = |\mathcal{S}_P(A_p(G))| .$$

As before, let $Z \subset A_p(P) \times \mathcal{S}_P(A_p(G))$ be the closed subset consisting of pairs (H, x) where $x \in X^H$ or equivalently $H \subseteq P_x$.

By definition of the singular set $P_x \neq \{1\}$ for all $x \in \mathcal{S}_P(A_p(G))$; hence $Z_y = A_p(P_x)$ is contractible by 2.4. Also, if $A \in A_p(P)$ then $Z_A = X^A$, the poset of simplices in $|A_p(G)^A|$. However, we have seen in 2.8 that this is contractible. Hence we have shown

$$|\mathcal{S}_P(A_p(G))| \simeq \mathcal{S}_P(|A_p(G)|) \simeq |A_p(P)| \simeq * .$$ □

Corollary 2.10 $\chi(|A_p(G)|) \equiv 1$ *mod* $|\mathrm{Syl}_p(G)|$.

Proof. For any finite G-complex we have

$$\chi(X) \equiv \chi(\mathcal{S}_G(X)) \bmod |G| .$$

The result follows from applying this to $X = |A_p(G)|$, $G = \mathrm{Syl}_p(G)$. □

To conclude this section we provide some examples of posets $A_p(G)$.

Examples 2.11

$G = \mathcal{S}_4$, the Symmetric Group on 4-Letters,

$A_2(G)/G$:

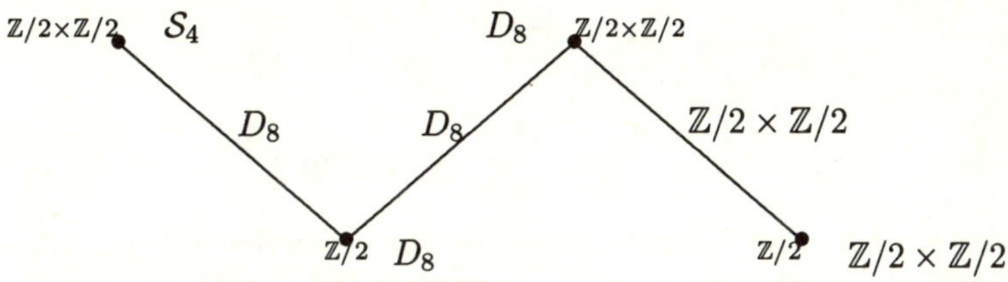

$G = L_3(\mathbb{F}_2)$, $p = 2$.

In this case $|T(\mathbb{F}_2^3)| \simeq |A_2(G)| \simeq \bigvee_1^8 S^1$ is a trivalent graph. The parabolics are

$$P_1 = \left\{\begin{pmatrix} 1 & * & * \\ 0 & * & * \\ 0 & * & * \end{pmatrix}\right\} \cong \mathcal{S}_4 \cong \left\{\begin{pmatrix} * & * & * \\ * & * & * \\ 0 & 0 & 1 \end{pmatrix}\right\} = P_2$$

and the Borel subgroup is

$$B = \left\{\begin{pmatrix} 1 & * & * \\ 0 & 1 & * \\ 0 & 0 & 1 \end{pmatrix}\right\} \cong D_8 .$$

G acts edge transitively on Π with quotient

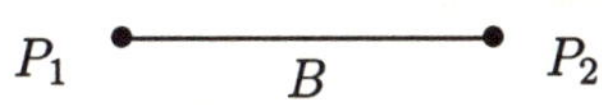

It is known that $H_1(T;\mathbb{Z}/2) \cong St$ the Steinberg module, which is projective of rank 8 as an $\mathbb{F}_2G$-module.

$G = M_{11}$, $p = 2$,

$A_2(M_{11})/M_{11}$:

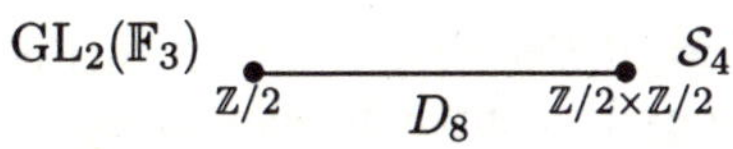

$G = J_1$, $p = 2$,

$A_2(J_1)/J_1$:

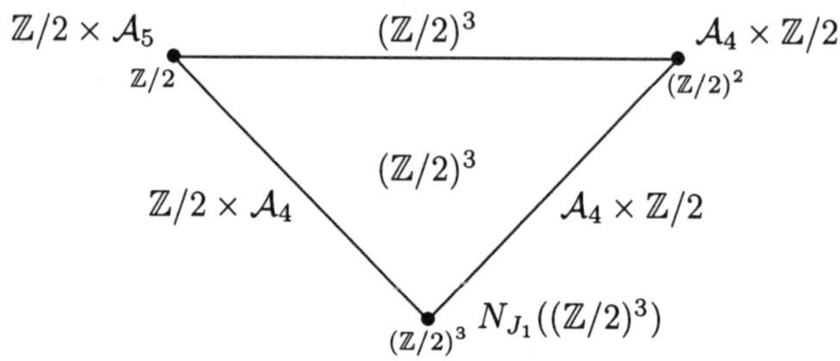

We point out here that from the above one can verify that this poset has negative Euler characteristic, hence its 1–dimensional homology is non–trivial. It is an example of a non–spherical poset space.

3. Applications to Cohomology

We have seen in §2 that we can associate a finite G-CW-complex $X = |A_p(G)|$ to any finite group which satisfies the condition that the fixed point set $|A_p(G)|^H$ is contractible for any p-subgroup $H \subseteq G$. In this section we will show how this very strong condition allows one to extract important and useful cohomological information about G as long as we concentrate only on the prime p.

To begin we have a result of K. Brown,, [Brown, pg. 293, Theorem X.7.3],

Theorem 3.1 *The map $\hat{H}^*(G;\mathbb{F}_p)\longrightarrow\hat{H}^*_G(|A_p(G)|)$ induced by $|A_p(G)|\longrightarrow$pt is an isomorphism for all $* \in \mathbb{Z}$.*

Proof. We have a short exact sequence of G-cochain complexes

$$0\longrightarrow\mathbb{F}_p \xrightarrow{\epsilon} C^*(|A_p(G)|;\mathbb{F}_p)\longrightarrow\tilde{C}^*(|A_p(G)|;\mathbb{F}_p)\longrightarrow 0\,.$$

This induces a long exact sequence

$$\cdots\longrightarrow\hat{H}^i(G;\mathbb{F}_p)\xrightarrow{\epsilon^*_G}\hat{H}^i_G(X;\mathbb{F}_p)\longrightarrow\hat{H}^i_G(\tilde{C}^*)\longrightarrow\hat{H}^{i+1}(G;\mathbb{F}_p)\longrightarrow\cdots\,,$$

where $X = |A_p(G)|$. Now, if $P = \mathrm{Syl}_p(G)$ recall that $\mathcal{S}_P(X) \simeq *$ and furthermore, as Tate cohomology is identically zero on free complexes we have

$$\hat{H}^*_P(X;\mathbb{F}_p) \cong \hat{H}^*_P(\mathcal{S}_P(X);\mathbb{F}_p) \cong \hat{H}^*(P;\mathbb{F}_p)\,.$$

As this isomorphism is clearly induced by the augmentation we deduce that

$$\hat{H}^*_P(\tilde{C}(X);\mathbb{F}_p) \equiv 0 \qquad \text{for all } * \in \mathbb{Z}\,.$$

However, using restriction-transfer we have

$$\hat{H}^*_G(\tilde{C}(X);\mathbb{F}_p) \underset{\text{res}}{\hookrightarrow} \hat{H}^*_P(\tilde{C}(X);\mathbb{F}_p)$$

($[G{:}\,P]$ prime to p). Hence $\hat{H}^*(G;\tilde{C}^*) \equiv 0$ and so ϵ^*_G is an isomorphism. □

What this illustrates is that the equivariant structure of $A_p(G)$ determines the cohomology of G for p-torsion coefficients. The next result, a theorem of P. Webb,, [We], makes this more precise.

Theorem 3.2 *In the Leray spectral sequence associated to* $|A_p(G)| \times_G EG \longrightarrow |A_p(G)|/G$ *with* $\mathbb{F}_p$ *coefficients we have*

$$E_2^{p,q} \cong \begin{cases} 0 & p > 0, \\ H^q(G;\mathbb{F}_p) & p = 0. \end{cases}$$

Proof. Our proof of this result requires standard techniques in equivariant topology; a good general reference for this material is [Bre2]. Recall that $|A_p(G)|$ is a finite G-CW complex with constant isotropy on each cell. Under these conditions the E_1-term of the above spectral sequence can easily be identified with

$$E_1^{s,q} = H^q(G;C^s(|A_p(G)|;\mathbb{F}_p)) = \bigoplus_{\sigma_s \in |A_p(G)|/G} H^q(G_\sigma)$$

and d_1 is the differential induced by the coboundary map on C^*. Define for any $H \subseteq G$

$$D^*_H(q) = H^q(H;C^*(|A_p(G)|;\mathbb{F}_p)) \,.$$

This is a finite cochain complex and (3.2) is equivalent to proving that for all $q \geq 0$

$$H^i(D^*_G(q)) = \begin{cases} H^q(G;\mathbb{F}_p) & i = 0, \\ 0 & \text{otherwise.} \end{cases}$$

Now we have (for all $q > 0$) a split monomorphism of cochain complexes

$$D^*_G(q) \hookrightarrow D^*_P(q) \qquad (P = \mathrm{Syl}_p(G))$$

and so $D^*_P(q) \simeq D^*_G(q) \oplus D'$. However, if we define

$$E^*_P(q) = H^q(P;C^*(\mathcal{S}_P(|A_p(G)|);\mathbb{F}_p)) \,,$$

we see that $E^*_P(q) \cong D^*_P(q)$, hence $D^*_G(q)$ is a direct summand in $E^*_P(q)$. It is now clear that it suffices to prove the claim for the cochain complex $E^*_P(q)$. For this note that

$$\mathcal{S}_P(|A_p(G)|) \longrightarrow *$$

induces homotopy equivalences

$$\mathcal{S}_P(|A_p(G)|)^H \longrightarrow *$$

for **all** $H \subseteq P$. Hence this complex is **equivariantly** contractible and so we have an isomorphism of spectral sequences at the E_2-level

$$\begin{array}{ccc} E_2^{p,q} & = & H^p(\mathcal{S}_P(|A_p(G)|)/P; \mathcal{H}^q) \\ & & \big\downarrow \cong \\ E_2^{p,q} & = & H^q(*; \mathcal{H}^q) \end{array}$$

which completes the proof for $q > 0$. However one observes that the argument above holds if we use Tate Cohomology and the corresponding spectral sequence, for all $q \geq 0$. Using the fact that for any finite group G with order divisible by p we have $H^0(G, \mathbb{F}_p) \cong \widehat{H}^0(G, \mathbb{F}_p)$ completes the proof of (3.2).

□

Corollary 3.3

$$H^*(G; \mathbb{F}_p) \oplus \left(\bigoplus_{\substack{\sigma_i \in A_p(G)/G \\ i \text{ odd}}} H^*(G_{\sigma_i}; \mathbb{F}_p) \right) \cong \bigoplus_{\substack{\sigma_i \in A_p(G)/G \\ i \text{ even}}} H^*(G_{\sigma_i}; \mathbb{F}_p)$$

for all $* \geq 0$.

Remark. This extends easily to arbitrary twisted coefficients M and all $* \in \mathbb{Z}$ using Tate cohomology instead in the proof.

We may apply this to the examples in §2.

Examples 3.4 (all for p=2)

$\mathcal{S}_4$

For $\mathcal{S}_4$ we obtain nothing new

$$\begin{aligned} &H^*(\mathcal{S}_4) \oplus H^*(D_8) \oplus H^*(D_8) \oplus H^*((\mathbb{Z}/2)^2) \\ &\qquad \cong H^*(\mathcal{S}_4) \oplus H^*(D_8) \oplus H^*(D_8) \oplus H^*((\mathbb{Z}/2)^2). \end{aligned}$$

$\mathbf{SL_3(\mathbb{F}_2)}$

For $G = \mathrm{SL}_3(\mathbb{F}_2)$ we get

$$H^*(\mathrm{SL}_3(\mathbb{F}_2)) \oplus H^*(D_8) \cong H^*(\mathcal{S}_4) \oplus H^*(\mathcal{S}_4)\ .$$

The Sporadic Group M_{11}

For $G = M_{11}$ we get

$$H^*(M_{11}) \oplus H^*(D_8) \cong H^*(\mathrm{GL}_2(\mathbb{F}_3)) \oplus H^*(\mathcal{S}_4) .$$

The Sporadic Group J_1

For $G = J_1$ we get

$$\begin{aligned} &H^*(J_1) \oplus H^*((\mathbb{Z}/2)^3) \oplus H^*(\mathbb{Z}/2 \times \mathcal{A}_4) \oplus H^*(\mathbb{Z}/2 \times \mathcal{A}_4) \\ &\quad \cong H^*((\mathbb{Z}/2)^3) \oplus H^*(\mathbb{Z}/2 \times \mathcal{A}_5) \oplus H^*(\mathcal{A}_4 \times \mathbb{Z}/2) \oplus H^*(N_{J_1}((\mathbb{Z}/2)^3)). \end{aligned}$$

Using the fact that $H^*(\mathcal{A}_4) \cong H^*(\mathcal{A}_5)$ at $p = 2$ we recover the result of II.6.9,

$$H^*(J_1) \cong H^*(N_{J_1}((\mathbb{Z}/2)^3)) ,$$

where $N_{J_1}((\mathbb{Z}/2)^3)$ is a group of order 168.

Chapter VI. The Cohomology of Symmetric Groups

0. Introduction

There are intimate connections between the homology and cohomology of the symmetric groups and algebraic topology. The first of these is their connection with the structure of cohomology operations. This arises through Steenrod's definition of the Pth power operations in terms of properties of certain elements in the groups $H_*(\mathcal{S}_p; \mathbb{F}_p)$. Indeed, the original calculation of $H_*(\mathcal{S}_n; \mathbb{F}_p)$ by Nakaoka was motivated by this connection.

These connections were exploited and developed from about 1952 – 1964 in work of J. Adem, N. Steenrod, A. Dold, M. Nakaoka and others. In particular, Adem used properties of the groups $H_*(\mathcal{S}_{p^2}; \mathbb{F}_p)$ to determine the relations in the Steenrod algebras. The complete cohomology rings, $H^*(\mathcal{S}_{p^2}; \mathbb{F}_p)$, were then determined by H. Cardenas, [Card], and it is here that the Cardenas-Kuhn theorem first appears.

Then, in the period from 1959 – 1961 E. Dyer and R. Lashof discovered a second connection of the symmetric groups with topology: a fundamental relationship between $H_*(\mathcal{S}_n; \mathbb{F}_p)$ and the structure of the homology of infinite loop spaces. The particular relation that expresses the spirit of their results best is a remarkable map of spaces

$$f\colon \mathbb{Z} \times B_{\mathcal{S}_\infty} \longrightarrow \lim_{n\to\infty} \Omega^n S^n = QS^0$$

which induces isomorphism in homology, but it is not a homotopy equivalence.

In the late 1960's and early 1970's this isomorphism formed the starting point for much of Quillen's work relating the classifying spaces of finite groups of Lie type to stable homotopy theory. We discuss some aspects of this in Chapter VII.

The original calculations of the groups $H_*(\mathcal{S}_n; \mathbb{F}_p)$ was based on an important connection between these groups and the cohomology of symmetric products of spheres $SP^n(S^{2m})$,

$$H^{2mk-s}(SP^n(S^{2m}); \mathbb{F}_p) \cong H_s(\mathcal{S}_n; \mathbb{F}_p)$$

for $s < 2m$ (recall from Chapter II that $SP^n(X) = X^n/\mathcal{S}_n$ where $\mathcal{S}_n$ acts on X^n by permuting coordinates). In turn, the spaces $SP^n(S^{2m})$ were identified with subspaces of Eilenberg-MacLane spaces by the fundamental theorem of Dold and Thom, [DT],

$$SP^{\infty}(X) \simeq \prod_{1}^{\infty} K(\tilde{H}_i(X;\mathbb{Z}), i)$$

for all connected CW complexes X. From this, Steenrod, in unpublished work, Milgram, [M3], and Dold, [Do], determined the homology groups of the $SP^n(X)$.

This work has led to recent connections between the homology of symmetric groups, certain moduli spaces of holomorphic maps from the Riemann sphere to symmetric spaces, the geometry of spaces of instantons and monopoles, and even results on linear control theory. Some of this work is described in the papers [C²M²], [MM1], [MM2], [BHMM], [Segal].

In this chapter we give a fairly complete and self-contained exposition of the structure of the homology and cohomology of the symmetric groups, $\mathcal{S}_n$, for all n. The connection with infinite loop space theory is summarized in Theorem (3.5), but we do not discuss the other applications.

We now describe the form of the answer. The homology of $\mathcal{S}_n$ injects with any untwisted coefficients, $\mathbb{A}$, onto a direct summand in the homology of $\mathcal{S}_\infty$ for each n via the homology map induced from the usual inclusion of groups. In particular this implies that $H_*(\mathcal{S}_\infty;\mathbb{A}) \cong \coprod_1^\infty H_*(B_{\mathcal{S}_n}, B_{\mathcal{S}_{n-1}};\mathbb{A})$ for all untwisted coefficients $\mathbb{A}$. On the other hand, if $\mathbb{A}$ is a ring there is a ring structure induced on $H_*(\mathcal{S}_\infty;\mathbb{A})$ from fitting together the maps $\mathcal{S}_n \times \mathcal{S}_m \longrightarrow \mathcal{S}_{n+m}$. This induces pairings

$$H_*(B_{\mathcal{S}_n}, B_{\mathcal{S}_{n-1}};\mathbb{A}) \otimes H_{*'}(B_{\mathcal{S}_m}, B_{\mathcal{S}_{m-1}};\mathbb{A}) \longrightarrow H_{*+*'}(B_{\mathcal{S}_{n+m}}, B_{\mathcal{S}_{n+m-1}};\mathbb{A})$$

which makes $H_*(\mathcal{S}_\infty;\mathbb{A})$ into a *bigraded* ring. Then the main structure theorems have the form

Theorem. *$H_*(\mathcal{S}_\infty;\mathbb{F}_p)$ is a tensor product of exterior algebras on odd dimensional generators and polynomial algebras on even generators where each generator has a known bidegree when p is any odd prime. When $p = 2$ it has the form of a polynomial algebra on generators of known bidegrees.*

As an example, here is the answer for $p = 2$. An admissible sequence, of length n, $I = (i_1, i_2, \ldots, i_n)$, is any non-decreasing sequence of positive integers $1 \le i_1 \le i_2 \le \cdots \le i_n$. The dimension of the sequence, $d(I) = i_1 + 2i_2 + 4i_3 + \cdots + 2^{n-1}i_n$, and the bidegree of I, $b(I) = 2^n$. Then

$$H_*(\mathcal{S}_\infty;\mathbb{F}_2) \cong \mathbb{F}_2[x_1, x_2, \ldots, x_I, \ldots]$$

as I runs over all admissible sequences. These admissible sequences are given by certain constructions called (iterated) Dyer-Lashof operations, [DL], applied to $H_*(\mathbb{Z}/2;\mathbb{F}_2)$. (We will discuss this further in §3.) In any case

$H_*(\mathcal{S}_n;\mathbb{F}_2) \subset H_*(\mathcal{S}_\infty;\mathbb{F}_2)$ is generated by all the monomial products of the generators with bidegree $\leq n$. Precisely, these monomials have the form

$$x_{I_1}^{i_1}\cdots x_{I_r}^{i_r}\ .$$

Such a monomial has bidegree $\sum_1^r i_j b(I_j)$ and dimension $\sum_1^r i_j d(I_j)$. In particular $\tilde{H}_*(\mathcal{S}_2;\mathbb{F}_2)$ has generators x_i, $1 \leq i < \infty$, where x_i has dimension i and bidegree 2. Similarly $H_*(\mathcal{S}_4;\mathbb{F}_2)$ has these generators, their products $x_i x_j$, $i \leq j$, and further generators corresponding to the admissible sequences (i_1, i_2) of dimension $i_1 + 2i_2$.

The x_i's are used by Steenrod to construct the basic Steenrod squaring operations, Sq^j, while the x_I with $I = (i_1, i_2)$ are used by J. Adem to construct the iterates Sq^iSq^j and determine the relations between them, such as the basic relations $Sq^{2n-1}Sq^n = 0$.

The first section of this chapter is primarily algebraic. We determine the groups $\mathrm{Syl}_p(\mathcal{S}_n)$ and show that $H^*(\mathcal{S}_n;\mathbb{F}_p)$ is detected by restriction to elementary abelian p-groups. Then we determine the conjugacy classes of these groups in $\mathcal{S}_n$ and show that certain key conjugacy classes satisfy weak closure conditions in $\mathcal{S}_{p^m}$. This allows us to determine the image of the restriction homomorphism for these groups and at this point we are able to write down the groups $H^*(\mathcal{S}_n;\mathbb{F}_p)$ for $n \leq p^2$.

The hard work in §1 is the determination of the image of restriction

$$H^*(\mathcal{S}_{p^n};\mathbb{F}_p) \longrightarrow H^*(V_n(p);\mathbb{F}_p)$$

where $V_n(p) \cong (\mathbb{Z}/p)^n$ and the inclusion is the regular representation. It turns out that the classes detected by these groups construct every cohomology class in $H^*(\mathcal{S}_n;\mathbb{F}_p)$ via certain composition pairings. Here, for the most part, we follow the exposition of [Ma], reporting on work done in the early 1970's. We extend his ideas in a few places to make things more self-contained.

To understand these groups and the composition pairing which builds them for all n we must introduce a more global point of view. This is accomplished in the second section where we introduce the techniques of Hopf algebras. We first introduce the notions of Hopf algebras, then prove the basic theorems of Borel and Hopf on the structure of commutative and cocommutative Hopf algebras.

In §3 we apply the Borel-Hopf theorems to the work of §1 to quickly obtain the Hopf algebras $H_*(\mathcal{S}_\infty;\mathbb{F}_p)$ for all primes p.

In §4 we discuss the structure of some rings of invariants. This is in preparation for §5 where we give complete calculations of $H^*(\mathcal{S}_n;\mathbb{F}_2)$ for $n = 6, 8, 10, 12$. Finally, in §6 we discuss the cohomology groups $H^*(\mathcal{A}_n;\mathbb{F}_2)$. The work in §4 is largely incomplete and the complete answers should be of considerable interest when they are finally understood.

1. Detection Theorems for $H^*(\mathcal{S}_n; \mathbb{F}_p)$ and Construction of Generators

In this section we concentrate on the structure of the groups $\mathcal{S}_{p^n}$. In particular, we construct a great many cohomology classes which are non-trivial in these groups using the Cardenas-Kuhn theorem and restriction to some of the more important maximal elementary p-subgroups of $\mathcal{S}_{p^n}$. These subgroups are classified in (1.3). Then, after we have discussed Hopf algebras in §2 we will show that the elements constructed here generate $H^*(\mathcal{S}_n; \mathbb{F}_p)$ for all n.

The Sylow p-Subgroups of $\mathcal{S}_n$

Recall that for any $G \subseteq \mathcal{S}_n$ and any other group H, the wreath product $H \wr G$ is defined as the product $H^n \times G$ with multiplication

$$(h_1, \ldots, h_n, g)(h'_1, \ldots, h'_n, g') = (h_1 h'_{g^{-1}(1)}, \ldots, h_n h_{g^{-1}(n)}, gg') .$$

In particular, $\mathcal{S}_m \wr \mathcal{S}_n$ may be thought of as the set of permutations of pairs (i, j), $1 \le i \le m$ and $1 \le j \le n$ by defining

$$(h_1, \ldots, h_n, g)(i, j) = (h_j(i), g^{-1}(j)) .$$

Then, using lexicographic ordering, this provides an embedding

$$J_{n,m}: \mathcal{S}_m \wr \mathcal{S}_n \longrightarrow \mathcal{S}_{nm} .$$

Note that, in particular, we have

$$J: \mathbb{Z}_2 \wr \mathbb{Z}_2 \hookrightarrow \mathcal{S}_4$$

and, iterating this, we obtain

$$J: \underbrace{\mathbb{Z}_2 \wr \cdots \wr \mathbb{Z}_2}_{n \text{ times}} \hookrightarrow \mathcal{S}_{2^n} .$$

Lemma 1.1
1. *The power of 2 which divides $n!$ is $n - \alpha(n)$ where $\alpha(n)$ is the number of 1's in the dyadic expansion of n.*
2. *Let p be an odd prime, then the power of p which divides $n!$ is $\frac{n-\alpha_p(n)}{p-1}$ where $\alpha_p(n)$ is the sum of the coefficients in the p-adic expansion of n.*

Proof. The power of p which divides $n!$ is

$$\left[\frac{n}{p}\right] + \left[\frac{n}{p^2}\right] + \cdots$$

so, if $n = \sum_0^v a_r p^r$ then $\left[\frac{n}{p^t}\right] = \sum_t^v a_r p^{r-t}$ and the sum above is

$$\sum a_r(p^{r-1} + p^{r-2} + \cdots + 1) = \sum a_r \left(\frac{p^r - 1}{p-1}\right) = \frac{1}{p-1}(n - \alpha_p(n)) . \quad \square$$

It follows that J embeds the wreath product as the 2-Sylow subgroup of $\mathcal{S}_{2^n}$.

Now let m be an arbitrary integer. Write it out in terms of its dyadic expansion

$$m = 2^{i_1} + \cdots + 2^{i_r} \quad i_1 < i_2 < \cdots < i_r .$$

Then

$$\mathcal{S}_{2^{i_1}} \times \mathcal{S}_{2^{i_2}} \times \cdots \times \mathcal{S}_{2^{i_r}} \hookrightarrow \mathcal{S}_m,$$

with *odd* index. hence

$$\mathrm{Syl}_2(\mathcal{S}_n) = \mathrm{Syl}_2(\mathcal{S}_{2^{i_1}}) \times \cdots \times \mathrm{Syl}_2(\mathcal{S}_{2^{i_r}}),$$

from which we conclude that the Sylow 2-subgroup of any finite symmetric group is a product of iterated wreath products of $\mathbb{Z}/2$. A similar analysis applies for p odd. One checks as before that

$$\underbrace{(\mathbb{Z}/p) \wr (\mathbb{Z}/p) \wr \cdots \wr (\mathbb{Z}/p)}_{r-times} \subset \mathcal{S}_{p^r}$$

is $\mathrm{Syl}_p(\mathcal{S}_{p^r})$ and a product, depending on the p-adic expansion of n, of these wreath products is $\mathrm{Syl}_p(\mathcal{S}_n)$ for general n.

From (IV.4.3) we obtain an important detection theorem for symmetric groups.

Theorem 1.2 *The mod p cohomology of $\mathcal{S}_n$ is detected by its elementary abelian p-subgroups.*

The Conjugacy Classes of Elementary p-Subgroups in $\mathcal{S}_n$

Let $V_n(p) = (\mathbb{Z}/p)^n \hookrightarrow \mathcal{S}_{p^n}$ be the regular representation, (the permutation representation on the cosets of the identity).

Theorem 1.3 *Write $n = \alpha + i_1 p + i_2 p^2 + \cdots + i_r p^r$ with $0 \leq \alpha < p$, $i_j \geq 0$ for $1 \leq j \leq r$. (Note that the i_j can be greater than p in this decomposition.) Then there is a maximal p-elementary subgroup of $\mathcal{S}_n$ corresponding to this decomposition*

$$\underbrace{V_1(p) \times \cdots \times V_1(p)}_{i_i} \times \cdots \times \underbrace{V_r(p) \times \cdots \times V_r(p)}_{i_r}$$

$$\subset \underbrace{\mathcal{S}_p \times \cdots \times \mathcal{S}_p}_{i_1} \times \cdots \times \underbrace{\mathcal{S}_{p^r} \times \cdots \times \mathcal{S}_{p_r}}_{i_r} \subset \mathcal{S}_n$$

and as we run over distinct decompositions $(i_1, \ldots, i_r)$ these give the distinct conjugacy classes of maximal elementary p-subgroups of $\mathcal{S}_n$.

Proof. Let $H = (\mathbb{Z}/p)^i \subset \mathcal{S}_n$ be any p-elementary subgroup. The action of H on $(1, \ldots, n)$ breaks this set into orbits, each of length a power of p. Now restrict H to its action on a single orbit. This gives a homomorphism $H \longrightarrow \mathcal{S}_{p^t}$ with image conjugate to $V_t(p)$. It follows that H is contained in one of the groups of the theorem. □

Corollary 1.4 *$H^*(\mathcal{S}_{p^n}; \mathbb{F}_p)$ is detected by $H^*(V_n(p); \mathbb{F}_p)$ and*

$$H^*(\underbrace{\mathcal{S}_{p^{n-1}} \times \cdots \times \mathcal{S}_{p^{n-1}}}_{p}; \mathbb{F}_p)\ .$$

Proof. Every elementary p-subgroup of $\mathcal{S}_{p^n}$ is conjugate to one contained in either $\underbrace{\mathcal{S}_{p^{n-1}} \times \cdots \times \mathcal{S}_{p^{n-1}}}_{p}$ or in $V_n(p)$ and $H^*(\mathrm{Syl}_p(\mathcal{S}_n); \mathbb{F}_p)$ is detected by p-elementary subgroups. □

Weak Closure Properties for $V_n(p) \subset \mathrm{Syl}_p(\mathcal{S}_{p^n})$ and $(V_{n-i}(p))^{p^i} \subset \mathcal{S}_{p^{n-1}} \wr \mathbb{Z}/p$

Recall that $N \subset H \subset G$ is said to be weakly closed in H if every subgroup of H which is conjugate to N in G is already conjugate to N in H.

Theorem 1.5 *For all $n > 0$ and each prime p the subgroup $V_n(p) \subset \mathcal{S}_{p^n}$ obtained via the regular representation of $(\mathbb{Z}/p)^n$ is weakly closed in* $\mathrm{Syl}_p(\mathcal{S}_{p^n})$.

Proof. The regular representation $G \hookrightarrow \mathcal{S}_{|G|}$ is defined by regarding the points of G as the elements being permuted and the embedding as permutations is $g(\{h\}) = \{gh\}$. Then the centralizer of $G \hookrightarrow \mathcal{S}_{|G|}$ is a second copy of G acting from the right, $c(g)(\{h\}) = \{hg^{-1}\}$. (This is well known, but the proof is easy: if $xg = gx$ for all $g \in G$, we have $xg\{h\} = gx\{h\} = g\{h\lambda\}$ for some λ, and $xg\{h\} = x(\{gh\}) = \{gh\lambda\}$.)

In particular, in the case where G is abelian, it follows that G is its own centralizer in $\mathcal{S}_{|G|}$ when G is embedded via the regular representation.

Next, choose $e \in V_n(p)$, $e \neq 0$, and let $C(e) \cong \mathbb{Z}/p \wr \mathcal{S}_{p^{n-1}}$ be the centralizer of e in $\mathcal{S}_{p^n}$. In particular, $V_n(p) \subset C(e)$. Then we also have $C(\mathbb{Z}/p \wr \mathcal{S}_{p^{n-1}}) = \langle e \rangle \cong \mathbb{Z}/p$, and we suppose $f \in \mathcal{S}_{p^n}$ given so that $fV_n(p)f^{-1} \subset C(e)$. We claim that there is a $g \in C(e)$ so that $fV_n(p)f^{-1} = gV_n(p)g^{-1}$ so that $V_n(p)$ is weakly closed in $C(e)$.

To see this note that $e \in C(fV_n(p)f^{-1}) = fV_n(p)f^{-1}$ since, as we have seen, the centralizer of the regular representation of an abelian group is the group itself. Thus there is $h \in V_n(p)$ so that $fhf^{-1} = e$ and there is an element $\lambda \in \mathrm{Aut}(V_n(p)) \cong N_{\mathcal{S}_{p^n}}(V_n(p))$ so that $f \cdot \lambda e(f \cdot \lambda)^{-1} = e$ and, of course, $f \cdot \lambda V_n(p)(f \cdot \lambda)^{-1} = fV_n(p)f^{-1}$. But this implies that $f \cdot \lambda \in C(e)$ and the claim is verified.

Now we are ready to prove the theorem. The proof is by induction, so we assume it is true for $n-1$. Suppose that we have $V' = gV_n(p)g^{-1} \subset \mathrm{Syl}_p(\mathcal{S}_{p^n})$. We wish to show that there is an $h \in \mathrm{Syl}_p(\mathcal{S}_{p^n})$ so that $hV_n(p)h^{-1} = gV_n(p)g^{-1}$. To begin we assume that $g \in \mathbb{Z}/p \wr \mathcal{S}_{p^{n-1}}$ by the remarks above. Then we project onto $\mathcal{S}_{p^{n-1}}$. Note that, if π denotes the projection, then $\pi(V_n(p)) = V_{n-1}(p)$. Thus $\pi(g)V_{n-1}(p)\pi(g)^{-1} \subset \mathrm{Syl}_p(\mathcal{S}_{p^{n-1}})$ and, by the inductive assumption, there is some $\lambda \in \mathrm{Syl}_p(\mathcal{S}_{p^{n-1}})$ so $\lambda\pi(g)V_{n-1}(p)(\lambda\pi(g))^{-1}$-$= V_{n-1}(p)$ and $\lambda\pi(g)$ is an automorphism of $V_{n-1}(p)$. Hence, there is some

$$\phi \in \mathcal{S}_{p^{n-1}} \subset \mathbb{Z}/p \wr \mathcal{S}_{p^{n-1}} \cap N_{\mathcal{S}_{p^n}}(V_n(p))$$

so that $\lambda\pi(g\phi)$ commutes with $V_{n-1}(p)$. But then, since $V_{n-1}(p)$ is its own centralizer in $\mathcal{S}_{p^{n-1}}$, it follows that $\lambda\pi(g\phi) \in \mathrm{Syl}_p(\mathcal{S}_{p^{n-1}})$, so $\pi(g\phi) \in \mathrm{Syl}_p(\mathcal{S}_{p^{n-1}})$ as well, and $g\phi \in \pi^{-1}(\mathrm{Syl}_p(\mathcal{S}_{p^{n-1}}) = \mathrm{Syl}_p(\mathcal{S}_{p^n})$ and the induction is complete. □

Using this result we have a precise determination of the image of the restriction homomorphism from $H^*(\mathcal{S}_{p^n};\mathbb{F}_p)$ to $H^*(V_n(p);\mathbb{F}_p)$.

Corollary 1.6 *The image of the restriction homomorphism*

$$\mathrm{res}^*\colon H^*(\mathcal{S}_{p^n};\mathbb{F}_p) \longrightarrow H^*(V_n(p);\mathbb{F}_p)$$

is precisely equal to

$$\mathrm{im}(\mathrm{res}^*\colon\ H^*(\underbrace{\mathbb{Z}/p \wr \cdots \wr \mathbb{Z}/p}_{n-times};\mathbb{F}_p) \longrightarrow H^*(V_n(p);\mathbb{F}_p))$$

$$\bigcap H^*(V_n(p);\mathbb{F}_p)^{\mathrm{GL}_n(p)}.$$

Proof. First, the normalizer of the regular representation of $H \subset \mathcal{S}_{|H|}$ is always $\mathrm{Aut}(H)$ so the Weyl group is $\mathrm{Out}(H)$. In the case of $(\mathbb{Z}/p)^n$ this is $\mathrm{GL}_n(\mathbb{F}_p)$. Now (1.5) implies that we can apply the Cardenas-Kuhn theorem to $V_n(p) \subset \mathrm{Syl}_p(\mathcal{S}_{p^n})$ and (1.6) follows. □

We also have further weak closure properties which are very useful in understanding the groups $H^*(\mathcal{S}_{p^n};\mathbb{F}_p)$.

Theorem 1.7 *For all primes p and $1 \leq i \leq n-1$, the subgroup $(V_{n-i}(p))^{p^i} \subset \mathcal{S}_{p^{n-1}} \wr \mathbb{Z}/p$ is weakly closed in $\mathcal{S}_{p^n}$.*

Proof. Let $T = (1, 1, \ldots, 1, T) \in H \wr \mathbb{Z}/p$ where T acts by

$$T(h_1, \ldots, h_p, 1)T^{-1} = (h_2, h_3, \ldots, h_p, h_1, 1) ,$$

and suppose that $\tau \in H \wr \mathbb{Z}/p$ and T have the same image under the projection $\pi: H \wr \mathbb{Z}/p \longrightarrow \mathbb{Z}/p$. Then, we can write $\tau = (\tau_1, \ldots, \tau_p, T)$, and a direct calculation shows that

$$\tau^p = (\tau_1\tau_2 \cdots \tau_p, \tau_2 \cdots \tau_p\tau_1, \ldots, \tau_p\tau_1 \cdots \tau_{p-1}, 1)$$

so $\tau^p = 1$ if and only if $\tau_1 \cdots \tau_p = 1$.

Proposition. *Suppose that τ and T have the same image under π in $\mathbb{Z}/p$, then, if $\tau^p = 1$ it follows that T and τ are conjugate in $H \wr \mathbb{Z}/p$.*

(Write $\theta = (\tau_1, \tau_2\tau_1, \ldots, \tau_{p-1} \ldots \tau_1, 1, 1) \in H \wr \mathbb{Z}/p$. Then directly $\theta T \theta^{-1} = \tau$ and the proposition follows.)

Now we turn to the proof of the theorem. We introduce the following notation. Let $M = g(V_{n-i}(p))^{p^i} g^{-1} \subset \mathcal{S}_{p^{n-i}} \wr \mathbb{Z}/p$ for some $g \in \mathcal{S}_{p^n}$. Suppose that the projection π on M is not $\{1\}$. Then the proposition shows that we may assume $T \in M$. The centralizer of $\langle T \rangle$ in $\mathcal{S}_{p^{n-1}} \wr \mathbb{Z}/p$ is $\Delta^p(\mathcal{S}_{p^{n-1}}) \times \langle T \rangle$ and consequently,

$$M = \left(M \cap \Delta^p(\mathcal{S}_{p^{n-1}})\right) \times \langle T \rangle .$$

Thus, the intersection with $\Delta(\mathcal{S}_{p^{n-1}})$, now regarded as the permutation group on p^{n-1} points, must have p^i orbits, each of length p^{n-i-1} to give p^i orbits each of length p^{n-i} under under the action of the entire subgroup in $\mathcal{S}_{p^{n-1}} \wr \mathbb{Z}/p$. It follows that

$$M \cap \Delta^p(\mathcal{S}_{p^{n-1}}) \subset (V_{n-i-1}(p))^{p^i}$$

and the order of the resulting group $\langle M \cap \mathcal{S}^p_{p^{n-1}}, T \rangle$ is at most $p^{(n-i-1)p^i+1} < p^{(n-i)p^i}$ so it follows that $\pi(M) = \{1\}$. Consequently, $M \subset \mathcal{S}^p_{p^{n-1}}$ and the result now follows by checking orbits. □

Corollary 1.8 *The image of the restriction map*

$$H^*(\mathcal{S}_{p^n}; \mathbb{F}_p) \longrightarrow H^*((V_{n-i}(p))^{p^i}; \mathbb{F}_p)$$

is

$$\left(\mathrm{im}(\mathrm{res}^*: H^*(\mathcal{S}_{p^{n-i}}; \mathbb{F}_p) \longrightarrow H^*(V_{n-i}(p); \mathbb{F}_p)\right)^{p^i} \cap \left(H^*(V_{n-i}(p); \mathbb{F}_p)^{p^i}\right)^N$$

where N is the Weyl group $\mathrm{GL}_{n-i}(p) \wr \mathcal{S}_{p^i}$ of $V_{n-i}(p)^{p^i}$ in $\mathcal{S}_{p^n}$.

There are two special cases for the result above. Let

$$v \in \mathrm{im}(\mathrm{res}^*(H^*(\mathcal{S}_{p^n}; \mathbb{F}_p) \longrightarrow H^*(V_n(p); \mathbb{F}_p))) ,$$

then

$$\underbrace{v\otimes 1\otimes\cdots\otimes 1}_{p^i\ times}+\underbrace{1\otimes v\otimes\cdots\otimes 1}_{p^i\ times}+\cdots+\underbrace{1\otimes\cdots\otimes 1\otimes v}_{p^i\ times}$$

is in $H^*(V_n(p)^{p^i};\mathbb{F}_p)^N$. Also, when $\dim(v)$ is even

$$\underbrace{v\otimes v\otimes\cdots\otimes v}_{p^i\ times}\in H^*(V_n(p)^{p^i};\mathbb{F}_p)^N .$$

Consequently we have

Corollary 1.9

1. *For all integers $n, i > 0$ we have that the composite restriction map*

$$H^*(\mathcal{S}_{p^{n+i}};\mathbb{F}_p)\xrightarrow{\mathrm{res}^*_{i,n}}H^*(\mathcal{S}_{p^n};\mathbb{F}_p)\xrightarrow{\mathrm{res}^*}H^*(V_n(p);\mathbb{F}_p)$$

is onto the image of res^*.

2. *Let $v\in\mathrm{im}(\mathrm{res}^*\colon H^{2*}(\mathcal{S}_{p^n};\mathbb{F}_p)\longrightarrow H^{2*}(V_n(p);\mathbb{F}_p))$, then for all $i>0$,*

$$\underbrace{v\otimes v\otimes\cdots\otimes v}_{p^i\ times}$$

is in the image of the restriction

$$\mathrm{res}^*\colon\ H^*(\mathcal{S}_{p^{n+i}};\mathbb{F}_p)\longrightarrow H^*(V_n(p)^{p^i};\mathbb{F}_p) .$$

(1.9) is a key tool in our final description of the homology structure of the symmetric groups which will be completed in the next 2 sections using the methods of Hopf algebras. We now determine the image of the restriction map in $H^*(V_n(p);\mathbb{F}_p)$.

The Image of res^*: $H^*(\mathcal{S}_{p^n};\mathbb{F}_p)\longrightarrow H^*(V_n(p);\mathbb{F}_p)$.

We apply the results of (IV.1) on the homology of wreath products, to further understand how the mod (p) homology and cohomology of $\mathcal{S}_n$ occurs. Clearly, the critical case is when $n = p^m$. We have a factorization of the inclusion $\mathrm{Syl}_p(\mathcal{S}_{np})\hookrightarrow\mathcal{S}_{np}$ as follows

$$\mathrm{Syl}_p(\mathcal{S}_{np})\hookrightarrow\mathrm{Syl}_p(\mathcal{S}_n)\wr\mathbb{Z}/p\hookrightarrow\mathcal{S}_n\wr\mathbb{Z}/p\hookrightarrow\mathcal{S}_n\wr\mathcal{S}_p\hookrightarrow\mathcal{S}_{np} .$$

In (IV.1) we determined $H^*(H\wr\mathbb{Z}/p;\mathbb{F}_p)$. It is generated by elements of two types. First there are the $(1+T+\cdots T^{p-1})(\lambda_1\otimes\cdots\lambda_p)\in H^*(H^p;\mathbb{F}_p)$ where the λ_j are not all equal and T acts to shift the elements $T(\lambda_1\otimes\cdots\otimes\lambda_p) = (\lambda_p\otimes\lambda_1\otimes\cdots\otimes\lambda_{p-1})$. These are all in the image of the composite $\mathrm{res}\cdot\mathrm{tr}$. Then there are the elements of the form

(1.10) $$\underbrace{(\lambda\otimes\lambda\otimes\cdots\otimes\lambda)}_{p\ times}\cup\pi^*(\gamma_j)$$

where $(\gamma_j)\in H^j(\mathbb{Z}/p;\mathbb{F}_p)$ is a generator and $\pi\colon H\wr\mathbb{Z}/p\longrightarrow\mathbb{Z}/p$ is the projection.

Precisely, γ_j is given as follows. Let $e\in H^1(B_{\mathbb{Z}/p};\mathbb{F}_p)$ be the fundamental class which corresponds to the identity homomorphism

$$\mathrm{id}\in \mathrm{Hom}(H_1(B_{\mathbb{Z}/p};\mathbb{Z});\mathbb{F}_p)=H^1(B_{\mathbb{Z}/p};\mathbb{F}_p)$$

and suppose $b=\beta(e)$ is the Bockstein. Then set $\gamma_{2j}=b^j$, $\gamma_{2j+1}=b^je$ for p odd, and $\gamma_j=e^j$ when $p=2$. For notational convenience we will sometimes write $\pi^*(\gamma_j)=e^j$, $\pi^*(\gamma_{2j+\epsilon})=b^je^\epsilon$ as well in what follows.

There are further restrictions that we can put on the possible classes in (1.10) which are in the image from $H^*(\mathcal{S}_{np};\mathbb{F}_p)$ when p is odd. The normalizer of $\mathbb{Z}/p\subset\mathcal{S}_p$ is $\mathbb{Z}/p\times_T\mathbb{Z}/(p-1)$. The action of $\theta\in\mathbb{Z}/(p-1)$ on $(\lambda\otimes\cdots\otimes\lambda)$ is trivial if $\dim(\lambda)$ is even, but is given by the sign representation if $\dim(\lambda)$ is odd. Likewise, we can make $\mathrm{Aut}(\mathbb{Z}/p)\cong\mathbb{Z}/(p-1)$ correspond to the units $\mu\in\mathbb{Z}/p$ and under this correspondence $\mu(\gamma_j)=\mu^{[j+1/2]}\gamma_j$. Thus, the invariants under this action in $H^*(\mathcal{S}_n\wr\mathbb{Z}/p;\mathbb{F}_p)$ among the elements of the form (1.10) are the elements $(\lambda\otimes\cdots\otimes\lambda)\cup\gamma_k$ with $k=2j(p-1)$ or $k=2j(p-1)-1$ when $\dim(\lambda)$ is even and $(\lambda\otimes\cdots\otimes\lambda)\cup b^{(2j+1)(p-1)/2}$ or $(\lambda\otimes\cdots\otimes\lambda)\cup b^{(2j+1)(p-1)/2-1}e$ when $\dim(\lambda)$ is odd.

Lemma 1.11 *Let p be a prime. The composite*

$$H^*(\mathcal{S}_{p^{n-1}}\times\cdots\times\mathcal{S}_{p^{n-1}};\mathbb{F}_p)\xrightarrow{\mathrm{tr}}H^*(\mathcal{S}_{p^{n-1}}\wr\mathbb{Z}/p;\mathbb{F}_p)\xrightarrow{\mathrm{res}}H^*(V_n(p);\mathbb{F}_p)$$

is identically zero.

Proof. We use the double coset formula for $\mathcal{S}^p_{p^{n-1}}$ and $V_n(p)$. There is only one double coset in the wreath product. Hence the composite factors as

$$\mathrm{tr}\cdot\mathrm{res}\colon H^*(\mathcal{S}^p_{p^{n-1}};\mathbb{F}_p)\longrightarrow H^*(V_n(p)\cap\mathcal{S}_{p^{n-1}};\mathbb{F}_p)\longrightarrow H^*(V_n(p);\mathbb{F}_p)$$

and since the index of this intersection in $V_n(p)$ is p the result follows. □

In view of (1.11), to understand the restriction map

$$H^*(\mathcal{S}_{p^{n-1}}\wr\mathbb{Z}/p;\mathbb{F}_p)\longrightarrow H^*(V_n(p);\mathbb{F}_p)$$

it suffices to study it on elements of the form $\Gamma(\alpha)$ where $\Gamma(\alpha)$ is described in the proof of (IV.1.5). From this proof we see that, in fact, $\Gamma(\alpha)$ is defined and non-zero in $H^{pt}(X^p\times_T E_{\mathbb{Z}/p};\mathbb{F}_p)$ for any cohomology class $\alpha\in H^t(X;\mathbb{F}_p)$ where X is an arbitrary CW complex. $\Gamma(\alpha)$ satisfies the following four properties, all of which follow directly from its construction in (IV.1.5):

1. $\mathrm{res}(\Gamma(\alpha))=\underbrace{\alpha\otimes\cdots\otimes\alpha}_{p\ times}$ in $H^*(\mathcal{S}^p_{p^{n-1}};\mathbb{F}_p)$,

2. the construction is natural in the sense that if $f: X \longrightarrow Y$ is a continuous map and $f^*(\tau) = \alpha$ then $(f^p \times \mathrm{id})^*(\Gamma(\tau)) = \Gamma(\alpha)$,
3. $\Gamma(\alpha \cup \beta) = \pm\Gamma(\alpha) \cup \Gamma(\beta)$,
4. $\Gamma(\alpha + \beta) = \Gamma(\alpha) + \Gamma(\beta) + \mathrm{tr}(w)$ for some $w \in H^t(X^p; \mathbb{F}_p)$.

From the appendix to (IV) we have the formulae for the restriction map

$$(\Delta^p \times 1)^*: \Gamma(\alpha) \longrightarrow H^*(X \times B_{\mathbb{Z}/p}; \mathbb{F}_p):$$

(1.12)

$$(\Delta^2 \times 1)^*(\Gamma(\alpha)) = \sum_0^t Sq^j(\alpha) \otimes e^{t-j} \text{ when } p = 2,$$

$$(\Delta^p \times 1)^*(\Gamma(\alpha)) = \lambda_t \sum_0^{[t/2(p-1)]} (-1)^i [P^i(\alpha) \otimes b^{m(t-2i)} + (-1)^t \beta P^i(\alpha) \otimes b^{m(t-2i)-1} e],$$

where $m = (p-1)/2$ and

$$\lambda_t = \frac{(-1)^{mt(t-1)/2}}{(m!)^t}.$$

Note that since $(m!)^2 = (-1)^{(p+1)/2}$, it follows that λ_t is ± 1 for t even.

This can be applied inductively to understand the restriction map to $H^*(V_n(p))$ from $H^*(\mathcal{S}_{p^n})$ since the inclusion $V_n(p) \hookrightarrow \mathcal{S}_{p^n}$ factors as follows

$$V_n(p) = V_{n-1}(p) \times \mathbb{Z}/p \xrightarrow{\Delta^p \times 1} \mathcal{S}_{p^{n-1}} \wr \mathbb{Z}/p \longrightarrow \mathcal{S}_{p^n}.$$

When $p = 2$ and $n = 2$ this gives

$$\mathrm{res}^*(\Gamma(e)) = e^2 \otimes 1 + e \otimes e \in H^*(V_2(2)).$$

This element is not invariant under $\mathrm{GL}_2(2)$ but $\mathrm{res}^*(\Gamma(e)) + 1 \otimes e^2 = d_2$ is invariant, and since $\mathrm{res}^*(e) = 1 \otimes e$, it is also in the image of res. Consequently, d_2 is in the image of $H^*(\mathcal{S}_4; \mathbb{F}_2)$, and since $d_3 = Sq^1(d_2)$ it follows that d_3 is also in the image. We thus have a proof that $\mathbb{F}_2[x_1, x_2]^{L_2(2)}$ is in the image of restriction from $H^*(\mathcal{S}_4; \mathbb{F}_2)$.

Example 1.13 $H^*(\mathcal{S}_4; \mathbb{F}_2)$

In this case we need only 2 detecting subgroups, $V_1(2)^2 = \mathcal{S}_2 \times \mathcal{S}_2 \subset \mathcal{S}_4$, and $V_2(2)$. Then we have

$$\begin{array}{ccc} H^*(\mathcal{S}_4) & \xrightarrow{\phi} & H^*(V_1(2)^2) \oplus H^*(V_2) \\ & & \Big\downarrow \varphi \\ & & H^*(V_1(2)^2 \cap V_2) \end{array}$$

where $\phi = (\mathrm{res}_1, \mathrm{res}_2)$, and $\varphi = \mathrm{res}^{V_1(2)^2}_{V_1(2)^2\cap V_2} + \mathrm{res}^{V_2}_{V_1(2)^2\cap V_2}$. Note that $\mathrm{im}\,\phi \subseteq \ker\varphi$; now $V_1(2)^2 = \langle(12),(34)\rangle$, $V_2(2) = \langle(12)(34),(14)(23)\rangle$, hence $V_1(2)^2 \cap V_2(2) = \langle(12)(34)\rangle = \Delta(V_2(1))$. The invariant subrings are given as follows:

$$\begin{aligned} H^*(V_1(2)^2)^{\mathbb{Z}_2} &\cong \mathbb{F}_2[\sigma_1,\sigma_2], \\ H^*(V_2(2))^{\mathrm{GL}_2(\mathbb{F}_2)} &\cong \mathbb{F}_2[d_2,d_3], \\ H^*(V_1(2)^2\cap V_2) &\cong \mathbb{F}_2[e], \end{aligned}$$

and

$$\varphi(\sigma_1,0)=0,\ \varphi(\sigma_2,0)=e^2,\ \varphi(0,d_2)=e^2,\ \varphi(0,d_3)=0.$$

From the above we deduce the existence of classes $\sigma_1,\sigma_2,c_3 \in H^*(\mathcal{S}_4)$, with

$$\phi(\sigma_1)=(\sigma_1,0),\ \phi(\sigma_2)=(\sigma_2,d_2),\ \phi(c_3)=(0,d_3)$$

and so

$$H^*(\mathcal{S}_4;\mathbb{F}_2) \cong \mathbb{F}_2[\sigma_1,\sigma_2,c_3]/(\sigma_1 c_3)\,.$$

This approach can be iterated. First, we need an inductive formula for constructing $d_{(p-1)p^{n-1}} \in H^*(V_n(p))^{\mathrm{GL}_n(p)}$. We fix the generators $d_{p^n-p^j}$ for the Dickson algebra discussed in (III.2) as follows. Write $V_n(p) = (\mathbb{Z}/p)^{n-1}\times \mathbb{Z}/p$ where the elements in the summand $\mathbb{Z}/p$ are written λe, $0\le\lambda\le p-1$. Then the restriction of $d_{p^n-p^j}$ to the subalgebra for $(\mathbb{Z}/p)^{n-1}$ is $(d_{p^{n-1}-p^{j-1}})^p$ for $j>1$, and when $j=1$ we have $d_{p^n-1} = d_{p^{n-1}-1}e^{p^{n-1}(p-1)}$ where $d_{p^{n-1}-*}$ is the Dickson invariant for $(\mathbb{Z}/p)^{n-1}$.

Lemma 1.14 *The following expansion is valid for all primes p.*

$$x^{p^n} - \sum_{j=0}^{n-1} \lambda_j x^{p^j} d_{p^n-p^j} = \prod_{v\in V_n(p)} (x-v)\,,$$

for each integer n, $1\le n<\infty$.

Proof. To begin consider the case $n=1$. We have $\prod_{i=0}^{p-1}(x-ie) = x^p - xe^{p-1}$ so the result is true here. Assume the result true for n, we show this implies it for $n+1$. Write $V_n(p) = V_{n-1}(p)\times\mathbb{Z}/p$, so

$$\begin{aligned}
\prod_{v\in V_n(p)}(x-v) &= \prod_{\lambda=0}^{p-1}\left(\prod_{w\in V_{n-1}(p)}(x+\lambda e+w)\right)\\
&= \prod_{\lambda=0}^{p-1}\left(\sum_{j=0}^{n-1}\lambda_j(x+\lambda e)^{p^j}d_{p^{n-1}-p^j}\right)\\
&= \prod_{\lambda=0}^{p-1}\left(\sum_{j=0}^{n-1}\lambda_j x^{p^j}d_{p^{n-1}-p^j}+\lambda\sum_{j=0}^{n-1}\lambda_j e^{p^j}d_{p^{n-1}-p^j}\right)\\
&= \left(\sum_{j=0}^{n-1}\lambda_j x^{p^j}d_{p^{n-1}-p^j}\right)^p+\\
&\qquad \left(\sum_{j=0}^{n-1}\lambda_j x^{p^j}d_{p^{n-1}-p^j}\right)\left(\sum_{j=0}^{n-1}\lambda_j e^{p^j}d_{p^{n-1}-p^j}\right)^{p-1}\\
&= \sum_{j=0}^{n}\lambda_j x^{p^j}((d_{p^{n-1}-p^{j-1}})^p+B_j e^{p-1}).
\end{aligned}$$

In particular, by restricting to $\mathbb{F}_p[V_{n-1}(p)]$ we see that when $j\neq 0$ the restriction of the coefficient of x^{p^j} is $-d^p_{p^{n-1}-p^{j-1}}$. Thus, since there is only a one dimensional set of invariants in $\mathbb{F}_p[V_n(p)]$ with degree p^n-p^j we obtain the result for all the lower coefficients. But the top coefficient is $\pm\prod v$ where $v\in V_n(p)$ and $v\neq 0$. The inductive step follows. □

Corollary 1.15 *Write $V_n(p)=V_{n-1}(p)\times\mathbb{Z}/p$, then the lowest dimensional* $\mathrm{GL}_n(p)$ *invariant can be written* $d_{p^n-p^{n-1}}=\sum_{j=0}^{n-1}e^{p^j}d_{p^{n-1}-p^j}+d^p_{p^{n-1}-p^{n-2}}$.

(Just expand out the coefficient of $x^{p^{n-1}}$ in the expression above.)

Corollary 1.16 $d_{p^n-p^{n-1}}$ *is contained in the image of the restriction map,* $\mathrm{im}(\mathrm{res}^*)$, *from* $H^*(\mathrm{Syl}_p(\mathcal{S}_{p^n}))$.

Proof. Indeed, $d_{p^n-p^{n-1}}=\mathrm{res}^*(\Gamma(d_{p^{n-1}-p^{n-2}})\pm 1\otimes b^{p^n-p^{n-1}}$, (or $1\otimes e^{2^{n-1}}$ when $p=2$) but if $\pi\colon\mathrm{Syl}_p(\mathcal{S}_{p^{n-1}})\wr\mathbb{Z}/p\to\mathbb{Z}/p$ is the projection onto the new $\mathbb{Z}/p$ and $e=\pi^*(e)$ then $\mathrm{res}^*(e)=e$ in the expression above for $p=2$ and similarly for b when p is odd. □

This shows, since the $d_{p^n-p^j}$ are all Steenrod P^{th} powers of $d_{p^n-p^{n-1}}$, that when $p=2$ the entire invariant subalgebra is in the image of the restriction map and when p is odd, the entire invariant **polynomial** subalgebra is also in the image of restriction, but it leaves open the question of the part which involves the exterior terms. Here we follow the discussion in [Ma]. Define

(1.17)
$$L_i = \begin{vmatrix} b_1^{p^{i-1}} & \dots & b_i^{p^{i-1}} \\ \vdots & \ddots & \vdots \\ b_1^{p^r} & \dots & b_i^{p^r} \\ \vdots & \ddots & \vdots \\ b_1 & \dots & b_i \end{vmatrix}$$
i.e. the k, j entry of L_i is $b_j^{p^k}$, $(\leq r \leq i-1)$.

Similarly, set

$$M_{j,i} = \begin{vmatrix} b_1^{p^{i-1}} & \dots & b_i^{p^{i-1}} \\ \vdots & \ddots & \vdots \\ \widehat{b_1^{p^j}} & \dots & \widehat{b_i^{p^j}} \\ \vdots & \ddots & \vdots \\ b_1 & \dots & b_i \\ e_1 & \dots & e_i \end{vmatrix}$$
i.e. the b^{p^j} row is omitted $(1 \leq j \leq i-1)$.

We have $P^{p^j}(b^{p^j}) = b^{p^{j+1}}$ but $P^{p^j}(b^r) = 0$ for $r < p^j$. From this we find

$$\begin{aligned} M_{j,i} &= P^{p^{j-1}+p^j+\cdots+p^{i-1}}(M_{i-1,i}), \\ L_i &= \beta P^{1+p+\cdots+p^{i-1}}(M_{i-1,i}), \end{aligned}$$

while

$$L_{j,i} = \begin{vmatrix} b_1^{p^i} & \dots & b_i^{p^i} \\ \vdots & \ddots & \vdots \\ \widehat{b_1^{p^j}} & \dots & \widehat{b_i^{p^j}} \\ \vdots & \ddots & \vdots \\ b_1 & \dots & b_i \end{vmatrix} = P^{p^j+p^{j+1}+\cdots+p^{i-1}}(L_i) \,.$$

Note that $\dim(M_{i,i+1}) = p(\dim(M_{i-1,i})) - p + 3$ so we have, setting $q = (p-3)/2 = m-1$,
(1.18)

$$\operatorname{res}^*(\Gamma(M_{i-1,i})) = \pm \left(\sum_{j=0}^{i-1} (-1)^j M_{j,i} b_{i+1}^{p^j+q} - L_i b_{i+1}^q e \right) = M_{i,i+1} b_{i+1}^q$$

up to a sign. Here the restriction is from $S_{p^i} \wr \mathbb{Z}/p$ to $V_{i+1}(p) = \Delta^p(V_i(p)) \times \mathbb{Z}/p$ and the classes b_{i+1}, e are associated to the rightmost $\mathbb{Z}/p$. Moreover, $P^{p^i} M_{i,i+1} b_{i+1}^q = M_{i-1,i+1} b_{i+1}^q$. A similar calculation shows that $\operatorname{res}^*(\Gamma(L_i)) \cup b_{i+1} = L_{i+1}$. Also we should note that each of the matrices above is invariant under $\mathrm{SL}_n(p)$ but that $gJ = det(g)J$ for $g \in \mathrm{GL}_n(p)$, so any product of $p-1$ of them is invariant under $\mathrm{GL}_n(p)$.

Example 1.19 $H^*(\mathcal{S}_{p^2};\mathbb{F}_p)$. The following calculation is due to H. Cardenas. We have

$$\begin{aligned}
M_{1,2}M_{0,2}L_2^{p-3} &= \begin{vmatrix} b_1 & b_2 \\ e_1 & e_2 \end{vmatrix}\begin{vmatrix} b_1^p & b_2^p \\ e_1 & e_2 \end{vmatrix}\begin{vmatrix} b_1^p & b_2^p \\ b_1 & b_2 \end{vmatrix}^{p-3} \\
&= (b_1e_2 - b_2e_1)(b_1^pe_2 - e_1b_2^p)(b_1^pb_2 - b_1b_2^p)^{p-3} \\
&= (b_1b_2^p - b_1^pb_2)^{p-2}e_1e_2 \\
&= (b_1^p - b_1b_2^{p-1})^{p-2}e_1b_2^mb_2^{m-1}e_2 \\
&= \pm\mathrm{res}^*(\Gamma(b))^{p-2}\mathrm{res}^*(\Gamma(e))b_2^{m-1}e_2
\end{aligned}$$

and we see that the $\mathrm{GL}_2(p)$-invariant class $M_{1,2}M_{0,2}L_2^{p-3}$ is in the image of restriction from $H^*(\mathrm{Syl}_p(\mathcal{S}_{p^2});\mathbb{F}_p)$. Thus, applying β we have that $M_{1,2}L_1^{p-2}$ is in the restriction image, and applying $P^1\beta$ we have $M_{0,2}L_1^{p-2}$ is also in this image. The polynomial algebra $\mathbb{F}_p[d_{p^2-p}, d_{p^2-1} = L_2^{p-1}]$ is also in the restriction image, and, from (III.2.9), we see that this is the entire $\mathrm{GL}_2(p)$ invariant subalgebra of $H^*(V_2(p);\mathbb{F}_p)$.

The groups $H^*(\mathcal{S}_{p^2};\mathbb{F}_p)$ are detected by restriction to $H^*(V_2(p);\mathbb{F}_p)$, $H^*(V_1(p)^p;\mathbb{F}_p)$. From (III.4.2) and (1.8) above it follows that the image of this restriction for $V_1(p)^p$ is

$$\mathbb{F}_p[s_1,\ldots,s_p]\otimes E(f_1,\ldots,f_p)$$

where $\dim(s_i) = 2i(p-1)$ and $\dim(f_i) = 2i(p-1)-1$. Also, the only generating class which restricts to $H^*(V_2(p);\mathbb{F}_p)$ non-trivially and also restricts to $H^*(V_1(p)^p;\mathbb{F}_p)$ gives the pair (σ_p, d_{p^2-p}).

The remarks preceeding (1.19) show that more generally

$$\Gamma(M_{i-1,i}M_{i-2,i}L_i^{p-3}) = \pm M_{i,i+1}M_{i-1,i+1}L_{i+1}^{p-3}$$

and so this element is in the image from the restriction of $H^*(\mathcal{S}_{p^i};\mathbb{F}_p)$ for each i. Applying Steenrod p-power operations and Bocksteins we obtain the following table of further elements:

$$
\begin{array}{ccccccc}
M_{i-1}M_{i-2}L^{p-3} & & & & & & \\
\Big\downarrow P^{p^{i-3}} & & & & & & \\
M_{i-1}M_{i-3}L^{p-3} & \xrightarrow{P^{p^{i-2}}} & M_{i-2}M_{i-3}L^{p-3} & & & & \\
\Big\downarrow P^{p^{i-4}} & & \Big\downarrow P^{p^{i-4}} & & & & \\
M_{i-1}M_{i-4}L^{p-3} & \xrightarrow{P^{p^{i-2}}} & M_{i-2}M_{i-4}L^{p-3} & \xrightarrow{P^{p^{i-3}}} & \cdots & & \\
\Big\downarrow P^{p^{i-5}} & & \Big\downarrow P^{p^{i-5}} & & & & \\
\vdots & & \vdots & & \cdots & \xrightarrow{P^{p^2}} & M_2M_1L^{p-3} \\
\Big\downarrow P^{1} & & \Big\downarrow P^{1} & & & & \Big\downarrow P^{1} \\
M_{i-1}M_0L^{p-3} & \xrightarrow{P^{p^{i-2}}} & M_{i-2}M_0L^{p-3} & \xrightarrow{P^{p^{i-3}}} & \cdots & \xrightarrow{P^{p^2}} & M_2M_0L^{p-3} \\
\Big\downarrow \beta & & \Big\downarrow \beta & & & & \Big\downarrow \beta \\
M_{i-1}L^{p-2} & \xrightarrow{P^{p^{i-2}}} & M_{i-2}L^{p-2} & \xrightarrow{P^{p^{i-3}}} & \cdots & \xrightarrow{P^{p^2}} & M_2L^{p-2}
\end{array}
$$

Further, one can verify that $M_j^2 = 0, 0 \leq j \leq i-1$ but $M_{i-1}M_{i-2}\cdots M_0 \neq 0$. Thus, there are further products that we can write down and Mann shows, using a counting argument depending on the homology of symmetric products of spheres, that the image of the restriction map is exactly the free module over the Dickson algebra of $\mathrm{GL}_n(p)$-polynomial invariants generated by 1 and the elements together with their distinct products above. It is possible to give an algebraic proof of these results thus avoiding any reference to topology. Here are details for these last two steps.

Lemma 1.20 *For each $i > 0$ and every odd prime p we have*

$$M_{i-1}M_{i-2}\cdots M_0 = \pm e_1e_2\cdots e_iL_i^{i-1}\,.$$

Proof. One has an expansion for each j,

$$M_j \;=\; \pm\sum_{r=0}^{i} e_rL^{j,r}$$

where $L^{j,r}$ is the (j,r) minor of L given as $(-1)^{r+j}\text{Det}(V^{j,r})$ where $V^{j,r}$ is the matrix obtained from the matrix, $\mathcal{L}$, with determinant L by deleting the $(i-j)^{th}$ row and the rth column. But then it is easy to verify that

$$M_{i-1}\cdots M_0 \;=\; e_1e_2\cdots e_i\text{Det}\left(L^{r,j}\right)\ .$$

On the other hand

$$\mathcal{L}\cdot\left(L^{r,j}\right)=\begin{pmatrix} L & 0 & \dots & 0\\ 0 & L & \dots & 0\\ \vdots & \vdots & \ddots & \vdots\\ 0 & 0 & \dots & L\end{pmatrix}$$

so taking determinants of both sides (1.20) follows. □

Finally, to complete the demonstration that the elements above generate the entire image of res* we proceed by induction using the factorization

$$V_i(p)\xrightarrow{\Delta^p\times\text{id}}V_{i-1}(p)\wr\mathbb{Z}/p\xrightarrow{(\text{res})\wr\text{id}}\mathcal{S}_{p^{i-1}}\wr\mathbb{Z}/p\hookrightarrow\mathcal{S}_{p^i}$$

for the restriction map. The details are direct.

Remark 1.21 The entire ring of invariants has been discussed in (III.2.9), and for $p>3$ and $n>3$ also, [Ma] shows that the classes above generate a **proper** subalgebra of this ring.

2. Hopf Algebras

An augmented, graded, algebra over a field $\mathbb{F}$ is an associative, graded, unitary $\mathbb{F}$-module together with a ring homomorphism $\epsilon\colon A\to\mathbb{F}$, where $\mathbb{F}$ is thought of as concentrated in degree 0. Here, to be precise the multiplication $\mu\colon A\otimes_{\mathbb{F}}A\to A$ is given and the unit corresponds to a graded homomorphism $1\colon\mathbb{F}\to A$ so that the compositions $\mathbb{F}\otimes_{\mathbb{F}}A\xrightarrow{\epsilon\otimes 1}A\otimes_{\mathbb{F}}A\xrightarrow{\mu}A$ and $A\otimes_{\mathbb{F}}\mathbb{F}\xrightarrow{1\otimes\epsilon}A\otimes_{\mathbb{F}}A\xrightarrow{\mu}A$ are both the identity maps, as is the composition $\epsilon 1\colon\mathbb{F}\to A\to\mathbb{F}$. The associative condition can be written diagramatically as saying that the diagram

$$\begin{array}{ccc} A\otimes A\otimes A & \xrightarrow{\mu\otimes 1} & A\otimes A\\ \Big\downarrow{\scriptstyle 1\otimes\mu} & & \Big\downarrow{\scriptstyle\mu}\\ A\otimes A & \xrightarrow{\mu} & A\end{array}$$

commutes.

A graded coassociative, counitary, coaugmented, coalgebra over a field $\mathbb{F}$ is a graded $\mathbb{F}$-module, B, together with a coproduct map

$$\phi : B \longrightarrow B \otimes_{\mathbb{F}} B\,,$$

counit, i.e. a graded $\mathbb{F}$-module morphism

$$1^* : B \longrightarrow \mathbb{F}$$

so that $(\mathrm{id} \otimes 1^*)\phi = (1^* \otimes \mathrm{id})\phi = \mathrm{id}$, and coaugmentation map

$$\epsilon^* : \mathbb{F} \rightarrow B$$

which is also graded and satisfies the condition that the composition

$$(\epsilon^* \otimes_{\mathbb{F}} \epsilon^*)\mathrm{id} : \mathbb{F} \xrightarrow{=} \mathbb{F} \otimes_{\mathbb{F}} \mathbb{F} \rightarrow B \otimes_{\mathbb{F}} B$$

is just $\phi\epsilon^*$. As before the coaugmentation and counit are related by $1^*\epsilon^* = \mathrm{id}\colon \mathbb{F} \rightarrow \mathbb{F}$. The coassociative condition is that the diagram below commutes

$$\begin{array}{ccc} B & \xrightarrow{\phi} & B \otimes_{\mathbb{F}} B \\ \Big\downarrow{\scriptstyle \phi} & & \Big\downarrow{\scriptstyle \phi\otimes \mathrm{id}} \\ B \otimes_{\mathbb{F}} B & \xrightarrow{\mathrm{id}\otimes\phi} & B \otimes_{\mathbb{F}} B \otimes_{\mathbb{F}} B\,. \end{array}$$

Definition 2.1 A *Hopf algebra* $(A, \mu, \phi, \epsilon, 1)$ over a field $\mathbb{F}$ is a graded, associative, unitary, augmented algebra, together with a counitary, coaugmented, coassociative, coalgebra map ϕ so that ϕ is a homomorphism of graded algebras for which the counit is ϵ and the coaugmentation is 1.

Example 2.2 Let A be the polynomial algebra on an single variable x of dimension $2i$ over the field $\mathbb{F}$. Then the coproduct structure is determined by $\phi(x) = x \otimes 1 + 1 \otimes x$ so $\phi(x^i) = \sum_{ji=0}^{i} \binom{i}{j} x^j \otimes x^{i-j}$. Thus, in the dual algebra the product is given by the rule

$$(x^j)^* \cdot (x^k)^* = \binom{j+k}{j} (x^{j+k})^*\,.$$

It follows that if $\mathbb{F}$ has characteristic 0 then A^* is also isomorphic to the polynomial algebra on one variable x^* in dimension $2i$, but if $char(\mathbb{F}) = p$ then a direct exercise with binomial coefficients $\mathrm{mod}(p)$ shows that

$$A^* = \mathbb{F}[x^*, \gamma_p, \ldots, \gamma_{p^i}, \ldots]/(x^{*p} = \ldots = \gamma_{p^i}^p = \ldots = 0), \quad \dim(\gamma_{p^j}) = 2ip^j\,.$$

In case A is of finite type, i.e. A_i is finite dimensional over $\mathbb{F}$ for all i these notions are self dual, $A_i^* = \mathrm{Hom}_{\mathbb{F}}(A_i, \mathbb{F})$, $(A^*, \phi^*, \epsilon^*, 1^*)$ becomes an associative, unitary, augmented algebra and $(A^*, \mu^*, \epsilon^*, 1^*)$ becomes a coassociative, coaugmented, counitary coalgebra. In particular, $(A^*, \phi^*, \mu^*, 1^*, \epsilon^*)$ is also a Hopf algebra and $(A^{**}, \mu^{**}, \phi^{**}, \epsilon^{**}, 1^{**}) = (A, \mu, \phi, \epsilon, 1)$.

The Hopf algebra $(A, \mu, \phi, \epsilon, 1)$ is said to be connected if $A^i = 0$ for $i < 0$ and $1: \mathbb{F} \longrightarrow A_0$ is an isomorphism. A is said to be commutative in case $\mu: A \otimes A \longrightarrow A$ satisfies $\mu(a \otimes b) = (-1)^{|a||b|}\mu(b \otimes a)$ for all $a \in A_{|a|}$, $b \in A_{|b|}$.

Example 2.3 Let G be a group and $\mathbb{F}(G)$ its group ring, then

$$\phi: \mathbb{F}(G) \longrightarrow \mathbb{F}(G) \otimes \mathbb{F}(G)$$

is defined by $\phi(g) = (g \otimes g)$ for all $g \in G$, so $\phi \sum f_i g_i = \sum f_i(g_i \otimes g_i)$. This is also a unitary, coconnected Hopf algebra, but it is not connected unless $G = 1$.

In the sequel all the Hopf algebras we consider will be connected, commutative, and even cocommutative of finite type. So we assume these conditions in the remainder of the section. Hopf algebras arise for us as follows. We are given a sequence of groups and injective homomorphisms

$$G_n, \ 0 \le n < \infty, \ G_0 = \{1\}, \qquad \mu(n,m): G_n \times G_m \longrightarrow G_{n+m}$$

for which the $\mu(n,m)$ are associative up to congugation, i.e. for any triple (n,m,l) we have that there is an element $g(n,m,l) \in G_{n+m+l}$ so that

$$\begin{aligned} g(n,m,l)^{-1}(\mu(n+m,l)(\mu(n,m)(a,b),c)g(n,m,l) \\ = \mu(n,m+l)(a,\mu(m,l)(b,c)) \end{aligned}$$

for all $(a,b,c) \in G_n \times G_m \times G_l$. We also assume the $\mu(n,m)$ are commutative up to conjugation. This means that for each pair (n,m) there is $g(n,m) \in G_{n+m}$ so that

$$g(n,m)^{-1}\mu(n,m)(a,b)g(n,m) = \mu(m,n)(b,a) \qquad \forall (a,b) \in G_n \times G_m \, .$$

For example if $G_n = \mathcal{S}_n$, the symmetric group and the pairing $\mu(n,m)$ is the usual inclusion of $\mathcal{S}_n \times \mathcal{S}_m \subset \mathcal{S}_{n+m}$, with $\mathcal{S}_n$ acting on the first n elements and $\mathcal{S}_m$ acting on the last m, then these conditions are satisfied. Here the $g(n,m,l) = \text{id}$ but the $g(n,m)$ are non-trivial.

Of course $\coprod_1^\infty H_*(G_n; \mathbb{F}_p)$ is only an associative, commutative, graded algebra, but not a Hopf algebra since the diagonal map $\Delta: B_G \longrightarrow B_G \times B_G$ which, on passing to cohomology, is supposed to give the coproduct, ϕ, has the form

$$\Delta_*(\alpha_n) = \alpha_n \otimes t^n + t^n \otimes \alpha_n + \sum_i a_i' \otimes a_i''$$

where $\alpha_n \in H_*(G_n; \mathbb{F}_p)$ and $t^n \in H_0(G_n; \mathbb{F}_p)$ is the element given as the image of a chosen generator $t \in H_0(G_1; \mathbb{F}_p)$ under the iterate of the multiplication map,

$$\underbrace{G_1 \times \cdots \times G_1}_{n\ times} \longrightarrow G_n\ .$$

Next, adjoin a unit 1 to the algebra and note that the quotient of this algebra by the ideal $(1-t)$ is a Hopf algebra. In dimension i we have, clearly, that this quotient is given as $\lim_n(H_i(G_n;\mathbb{F}_p))$.

Lemma 2.4 *Let* $G = \lim\limits_{n\to\infty} G_n$ *where the inclusion* $G_n \subset G_{n+1}$ *is* $\mu(n,0)$, *then*

$$H_*(G;\mathbb{Z}/p) = \lim_{n\to\infty} H_*(G_n;\mathbb{Z}/p)\ .$$

(Any chain in the bar construction for G appears in the bar construction of G_m for some sufficiently large m.)

In particular, while the maps $\mu(n,m)$ need not pass to limits to define a "reasonable" homomorphism $G\times G\longrightarrow G$, in homology the maps do pass to limits to define a pairing

$$\mu\colon H_*(G;\mathbb{Z}/p)\otimes H_*(G;\mathbb{Z}/p)\longrightarrow H_*(G;\mathbb{Z}/p)$$

which, together with the diagonal map

$$\Delta\colon H_*(G;\mathbb{Z}/p)\longrightarrow H_*(G;\mathbb{Z}/p)\otimes H_*(G;\mathbb{Z}/p)$$

gives $H_*(G;\mathbb{Z}/p)$ the structure of a commutative, cocommutative, connected Hopf algebra. Of course, it is not always going to be true that the Hopf algebra is of finite type but this does turn out to be the case for the symmetric groups above and the other cases we consider, finite groups of Lie type which are studied in Chapter VII.

Thus, associated to $\coprod \mathcal{S}_n$ is the Hopf algebra $H_*(\mathcal{S}_\infty;\mathbb{F}_p)$. In §3 we will show that the restriction map $\text{res}_*\colon H_*(\mathcal{S}_n;\mathbb{F}_p)\longrightarrow H_*(\mathcal{S}_\infty;\mathbb{F}_p)$ is injective for all n, and we will determine the Hopf algebra structure on $H_*(\mathcal{S}_\infty;\mathbb{F}_p)$ completely. As a result we will obtain the groups $H_*(\mathcal{S}_n;\mathbb{F}_p)$ by simply describing the image of the restriction map in the resulting Hopf algebra.

Before we can do this we need to discuss some basic structure theorems for commutative and cocommutative Hopf algebras and for this we need to discuss sub-Hopf algebras and quotient Hopf algebras.

The notion of sub-Hopf algebra is evident, as is the notion of quotient Hopf algebra and Hopf algebra map $\lambda\colon A\longrightarrow B$ of Hopf algebras . Duality connects these notions together very nicely. Given a sub-Hopf algebra $A'\subset A$, note that $A'_0 = A_0 = \mathbb{F}$ by our assumptions and let $IA' = \text{Ker}(\epsilon|A')$. ($IA$ is called the augmentation ideal in A.) Then $AIA'A$ is a two sided ideal in A and $B = A/AIA'A$ is written $A//A'$. B inherits a unit, augmentation, product and coproduct from A, and so becomes a Hopf algebra. It is the quotient of A by the Hopf subalgebra A'. Dually, $B^*\subset A^*$ is a Hopf subalgebra of A^* and $A^*//B^* = A'^*$.

Proposition 2.5 *Let $B \subset A$ be a sub-Hopf algebra of the commutative Hopf algebra A, and $C = A//B$ be the quotient. Let $e\colon A \longrightarrow A \otimes_{\mathbb{F}} C$ be the composition*

$$A \xrightarrow{\phi} A \otimes A \xrightarrow{1 \otimes p} A \otimes_{\mathbb{F}} C$$

where $p\colon A \longrightarrow C$ is the projection. Then $e(\theta) = \theta \otimes 1$ if and only if $\theta \in B$.

(The kernel of p has the form $I(B)A$ and can be written $\sum I(B)w_j$ where the w_j map onto an $\mathbb{F}$-basis for C since A is commutative. But from this the result is direct.)

An element $\alpha \in A_i$ is said to be primitive if $\phi(\alpha) = \alpha \otimes 1 + 1 \otimes \alpha$. The set of primitives, $\mathcal{P}(A)$ is a graded $\mathbb{F}$-submodule of A and is closed under the pth-power operation, $\xi\colon \alpha \mapsto \alpha^p$, if $char(\mathbb{F}) = p$. It is always closed under the Lie bracket operation $[\alpha, \beta] = \alpha\beta - (-1)^{|\alpha||\beta|}\beta\alpha$. Thus the set $\mathcal{P}(A)$ is a (restricted) Lie algebra where "restricted" means the existence of the operation ξ. A very important property of $\mathcal{P}(A)$ is that $\mathcal{P}(A \otimes_{\mathbb{F}} B) = \mathcal{P}(A) \oplus \mathcal{P}(B)$ where A and B are Hopf algebras.

A dual notion to being primitive is being indecomposable. The set of indecomposables $\mathcal{Q}(A)$ for A is the quotient $IA/(IA)^2$ of IA by the square of its augmentation ideal. The lifts of a basis for $\mathcal{Q}(A)$ to A form a set of algebra generators for A.

Proposition 2.6 *Let $\nu\colon A \longrightarrow B$ be a surjection of Hopf algebras with A connected. Let $\mu \in \mathrm{Ker}(\nu)$ have minimal dimension then μ is primitive.*

$(\nu \otimes \nu(\phi(\mu)) = 0$ since ν is a map of Hopf algebras, but $\phi(\mu) = \mu \otimes 1 + 1 \otimes \mu + \sum \mu' \otimes \mu''$ and the connectedness of A implies that $\dim(\mu')$, $\dim(\mu'')$ are both less than $\dim(\mu)$, thus the sum term is identically zero.)

The Theorems of Borel and Hopf

A Hopf algebra is said to be primitively generated if the lifts of its indecomposables can be chosen to be primitives. For example, this is the case with $\mathbb{F}_p[x]$ where $\phi(x) = x \otimes 1 + 1 \otimes x$. But note that it is not the case for $\mathbb{F}_p[x]^*$ where only x^*, among the indecomposables, can be taken to be primitive. We have

Proposition 2.7 *Let A be a connected Hopf algebra of finite type which is commutative and primitively generated. Suppose also that $\mathbb{F}$ is perfect if $char(\mathbb{F}) < \infty$. Then as an algebra, $A = \bigotimes_i \mathcal{M}(i)$ where each $\mathcal{M}(i)$ has the form $\mathbb{F}[x]/(x^r)$ and r is constrained by the following conditions:*

1. *$r = 2$ if $\dim(x)$ is odd and $char(\mathbb{F}) \neq 2$,*
2. *when $\dim(x)$ is even $r = 0$ if $char(\mathbb{F}) = 0$ and has the form p^l (or zero) if $char(\mathbb{F}) = p$.*

Proof. Order the generators of A, $f_1, \dots, f_n, \dots$ so that if $i < j$ then $\dim(f_i) \le \dim(f_j)$. We prove the result by induction. Set $A(i)$ to be the subalgebra of A generated by $f_1, \dots, f_i$. Since the f_i are primitive $A(i)$ is also a Hopf subalgebra of A. Clearly, the proposition holds for $A(1)$. We assume it for $A(i)$ and consider the surjection

$$A(i) \otimes_{\mathbb{F}} \mathbb{F}[\bar{f}_{i+1}] \longrightarrow A(i+1)$$

where $\bar{f}_{i+1} \mapsto f_{i+1}$. The primitives in $A(i) \otimes \mathbb{F}[\bar{f}_{i+1}]$ are $\mathcal{P}(A(i)) \oplus \langle \bar{f}_{i+1}^{p^j} \mid j = 0, 1, \dots \rangle$. Let θ be an element of minimal degree in the kernel. Then θ is primitive and so can be written as $(\bar{f}_{i+1}^{p^j} + \sum_{t<i} \lambda_t f_t^{p^{s(t)}})$. By the ordering condition $s(t) \ge j$ for all t. Since $\mathbb{F}$ is perfect it follows that there is ρ_t so that $\rho_t^{p^{s(t)}} = \lambda_t$ for each t. Replace f_{i+1} by $f'_{i+1} = f_{i+1} - \sum \rho_t f_t^{p^{s(t)-j}}$. Then $f'^{p^j}_{i+1} = 0$, $A(i+1)$ is the surjective image of $A(i) \otimes_{\mathbb{F}} \mathbb{F}[\bar{f}_{i+1}]/(\bar{f}_{i+1}^{p^j})$ and there is no longer any primitive in the kernel so the surjection is an isomorphism. □

This last result generalizes to the important theorem of Borel and Hopf.

Theorem 2.8 *Let A be a connected Hopf algebra of finite type over the perfect field $\mathbb{F}$. If the multiplication in A is commutative then, as an algebra A is a tensor product $A = \otimes_i \mathcal{M}_i$ where each $\mathcal{M}_i$ is of the type above.*

Proof. Define $A(i) \subset A$ to be the subalgebra of A generated by all the elements of dimension $\le i$ in A. It is a sub-Hopf algebra. Since $A(1)$ is primitively generated we can begin an induction. Assume the result proved for $A(i)$. Set $W \subset A_{i+1}$ to be a complementary $\mathbb{F}$-subspace to $A(i) \cap A_{i+1}$, and choose a basis $x_1, \dots, x_r$ for W. We have the sequence of Hopf-subalgebras

$$A(i) \subset \langle A(i), x_1 \rangle \subset \langle A(i), x_1, x_2 \rangle \subset \cdots \subset A(i+1)$$

and we assume the result for $\langle A(i), x_1, \dots, x_j \rangle$. Thus we are reduced to considering the situation $\langle A, b \rangle$ where the indecomposables of A all have dimension $\le \dim(b)$ and there is an exact sequence $Q(A) \longrightarrow Q(\langle A, b \rangle) \longrightarrow \langle b \rangle$.

We have that $\langle A, b \rangle // A$ is primitively generated and thus has the form $\mathbb{F}[b]/(b^{p^f})$ (or $\mathbb{F}[b]/(b^2)$ when $\dim(b)$ is odd, but since this case is easy we assume for the rest of the proof that $\dim(b)$ is even). Consider, as before the surjection $A \otimes_{\mathbb{F}} \mathbb{F}[\bar{b}] \xrightarrow{e} \langle A, b \rangle$ and let $\bar{b}^{p^r} + w$ be the least dimensional element in the kernel. If w is zero we are finished since the composite map

$$A \otimes \mathbb{F}[\bar{b}]/(\bar{b}^{p^f}) \xrightarrow{e} \langle A, \bar{b} \rangle \xrightarrow{\nu} \langle A, \bar{b} \rangle \otimes \langle A, \bar{b} \rangle // A$$

is clearly injective so e is an isomorphism of algebras.

In case $w \ne 0$ we proceed as follows. Let $B \subset A$ be the sub-Hopf algebra of A consisting of p^f-powers. (This is where we use the fact that $\mathbb{F}$ is perfect.) Consider $\langle A, \bar{b} \rangle // B = C$. Since $\phi(\bar{b}) = \bar{b} \otimes 1 + 1 \otimes \bar{b} + \sum v_i \otimes v'_i$ with $\sum v_i \otimes v'_i \in$

$A\otimes A$ we have that $\phi(\xi^f(\bar{b}))=\xi^f(\bar{b})\otimes 1+1\otimes\xi^f(\bar{b})$ in C. We need to show that this implies $\xi^f(\bar{b})=0$ in C and this follows from

Proposition 2.9 *If A is a connected Hopf algebra over a field of characteristic $p>0$ and the multiplication in A is commutative then if*
1. $Q(A)_r=0$ for $r>n$, and
2. $\xi^f(IA)=0$ for some integer f,
it follows that $P(A)_r=0$ for $r>p^{f-1}n$.

Proof of 2.9 Suppose that the assertion is proved for Hopf algebras with q or less generators as an algebra and A has $q+1$ generators. Choose a set of $q+1$ generators for A and let X be one having highest degree. Let A' be the sub-Hopf algebra generated by the remaining generators, and let $A''=A//A'$, so A'' is a Hopf algebra with 1 generator that satisfies (1) and (2). We have an exact sequence $P(A')\longrightarrow P(A)\longrightarrow P(A'')$. Now we can apply the inductive hypothesis. □

Returning to the proof of the theorem, we have that $\xi^f(b)=b^{p^f}=0$ under projection onto C. Thus $(\mathrm{id}\otimes p)\phi(b^{p^f})=\xi^f(b)\otimes 1$. Consequently $b^{p^f}\in B$ and there is an element $v\in A$ so $b^{p^f}=v^{p^f}$ and if we set $b'=b-v$ we are in the previous case and the result follows. □

3. The Structure of $H_*(\mathcal{S}_n;\mathbb{F}_p)$

The element in $H^*(\mathcal{S}_{p^{i-1}}\wr\mathbb{Z}/p;\mathbb{F}_p)$ which restricts to L_i (defined in (1.17)), which we write as L when there is no possibility of confusion, has the form $\Gamma(L_{i-1})\cup\pi^*(b)$. Consequently it restricts to 0 in $H^*(\mathcal{S}^p_{p^{i-1}};\mathbb{F}_p)$. It follows that all the elements described in (1.14) – (1.20) which restrict to the ideal

(3.1)
$$\begin{aligned}\mathcal{I}_i &= \mathbb{F}_p[d_{p^i-p^{i-1}},\ldots,d_{p^i-1}](d_{p^i-1},M_0L^{p-2},\cdots,M_jL^{p-2},\\ &\qquad\qquad\ldots,M_0\cdots M_{i-1}L^{\gamma})\end{aligned}$$

all restrict to 0 in $H^*(\mathcal{S}_{p^{i-1}};\mathbb{F}_p)$. Also, by (1.9(1)) they survive non-trivially to $H_*(\mathcal{S}_\infty;\mathbb{F}_p)$ as do the classes

(3.2)
$$\underbrace{\theta\otimes\theta\otimes\cdots\otimes\theta}_{p^j-times}$$

for $\theta\in\mathcal{I}_i$ and $\dim(\theta)$ even (by (1.9(2))). In particular, if θ_* is the dual of any such even dimensional element in $\mathcal{I}_i$ it has image an indecomposable in the Hopf algebra $H_*(\mathcal{S}_\infty;\mathbb{F}_p)$, $\xi^j(\theta_*)\neq 0$ for $j\geq 1$, and is independent of $\xi^l(\lambda_*)$ for any other $\lambda\in\mathcal{I}_k$. Thus the dual groups $\mathcal{N}_i=\mathcal{I}_i^*$ generate a polynomial algebra on even dimensional generators tensored with an exterior algebra on odd dimensional elements which is a split summand of $H_*(\mathcal{S}_\infty;\mathbb{F}_p)$. Letting $\Lambda(\mathcal{N}_i)$ denote this algebra we have from (1.8), the discussion after (1.19), and the Borel-Hopf theorem (2.8), that

$$H_*(\mathcal{S}_\infty;\mathbb{F}_p) \cong \bigotimes_0^\infty \Lambda(\mathcal{N}_i)\,. \tag{3.3}$$

Moreover, applying (1.8) again we see that the map

$$\text{res}_*\colon H_*(\mathcal{S}_n;\mathbb{F}_p) \longrightarrow H_*(\mathcal{S}_\infty;\mathbb{F}_p)$$

must be injective since the Weyl group of a group of the form appearing in (1.3) is a product of the Weyl groups for the

$$\underbrace{V_r(p)\times\cdots\times V_r(p)}_{i_r}$$

separately and this Weyl group is $\mathrm{GL}_r(p)\wr\mathcal{S}_{i_r}$. Thus the homology images of tensor products of elements in $\mathcal{N}_{i_r}$ surject onto $H_*(\mathcal{S}_n;\mathbb{F}_p)$ and, since these elements last to $H_*(\mathcal{S}_\infty;\mathbb{F}_p)$ subject only to the same symmetrizing conditions imposed in $H_*(\mathcal{S}_n;\mathbb{F}_p)$ – namely symmetry under $\mathcal{S}_{i_r}$ – the injectivity follows.

We specify the image of $H_*(\mathcal{S}_n;\mathbb{F}_p)$ more precisely by giving a second degree to each monomial generator of $\mathcal{N}_i$, $deg_2(\theta)=p^i$. Then to each monomial in (3.2), $\theta_1\otimes\cdots\otimes\theta_l\in H_*(\mathcal{S}_\infty;\mathbb{F}_p)$, we assign the second degree

$$deg_2(\theta_1\otimes\cdots\otimes\theta_l)=deg_2(\theta_1)+\cdots+deg_2(\theta_l)\,,$$

and it follows that $H_*(\mathcal{S}_n;\mathbb{F}_p)\subset H_*(\mathcal{S}_\infty;\mathbb{F}_p)$ is the **additive** subgroup spanned by the tensor product monomials having second degree $\le n$.

When $p=2$ we have a very convenient method for indexing these classes. Let $I=(i_1\le i_2\le\cdots\le i_r)$ be a sequence of non-negative integers, not all of them zero. Then there is a unique generator $Q_I=Q_{i_1}Q_{i_2}\cdots Q_{i_r}\in\mathcal{N}_r$ having dimension $i_1+2i_2+\cdot+2^{r-1}i_r$ corresponding to it. It is the image of the class $e_*^{i_1}\otimes\cdots\otimes e_*^{2^{r-1}i_r}$. There are similar descriptions for the generators of $\mathcal{N}_i$ for odd primes, but the conditions on the sequence $J=(i_1\le i_2\le\cdots\le i_r)$ are more complex.

For notational convenience, when $p=2$ we write the product in homology as $\alpha*\beta$. thus a typical monomial in $H_*(\mathcal{S}_\infty;\mathbb{F}_2)$ will have the form

$$Q_{I_1}*Q_{I_2}*\cdots*Q_{i_r}\,.$$

It should also be pointed out that there are significant differences between the dual algebras. The following result is direct when we check the pth power operation on the subrings of the $H^*(V_i(p);\mathbb{F}_p)$ given as the images of res*.

Theorem 3.4

1. *$H^*(\mathcal{S}_\infty;\mathbb{F}_2)$ is a polynomial algebra on generators in one to one correspondence with the generators for $H_*(\mathcal{S}_\infty;\mathbb{F}_p)$.*

2. *For each odd prime p* $H^*(\mathcal{S}_\infty;\mathbb{F}_p)$ *is a tensor product of a polynomial algebra on classes indexed by r-tuples* $(i_1,i_2,\ldots,i_r)$ *with* $i_1 \le i_2 \le \cdots i_r$ *and dimension* $2(p-1)(i_1+pi_2+\cdots+p^{r-1}i_r)$ *together with an exterior algebra on odd generators dual to the odd generators for* $H_*(\mathcal{S}_\infty;\mathbb{F}_p)$ *together with a divided power algebra on generators indexed by the remaining elements in the union of the* $\mathcal{N}_j$.

Remark F or $p = 2$ the cohomology generator which corresponds to D^{2^r} where $D = d^{j_1}_{2^i-1}\cdots d^{j_i}_{2^i-1} \in \mathcal{I}_i$ is actually the element which restricts to

$$\underbrace{D\otimes D\otimes\cdots\otimes D}_{2^r\ times} \in H^*(V_i(p)^{2^r};\mathbb{F}_2)\ ,$$

and a similar discription is given for the elements corresponding to p^{th} powers of Dickson elements at odd primes.

By divided power algebra on an even dimensional generator γ, we mean the tensor product

$$\mathbb{F}_p[\gamma]/(\gamma^p)\otimes\mathbb{F}_p[\gamma_1]/(\gamma_1^p)\otimes\cdots\otimes\mathbb{F}_p[\gamma_r]/(\gamma_r^p)\otimes\cdots$$

where $\dim(\gamma_r) = p^r\cdot\dim(\gamma)$. Then classes of the form e.g.,

$$\gamma = \{d^{j_1}_{p^i-p^{i-1}}\cdots d^{j_i}_{p^i-1}M_rM_sL^{p-3}\}$$

together with the classes restricting to

$$\underbrace{\gamma\otimes\cdots\gamma}_{p^r\ times}$$

generate the divided power algebras.

The structure of $H_*(\mathcal{S}_n;\mathbb{F}_p)$ was first discussed by Nakaoka [Na]. Later H. Cardenas [Card] determined the cohomology rings $H^*(\mathcal{S}_{p^2};\mathbb{F}_p)$. Nakaoka's idea was to determine $H_*(\mathcal{S}_n;\mathbb{F}_p)$ using the cohomology of symmetric products of spheres which he had determined previously, but which were also determined by Steenrod (unpublished) at about the same time using a key result of Dold and Thom, [DT]. Cardenas' ideas were very much in the spirit of the development in §1. Perhaps the interest in these groups was prompted during the 1950's by J. Adem's thesis [Adem], where he used preliminary information about $H^*(\mathcal{S}_{p^2};\mathbb{F}_p)$ to obtain the Adem relations in the Steenrod algebra.

From another point of view $B_{\mathcal{S}_\infty}$ occurs in an essential way in the study of infinite loop spaces. Recall that the n-fold loop space Ω^nX is the space of based continuous maps $f\colon(S^n,\infty)\longrightarrow(X,*)$ with the compact open topology. If X is a space with basepoint, the natural inclusions

$$\Omega^nS^n(X)\longrightarrow\Omega^{n+1}S^{n+1}(X)$$

on passing to the limit define the space $Q(X) = \lim_{n\to\infty} \Omega^n S^n(X)$. $Q(X)$ depends functorially on X and $\pi_i(Q(X)) = \pi_i^s(X)$, the i–th stable homotopy group of X.

Now $Q(S^0)$ has a "loop sum product", which is given as the limit of the $\Omega^{n-1}(*)\colon \Omega^n S^n \times \Omega^n S^n \longrightarrow \Omega^n S^n$ where $*\colon \Omega S^n \times \Omega S^n \longrightarrow \Omega S^n$ is the loop sum. We have the following fundamental result of E. Dyer and R. Lashof

Theorem 3.5 *There is a map $e\colon \mathbb{Z} \times B_{\mathcal{S}_\infty} \longrightarrow Q(S^0)$ which induces isomorphisms in homology with all (untwisted) coefficients and which is additive in the sense that the diagram*

$$\begin{array}{ccc} (\mathbb{Z} \times B_{\mathcal{S}_\infty})^2 & \xrightarrow{e \times e} & Q(S^0) \times Q(S^0) \\ \downarrow \mu & & \downarrow * \\ \mathbb{Z} \times B_{\mathcal{S}_\infty} & \xrightarrow{e} & Q(S^0) \end{array}$$

homotopy commutes where the product on the left is addition in $\mathbb{Z}$ times the product discussed in §2 on $B_{\mathcal{S}_\infty}$.

Although the language was different, this result first appeared in the original 1960 preprint of [D-L] which was widely circulated at the time but not published. It later was independently rediscovered by Barratt-Priddy, [BP], and Quillen, also unpublished, and has been of fundamental importance in algebraic topology since.

4. More Invariant Theory

As we saw in §3, the additive structure of $H^*(\mathcal{S}_n)$ has been completely understood for over 20 years. In (1.13) the ring structure of $H^*(\mathcal{S}_4)$ was determined to be $H^*(\mathcal{S}_4) \cong P[\sigma_1, \sigma_2, c_3]/\langle \sigma_1 c_3 = 0\rangle$.

It was assumed that the same type of simple relation would hold for symmetric groups of higher degree. In this section we will use invariant theory to show otherwise. In fact we will exhibit numerical evidence that indicates the presence of very complicated relations, rich with symmetry, that build up successively. We will use these to give complete descriptions of the *rings* $H^*(\mathcal{S}_n)$ for $n = 6, 8, 10, 12$ in §5.

As in §1, let $j\colon V_n(2) = (\mathbb{Z}/2)^n \hookrightarrow \mathbb{Z}/2 \wr \cdots \wr \mathbb{Z}/2 \hookrightarrow \mathcal{S}_{2^n}$ denote the embedding where we consider $\mathcal{S}_{2^n}$ as the automorphism group of the set $\oplus_1^n \mathbb{Z}/2$ and $V_n(2)$ as the set of translations. Its normalizer is the group of affine transformations, $Aff_n(\mathbb{Z}/2)$. Therefore $N(V_n(2))/V_n(2) = \mathrm{GL}_n(\mathbb{F}_2)$ and we have seen that in fact $\mathrm{im} j^* = H^*(V_n(2))^{\mathrm{GL}_n(\mathbb{F}_2)}$, the ring of invariants. For $n = 2$, we obtain that $\mathrm{im} j^* \cong P[x_0, x_1]$, $\deg x_0 = 2$, $\deg x_1 = 3$.

For $n = 4k$, we have a natural inclusion

$$i\colon (V_2)^k \hookrightarrow (\mathcal{S}_4)^k \hookrightarrow \mathcal{S}_n \ .$$

In cohomology, the image of i^* lies in $\left(\otimes_{i=1}^k P[x_0, x_1]\right)^{\mathcal{S}_k}$, where $\mathcal{S}_k$ acts by permuting the polynomial generators. We will now analyze this ring of invariants, which is, as we have seen in §3, a homomorphic image of $H^*(\mathcal{S}_n)$.

We start with the case $k = 2$:

Theorem 4.1 *The Poincaré series of* $(P[x_0, x_1] \otimes P[x_0, x_1])^{\mathcal{S}_2}$ *is*

$$P_2(t) = \frac{1 + t^5}{(1 - t^2)(1 - t^3)(1 - t^4)(1 - t^6)} .$$

Proof. As we are taking two–fold symmetric tensors, there are two types of invariants, $x \otimes y + y \otimes x$ and $x \otimes x$. The latter are counted by the Poincaré series $\frac{1}{(1-t^4)(1-t^6)}$. Consequently, the former are counted by the series $\frac{1}{2}\left[\frac{1}{(1-t^2)(1-t^3)} - \frac{1}{(1-t^4)(1-t^6)}\right]$, and doing an easy manipulation we obtain :

$$\begin{aligned} P_2(t) &= \frac{1}{2}\left[\left(\frac{1}{(1 - t^2)(1 - t^3)}\right)^2 + \frac{1}{(1 - t^4)(1 - t^6)}\right] \\ &= \frac{1}{2}\left(\frac{1}{(1 - t^2)(1 - t^3)}\right)\left(\frac{1}{(1 - t^2)(1 - t^3)} + \frac{1}{(1 + t^2)(1 + t^3)}\right) \\ &= \frac{1}{2}\left(\frac{1}{(1 - t^2)(1 - t^3)}\right)\left(\frac{2 + 2t^5}{(1 - t^4)(1 - t^6)}\right) \\ &= \frac{1 + t^5}{(1 - t^2)(1 - t^3)(1 - t^4)(1 - t^6)} . \end{aligned}$$

□

Examining this series, we deduce the existence of a five–dimensional class, which satisfies a *quadratic* relation. This is the first indication of the occurrence of relations which are not of the form seen before, the products of two generators being zero.

Theorem 4.2

$$(P[x_0, x_1] \otimes P[x_0, x_1])^{\mathcal{S}_2} \cong$$
$$P\left[d_{01}, d_{02}, d_{11}, d_{22}\right](x_5)/\langle x_5^2 + x_5 d_{01} d_{11} + d_{22} d_{01}^2 + d_{02} d_{11}^2 = 0\rangle$$

where

$$\begin{aligned} d_{01} &= x_0 \otimes 1 + 1 \otimes x_0, d_{11} = x_1 \otimes 1 + 1 \otimes x_1 \\ d_{02} &= x_0 \otimes x_0, d_{22} = x_1 \otimes x_1 \\ x_5 &= x_0 \otimes x_1 + x_1 \otimes x_0 . \end{aligned}$$

Proof. The elements $d_{01}, d_{02}, d_{11}, d_{22}$ are independent, generating a subpolynomial algebra. Verifying the relation, we conclude that together with x_5 they generate a subalgebra with the same Poincaré series as

$$(P[x_0, x_1] \otimes P[x_0, x_1])^{\mathcal{S}_2} . \qquad \square$$

The relation in 4.2 will pull back to $H^*(\mathcal{S}_8)$, as will be shown in §5. Next we analyze the case $k = 3$, denote

$$P_3 = (P[x_0, x_1] \otimes P[x_0, x_1] \otimes P[x_0, x_1])^{\mathcal{S}_3} .$$

Theorem 4.3 *The Poincaré series for P_3 is*

$$P_3(t) = \frac{1 + t^5 + t^7 + t^8 + t^{10} + t^{15}}{(1-t^2)(1-t^3)(1-t^4)(1-t^6)^2(1-t^9)} .$$

Proof. Denote

$$\begin{aligned} abc &= \left(1/(1-t^2)(1-t^3) - 1\right)^3 \\ a^2b &= \left(1/(1-t^4)(1-t^6) - 1\right)\left(1/(1-t^2)(1-t^3) - 1\right) \\ a^3 &= \left(1/(1-t^6)(1-t^9) - 1\right) . \end{aligned}$$

The letters denote the type of products they represent. Let

$$\begin{aligned} e_3 &= a^3 \leftarrow \text{cubes} \\ e_{12} &= a^2b - e_3 \leftarrow \text{products of type } a^2b \text{ which are not cubes} \\ e_{111} &= abc - 3e_{12} - e_3 \leftarrow \text{ products of type } abc \text{ not of two other types.} \end{aligned}$$

As we are taking symmetric invariants, the formula for the desired Poincaré series is

$$P_3(t) = \frac{1}{6}e_{111} + e_{12} + e_3 + \left(1/(1-t^2)(1-t^3)\right) .$$

The term on the extreme right records the 3–fold symmetrization of the generators of $P[x_0, x_1]$.

Using *macsyma*, this simplifies to the asserted Poincaré series. $\square$

As before, we use the numerator to deduce the relations. Let $S(u_1 \otimes u_2 \otimes u_3)$ denote symmetrization.

Theorem 4.4

$$P_3 \cong P[d_{01}, d_{02}, d_{03}, d_{11}, d_{22}, d_{33}, x_5, x_7, x_8]/\langle R \rangle$$

where

$$d_{01} = S(x_0 \otimes 1 \otimes 1),\ d_{02} = S(x_0 \otimes x_0 \otimes 1),\ d_{03} = x_0 \otimes x_0 \otimes x_0$$
$$d_{11} = S(x_1 \otimes 1 \otimes 1),\ d_{22} = S(x_1 \otimes x_1 \otimes 1),\ d_{33} = x_1 \otimes x_1 \otimes x_1$$
$$x_5 = S(x_0 \otimes x_1 \otimes 1),\ x_7 = S(x_0 \otimes x_0 \otimes x_1),\ x_8 = S(x_0 \otimes x_1 \otimes x_1)$$

Relations:

(1)
$$x_7^2 + d_{02}x_5^2 + d_{02}d_{01}x_8 + (d_{03}d_{11} + d_{02}d_{01}d_{11})x_5 + d_{02}^2d_{11}^2 + d_{01}^2d_{22}d_{02} + d_{03}d_{01}d_{22}$$

(2)
$$x_8^2 + d_{22}x_5^2 + d_{22}d_{11}x_7 + (d_{33}d_{01} + d_{22}d_{11}d_{01})\,x_5 + d_{22}^2d_{01}^2 + d_{11}^2d_{02}d_{22} + d_{33}d_{11}d_{02}$$

(3)
$$x_5x_7 + d_{01}x_5^2 + [d_{01}^2 + d_{02}]x_8 + d_{01}^2d_{11}x_5 + d_{01}^3d_{22} + d_{03}d_{11}^2 + d_{03}d_{22} + d_{01}d_{02}d_{11}^2$$

(4)
$$x_5x_8 + d_{11}x_5^2 + [d_{11}^2 + d_{22}]x_7 + d_{11}^2d_{01}x_5 + d_{11}^3d_{02} + d_{33}d_{01}^2 + d_{33}d_{02} + d_{11}d_{22}d_{01}^2$$

(5)
$$x_5^3 + d_{11}d_{01}x_5^2 + x_7x_8 + d_{11}d_{01}^2x_8 + d_{01}d_{11}^2x_7 + \left(d_{01}^2d_{22} + d_{02}d_{11}^2\right)x_5 + {} + d_{01}^3d_{33} + d_{11}^3d_{03} + d_{03}d_{33}\,.$$

Proof. The elements d_{ij} $i = 0, 1$, $j = 1, 2, 3$ are algebraically independent, and generate a polynomial subalgebra. We examine the numerator of the Poincaré series for P_3:

$$1 + t^5 + t^7 + t^8 + t^{10} + t^{15}\ .$$

The quintic term must come from $x_5 = S(x_0 \otimes x_1 \otimes 1)$ and similarly $x_7 = S(x_0 \otimes x_0 \otimes x_1)$, $x_8 = S(x_0 \otimes x_1 \otimes x_1)$ account for t^7, t^8. Likewise x_5^2 explains the occurrence of t^{10}. (One checks that x_5 does not satisfy a quadratic equation).

There must be relations involving x_5^3, x_7^2, x_8^2, x_5x_7 and x_5x_8. Finding the exact relation is a rather tedious task, which involves taking all possible products of the right dimension involving x_5, x_7, x_8, x_5^2 and expressing their spill over in terms of the d_{ij}. Knowing the exact formulas, their verification is a lengthy but straight forward calculation, which we leave as a horrible exercise to the reader. Note that relation (2) is obtained from (1) by exchanging x_0 and x_1; likewise we get (4) from (3) in this manner. Relation (5) is invariant under this exchange.

To complete the proof we just compare Poincaré series. □

For the cases $k = 4, 5, 6$ we do not have the algebra structure, only the Poincaré series. These can be obtained in a combinatorial way extending the methods used in 4.3; we used *macsyma* to perform the simplifications. As the denominators of these series are clear, we only give the numerators:

Theorem 4.5 *The invariant algebras P_4, P_5, P_6 have the following polynomials as numerators of their Poincaré series:*

$$
\begin{aligned}
(1)\qquad N_4(t) = {} & t^{30} + t^{25} + t^{23} + t^{22} + t^{21} + 2t^{20} + t^{19} + \\
& + t^{18} + t^{17} + t^{16} + 2t^{15} + t^{14} + t^{13} + t^{12} + t^{11} \\
& + 2t^{10} + t^9 + t^8 + t^7 + t^5 + 1
\end{aligned}
$$

$$
\begin{aligned}
(2)\qquad N_5(t) = {} & t^{50} + t^{45} + t^{43} + t^{42} + t^{41} + 2t^{40} + 2t^{39} + 2t^{38} \\
& + 2t^{37} + 3t^{36} + 3t^{35} + 3t^{34} + 3t^{33} + 4t^{32} + 4t^{31} \\
& + 4t^{30} + 5t^{29} + 5t^{28} + 5t^{27} + 5t^{26} + 6t^{25} + 5t^{24} \\
& + 5t^{23} + 5t^{22} + 5t^{21} + 4t^{20} + 4t^{19} + 4t^{18} + 3t^{17} \\
& + 3t^{16} + 3t^{15} + 3t^{14} + 2t^{13} + 2t^{12} + 2t^{11} + 2t^{10} \\
& + t^9 + t^8 + t^7 + t^5 + 1
\end{aligned}
$$

$$
\begin{aligned}
(3)\ N_6(t) = {} & t^{75} + t^{70} + t^{68} + t^{67} + t^{66} + 2t^{65} + 2t^{64} + 2t^{63} + 3t^{62} + 4t^{61} \\
& + 4t^{60} + 5t^{59} + 5t^{58} + 6t^{57} + 7t^{56} + 8t^{55} + 10t^{54} + 10t^{53} + 11t^{52} \\
& + 13t^{51} + 14t^{50} + 15t^{49} + 17t^{48} + 18t^{47} + 19t^{46} + 20t^{45} + 21t^{44} \\
& + 22t^{43} + 23t^{42} + 23t^{41} + 24t^{40} + 23t^{39} + 24t^{38} + 24t^{37} + 23t^{36} \\
& + 24t^{35} + 23t^{34} + 23t^{33} + 22t^{32} + 21t^{31} + 20t^{30} + 19t^{29} + 18t^{28} \\
& + 17t^{27} + 15t^{26} + 14t^{25} + 13t^{24} + 11t^{23} + 10t^{22} + 10t^{21} + 8t^{20} \\
& + 7t^{19} + 6t^{18} + 5t^{17} + 5t^{16} + 4t^{15} + 4t^{14} + 3t^{13} + 2t^{12} + 2t^{11} \\
& + 2t^{10} + t^9 + t^8 + t^7 + t^5 + 1\,.
\end{aligned}
$$

Let $P[x_1, \ldots, x_n]$ be a polynomial algebra which has an $\mathcal{S}_n$ action defined by permuting the n generators. A fundamental result in Galois Theory is that the ring of invariants of this action is also a polynomial ring, on the symmetric generators $\sigma_1, \ldots, \sigma_n$. We have shown that this result does not extend to invariants of the form $\left(\otimes_1^k P[x_0, x_1]\right)^{\mathcal{S}_k}$. As more variables are introduced, the complexity of this ring of invariants increases and this will be reflected by interesting relations in the cohomology of the symmetric groups.

5. $H^*(\mathcal{S}_n)$, $n = 6, 8, 10, 12$

In this section we will combine the well-known additive structure of $H_*(\mathcal{S}_n)$ with the invariant theory of §4 to obtain precise descriptions of the cohomology rings $H^*(\mathcal{S}_n)$ $n = 6, 8, 10, 12$, as well as the action of the Steenrod algebra on them. We start by describing $H^*(\mathcal{S}_6)$:

Theorem 5.1

$$H^*(\mathcal{S}_6) \cong P[\sigma_1, \sigma_2, \sigma_3, c_3]/\langle c_3(\sigma_3 + \sigma_1\sigma_2) = 0\rangle$$

$\deg \sigma_i = i$, $\deg c_3 = 3$ *where*

$$Sq^1\sigma_2 = \sigma_1\sigma_2 + \sigma_3 + c_3, \; Sq^1 c_3 = 0, \; Sq^2 c_3 = \sigma_2 c_3$$
$$Sq^1\sigma_3 = (c_3 + \sigma_3)\sigma_1, \; Sq^2\sigma_3 = \sigma_3\sigma_2 + c_3\sigma_1^2 .$$

Proof. From §1, we see that $H_*(\mathcal{S}_6)$ has a basis given by elements

$$\{Q_i * Q_j * Q_k, Q_{r,s} * Q_t\}$$

where $1 \leq r \leq s$, $\dim Q_{r,s} = 2s + r$ and $Q_{0,s} = Q_s * Q_s$. We now consider the sequence of inclusions

$$(\mathcal{S}_2)^3 \overset{i}{\hookrightarrow} \mathcal{S}_4 \times \mathcal{S}_2 \overset{j}{\hookrightarrow} \mathcal{S}_6 .$$

Using a basis dual to the one above, we have that if $k = j \cdot i$,

$$k^*((Q_1 * 1 * 1)^*) = \sigma_1$$
$$k^*((Q_1 * Q_1 * 1)^*) = \sigma_2$$
$$k^*((Q_1 * Q_1 * Q_1)^*) = \sigma_3 .$$

Using the symbol σ_i to denote the cohomology classes identified to symmetric classes, the above implies that

$$j^*((Q_1 * 1 * 1)^*) = \sigma_1 \otimes 1 + 1 \otimes \sigma_1$$
$$j^*((Q_1 * Q_1 * 1)^*) = \sigma_1 \otimes \sigma_1 + \sigma_2 \otimes 1$$
$$j^*((Q_1 * Q_1 * Q_1)^*) = \sigma_2 \otimes \sigma_1 .$$

Note that the 3–dimensional generator c_3 in $H^*(\mathcal{S}_4)$ restricts to zero. As $j_*(Q_{11} \otimes 1) = Q_{11} * 1$, we conclude that $j^*((Q_{11} * 1)^*) = Q_{11} \otimes 1 = c_3 \otimes 1$.

The elements $\sigma_1 \otimes 1 + 1 \otimes \sigma_1$, $\sigma_1 \otimes \sigma_1 + \sigma_2 \otimes 1$ and $\sigma_2 \otimes \sigma_1$ are independent, as they map to the symmetric generators under i^*. Recalling that in $H^*(\mathcal{S}_4)$ $c_3\sigma_1 = 0$ is the only relation, we obtain a unique relation in the subalgebra generated by the above elements together with $c_3 \otimes 1$:

$$(c_3 \otimes 1)\left[(\sigma_2 \otimes \sigma_1) + (\sigma_1 \otimes 1 + 1 \otimes \sigma_1)(\sigma_1 \otimes \sigma_1 + \sigma_2 \otimes 1)\right] = 0 .$$

Hence observing that j^* is a monomorphism and that this subalgebra has the same Poincaré series as $H^*(\mathcal{S}_6)$ (which is $1+x^3/(1-x)(1-x^2)(1-x^3)$) we conclude

$$H^*(\mathcal{S}_6) \cong P[\sigma_1, \sigma_2, \sigma_3, c_3]/\langle c_3(\sigma_3 + \sigma_1\sigma_2) = 0\rangle \ .$$

Here we identify $\sigma_1 \otimes 1 + 1 \otimes \sigma_1 \longrightarrow \sigma_1$, $\sigma_1 \otimes \sigma_1 + \sigma_2 \otimes 1 \longrightarrow \sigma_2$, $\sigma_2 \otimes \sigma_1 \longrightarrow \sigma_3$, $c_3 \otimes 1 \longrightarrow c_3$.

For the action of the Steenrod Algebra, we apply the Sq^i in $H^*(\mathcal{S}_4) \otimes H^*(\mathcal{S}_2)$ and use the relations there:

$$\begin{aligned} Sq^1\sigma_2 &= Sq^1 i^*(\sigma_2 \otimes 1 + \sigma_1 \otimes \sigma_1) = i^*(Sq^1\sigma_2 \otimes 1 + \sigma_1^2 \otimes \sigma_1 + \sigma_1 \otimes \sigma_1^2) \\ &= i^*(c_3 \otimes 1 + \sigma_1\sigma_2 \otimes 1 + \sigma_1^2 \otimes \sigma_1 + \sigma_1 \otimes \sigma_1^2) \\ &= i^*\left((\sigma_1 \otimes 1 + 1 \otimes \sigma_1)(\sigma_1 \otimes \sigma_1 + \sigma_2 \otimes 1) + \sigma_2 \otimes \sigma_1 + c_3 \otimes 1\right) \\ &= \sigma_1\sigma_2 + \sigma_3 + c_3 \end{aligned}$$

$$\begin{array}{rclcl} Sq^1c_3 &=& Sq^1 i^*(c_3 \otimes 1) &=& i^*(Sq^1(c_3 \otimes 1)) = 0 \\ Sq^2c_3 &=& i^*(Sq^2(c_3 \otimes 1)) &=& i^*(\sigma_2 c_3 \otimes 1)) = \sigma_2 c_3 \\ Sq^1\sigma_3 &=& Sq^1 i^*(\sigma_2 \otimes \sigma_1) &=& i^*(Sq^1\sigma_2 \otimes \sigma_1 + \sigma_2 \otimes \sigma_1^2) \\ &&&=& i^*(c_3 \otimes \sigma_1 + \sigma_1\sigma_2 \otimes \sigma_1 + \sigma_2 \otimes \sigma_1^2) \\ &&&=& (c_3 + \sigma_3)\sigma_1 \end{array}$$

$$\begin{aligned} Sq^2\sigma_3 &= i^*(Sq^2\sigma_2 \otimes \sigma_1 + Sq^1\sigma_2 \otimes Sq^1\sigma_1) \\ &= i^*(\sigma_2^2 \otimes \sigma_1 + c_3 \otimes \sigma_1^2 + \sigma_1\sigma_2 \otimes \sigma_1^2) \\ &= i^*((\sigma_1 \otimes \sigma_1)(\sigma_2 \otimes 1 + \sigma_1 \otimes \sigma_1) + (c_3 \otimes 1)(\sigma_1 \otimes 1 + 1 \otimes \sigma_1)^2) \\ &= \sigma_3\sigma_2 + c_3\sigma_1^2 . \end{aligned}$$

□

Next we consider $\mathcal{S}_8$: From this point forward we will omit details and just describe the results. Full details can be found in [AMM1].

Theorem 5.2

$$H^*(\mathcal{S}_8) \cong P[\sigma_1, \sigma_2, \sigma_3, c_3, \sigma_4, d_6, d_7](x_5)/\langle R\rangle$$

where $\deg \sigma_i = i$, $\deg c_3 = 3$, $\deg d_i = i$, $\deg x_5 = 5$, *and* R *is the following set of relations:*

$$\begin{aligned} &d_6\sigma_1 = d_6\sigma_3 = 0 \\ &d_7\sigma_1 = d_7\sigma_2 = d_7\sigma_3 = d_7c_3 = d_7x_5 = 0 \\ &x_5\sigma_3 + c_3\sigma_4\sigma_1 = 0 \\ &c_3(\sigma_3 + \sigma_1\sigma_2) + \sigma_1 x_5 = 0 \\ &x_5^2 + x_5\sigma_2 c_3 + d_6\sigma_2^2 + \sigma_4 c_3^2 = 0 . \end{aligned}$$

The following tables describe the action of the Steenrod algebra

	σ_2	σ_3	c_3
Sq^1	$\sigma_1\sigma_2 + \sigma_3 + c_3$	$\sigma_1(c_3 + \sigma_3)$	0
Sq^2	σ_2^2	$c_3\sigma_1^2 + \sigma_2\sigma_3 + \sigma_1\sigma_4$	$\sigma_2 c_3 + x_5$
Sq^3	0	σ_3^2	c_3^2

	σ_4	x_5	d_6	d_7
Sq^1	$x_5 + \sigma_1\sigma_4$	$\sigma_1 x_5$	d_7	0
Sq^2	$\sigma_2\sigma_4 + d_6 + x_5\sigma_1$	$\sigma_2 x_5$	$\sigma_2 d_6$	0
Sq^3	m_7	$(c_3 + \sigma_3)x_5$	$c_3 d_6$	0
Sq^4	σ_4^2	n_9	$\sigma_4 d_6$	$\sigma_4 d_7$
Sq^5	0	x_5^2	$x_5 d_6 + \sigma_4 d_7$	0
Sq^6	0	0	d_6^2	$d_6 d_7$

Here $m_7 = d_7 + c_3\sigma_4 + \sigma_3\sigma_4 + \sigma_2 x_5$, and $n_9 = \sigma_4 x_5 + c_3(d_6 + \sigma_1 x_5)$. From this a relatively direct calculation gives the cohomology of $\mathcal{S}_{10}$. Indeed, going from $\mathcal{S}_{4i}$ to $\mathcal{S}_{4i+2}$ only has the effect of freeing up products with σ_1, and adding a new generator σ_{2i+1}. We omit the $\mathcal{A}(2)$ action as it can easily be obtained using the embedding $H^*(\mathcal{S}_{10}) \longrightarrow H^*(\mathcal{S}_8 \times \mathcal{S}_2)$.

Theorem 5.3

$$H^*(\mathcal{S}_{10}) \cong P[\sigma_1, \sigma_2, \sigma_3, \sigma_4, \sigma_5, c_3, x_5, d_6, d_7]/\langle R\rangle$$

where R is the set of relations

$$\sigma_2 d_7 = \sigma_3 d_7 = c_3 d_7 = x_5 d_7 = 0$$
$$(\sigma_5 + \sigma_1\sigma_4)d_7 = 0$$
$$(\sigma_3 + \sigma_1\sigma_2)d_6 = (\sigma_5 + \sigma_1\sigma_4)d_6 = 0$$
$$x_5^2 = \sigma_2 c_3 x_5 + c_3^2\sigma_4 + \sigma_2^2 d_6$$
$$(\sigma_3 + \sigma_1\sigma_2)x_5 = c_3(\sigma_5 + \sigma_1\sigma_4 + \sigma_2\sigma_3 + \sigma_1\sigma_2^2)$$
$$(\sigma_5 + \sigma_1\sigma_4)x_5 = c_3(\sigma_3 + \sigma_1\sigma_2)\sigma_4 .$$

Theorem 5.4

$$H^*(\mathcal{S}_{12}) \cong P[\sigma_1, \sigma_2, \sigma_3, c_3, \sigma_4, \sigma_5, \sigma_6, d_6, d_7, d_9]\,(x_5, x_7, x_8)\,/\langle R\rangle$$

where deg $\sigma_i = i$, deg $c_3 = 3$, deg $d_i = i$, deg $x_i = i$ *and R is the set of relations*

$$x_7^2 + \sigma_4 x_5^2 + \sigma_2\sigma_4 x_8 + (\sigma_6 c_3 + \sigma_2\sigma_4 c_3)\,x_5 + \sigma_4^2 + c_3^2 + \sigma_2^2\sigma_4 d_6 + \sigma_6\sigma_2 d_6,$$
$$x_8^2 + d_6 x_5^2 + c_3 d_6 x_7 + (d_9\sigma_2 + c_3 d_6\sigma_2)x_5 + d_6^2\sigma_2^2 + c_3^2\sigma_4 d_6 + d_9 c_3\sigma_2,$$
$$x_5 x_7 + \sigma_2 x_5^2 + \left[\sigma_2^2 + \sigma_4\right] x_8 + \sigma_2^2 c_3 x_5 + \sigma_2^3 d_6 + \sigma_6 c_3^2 + \sigma_6 d_6 + \sigma_2\sigma_4 c_3^2,$$
$$x_5 x_8 + c_3 x_5^2 + \left[c_3^2 + d_6\right] x_7 + c_3^2\sigma_2 x_5 + c_3^3\sigma_4 + d_9\sigma_2^2 + d_9\sigma_4 + c_3 d_6\sigma_2^2,$$
$$x_5^3 + c_3\sigma_2 x_5^2 + x_7 x_8 + c_3\sigma_2^2 x_8 + \sigma_2 c_3^2 x_7 + \left(\sigma_2^2 d_6 + \sigma_4 c_3^2\right) x_5$$
$$+\sigma_2^3 d_9 + c_3^3\sigma_6 + \sigma_6 d_9,$$

$d_9\sigma_1, d_9\sigma_3, d_9\sigma_5$
$d_7\sigma_3, d_7x_5, d_7\sigma_1c_3, d_7(x_7+\sigma_4c_3), d_7(\sigma_5+\sigma_4\sigma_1), d_7(\sigma_6+\sigma_4\sigma_2),$
$d_7\left(x_8+d_6\sigma_2\right), d_7\left(d_9+d_6c_3\right),$

$x_7\sigma_1+x_5\left(\sigma_1\sigma_2+\sigma_3\right)+c_3\left(\sigma_2\sigma_3+\sigma_2^2\sigma_1+\sigma_1\sigma_4+\sigma_5\right),$
$x_7\sigma_3+x_5\left(\sigma_5+\sigma_1\sigma_4\right)+c_3\left(\sigma_1\sigma_6+\sigma_3\sigma_4+\sigma_1\sigma_2\sigma_4\right),$
$x_7\sigma_5+x_5\sigma_1\sigma_6+c_3\left(\sigma_3\sigma_6+\sigma_1\sigma_2\sigma_6\right),$
$x_8\sigma_1+d_6\left(\sigma_3+\sigma_1\sigma_2\right),$
$x_8\sigma_3+d_6\left(\sigma_5+\sigma_1\sigma_4\right),$
$x_8\sigma_5+d_6\sigma_1\sigma_6 .$

The action on the Steenrod Algebra on $H^*\left(\mathcal{S}_{12}\right)$ is determined by the following values:

$Sq^1\sigma_2=\sigma_1\sigma_2+\sigma_3+c_3,$
$Sq^1c_3=0, Sq^2c_3=\sigma_2c_3+x_5,$
$Sq^2\sigma_3=c_3\sigma_1^2+\sigma_2\sigma_3+\sigma_1\sigma_4+\sigma_5,$
$Sq^1\sigma_4=x_5+\sigma_1\sigma_4+\sigma_5, Sq^2\sigma_4=\sigma_2\sigma_4+\sigma_6+d_6+c_3\left(\sigma_3+\sigma_1\sigma_2\right),$
$Sq^2\sigma_5=d_6\sigma_1+x_5\sigma_1^2+c_3\left(\sigma_1\sigma_3+\sigma_1^2\sigma_2\right)+\sigma_2\sigma_5+\sigma_1\sigma_6,$
$Sq^4\sigma_5=c_3\sigma_3^2+\sigma_4\sigma_5+\sigma_1x_5\sigma_3+\sigma_3\sigma_6+c_3\sigma_1\sigma_2\sigma_3+c_3\sigma_1\sigma_5+d_7\sigma_1^2,$
$Sq^2x_5=\sigma_2x_5,$
$Sq^1\sigma_6=x_7+\sigma_1\sigma_6,\ Sq^2\sigma_6=x_8+x_7\sigma_1+\sigma_2\sigma_6,$
$Sq^4\sigma_6=\sigma_4\sigma_6+x_7\sigma_3+d_6\sigma_2^2+x_5^2+x_5\sigma_2c_3+c_3^2\sigma_4+c_3x_7+c_3\sigma_1\sigma_6,$
$Sq^1d_6=d_7,\ Sq^2d_6=x_8+\sigma_2d_6,$
$Sq^4d_6=\sigma_4\sigma_6+x_5\sigma_2c_3+d_6\sigma_2^2+\sigma_4c_3^2+x_5^2+x_7c_3+x_8\sigma_2,$
$Sq^1d_7=Sq^2d_7=0,\ Sq^4d_7=\sigma_4d_7,\ Sq^6d_7=d_6d_7,$
$Sq^1x_8=d_9+\sigma_2d_7+\sigma_3d_6+\sigma_1\sigma_2d_6,\ Sq^4x_8=x_8\left(\sigma_4+c_3\sigma_1\right)$
$Sq^1d_9=d_7c_3,\ Sq^2d_9=\sigma_2d_9,\ Sq^8d_9=d_7x_7c_3+x_8d_9 .$

6. The Cohomology of the Alternating Groups

An important class of simple groups whose cohomology rings have not been computed are the alternating groups A_n. The absence of proper normal subgroups makes the cohomology of simple groups particularly inaccessible, but in our situation we can make good use of $H^*(\mathcal{S}_n)$. The Gysin sequence, as in II.5, allows one to relate the cohomology of a group to that of an index 2 subgroup. For convenience we recall its statement and give an independent proof.

Lemma 6.1 *Let $v\in \mathrm{Hom}(G,\mathbb{Z}/2)$ non-trivial, with $H=\ker v$. Then there is a long exact sequence*

$$\cdots\longrightarrow H^i(G)\xrightarrow{\cup v}H^{i+1}(G)\xrightarrow{\text{res}}H^{i+1}(H)\xrightarrow{\text{tr}}H^{i+1}(G)\xrightarrow{\cup v}\cdots .$$

Proof. (Compare [DM2], Appendix 1). The fibration $\mathbb{Z}/2 \longrightarrow B_H \longrightarrow B_G$ is the sphere fibration of a line bundle. Consequently $MC(p) = B_G/B_H$ is the Thom space of the line bundle with first Stiefel– Whitney class v. Now apply the Gysin sequence. Transversality identifies

$$B_G/B_H \longrightarrow \mathcal{S}B_H$$

with the transfer. This can also be proved algebraically using the short exact sequence of G–modules

$$0 \longrightarrow \mathbb{F}_2 \longrightarrow \mathbb{F}_2[G/H] \longrightarrow \mathbb{F}_2 \longrightarrow 0.$$ □

We apply this lemma to the situation $A_n \subseteq \mathcal{S}_n$, with $[\mathcal{S}_n : A_n] = 2$. We know $H^1(\mathcal{S}_n) \cong \mathbb{F}_2$, generated by the symmetric class σ_1; hence

Theorem 6.2 *There is a short exact sequence of $H^*(\mathcal{S}_n)$–modules:*

$$0 \longrightarrow H^*(\mathcal{S}_n)/(\sigma_1) \xrightarrow{\text{res}} H^*(A_n) \xrightarrow{\text{tr}} Ann(\sigma_1) \longrightarrow 0\,.$$

From the considerations in §1, it is not hard to see that the ideal $J_n = Ann(\sigma_1)$ is generated by "pure" Dickson classes. These can be described as follows: let $n = 2^{j_1} + \ldots + 2^{j_r}$, where $1 \le j_1 \le \ldots \le j_r$ (the case $j_1 = 0$ can be reduced to this situation); the pure Dickson classes are those of the form $(Q_{I_1} * \ldots * Q_{I_r})^*$, where $2 \le length(I_k) = j_k$ for all $k = 1, \ldots, r$ and each $I_k = (1, \ldots, 1)$.

An immediate consequence of this is the following

Corollary 6.3 *If n is congruent to 2 or 3 mod 4, then the restriction map $H^*(\mathcal{S}_n) \to H^*(A_n)$ is onto, and*

$$H^*(A_n) \cong H^*(\mathcal{S}_n)/(\sigma_1)\,.$$

We remark that the results in §1 together with the preceding discussion can be used to completely determine the additive structure of $H^*(A_n)$, any n. The ring structure is of course much harder, but we can recover the rings $H^*(A_6)$, $H^*(A_8)$, $H^*(A_{10})$ and $H^*(A_{12})$ in what follows.

From 5.1 and 6.3 we have:

Corollary 6.4

$$H^*(A_6) \cong P[\sigma_2, \sigma_3, c_3]/\langle c_3\sigma_3 = 0\rangle$$

with $Sq^1\sigma_2 = \sigma_3 + c_3,\ Sq^1 c_3 = 0,\ Sq^2 c_3 = \sigma_2 c_3,\ Sq^1\sigma_3 = 0,\ Sq^2\sigma_3 = \sigma_3\sigma_2.$

Proof. By 5.1, $Ann(\sigma_1) = 0$, hence

$$H^*(A_6) \cong H^*(\mathcal{S}_6)/(\sigma_1). \qquad \square$$

Applying 6.2 to A_8, we obtain the following

Corollary 6.5

$$H^*(A_8) \cong P[\sigma_2, c_3, \sigma_3, \sigma_4, d_6, e_6, d_7, e_7](x_5)/\langle R\rangle$$

where $\deg \sigma_i = i$, $\deg c_3 = 3$, $\deg d_i = i$, $\deg e_i = i$ $\deg x_i = 5$, *and R is the following set of relations:*

$$\begin{aligned}
&d_6\sigma_3 = 0, d_6d_7 + d_6e_7 + e_6e_7 = 0, d_6^2 + d_6e_6 + e_6^2 = 0,\\
&d_7\sigma_2 = d_7\sigma_3 = d_7c_3 = d_7x_5 = 0, d_6d_7 + d_7e_6 + e_6e_7 = 0,\\
&d_7^2 + d_7e_7 + e_7^2 = 0, e_6\sigma_3 = 0\\
&e_7\sigma_2 = e_7\sigma_3 = e_7c_3 = e_7x_5 = 0\\
&x_5\sigma_3 = 0, c_3\sigma_3 = 0\\
&x_5^2 + x_5\sigma_2c_3 + (d_6 + e_6)\sigma_2^2 + \sigma_4c_3^2 = 0.
\end{aligned}$$

The action of the Steenrod algebra on the generators different from e_6, e_7 can be read from 5.2 using the restriction map; for e_6, e_7 the action is identical to that on d_6 and d_7 substituting e_6, e_7 for d_6, d_7 throughout.

Proof. Using 5.2 and 6.2, we have a short exact sequence

$$0 \to H^*(\mathcal{S}_8)/(\sigma_1) \xrightarrow{\text{res}} H^*(A_8) \xrightarrow{\text{tr}} (d_6, d_7) \to 0.$$

Here (d_6, d_7) is the ideal generated by d_6 and d_7 in $H^*(\mathcal{S}_8)$, and res is an isomorphism until degree 6, where a new class e_6 appears in $H^6(A_8)$; similarly there is a new class $e_7 \in H^7(A_8)$, and we have $\text{tr}(e_6) = d_6$, $\text{tr}(e_7) = d_7$. Note that $\text{tr}(xy) = \text{tr}(x)\cdot y$ for $y \in H^*(\mathcal{S}_8)/(\sigma_1)$, hence we need only adjoin e_6, e_7 and their products with elements in im res to obtain $H^*(A_8)$. It only remains to determine the multiplicative relations involving these two classes.

Let $N_{\mathcal{S}_8}(V_3)$, $N_{A_8}(V_3)$ be the normalizers of V_3 in $\mathcal{S}_8$ and A_8 respectively. Then $[\mathcal{S}_8 \colon N_{\mathcal{S}_8}(V_3)] = 2[A_8 \colon N_{A_8}(V_3)]$ as there are twice as many conjugates of V_3 obtained by using elements in $\mathcal{S}_8$ as those obtained by using only elements in A_8. Hence there exists $1 \neq \bar{g} \in \mathcal{S}_8/A_8$ such that V_3 and gV_3g^{-1} are not conjugate in A_8. The classes e_6, e_7 will be detected on gVg^{-1}. For this note that $N_{A_8}(V_3) = N_{\mathcal{S}_8}(V_3) = \text{Aff}_3(\mathbb{Z}/2)$; similarly $N_{A_8}(gV_3g^{-1}) = N_{\mathcal{S}_8}(V_3)$. We have a diagram of restriction maps

$$\begin{array}{ccccc}
H^*(\mathcal{S}_8) & \xrightarrow{i^*} & H^*(A_8) & \xrightarrow{\ell^*} & H^*(gV_3g^{-1})^{\mathrm{GL}_3(\mathbb{F}_2)}\\
 & & \big\downarrow{\scriptstyle j^*} & & \\
 & & H^*(V_3)^{\mathrm{GL}_3(\mathbb{F}_2)}. & &
\end{array}$$

The maps j^* and ℓ^* are onto: as noted before their images will be subpolynomial algebras generated by the Dickson invariants D_4, D_6, D_7 and E_4, E_6, E_7 respectively. The classes D_4, E_4 are identified with $\sigma_4 \in H^4(A_8)$ and we obtain $d_6, e_6, d_7, e_7 \in H^*(A_8)$ with $j^*(d_6) = D_6$, $j^*(d_7) = D_7$, $\ell^*(e_6) = E_6$, $\ell^*(e_7) = E_7$.

The element $\bar{g} \in \mathcal{S}_8/A_8$ induces an involution on $H^*(A_8)$ with $\bar{g}^* d_i = e_i, i = 6, 7$; also note that $\text{res}\, d_i = d_i + e_i$. Now $\text{tr}\,((e_i + d_i)e_i) = \text{tr}((\text{res}\, d_i)e_i) = d_i \cdot \text{tr} e_i = d_i^2$. Therefore

$$\text{tr}\ (e_i^2 + (e_i + d_i)e_i) = 0 \text{ and } e_i d_i \in \text{im res} \ .$$

Using the restriction to V_3 and gV_3g^{-1}, it follows that

$$e_i d_i = \text{res}\ d_i^2$$

yielding the relation

$$e_i^2 + e_i d_i + d_i^2 = 0\ .$$

The same procedure yields the relations

$$e_6e_7 + d_6e_7 + d_6d_7 = 0$$
$$e_6e_7 + d_7e_6 + d_6d_7 = 0\ .$$

The other relations, involving e_6, e_7 and the other generators, follow from the corresponding relations for d_6, d_7. The remaining relations follow from applying res to relations in $H^*(\mathcal{S}_8)$. □

The explicit calculation of $H^*(A_8)$ is particularly interesting in light of the fact that $A_8 \cong \text{GL}_4(\mathbb{F}_2)$. This isomorphism, due to Jordan and found in Dickson [D], can be described by the following correspondence of generators:

$$(23)(12) \mapsto \begin{pmatrix} 1 & 1 & 1 & 1 \\ 0 & 0 & 0 & 1 \\ 1 & 1 & 0 & 0 \\ 0 & 1 & 0 & 1 \end{pmatrix}, \quad (34)(12) \mapsto \begin{pmatrix} 0 & 1 & 0 & 1 \\ 0 & 0 & 1 & 0 \\ 0 & 1 & 0 & 0 \\ 1 & 0 & 1 & 0 \end{pmatrix}$$

$$(45)(12) \mapsto \begin{pmatrix} 0 & 1 & 1 & 1 \\ 0 & 1 & 0 & 1 \\ 1 & 1 & 0 & 0 \\ 0 & 0 & 0 & 1 \end{pmatrix}, \quad (56)(12) \mapsto \begin{pmatrix} 1 & 0 & 1 & 0 \\ 0 & 1 & 0 & 0 \\ 0 & 0 & 1 & 0 \\ 0 & 1 & 0 & 1 \end{pmatrix}$$

$$(67)(12) \mapsto \begin{pmatrix} 0 & 0 & 1 & 0 \\ 0 & 1 & 0 & 1 \\ 1 & 0 & 0 & 0 \\ 0 & 0 & 0 & 1 \end{pmatrix}, \quad (78)(12) \mapsto \begin{pmatrix} 0 & 1 & 1 & 1 \\ 0 & 0 & 1 & 0 \\ 0 & 1 & 0 & 0 \\ 1 & 1 & 1 & 0 \end{pmatrix}.$$

A calculation for $H^*(\text{GL}_4(\mathbb{F}_2))$ was first attempted in [T–Y]; there was an error which was corrected later. However, at best their approach led to a listing of elements without giving any insight as to their significance or explicit multiplicative relations.

Corollary 6.6 *The Poincaré Series for* $H^*(\mathrm{GL}_4(\mathbb{F}_2))$ *is*

$$P(t) = \frac{(t+1)(t^{15}-t^{14}+t^{13}-t^{12}+t^{11}-t^9+t^8+t^7+2t^6+2t^4-t^3+2t^2-t+1)}{(t^7-1)(t^6-1)(t^3-1)(t^4-1)}.$$

Using an analogous approach, we obtain

Corollary 6.7 $H^*(A_{10}) \cong P[\sigma_2, \sigma_3, \sigma_4, \sigma_5, c_3, x_5, d_6, d_7]/\langle R\rangle$ *where* R *is the set of relations*

$$\begin{aligned}
&\sigma_5 d_7 = \sigma_3 d_7 = \sigma_3 d_7 = c_3 d_7 = x_5 d_7 = 0\\
&\sigma_3 d_6 = \sigma_5 d_6 = 0\\
&x_5^2 = \sigma_2 c_3 x_5 + c_3^2 \sigma_4 + \sigma_2^2 d_6\\
&\sigma_5 x_5 = c_3 \sigma_3 \sigma_4\\
&\sigma_3 x_5 = c_3(\sigma_5 + \sigma_2 \sigma_3).
\end{aligned}$$

(As for A_6 the $\mathcal{A}(2)$-module structure follows directly from that of $H^*(\mathcal{S}_{10})$.)

Finally, we have

Corollary 6.8

$$H^*(A_{12}) \cong P[\sigma_2, \sigma_3, c_3, \sigma_4, \sigma_5, \sigma_6, d_6, d_7, d_9, e_9, e_{10}]\,(x_5, x_7, x_8)\,/\langle R\rangle$$

where deg $\sigma_i = i$, deg $c_3 = 3$, deg $d_i = i$, deg $x_i = i$, deg $e_i = i$ *and* R *is the set of relations*

$$\begin{aligned}
&x_7^2 + \sigma_4 x_5^2 + \sigma_2\sigma_4 x_8 + (\sigma_6 c_3 + \sigma_2\sigma_4 c_3)\,x_5 + \sigma_4^2 + c_3^2 + \sigma_2^2\sigma_4 d_6 + \sigma_6\sigma_2 d_6,\\
&x_8^2 + d_6 x_5^2 + c_3 d_6 x_7 + ([e_9 + d_9]\sigma_2 + c_3 d_6 \sigma_2)x_5 + d_6^2\sigma_2^2\\
&\quad + c_3^2\sigma_4 d_6 + [e_9 + d_9]c_3\sigma_2,\\
&x_5 x_7 + \sigma_2 x_5^2 + \left[\sigma_2^2 + \sigma_4\right] x_8 + \sigma_2^2 c_3 x_5 + \sigma_2^3 d_6 + \sigma_6 c_3^2 + \sigma_6 d_6 + \sigma_2\sigma_4 c_3^2,\\
&x_5 x_8 + c_3 x_5^2 + \left[c_3^2 + d_6\right] x_7 + c_3^2\sigma_2 x_5 + c_3^3\sigma_4 + [e_9 + d_9]\sigma_2^2\\
&\quad + [e_9 + d_9]\sigma_4 + c_3 d_6 \sigma_2^2,\\
&x_5^3 + c_3\sigma_2 x_5^2 + x_7 x_8 + c_3\sigma_2^2 x_8 + \sigma_2 c_3^2 x_7 + \left(\sigma_2^2 d_6 + \sigma_4 c_3^2\right) x_5 +\\
&\quad \sigma_2^3[e_9 + d_9] + c_3^3\sigma_6 + \sigma_6[e_9 + d_9],\\[4pt]
&d_9\sigma_3, d_9\sigma_5, e_9\sigma_3, e_9\sigma_5\\
&d_7\sigma_3, d_7 x_5, e_{10}\sigma_4 + d_7(x_7 + \sigma_4 c_3), d_7\sigma_5, d_7(\sigma_6 + \sigma_4\sigma_2), d_7\,(x_8 + d_6\sigma_2),\\
&d_7\,(e_9 + d_9) + d_6(d_7 c_3 + e_{10}),\\[4pt]
&x_5\sigma_3 + c_3\,(\sigma_2\sigma_3 + \sigma_5)\,, x_7\sigma_3 + x_5\sigma_5 + c_3\sigma_3\sigma_4, x_7\sigma_5 + c_3\sigma_3\sigma_6,\\
&d_6\sigma_3, x_8\sigma_3 + d_6\sigma_5, x_8\sigma_5,\\[4pt]
&d_9^2 + d_9 e_9 + e_9^2, (d_7 c_3)^2 + d_7 c_3 e_{10} + e_{10}^2,\\
&d_9 d_7 c_3 + d_9 e_{10} + e_9 e_{10}, d_9 d_7 c_3 + d_7 c_3 e_9 + e_9 e_{10}.
\end{aligned}$$

The action of the Steenrod Algebra on generators different from e_9, e_{10} can be obtained from 5.3 using the restriction map; for e_9, e_{10} the action is identical to that on d_9 and $d_7 c_3$ doing the appropriate substitutions.

Chapter VII. Finite Groups of Lie Type

1. Preliminary Remarks

One of the most remarkable results of this century in mathematics has been the classification – completed in 1980 – of all the finite simple groups. This took over 20 years and occupies almost 5000 pages in the literature. It is conceivable that there are some errors there, so the details of classification are not really available to us, but the main results can be summarized. There are 17 families of simple groups, the alternating groups and 16 families of "Lie type". These in turn are broken up into subfamilies in several different ways. There are, first, the historical breakdowns, 6 families of "classical groups", the projective special linear groups over finite fields, the orthogonal groups, (three types), unitary groups, and the symplectic groups. Then there are the 5 families of groups of "exceptional types", the groups $G_2(q)$, $F_4(q)$, $E_6(q)$, $E_7(q)$, and $E_8(q)$, the q denoting the order of the finite field over which they are defined. Finally, there are five special families, the groups of "twisted type", the Suzuki groups $Sz[2^{2n+1}]$, the Ree groups ${}^2G_2[3^{2n+1}]$ and ${}^2F_2[2^{2n+1}]$, the twisted exceptional groups ${}^2E_6[2^{2n+1}]$, and the groups ${}^3D_4(q)$, the "triality twisted $D_4(q)$'s". Of course, the most natural classification of these groups is via Lie theory and Dynkin diagrams. A complete discussion can be found in R.W. Carter's book [Carter].

In each family the first few groups are somewhat exceptional: they may not be simple, for example $\mathcal{A}_4$ or $\mathrm{GL}_2(\mathbb{F}_2) = \mathcal{S}_3$, $\mathrm{Sp}_4(\mathbb{F}_2) = \mathcal{S}_6$, or they are not distinct, $\mathrm{PSL}_2(\mathbb{F}_4) = \mathrm{PSL}_2(\mathbb{F}_5) = \mathcal{A}_5$, $\mathrm{PSL}_4(\mathbb{F}_2) = \mathcal{A}_8$. Additionally, the situation for some of these groups at the prime 2 seems to be much more involved than at other primes. For these reasons and probably others there arise 26 further simple groups, the sporadic groups. The first five of these, the Mathieu groups M_{11}, M_{12}, M_{22}, M_{23}, and M_{24}, have been known and explored for about 100 years. The remaining ones have been known for less than thirty years. The first of them J_1 – the first Janko group – has already been discussed in Chapters II, III and V. A good discussion of these groups, their construction and history, can be found in Gorenstein's book [Gor].

In this chapter we describe the main features of the classical groups of Lie type and discuss some aspects of Quillen's calculation of the cohomology of these groups. The most striking of his results appears in his paper in the Proceedings of the International Congress of Mathematics, Nice, 1970 [Q2], where he indicates the existence of a spectral sequence for groups sufficiently closely related to the continuous Lie groups G with E_2-term

$$E_2 = H^*(G;\mathbb{F}_p) \otimes H^*(B_G;\mathbb{F}_p) \Rightarrow H^*(G(q);\mathbb{F}_p)$$

where $G(q)$ is the group defined by the same (integral) equations over the finite field $\mathbb{F}_q$ and q is prime to p. He also indicates there how to use this result to get information about the groups of twisted type. However, the indicated proof, using Etale techniques from algebraic geometry is far removed from our discussion. Rather, we give proofs of most of his results here but use the more elementary techniques discussed here. An alternate source for much of this discussion is [FP].

It should be emphasized that the groups we are actually studying are not the simple groups, generally quotients of the groups here by their centers, but the full groups, described as the groups of automorphisms of a finite dimensional vector space over a finite field, in the case of the general linear groups, or which preserve a bilinear form in the case of the orthogonal and symplectic groups, or which preserve a Hermitian form in the case of the unitary groups. In brilliant work of Quillen in the early 1970's these groups and their cohomology played an essential role in vital questions in homotopy theory some of which we will also outline in the introduction to Chapter VIII and in Chapter IX.

2. The Classical Groups of Lie Type

Given a field K and a vector space $\mathbb{V}_K^n$ of dimension n over K we have the group $\mathrm{GL}_n(K)$ of invertible linear transformations $\alpha\colon \mathbb{V}_K^n \to \mathbb{V}_K^n$. $\mathrm{GL}_n(K)$ has a center $C(\mathrm{GL}_n(K)) = K^\bullet$ and there is a determinant homomorphism $\mathrm{Det}\colon \mathrm{GL}_n(K) \to K^\bullet \to 0$. The kernel of Det is the special linear group $\mathrm{SL}_n(K)$ and

$$\mathrm{SL}_n(K)/\mathrm{SL}_n(K) \cap C(\mathrm{GL}_n(K)) = \mathrm{PSL}_n(K)$$

is called the special projective linear group of rank n over K. It is simple except for $K = \mathbb{F}_2, \mathbb{F}_3$, and $n = 2$. As K runs over the finite fields $\mathbb{F}_q$ (where $q = p^n$, p a prime in $\mathbb{Z}^+$) we obtain in this way the finite simple groups $L_n(q) = \mathrm{PSL}_n(\mathbb{F}_q)$. This is the first family of finite simple groups of Lie type.

Next, suppose $\mathbb{V}_K^n$ is given together with a non-singular bilinear form b. A bilinear form can be thought of as a K-linear map $B\colon \mathbb{V}_K^n \to (\mathbb{V}_K^n)^*$ where $(\mathbb{V}_K^n)^* = \mathrm{Hom}_K(\mathbb{V}_K^n, K)$. There are natural isomorphisms $e\colon (\mathbb{V}_K^n)^{**} \to \mathbb{V}_K^n$ (for $n \neq \infty$, $e(f)$ is given by $w(e(f)) = f(w)$ for $w \in \mathbb{V}_K^n$), and given B there is the induced map $B^*\colon (\mathbb{V}_K^n)^{**} \to (\mathbb{V}_K^n)^*$ defined by $B^*(f)(w) = f(B(w))$. Consequently B induces a map $Be^{-1}\colon (\mathbb{V}_K^n)^* \to \mathbb{V}_K^n$ which we again write as B^* using a mild abuse of the language. B is symmetric or anti-symmetric according to whether $B^* = B$ or $B^* = -B$ respectively. B is said to be non-singular if B is an isomorphism.

B is equally well determined by the bilinear form $\langle\ ,\ \rangle_B\colon \mathbb{V}_K^n \times \mathbb{V}_K^n \to K$, $\langle\ ,\ \rangle_B(v,v') = \langle v,v'\rangle_B = B(v)(v')$. We have

Lemma 2.1

a. *B is symmetric if and only if $\langle v,v'\rangle_B = \langle v',v\rangle_B$ for all v, v' in $\mathbb{V}_K^n$.*
b. *B is skew symmetric if and only if if $\langle v,v'\rangle_B = -\langle v',v\rangle_B$ for all v, v' in $\mathbb{V}_K^n$.*
c. *B is non-singular if and only if, given any $v \in \mathbb{V}_K^n$ there is a $v' \in \mathbb{V}_K^n$ so that $\langle v,v'\rangle \neq 0$.*

Proof. (a), (b) are direct. To see (c) note that B is a linear map of finite dimensional vector spaces over K having the same dimension. Consequently B is invertible if and only if $\mathrm{Ker}(B) = 0$. But $v \in \mathrm{Ker}(B)$ if and only if $\langle v,v'\rangle_B = B(v)(v') = 0$ for all $v' \in \mathbb{V}_K^n$. □

Two non-singular bilinear forms B and B' are said to be isomorphic if there is an isomorphism $g\colon \mathbb{V}_K^n \to \mathbb{V}_K^n$ so that the diagram

$$(*)\qquad \begin{array}{ccc} \mathbb{V}_K^n & \xrightarrow{B} & (\mathbb{V}_K^n)^* \\ \downarrow{\scriptstyle g} & & \uparrow{\scriptstyle g^*} \\ \mathbb{V}_K^n & \xrightarrow{B'} & (\mathbb{V}_K^n)^* \end{array}$$

commutes. Isomorphism is an equivalence relation. Moreover, to each non-singular bilinear form B there is associated its group of isometries G_B:

$$G_B = \{g\colon\ \mathbb{V}_K^n \to \mathbb{V}_K^n \mid g \in \mathrm{GL}_n(K),\ \langle g(v), g(v')\rangle_B = \langle v,v'\rangle_B\}$$

for all $v, v' \in \mathbb{V}_K^n$. ($g, h \in G_B$ implies that $\langle gh(v), gh(v')\rangle_B = \langle h(v), h(v')\rangle_B = \langle v,v'\rangle_B$ for all v, v' in $\mathbb{V}_K^n$ so G_B is closed under composition. Also

$$\langle v,v'\rangle_B = \langle g(g^{-1}(v)), g(g^{-1}(v'))\rangle_B = \langle g^{-1}(v), g^{-1}(v')\rangle_B$$

so G_B is closed under taking inverses and G_B is a subgroup of $\mathrm{GL}_n(K)$.)

Remark 2.2 If B is isomorphic to B' then $G_B \cong G_{B'}$, since if g takes B to B' in the sense of $(*)$ then $gG_{B'}g^{-1} \subset G_B$, and $g^{-1}G_B g \subset G_{B'}$.

Examples 2.3 Over K any bilinear form B is given by $\langle k,k'\rangle_B = kk'b$ for some $b \in K$ using the usual identification $K^* = K$ ($b \in K^*$ is the map $k \mapsto kb$.) Then $(*)$ says that two forms $\times b$ and $\times b'$ are equivalent if and only if there is a $k \in K^\bullet$ so that $b' = k^2 b$ and the equivalence classes of non-singular forms on K (they are all symmetric) are the elements of $K^\bullet/(K^\bullet)^2$ where $(K^\bullet)^2$ denotes the multiplicative subgroup of squares in $K^\bullet$. The group G_B in each of these cases is $K^\bullet$.

In particular, if $K = \mathbb{F}_q$, $(q \neq 2^r)$ there are 2 equivalence classes of forms, the first represented by $\langle k, k'\rangle_1 = kk'$ and the second represented by $\langle k, k'\rangle_a = kk'a$ where $a \in K^\bullet$ is any non-square.

Next, passing to based vector spaces, given $\mathbb{V}_K^n$ together with a basis $(e_1, \ldots, e_n)$, there is a dual basis for $(\mathbb{F}_K^n)^*$, $(e^1, \ldots, e^n)$ where e^i is defined as the homomorphism $e^i(e_j) = \begin{cases} 0 & i \neq j \\ 1 & i = j.\text{cr} \end{cases}$ With respect to these choices for bases B is represented by a matrix (which we again denote B) where the entries $B_{i,j}$ satisfy $B_{i,j} = B_{j,i}$ if B is symmetric or $B_{i,j} = -B_{j,i}$ if B is skew symmetric. Also, if $g \in \mathrm{GL}(\mathbb{V}_K^n)$ is represented with respect to $(e_1, \ldots, e_n)$ by the matrix $(g_{i,j})$ then g^* is represented by the transpose of g with respect to the basis $(e^1, \ldots, e^n)$ for $(\mathbb{V}_K^n)^*$. It follows that in matrix notation our equivalence relation becomes

$$B \sim gBg^t$$

for $g \in \mathrm{GL}_n(K)$.

It follows that $g \in G_B$ if and only if $B = gBg^t$. In particular for $B = \langle 1\rangle \perp \langle 1\rangle$ the matrix is $\begin{pmatrix} 1 & 0 \\ 0 & 1 \end{pmatrix}$ so $g \in G_B$ if and only if $gg^t = I$. Consequently $\mathrm{Det}(g) = \pm 1$. In general, for 2×2 matrices L,

$$L^{-1} = \begin{pmatrix} a & b \\ c & d \end{pmatrix}^{-1} = \frac{1}{\mathrm{Det}(L)} \begin{pmatrix} d & -b \\ -c & a \end{pmatrix}$$

so $L^{-1} = L^t$ if and only if $c = -b$, $a = d$ if $\mathrm{Det}(L) = 1$, and $c = b$, $a = -d$ if $\mathrm{Det}(L) = -1$. Denote by $\mathrm{SO}_2^B(K)$ those elements of G_B with determinant $+1$ so, if $\mathrm{Char}(K)$ is not 2 $\mathrm{SO}_2^B(K)$ is normal in G_B with quotient $\mathbb{Z}/2$. Summarizing, $\theta \in \mathrm{SO}_2^B(K)$ if and only if $\theta = \begin{pmatrix} a & b \\ -b & a \end{pmatrix}$ with $\mathrm{Det}(\theta) = a^2 + b^2 = 1$.

If $B\colon \mathbb{V}_K^n \to \mathbb{V}_K^n$, $B'\colon \mathbb{V}_K^m \to (\mathbb{V}_K^m)^*$ are given non-singular bilinear forms then $B \oplus B'\colon \mathbb{V}_K^n \oplus \mathbb{V}_K^m \to (\mathbb{V}_K^n)^* \oplus (\mathbb{V}_K^m)^* = (\mathbb{V}_K^n \oplus \mathbb{V}_K^m)^*$ is a non-singular bilinear form on $\mathbb{V}_K^{n+m}$. It is symmetric or skew-symmetric according as B, B' are, and it has the property of orthogonality, $\langle v, v'\rangle_{B\oplus B'} \equiv 0$ if $v \in \mathbb{V}_K^n$ while $v' \in \mathbb{V}_K^m$. This direct sum form is usually written $B \perp B'$. If B is isomorphic to $\hat{B}$ then $B \perp B'$ is isomorphic to $\hat{B} \perp B'$, and similarly for $\bar{B}$ isomorphic to B', $B \oplus B'$ is isomorphic to $B \perp \bar{B}$.

Theorem 2.4 *Let $K = \mathbb{V}_q$, $(q \neq 2^r)$ then there are exactly two equivalence classes of non-singular symmetric bilinear forms on $\mathbb{V}_K^n$, the first equivalent to $\langle 1\rangle \perp \langle 1\rangle \perp \cdots \perp \langle 1\rangle$ and the second equivalent to $\langle 1\rangle \perp \cdots \langle 1\rangle \perp \langle a\rangle$ with a any non-square in K.*

Proof. We need a splitting principle.

Lemma 2.5 *Let $v \in \mathbb{V}_K^n$ and let B be a non-singular symmetric form on $\mathbb{V}_K^n$. Suppose $\langle v, v\rangle_B = b \neq 0$, then there is an $n-1$ dimensional subspace $W \subset \mathbb{V}_K^n$ so that $\mathbb{V}_K^n = \langle v\rangle \oplus W_K^{n-1}$ as K-vector spaces, and with respect to this splitting $B = \langle b\rangle \perp \bar{B}$.*

Proof of 2.5 Let $W = \langle v\rangle^\perp$. That is to say, $w \in W$ if and only if $\langle w, v\rangle_B = 0$. Let $l \in \mathbb{V}_K^n$ be any non-zero vector. Then

$$l = \left(l - \frac{\langle l, v\rangle_B}{b} v\right) + \frac{\langle l, v\rangle_B}{b} v$$

and $l - \frac{\langle l,v\rangle}{l} v$ is contained in $\langle v\rangle^\perp$. Consequently $\mathbb{V}_K^n = \langle v\rangle + W$ and, since $b \neq 0$ it is clear that $\langle v\rangle \cap W = 0$ so the sum is direct. □

(More generally, if $L \subset \mathbb{V}_K^n$ is a sub-vector space of $\mathbb{V}_K^n$ then the same proof shows that if the restriction of B to L is non-singular there is a splitting $\mathbb{F}_K^n = L \oplus L^\perp$ and B splits as $(B|L) \perp \bar{B}$.)

Lemma 2.6 *Let $\mathbb{V}_K^n$ be given with $K = \mathbb{F}_{p^r}$, p odd, and B symmetric and non-singular. Then there is a vector $v \in \mathbb{V}_K^n$ so that $\langle v, v\rangle_B \neq 0$.*

Proof of 2.6 Choose $v \neq 0$, $v \in \mathbb{V}_K^n$ and suppose that $\langle v, v\rangle_B = 0$. Then, by the non-singularity of B, there is a $v' \in \mathbb{V}_K^n$ with $\langle v, v'\rangle_B \neq 0$. If $\langle v', v'\rangle \neq 0$ we are done. If not, then

$$\begin{aligned}\langle v + v', v + v'\rangle_B &= \langle v, v\rangle_B + 2\langle v, v'\rangle_B + \langle v', v'\rangle_B \\ &= 2\langle v, v'\rangle_B \\ &\neq 0\end{aligned}$$

so the proof is complete. □

At this point we have proved that B is isomorphic to $\langle b_1\rangle \perp \cdots \perp \langle b_n\rangle$. It remains to show that all but one of the b_i can be chosen equal to 1. To do this we study non-singular forms on $\mathbb{V}_K^2$.

Lemma 2.7 *Let B be a non-singular symmetric form on $\mathbb{V}_K^n$ where $K = \mathbb{F}_{p^r}$ with p odd, then B is isomorphic to $\langle 1\rangle \perp \langle 1\rangle$ or $\langle 1\rangle \perp \langle a\rangle$ where a is a non-square in $K^\bullet$.*

Proof of 2.7 After diagonalization $B \sim \langle 1\rangle \perp \langle 1\rangle$, $\langle 1\rangle \perp \langle a\rangle$ or $\langle a\rangle \perp \langle a\rangle$. We now verify that $\langle a\rangle \perp \langle a\rangle$ is isomorphic to $\langle 1\rangle \perp \langle 1\rangle$. By the pigeon hole principle there are k, k' in K so that $k^2 + k'^2$ is a non-square in $K^\bullet$. Then $(k^2 + k'^2)a$ is a square in $K^\bullet$, and the two vectors (k, k'), $(k', -k)$ are mutually orthogonal in $\langle a\rangle \perp \langle a\rangle$ and provide an explicit equivalence with $\langle 1\rangle \perp \langle 1\rangle$. □

Iterating the lemma above shows that there are at most two isomorphism classes, $\langle 1\rangle \perp \cdots \perp \langle 1\rangle$, and $\langle 1\rangle \perp \cdots \perp \langle 1\rangle \perp \langle a\rangle$. It remains to show that these two forms are not isomorphic. Suppose then that we have an isomorphism between them for some $n \geq 2$, and suppose that g is a map which effects the equivalence. Then

$$gg^t = \begin{pmatrix} 1 & 0 & 0 & \dots & 0 & 0 \\ 0 & 1 & 0 & \cdots & 0 & 0 \\ 0 & 0 & 1 & \cdots & 0 & 0 \\ \vdots & \vdots & \vdots & \ddots & \vdots & \vdots \\ 0 & 0 & 0 & \cdots & 1 & 0 \\ 0 & 0 & 0 & \cdots & 0 & b \end{pmatrix}$$

and $\text{Det}(gg^t) = b = \text{Det}(g)\text{Det}(g^t) = (\text{Det}(g))^2$ which is impossible. The theorem follows. □

In a similar way we can analyze skew-symmetric forms. On $\mathbb{V}_K^1$ we see from our previous discussion that there are none. On $\mathbb{V}_K^2$ we have $\langle v, v\rangle_B = 0$ for every $v \in \mathbb{V}_K^2$, but by the non-singularity of B given $v \neq 0$ in $\mathbb{V}_K^2$ there is a v' so $\langle v, v'\rangle_B = 1$. It follows that, with respect to the basis v, v' the form B is given by the matrix $\begin{pmatrix} 0 & 1 \\ -1 & 0 \end{pmatrix}$. Iterating this argument we have

Theorem 2.8 *Let B be a non-singular skew-symmetric form on $\mathbb{F}_K^n$ where the characteristic of K is not 2, then n is even and*

$$B \sim \langle S\rangle \perp \cdots \perp \langle S\rangle$$

where S is the form on $\mathbb{V}_K^2$ given above.

Remark 2.9 Note that when $n = 2m+1$ is odd and K is a finite field of odd characteristic we have that

$$\underbrace{\langle a\rangle \perp \cdots \perp \langle a\rangle}_{n \text{ times}} = \underbrace{\langle 1\rangle \perp \cdots \perp \langle 1\rangle}_{n-1 \text{ times}} \perp \langle a\rangle\,.$$

But this implies that for such n, K, the groups for the two forms are isomorphic since, if A represents the form with all ones, while B represents the form with all a's, we have $\langle v, v'\rangle_B = a\langle v, v'\rangle_A$ for all v, v' in $\mathbb{V}_K^n$ so $g \in G_B$ if and only if $g \in G_A$.

Lemma 2.10 *Let $K = \mathbb{F}_q$ be a finite field of odd characteristic.*
a. If $q \equiv 3 \bmod 4$ then $\text{SO}_2^A(K) = \mathbb{Z}/(q+1)$.
b. If $q \equiv 1 \bmod 4$ then $\text{SO}_2^A(K) = \mathbb{Z}/(q-1)$.

Proof. Suppose that $q \equiv 3 \bmod 4$, then $-1 \in K$ is a non-square and $\mathbb{F}_{q^2} = \mathbb{F}_q(\sqrt{-1})$. Write $\alpha \in \mathbb{F}_{q^2}$ as $k + k'\sqrt{-1}$. Then we have that the norm of α is

$k^2+k'^2$. Moreover, since Norm is an onto homomorphism $\mathbb{F}^{\bullet}_{q^2} = \mathbb{Z}/(q^2-1) \to \mathbb{F}^{\bullet}_q = \mathbb{Z}/(q-1)$, its kernel is $\mathbb{Z}/(q+1)$. Finally, for this case, if $N(\theta) = 1$ where $\theta = k + k'\sqrt{-1}$, then

$$\begin{pmatrix} k & k' \\ -k' & k \end{pmatrix} \in \mathrm{SO}_2^A(K)$$

and conversely, so $\mathrm{SO}_2^A(K) = \mathbb{Z}/(q+1)$.

To prove (b) note first that now $-1 = k^2$ for some $k \in K$ so there are two lines $\langle 1, k\rangle = l_1$, $\langle 1, -k\rangle = l_2$ and only two so that $\langle l, l\rangle_A = 0$ for $l \in l_i$. Also, since $\langle 1\rangle \perp \langle 1\rangle = \langle a\rangle \perp \langle a\rangle$ it follows that there is a one to one correspondence between those lines l_j for which A restricts as $\langle 1\rangle$ and those for which it restricts as $\langle a\rangle$. Consequently, since $\mathbb{V}^2_K$ contains exactly $q+1$ lines there are $(q-1)/2$ lines, $l_{+,1}, \ldots, l_{+,(q-1)/2}$ where A restricts to $\langle 1\rangle$. Each such line has exactly two vectors α and $-\alpha$ so that $\langle \beta, \beta\rangle_A = 1$, and each vector can be written with respect to the original basis as (k_1, k_2) with $k_1^2 + k_2^2 = 1$. This gives that $|\mathrm{SO}_2^A(K)| = q - 1$.

It remains to verify that $\mathrm{SO}_2^A(K)$ is as claimed. To see this consider

$$g = \begin{pmatrix} k & 1 \\ \frac{1}{2k} & \frac{1}{2} \end{pmatrix}.$$

We have

$$gg^t = \begin{pmatrix} k & 1 \\ \frac{1}{2k} & \frac{1}{2} \end{pmatrix}\begin{pmatrix} k & \frac{1}{2k} \\ 1 & \frac{1}{2} \end{pmatrix} = \begin{pmatrix} 0 & 1 \\ 1 & 0 \end{pmatrix}$$

so g gives an explicit isomorphism $\langle 1\rangle \perp \langle 1\rangle \sim \left\langle \begin{pmatrix} 0 & 1 \\ 1 & 0 \end{pmatrix} \right\rangle$. Hence $\mathrm{SO}_2^A(K)$ is isomorphic to the subgroup of $\mathrm{Aut}\begin{pmatrix} 0 & 1 \\ 1 & 0 \end{pmatrix}$ having determinant 1. However, here we have

$$\begin{aligned} g\begin{pmatrix} 0 & 1 \\ 1 & 0 \end{pmatrix} g^t &= \begin{pmatrix} a & b \\ c & d \end{pmatrix}\begin{pmatrix} 0 & 1 \\ 1 & 0 \end{pmatrix}\begin{pmatrix} a & c \\ b & d \end{pmatrix} \\ &= \begin{pmatrix} 2ab & bc+ad \\ bc+ad & 2cd \end{pmatrix} \\ &= \begin{pmatrix} 0 & 1 \\ 1 & 0 \end{pmatrix} \end{aligned}$$

and this implies that $g \in \mathrm{Aut}\begin{pmatrix} 0 & 1 \\ 1 & 0 \end{pmatrix}$ if and only if $g = \begin{pmatrix} a & 0 \\ 0 & 1/a \end{pmatrix}$ for $a \in K^{\bullet}$ or $g = \begin{pmatrix} 0 & a \\ 1/a & 0 \end{pmatrix} = \begin{pmatrix} a & 0 \\ 0 & 1/a \end{pmatrix}\begin{pmatrix} 0 & 1 \\ 1 & 0 \end{pmatrix}$. But the latter terms have determinant -1 so they do not occur and the result follows. □

Corollary 2.11 $\mathrm{O}_2^A(K) = \begin{cases} D_{2(q+1)} & q \equiv 3 \bmod 4 \\ D_{2(q-1)} & q \equiv 1 \bmod 4. \end{cases}$

Proof. For $q \equiv 1 \bmod 4$ this follows from the calculation above with the group $\mathrm{Aut}\begin{pmatrix} 0 & 1 \\ 1 & 0 \end{pmatrix}$. For $q \equiv 3 \bmod 4$ the element $\begin{pmatrix} 1 & 0 \\ 0 & -1 \end{pmatrix}$ acts to invert $\begin{pmatrix} 0 & 1 \\ -1 & 0 \end{pmatrix}$ and fixes $\begin{pmatrix} k & 0 \\ 0 & k \end{pmatrix}$. Thus

$$\begin{pmatrix} 1 & 0 \\ 0 & -1 \end{pmatrix} (k + k'\sqrt{-1}) \begin{pmatrix} 1 & 0 \\ 0 & -1 \end{pmatrix} = (k - k'\sqrt{-1})$$

and this is just inversion on the elements of norm 1. The corollary follows. □

Next we consider the case $B = \langle 1 \rangle \perp \langle a \rangle$.

Lemma 2.12 *Let K have odd characteristic. Then*

$$\mathrm{SO}_2^B(K) \;=\; \begin{cases} \mathbb{Z}/(q+1) & \text{when } q \equiv 1 \bmod 4, \\ \mathbb{Z}/(q-1) & \text{when } q \equiv 3 \bmod 4. \end{cases}$$

Proof. If $q \equiv 1 \bmod 4$ then $\langle 1 \rangle \perp \langle a \rangle$ has no isotropy vectors and on every line the restriction of the form is either $\langle 1 \rangle$ or $\langle a \rangle$. But the interchange mapping $(k, k') \leftrightarrow (k', k)$ has the effect of exchanging such lines so there are exactly $(q+1)/2$ lines l with $B|l = \langle 1 \rangle$. Consequently $B|l^{\perp} = \langle a \rangle$, and as before there are 2 transformations β, $-\beta$ taking $(1,0) \mapsto L_0 \in l$, $(0,1) \mapsto L_1 \in l^{\perp}$ and preserving the form. It follows that $|\mathrm{SO}_2^B(K)| = q+1$ if $q \equiv 1 \bmod 4$. On the other hand if $J = K(\sqrt{a})$ and $k + k'\sqrt{a}$ nas norm 1, then

$$\begin{pmatrix} k & k' \\ -ak' & k \end{pmatrix} \in \mathrm{SO}_2^B(K)\,.$$

Moreover, the correspondence above is a homomorphism since

$$\begin{pmatrix} k & k' \\ -ak' & k \end{pmatrix} \begin{pmatrix} \kappa & \kappa' \\ -a\kappa' & \kappa \end{pmatrix} = \begin{pmatrix} k\kappa - k'\kappa' a & k\kappa' + k'\kappa \\ -a(k\kappa' + k'\kappa) & k\kappa - ak'\kappa' \end{pmatrix}.$$

The first statement follows. The proof for $q \equiv 3 \bmod 4$ follows the lines of proof of the previous lemma (for $\mathrm{SO}_2^A(K)$ in the case $q \equiv 1 \bmod 4$. □

Similar considerations show

Corollary 2.13 $\mathrm{O}_2^B(K) \;=\; \begin{cases} D_{2(q+1)} & \textit{when } q \equiv 1 \bmod 4, \\ D_{2(q-1)} & \textit{when } q \equiv 3 \bmod 4. \end{cases}$

(In the case $q \equiv 1 \bmod 4$ we have

$$\begin{pmatrix} -1 & 0 \\ 0 & 1 \end{pmatrix} \begin{pmatrix} k & k' \\ -ak' & k \end{pmatrix} \begin{pmatrix} -1 & 0 \\ 0 & 1 \end{pmatrix} = \begin{pmatrix} k & -k' \\ ak' & k \end{pmatrix} = \begin{pmatrix} k & k' \\ -ak' & k \end{pmatrix}^{-1}.$$

When $q \equiv 3 \bmod 4$ we have that $B = \langle 1 \rangle \perp \langle -1 \rangle$ and

$$\mathrm{O}_2^B(K) = \mathrm{Aut}\left\{\begin{pmatrix} 0 & 1 \\ 1 & 0 \end{pmatrix}\right\}$$

and now the calculation follows.)

This completes the discussion for the orthogonal groups of rank ≤ 2. We now turn to the symplectic groups.

Proposition 2.14 $\mathrm{Aut}_K\left\{\begin{pmatrix} 0 & -1 \\ 1 & 0 \end{pmatrix}\right\} = \mathrm{Sp}_2(K)$ *is the special linear group* $\mathrm{SL}_2(K)$.

Proof. We calculate

$$\begin{aligned}\begin{pmatrix} a & b \\ c & d \end{pmatrix}\begin{pmatrix} 0 & 1 \\ -1 & 0 \end{pmatrix}\begin{pmatrix} a & c \\ b & d \end{pmatrix} &= \begin{pmatrix} -b & a \\ -d & c \end{pmatrix}\begin{pmatrix} a & c \\ b & d \end{pmatrix} \\ &= \begin{pmatrix} 0 & ad-bc \\ bc-ad & 0 \end{pmatrix}\end{aligned}$$

which is $\begin{pmatrix} 0 & 1 \\ -1 & 0 \end{pmatrix}$ if and only if $\mathrm{Det}(\begin{pmatrix} a & b \\ c & d \end{pmatrix}) = 1$. □

3. The Orders of the Finite Orthogonal and Symplectic Groups

Definition 3.1 A quadratic form on K^n is *hyperbolic* if and only if $n = 2l$ and

$$A \sim \underbrace{\left\langle\begin{pmatrix} 0 & 1 \\ 1 & 0 \end{pmatrix}\right\rangle \perp \cdots \left\langle\begin{pmatrix} 0 & 1 \\ 1 & 0 \end{pmatrix}\right\rangle}_{l \text{ times}}$$

In what follows we denote this form as $\langle H_{2l} \rangle$.

Remark 3.2 $\langle H_2 \rangle = \langle 1 \rangle \perp \langle -1 \rangle = \begin{cases} \langle 1 \rangle \perp \langle 1 \rangle & \text{for } q \equiv 1 \bmod 4 \\ \langle 1 \rangle \perp \langle \alpha \rangle & \text{for } q \equiv 1 \bmod 4. \end{cases}$

Assume that K is finite of odd characteristic. When n is odd we know there is only one orthogonal group so we may as well study it in the case of the form $H_{2n} \perp \langle 1 \rangle = \langle A \rangle$.

Proposition 3.3 *Let* $|K| = q < \infty$ *be of odd characteristic. Then we have*
a. The set of $v \in K^{2n}$ *so* $\langle v, v \rangle_{\langle H_{2n} \rangle} = \alpha$ *has cardinality*

$$\begin{cases} q^{2n-1} + q^n - q^{n-1} & \textit{when } \alpha = 0, \\ q^{n-1}(q^n - 1) & \textit{otherwise}. \end{cases}$$

b. *The set of* $v \in K^{2n+1}$ *so that* $\langle v, v \rangle_{\langle A \rangle} = \alpha$ *has cardinality*

$$\begin{cases} q^{2n} - q^n & \textit{when } \alpha \in K^\bullet \textit{ is a non-square}, \\ q^{2n} + q^n & \textit{when } \alpha \in K^\bullet \textit{ is a square}, \\ q^{2n} & \textit{when } \alpha = 0. \end{cases}$$

Proof. We begin with the case $n = 1$. Then (a, a) $(a, -a)$ are the only vectors of length zero (called isotropy vectors). Hence there are $2q-1$ isotropy vectors. Next, since

$$\langle \alpha \rangle \perp \langle -\alpha \rangle \ \sim \ \langle 1 \rangle \perp \langle -1 \rangle$$

the cardinality of the set of v so that $\langle v, v \rangle_{\langle H_2 \rangle} = \lambda$ is the same as that for $\langle v, v \rangle_{\langle H_2 \rangle} = \tau$ provided λ, τ are both non-zero, it follows that this cardinality is

$$\frac{q^2 - 2q + 1}{q - 1} = q - 1 .$$

Thus (a) is verified in the first case. Now we proceed by induction. First we show (a) for H_j with $j \leq n$ implies (a) for H_{2n+2}.

Since $\langle H_{2n+2} \rangle = \langle H_{2n} \rangle \perp \langle H_2 \rangle$ we can count isotropy vectors as

$$(q^{2n_1} + q^n - q^{n-1})(2q - 1) + q^{n-1}(q^n - 1)(q - 1)^2$$

which directly expands out to $q^{2n+1} + q^{n+1} - q^n$. As before the cardinality of the sets so $\langle v, v \rangle_{\langle H_{2n+2} \rangle} = \lambda$ with $\lambda \neq 0$ are all equal, so this cardinality is

$$\frac{(q^{2n+2} - q^{2n+1} - q^{n+1} + q^n)}{q - 1} = q^{2n+1} - q^n = q^n(q^{n+1} - 1) ,$$

and this case is complete.

We now verify that (a) for $\langle H_{2n} \rangle$ implies (b) for $\langle A \rangle = \langle H_{2n} \rangle \perp \langle 1 \rangle$. First we obtain the count of isotropy vectors as

$$\begin{aligned} (q^{2n-1} + q^n - q^{n-1}) + (q - 1)q^{n-1}(q^n - 1) &= q^{2n} - q^{2n-1} - q^n + q^{n-1} \\ &\quad + q^{2n-1} + q^n - q^{n-1} \\ &= q^{2n}. \end{aligned}$$

Suppose next that α is a non-square, then $\alpha = (\alpha - k^2) + k^2$ and $\alpha - k^2 \neq 0$, so as k runs over K this gives $q \times q^{n-1}(q^n - 1) = q^n(q^n - 1)$ solutions.

Finally, suppose that α is a square. Then $\alpha - k^2 = 0$ for two values of k so the count is

$$\begin{aligned} (q - 2)q^{n-1}(q^n - 1) + 2(q^{2n-1} + q^n - q^{n-1}) &= q^{2n} + q^n \\ &= q^n(q^n + 1)l \end{aligned}$$

so the induction is complete. □

We now turn to the calculation of these numbers for the other even dimensional form

$$\langle B\rangle = \langle H_{2n-2}\rangle \perp \left\langle \begin{pmatrix} 1 & 0 \\ 0 & -\alpha \end{pmatrix} \right\rangle$$

where $\alpha \in K^\bullet$ is a non-square.

Proposition 3.4 *The form $\langle B\rangle$ satisfies the condition that the cardinality of the set of v so that $\langle v, v\rangle_{\langle B\rangle} = \lambda$ is* $\begin{cases} q^{2n+1} - q^{n+1} + q^n & \textit{when } \lambda = 0, \\ q^n(q^{n+1}+1) & \textit{when } \lambda \neq 0. \end{cases}$

Proof. Clearly $\left\langle \begin{pmatrix} 1 & 0 \\ 0 & -\alpha \end{pmatrix} \right\rangle$ has only the 0-vector as an isotropy vector. Consequently, the remaining cases, all having the same cardinality, are $q+1$. Now consider the general case. Here the number of isotropy vectors is

$$\begin{aligned} q^{2n-1} + q^n - q^{n-1} + q^{n-1}(q^n-1)(q+1)(q-1) & \\ &= q^{2n-1} + q^n - q^{n-1} \\ &\quad + q^{n-1}(q^n - 11)(q^2-1) \\ &= q^{2n-1} + q^n - q^{n-1} \\ &\quad + q^{n-1}(q^{n+2} - q^n - q^2 + 1) \\ &= q^{2n+1} - q^{n+1} + q^n. \end{aligned}$$

From this the number when $\lambda \neq 0$ is

$$\frac{q^{2n+2} - (q^{2n+1} - q^{n+1} + q^n)}{q-1} = q^{2n+1} + q^n$$

and the result follows. □

Now write $SO^+_{2n}(K)$ for the group associated to the hyperbolic form $\langle H_{2n}\rangle$ and $SO^-_{2n}(K)$ for the group associated to the form $\langle H_{2n-2}\rangle \perp \left\langle \begin{pmatrix} 1 & 0 \\ 0 & -\alpha \end{pmatrix} \right\rangle$. Then we have

Theorem 3.5
a. $|SO^+_{2n}(K)| = q^{n(n-1)}(q^2-1)\cdots(q^{2n-2}-1)(q^n-1)$,
b. $|SO^-_{2n}(K)| = q^{n(n-1)}(q^2-1)\cdots(q^{2n-2}-1)(q^n+1)$,
c. $|SO_{2n+1}(K)| = q^{n^2}(q^2-1)(q^4-1)\cdots(q^{2n}-1)$.

Proof. We prove (a), (c), together. Clearly

$$|SO_{2n+1}(K)| = q^n(q^n+1)|SO^+_{2n}(K)|$$

since it suffices to map the generator i of $\langle 1\rangle$ to any element v of length 1 in $K^{2n+1} = \langle H_{2n}\rangle \perp \langle 1\rangle$. Then $\langle v\rangle^\perp \cong \langle H_{2n}\rangle$. Next note that $\langle H_2\rangle = \langle 1\rangle \perp \langle -1\rangle$ so $\langle H_{2n}\rangle = \langle -1\rangle \perp (\langle 1\rangle \perp \langle H_{2n-2}\rangle)$ so

$$|\mathrm{SO}_{2n}^+(K)| = q^{n-1}(q^n - 1)|\mathrm{SO}_{2n-1}(K)| .$$

From these two formulae (a) and (c) follow.

Finally we prove (b). Since

$$\langle H_{2n-2}\rangle \perp \left\langle \begin{pmatrix} 1 & 0 \\ 0 & -\alpha \end{pmatrix} \right\rangle = \langle H_{2n-2}\rangle \perp \langle 1\rangle \perp \langle -\alpha\rangle$$

we have $|\mathrm{SO}_{2n}^-(K)| = q^n(q^{n-1}+1)|\mathrm{SO}_{2n-1}(K)|$ and the theorem follows. □

Let K be any finite field (here characteristic 2 is allowed). We consider K^{2n} with the skew form

$$B = \langle S\rangle \perp \cdots \perp \langle S\rangle$$

where $S = \begin{pmatrix} 0 & 1 \\ -1 & 0 \end{pmatrix}$. In particular, we have a basis $e_1, \ldots, e_n, f_1, \ldots, f_n$ for K^{2n} so $\langle e_1, e_j\rangle_B = \langle f_i, f_j\rangle_B = 0$, $\langle e_1, f_j\rangle = -\langle f_j, e_i\rangle = \delta_{ij}$.

Lemma 3.6 *Let $v \in K^{2n}$ with $v \neq 0$. Then $\langle v, v\rangle_B = 0$ and there are precisely q^{2n-1} vectors $w \in K^{2n}$ so that $\langle v, w\rangle_B = 1$.*

Proof. Write $v = \sum v_i e_i + \sum \lambda_j f_j$, $w = \sum w - ie_i + \sum \theta_j f_j$. Then

$$\langle v, v\rangle_B = \sum (v_i\lambda_i - \lambda_i v_i) = 0$$
$$\langle v, w\rangle_B = \sum v_i\theta_i - \sum \lambda_i w_i$$

and since $v \neq 0$ there is a v_i or λ_i which is not 0. Say $v_i \neq 0$, then $\langle v, \frac{1}{v_i} f_i\rangle_B = 1$ so the set of w with $\langle v, w\rangle = 1$ is non-empty. On the other hand if $\langle v, w\rangle_B = \langle v, w'\rangle_B = 1$ then $\langle v, w - w'\rangle = 0$ and $w - w'$ is contained in the hyperplane

$$\sum v_i\theta_i - \sum \lambda_i w_i = 0 .$$

Conversely, if γ is contained in this hyperplane, then $\langle v, w + \gamma\rangle_B = 1$. Since any hyperplane has q^{2n-1} elements the lemma follows. □

Corollary 3.7 *Let $\mathrm{Sp}_{2n}(K)$ be the group of B, those $g \in \mathrm{GL}_{2n}(K)$ so that*

$$\langle gv, gw\rangle_B = \langle v, w\rangle_B$$

for all v, w, in K^{2n}. Then

$$|\mathrm{Sp}_{2n}(K)| = q^{n^2}(q^{2n} - 1)(q^{2n-2} - 1)\cdots(q^2 - 1) .$$

Proof. Given $g \in \mathrm{Sp}_{2n}(K)$, g can take the non-singular subspace $\langle e_1, f_1\rangle$ to a new non-singular subspace (in a form preserving manner) in exactly

$$q^{2n-1}(q^{2n}-1)$$

ways. Hence, since it also maps $\langle e_1, f_1\rangle^{\perp}$ to the perpendicular complement of $g(\langle e_1, f_1\rangle)$ we have

$$|\mathrm{Sp}_{2n}(K)| = q^{2n-1}(q^{2n}-1)|\mathrm{Sp}_{2n-2}(K)| .$$

Also, as we have already seen, $\mathrm{Sp}_2(K) = \mathrm{SL}_2(K)$ which has $q(q^2-1)$ elements. Now, note that

$$\begin{aligned}(2n-1)+(2n-3)+\cdots+1 &= \frac{2n(2n+1)}{2} - 2\frac{n(n+1)}{2} \\ &= n^2,\end{aligned}$$

so the corollary follows. □

Remark 3.8 Let $K/\mathbb{F}$ be an extension of degree d, then

$$\mathrm{Sp}_{2k}(K) \hookrightarrow \mathrm{Sp}_{2dk}(\mathbb{F})$$

by forgetting the K-structure. As an example

$$\mathrm{SL}_2(\mathbb{F}_4) = \mathrm{Sp}_2(\mathbb{F}_4) = \mathcal{A}_5 \subset \mathrm{Sp}_4(\mathbb{F}_2)$$

and we note that the index of this inclusion is 12. But the normalizer of $\mathcal{A}_5$ in $\mathrm{Sp}_4(\mathbb{F}_2)$ is generated by the Galois automorphism of $\mathbb{F}_4$ over $\mathbb{F}_2$, and this gives a copy of $\mathcal{S}_5 \subset \mathrm{Sp}_4(\mathbb{F}_2)$ having index 6. In turn, this gives, via the action of $Sp_4(\mathbb{F}_2)$ on cosets, a homomorphism $\mathrm{Sp}_4(\mathbb{F}_2) \to \mathcal{S}_6$. Moreover, the kernel is contained in $\mathcal{S}_5$ and hence is either $\mathcal{A}_5$ or 1. But since the normalizer of $Sp_2(\mathbb{F}_r)$ in $Sp_4(\mathbb{F}_2)$ is $\mathcal{S}_5$ it follows that the kernel is a proper subgroup of $\mathcal{A}_5$ and is thus 1 (since it is normal). Finally $|\mathrm{Sp}_4(\mathbb{F}_2)| = |\mathcal{S}_6|$ so the coset representation constructs an isomorphism $Sp_4(\mathbb{F}_2) \cong \mathcal{S}_6$.

4. The Cohomology of the Groups $\mathrm{GL}_n(q)$

The group $\mathrm{GL}_n(q)$ is a model for our calculations, so we study this case in detail now.

Theorem 4.1

a. *Suppose that* p *divides* $q-1$. *Then* $Syl_p(\mathrm{GL}_n(q))$ *is the* p*-Sylow subgroup of*

$$\mathbb{Z}/p^r \wr \mathcal{S}_n \subset \mathrm{GL}_n(q)$$

where $q-1 = p^r\theta$ *with* $(\theta, p) = 1$ *and* $r \geq 1$.

b. *If $p|q$ then*

$$\mathrm{Syl}_p(\mathrm{GL}_n(q)) = UT_n(q)$$

is the subgroup of upper triangular matrices with ones along the diagonal in $\mathrm{GL}_n(q)$.

c. *If $(q,p)=1$ but p does not divide $q-1$ then there is a unique non-negative integer s, $s \le p-1$ so that $q^s \equiv 1 \bmod (p)$ and*

$$\mathrm{Syl}_p(\mathrm{GL}_n(q)) = \mathrm{Syl}_p(\mathrm{GL}_{[\frac{n}{s}]}(q^s)) .$$

Moreover, if $q^s - 1 = p^r\theta$ then

$$\mathrm{Syl}_p(\mathrm{GL}_n(q)) \subset (\mathbb{Z}/p^r \times_T (\mathbb{Z}/s)) \wr \mathcal{S}_{[\frac{n}{s}]}$$

where $\mathbb{Z}/s$ acts on $\mathbb{Z}/p^r$ by $T(x) = x^q$ for T a generator of $\mathbb{Z}/s$.

Proof. We need a lemma.

Lemma 4.2 *Let $q-1 = p^r w$, $r \ge 1$ with $(w,p)=1$ and suppose either p is odd or $p=2$ and $r \ge 2$. Then $q^j - 1 = p^{r+s}w'$ where $(w',p)=1$ and $j = p^s v$ with $(v,p)=1$.*

Proof of 4.2 Write $(q^{p^s v} - 1) = (q^{p^s} - 1)(1 + q^{p^s} + \cdots + q^{p^s(v-1)})$. Since $q \equiv 1 \bmod (p)$ the second term is congruent to $v \bmod (p)$ and so is prime to p. Thus we need only analyze $(q^{p^s}-1)$, and this, in turn, reduces to studying $q^p - 1$. Write

$$q = 1 + p^r A$$

with $(A,p)=1$. Then

$$\begin{aligned} q^p &= 1 + p\cdot p^r A + \binom{p}{2} p^{2r}A^2 + \cdots + p^{rp}A^p \\ &= 1 + p^{r+1}A + p^{2r+1}B \\ &= 1 + p^{r+1}(A + p^r B) \end{aligned}$$

provided p is odd. If $p=2$ then $\binom{p}{2} = 1$ and

$$1 + 2^{r+1}A + 2^{2r}A^2 = 1 + 2^{r+1}(A + 2^{r-1}A^2)$$

and if $r=1$ this is $1 + 2^{r+2}W$ while $r>1$ implies that $(A + 2^{r-1}A^2)$ is odd. The lemma follows. □

We now prove (a) of the theorem. Note that

$$\mathbb{Z}/p^r \subset \mathbb{Z}/q-1 = \mathrm{GL}_1(q)$$

so $(\mathbb{Z}/p^r)^n \subset \mathrm{GL}_n(q)$ as a subgroup of the diagonal matrices. The normalizer of this group in $\mathrm{GL}_n(q)$ is $\mathbb{Z}/(q-1) \wr \mathcal{S}_n$ where $\mathcal{S}_n$ is embedded as the permutation matrices and $\mathbb{Z}/(q-1)$ embeds as the diagonal matrices. Now the power of p which divides the order of this group is

$$p^{nr+([n/p]+[n/p^2]\cdots])}$$

but from the lemma this is the p-order of

$$q^{n(n-1)/2}(q^n-1)q^{n-1}-1)\cdots(q-1)\,.$$

(b) and (c) are similar. □

Corollary 4.3 *Let $q-1=p^r w$, $r\geq 1$ if p is odd or $r\geq 2$ if $p=2$. Then*

$$\mathrm{Syl}_p(\mathrm{GL}_n(q)) \;=\; \coprod_{i\geq 0}\left(\underbrace{\mathbb{Z}/p^r\wr\mathbb{Z}/p\wr\cdots\wr\mathbb{Z}/p}_{i}\right)^{\alpha_i}$$

where α_i is the coefficient of p^i in the p-adic expansion of n; $n=\alpha_0+\alpha_1 p+\cdots+\alpha_i p^i+\cdots$.

The cohomology of $\mathbb{Z}/p^r\wr\mathbb{Z}/p\wr\cdots\mathbb{Z}/p$ is detected by groups of the form $(\mathbb{Z}/p^r)^s\times(\mathbb{Z}/p)^l$ as we see from the discussion in Chapter IV. On the other hand we have

Theorem 4.4 *Let $E=\prod_i(\mathbb{Z}/p^{j_i})\subset\mathrm{GL}_n(q)$ with $q-1=p^r\theta$ and each $j_i\leq r$. Then E is conjugate to a subgroup of the diagonal $(\mathbb{Z}/(q-1))^n$. In particular, $H^*(\mathrm{GL}_n(q);\mathbb{F}_p)$ is detected by restriction to the diagonal subgroup $(\mathbb{Z}/p^r)^n$.*

Proof. Since $p^r|q-1$, a primitive $p^r th$ root of unity is present in $\mathbb{F}_q$ so each element $x\in E$ is conjugate to a diagonal matrix. Next, since E is commutative, it follows that we can simultaneously diagonalize all the elements of E, and the result follows. □

Corollary 4.5 *Let $q-1=p\theta$ with p an odd prime and θ prime to p, then*

$$H^*(\mathrm{GL}_n(q);\mathbb{F}_p)=H^*((\mathbb{Z}/p)^n;\mathbb{F}_p)^{\mathcal{S}_n}\,.$$

Proof. Note first that the subgroup of the diagonal $(\mathbb{Z}/p)^n\subset\mathrm{Syl}_2(\mathrm{GL}_n(q))$ is normal in $\coprod\mathbb{Z}/(q-1)\wr\mathcal{S}_n$, and, indeed, it is the only elementary subgroup of this order in this wreath product. (This depends on the fact that p is odd.) Consequently $(\mathbb{Z}/p)^n\subset(\mathbb{Z}/p)\wr\mathcal{S}_n\subset\mathrm{GL}_n(q)$ is a closed system. From our discussion of the cohomology of wreath products with $\mathcal{S}_n$ it follows that the restriction map from $H^*(\mathbb{Z}/(q-1)\wr\mathcal{S}_n;\mathbb{F}_p)$ to $H^*((\mathbb{Z}/p)^n;\mathbb{F}_p)$ is surjective onto the ring of $\mathcal{S}_n$ invariants. The Cardenas-Kuhn theorem now gives the result. □

Remark. The condition that p^2 not divide $q-1$ in (4.5) is, in fact, satisfied in the major application of this calculation in Quillen's original proof of the Adams conjecture [Q6]. However, it is clearly desirable to have the general result. In fact we have the following theorem of Quillen,

Theorem 4.6 *Let p be an odd prime and suppose $q-1=p^r\theta$ with $r\geq 1$ and θ prime to p. Then the restriction map*

$$H^*(\mathrm{GL}_n(q);\mathbb{F}_p)\xrightarrow{\text{res}} H^*((\mathbb{Z}/p^r)^n;\mathbb{F}_p)^{\mathcal{S}_n}$$

is an isomorphism onto the entire ring of invariants.

(The proof of this result does not appear to be a formal consequence of the Cardenas-Kuhn theorem when r is greater than 1. Quillen's original method involved the use of the Adams operations in K-theory and is described in considerable detail in [FP]. However, it is also possible, using 4.4, III.4.3, and the results on Hopf algebras of VI.2, to give another proof, starting with $GL_1(q)$, and using the Hopf algebra structure of the disjoint union of the $H^*(GL_n(q);\mathbb{F}_p)$.)

Remark 4.7 The situation where p does not divide $q-1$ is not much different. Here the action of $(\mathbb{Z}/s)$ described in (4.1) gives that the invariants are in

$$\left\{\otimes_1^n H^*((\mathbb{Z}/p^r);\mathbb{F}_p)^{\mathbb{Z}/s}\right\}^{\mathcal{S}_n}.$$

But

$$H^*(\mathbb{Z}/p^r;\mathbb{F}_p)^{\mathbb{Z}/s}=\mathbb{F}_p[b^s]\otimes E(b^{s-1}e)$$

so the rings are formally the same as those before and, except for the evident change in dimensions of generators the arguments go as before. Thus we have

Corollary 4.8 *Suppose p does not divide q and $q^s-1=p^r\theta$ as in (4.1(c)). Then $H^*(\mathrm{GL}_n(q);\mathbb{F}_p)\cong H^*((\mathbb{Z}/p^r)^{[n/s]};\mathbb{F}_p)^{(\mathbb{Z}/s)\wr\mathcal{S}_{[n/s]}}$ and this invariant algebra is the tensor product of a polynomial algebra on generators in dimensions $2si$ and an exterior algebra on generators in dimensions $2si-1$, $1\leq i\leq[n/s]$.*

(Of course, we have only given a proof in case p^2 does not divide q^s-1, but (4.6) gives the proof in the general case.)

5. The Cohomology of the Groups $O_n^*(q)$ for q Odd

Lemma 5.1 *For the orthogonal groups $O_{2n}^{\pm}(q)$ and $O_{2n+1}(q)$ with q odd a copy of $\mathrm{Syl}_2(G)$ is always contained in the normalizer of a maximal torus. These tori normalizers are given as follows.*

1. For $q \equiv 1 \bmod (4)$ the maximal tori normalizers are

$$\begin{aligned} O_2^+(q) \wr \mathcal{S}_m &\subsetneq O_{2m}^+(q) \\ O_2^- \times \left(O_2^+ \wr \mathcal{S}_{m-1}\right) &\subsetneq O_{2m}^-(q) \\ \mathbb{Z}/2 \times \left(O_2^+ \wr \mathcal{S}_m\right) &\subsetneq O_{2m+1}(q). \end{aligned}$$

2. For $q \equiv 3 \bmod (4)$ and m odd the maximal tori normalizers are

$$\begin{aligned} O_2^+(q) \times \left(O_2^-(q) \wr \mathcal{S}_{m-1}\right) &\subsetneq O_{2m}^+(q) \\ O_2^- \wr \mathcal{S}_m &\subsetneq O_{2m}^-(q). \end{aligned}$$

3. For $q \equiv 3 \bmod (4)$ and m even the maximal tori normalizers are

$$\begin{aligned} O_2^-(q) \wr \mathcal{S}_m &\subsetneq O_{2m}^+(q) \\ O_2^+(q) \times \left(O_2^-(q) \wr \mathcal{S}_{m-1}\right) &\subsetneq O_{2m}^-(q). \end{aligned}$$

4. For $q \equiv 3 \bmod (4)$ the maximal tori normalizers are $O_2^- \wr \mathcal{S}_m \times \mathbb{Z}/2 \subsetneq O_{2m+1}(q)$.

(This is a direct counting argument.)

Corollary 5.2 *The groups $H^*(O_{2m}^{\pm}(q); \mathbb{F}_2)$ and $H^*(O_{2m+1}(q); \mathbb{F}_2)$ are detected by restriction to elementary 2-subgroups for q odd.*

Proof. From 2.13 and 5.1 we see that

$$\mathrm{Syl}_2(O_{2m}^{\pm}(q)) \cong \prod_l D_{2^j} \wr \underbrace{\mathbb{Z}/2 \wr \mathbb{Z}/2 \wr \cdots \wr \mathbb{Z}/2}_{l}$$

where the l run over the powers of 2 appearing in the dyadic expansion of m or such a group times $(\mathbb{Z}/2)^2$. Similarly, $\mathrm{Syl}_2(O_{2m+1}(q)) \cong \mathbb{Z}/2 \times \mathrm{Syl}_2(O_{2m}^{\pm}(q))$. But we have already seen in Chapter IV that the cohomology of such groups is detected by restriction to elementary 2-subgroups. □

Now we want to understand the conjugacy classes of elementary 2-subgroups in $O_{2m}^{\pm}(q)$ and $O_{2m+1}(q)$. Consider an elementary 2-subgroup $(\mathbb{Z}/2)^r = \langle t_1, \ldots, t_r \rangle$ contained in one of these orthogonal groups. For ease of notation we assume for the moment that we are dealing with one of the groups $O_{2m}^{\pm}(q)$. Since q is odd we can choose an orthogonal basis $e_1, \ldots, e_{2m}$ so that the t_i become diagonal with only ± 1's along the diagonal. Suppose $t_1(e_1) = -e_1$. Then we can choose new basis elements for $(\mathbb{Z}/2)^n$,

$t_1, t'_2, \ldots, t'_n$, so that $t'_i(e_1) = e_1$, $2 \le i \le n$. Then $V = \langle e_2, \ldots, e_{2m}\rangle$ is taken to itself by $(\mathbb{Z}/2)^n$. Write $V = V^+ \oplus V^-$ where $x \in V$ is contained in $V^\pm$ if and only if $t_1(x) = \pm x$. Then $\langle t'_2, \ldots, t'_n\rangle(V^\pm) = V^\pm$ and if we define τ by

$$\tau(x) = \begin{cases} e_1 & \text{if } x = e_1, \\ x & \text{if } x \in V^+, \\ -x & \text{if } x \in V^-, \end{cases}$$

then $\tau \in \mathrm{O}^\pm_{2m}(q)$ and τ commutes with $(\mathbb{Z}/2)^n$. Consequently either $\tau \in (\mathbb{Z}/2)^n$ or we can adjoin it obtaining an extension $(\mathbb{Z}/2)^{n+1} \subset \mathrm{O}^\pm_{2n}(q)$. By repeating this construction we obtain

Lemma 5.3 *Let $G = (\mathbb{Z}/2)^n \subset \mathrm{O}^\pm_{2m}(q)$ or $\mathrm{O}_{2m+1}(q)$, then G is contained in a maximal subgroup $(\mathbb{Z}/2)^{2m}$ or $(\mathbb{Z}/2)^{2m+1}$. Such a group splits the form uniquely as a sum of 1-forms $\langle 1\rangle^s \perp \langle\beta\rangle^{2m+\epsilon-s}$ and s is a complete characterization of the conjugacy class of a maximal 2-elementary subgroup of $\mathrm{O}^\pm_{2m}(q)$ or $\mathrm{O}_{2m+1}(q)$.*

Proof. The argument above shows that $(\mathbb{Z}/2)^n$ is contained in a subgroup $(\mathbb{Z}/2)^{2m+\epsilon} = \langle t_1, \ldots, t_{2m+\epsilon}\rangle$ and there is an orthogonal basis $e_1, \ldots, e_{2m+\epsilon}$ for the vector space so that $t_i(e_j) = e_j$ if $j \neq i$, and $t_i(e_i) = -e_i$. Since the (-1) eigenspace for t_i is orthogonal to the $(+1)$ eigenspace it follows that e_i is not an isotropy vector for the form and so the e_i give an orthogonal splitting of the form into a sum of 1-forms. On the other hand it is direct to see that among the $2^{2m+\epsilon}$ elements of this group exactly $2m + \epsilon$ of them have 1-dimensional (-1) eigenspaces. Hence, this splitting is intrinsic to the group and the result follows. □

In particular there are exactly two conjugacy classes of elementary 2-groups $(\mathbb{Z}/2)^2 \subset \mathrm{O}^\pm_2(q)$. Indeed, we have

Lemma 5.4 *The dihedral group $D_{2^n} = \{x, y \mid x^2 = (xy)^2 = y^{2^{n-1}} = 1\}$ has precisely 2 conjugacy classes of maximal elementary 2-groups $(\mathbb{Z}/2)^2$, the first conjugate to $\langle x, y^{2^{n-2}}\rangle$ and the second conjugate to $\langle xy, y^{2^{n-2}}\rangle$.*

Proof. The elements of order 2 are precisely the classes xy^i, $0 \le i \le 2^{n-1} - 1$ and $y^{2^{n-2}}$. $y^{2^{n-2}}$ is central and $xxy^ix = xy^{-i}$, $y^{-j}xy^iy^j = xy^{i+2j}$. Thus there are precisely 2 conjugacy classes of non-central elements of order two, the first represented by x and the second by xy. Also, $xy^ixy^j = y^{j-i}$ and we see that xy^i, xy^j commute if and only if $j = i \pm 2^{n-2}$. Hence there are two possibilties for conjugacy classes of $(\mathbb{Z}/2)^2$, $\langle x, y^{2^{n-2}}\rangle = \{1, x, xy^{2^{n-2}}, y^{2^{n-2}}\}$ and $\langle xy, y^{2^{n-2}}\rangle = \{1, xy, xy^{2^{n-2}+1}, y^{2^{n-2}}\}$ and these are clearly not conjugate. □

As an example of how 5.3 works we have

Example 5.5 Let β be a non-square in $\mathbb{F}_q$ and suppose $q \equiv 1 \bmod (4)$. Then the two subgroups $(\mathbb{Z}/2)^2 \subset O_2^+(q)$ described above are

$$I = \left\langle \begin{pmatrix} 0 & \beta \\ \beta^{-1} & 0 \end{pmatrix}, -I \right\rangle, \quad II = \left\langle \begin{pmatrix} 0 & 1 \\ 1 & 0 \end{pmatrix}, -I \right\rangle.$$

For II the eigenvector splitting is $\langle e_1 - e_2, e_1 + e_2 \rangle$ which corresponds to the splitting of the quadratic form as $\langle -2 \rangle \perp \langle 2 \rangle$. For I the eigenvector splitting is $\langle \beta e_1 + e_2, \beta e_1 - e_1 \rangle$ which gives the form $\langle 2\beta \rangle \perp \langle -2\beta \rangle$.

Corollary 5.6 *The normalizer of the maximal elementary 2-subgroup of* $O_{2m}^{\pm}(q)$ *or* $O_{2m+1}(q)$ *with splitting space* $\langle 1 \rangle^s \perp \langle \beta \rangle^{2m+\epsilon-s}$ *is*

$$(\mathbb{Z}/2 \wr \mathcal{S}_s) \times (\mathbb{Z}/2 \wr \mathcal{S}_{2m+\epsilon-s})$$

so the Weyl group is $\mathcal{S}_s \times \mathcal{S}_{2m+\epsilon-s}$.

Corollary 5.7 *The restriction map taking* $H^*(O_{2m}^{\pm}(q); \mathbb{F}_2)$ *to*

$$\begin{cases} H^*(O_2^{\pm}(q) \times (O_2^+(q))^{m-1}; \mathbb{F}_2) & \text{when } q \equiv 1 \bmod (4) \\ H^*(O_2^{\pm}(q) \times (O_2^-(q))^{m-1}; \mathbb{F}_2) & q \equiv 3 \bmod (4),\ m \text{ odd}, \\ H^*(O_2^{\mp}(q) \times (O_2^-(q))^{m-1}; \mathbb{F}_2) & q \equiv 3 \bmod (4),\ m \text{ even}, \end{cases}$$

or $H^*(O_{2m+1}(q); \mathbb{F}_2)$ *to*

$$\begin{cases} H^*(O_2^{\pm}(q)^m \times \mathbb{Z}/2; \mathbb{F}_2) & \text{for } q \equiv 1 \bmod (4), \\ H^*(O_2^-(q)^m \times \mathbb{Z}/2; \mathbb{F}_2) & \text{for } q \equiv 3 \bmod (4) \end{cases}$$

is injective in cohomology.

We now wish to precisely determine these images. The result will be direct from the Cardenas-Kuhn theorem once we have proved

Lemma 5.8 *Let* $G = (\mathbb{Z}/2)^{2m} \subset D_{2^n} \wr \mathcal{S}_m$. *Then* $G \subset (D_{2^n})^m \subset D_{2^n} \wr \mathcal{S}_m$. *In particular* $G \subset O_2^{\pm}(q) \wr \mathcal{S}_m \subset O_{2m}^{\pm}(q)$ *when* $q \equiv 1 \bmod (4)$ *and* $\pm$ *is* $+$ *or* $q \equiv 3 \bmod (4)$, $\pm = -$ *and* m *is odd, or* $\pm = +$ *and* m *is even.*

Proof. Consider the projection

$$\pi\colon\ D_{2^n} \wr \mathcal{S}_m \longrightarrow \mathcal{S}_{nm}$$

and look at $\pi(G)$. It is clear that any 2-elementary subgroup $(\mathbb{Z}/2) \subset \mathcal{S}_m$ must have $r \leq \left[\frac{m}{2}\right]$. Thus $\mathrm{Ker}(\pi) \cap G = (\mathbb{Z}/2)^v$ with $v \geq m + \left[\frac{m}{2}\right]$. In particular $G \cap (D_{2^n})^m$ must contain a subgroup of the form $(\mathbb{Z}/2)^m = \langle t_1, \ldots, t_m \rangle$ with t_i contained in the ith copy of D_{2^n} in $(D_{2^n})^m$ and $\pi(G)$ must centralize this group. But this is impossible unless $\pi(G) = \{1\}$ and the first statement of (5.8) follows.

To see the second statement we use (5.3), (5.4), (5.5) to assert that G has the form $V_1 \times \cdots \times V_m \subset (D_{2^n})^m$ where each V_i is either conjugate in D_{2^n} to $\langle x, y^{2^{n-2}}\rangle$ or $\langle xy, y^{2^{n-2}}\rangle$. Moreover, within $(D_{2^n}) \wr S_m$ each such group is conjugate to

$$\underbrace{V_I \times \cdots \times V_I}_{s} \times \underbrace{V_{II} \times \cdots \times V_{II}}_{m-s}$$

for a unique s, and (5.3) shows these groups are not conjugate in $O^{\pm}_{2m}(q)$ for distinct s. Thus the closure conditions are satisfied. □

Corollary 5.9 *Under the conditions of (5.8)*

$$H^*(O^{\pm}_{2m}(q); \mathbb{F}_2) \subset H^*(D_{2^n}; \mathbb{F}_2)^{S_m}$$

consists of exactly those elements θ which satisfy

$$\operatorname{res}^*_s(\theta) \in H^*((V_I)^s \times (V_{II})^{m-s}; \mathbb{F}_2)^{S_{2s} \times S_{2(m-s)}}, \qquad 0 \le s \le m\,.$$

Example. The case $q \equiv 1 \bmod (4)$

We now assume that $q \equiv 1 \bmod (4)$ and β will represent the a non-square in $\mathbb{F}_q$. The inclusions

$$j\colon\ O^+_{2m}(q) \longrightarrow O_{2m+1}(q)$$
$$i_{\langle 1\rangle},\ i_{\langle\beta\rangle}\colon\ O_{2m+1} \longrightarrow O^+_{2m+2}(q)$$

induce injections in homology and surjections in cohomology. Here $i_{\langle 1\rangle}$ is the inclusion induced by regarding $O_{2m+1}(q)$ as the group of the form $\underbrace{\langle 1\rangle\perp\ldots\perp\langle 1\rangle}_{2m+1}$ and

$$\langle 1\rangle\perp\left(\underbrace{\langle 1\rangle\perp\ldots\perp\langle 1\rangle}_{2m+1}\right) = \underbrace{\langle 1\rangle\perp\ldots\perp\langle 1\rangle}_{2m+2}$$

while $i_{\langle\beta\rangle}$ is the inclusion obtained by regarding $O_{2m+1}(1)$ as the group of the form $\underbrace{\langle\beta\rangle\perp\ldots\perp\langle\beta\rangle}_{2m+1}$ and noting

$$\underbrace{\langle\beta\rangle\perp\ldots\perp\langle\beta\rangle}_{2m+2} = \underbrace{\langle 1\rangle\perp\ldots\perp\langle 1\rangle}_{2m+1}\,.$$

$i_{\langle\beta\rangle}j$ is conjugate to $i_{\langle 1\rangle}j$ but $ji_{\langle\beta\rangle}$ need not be conjugate to $ji_{\langle 1\rangle}$. Hence, to be definite we define $O(q) = \lim_{\to}(O_n(q))$ where $O_n(q) = \begin{cases} O^+_n(q) & n \text{ even} \\ O_n(q) & n \text{ odd,}\end{cases}$ and the inclusion $O_n(q) \longrightarrow O_{n+1}(q)$ is j if n is odd and $i_{\langle 1\rangle}$ if n is even. We have

Theorem 5.10

a. $H_*(O(q);\mathbb{F}_2) = \coprod_{n=1}^{\infty} H_*(O_n(q), O_{n-1}(q);\mathbb{F}_2)$ *so the map taking*

$$H_*(O_n(q);\mathbb{F}_2) \longrightarrow H_*(O(q);\mathbb{F}_2)$$

is an inclusion for all n.

b. The inclusions $O_n(q) \times O_m(q) \longrightarrow O_{n+m}(q)$ *induced by orthogonal sums of vector spaces fit together on passing to limits to define a unitary, commutative and associative multiplication*

$$H_*(O(q);\mathbb{F}_2) \otimes H_*(O(q);\mathbb{F}_2) \longrightarrow H_*(O(q);\mathbb{F}_2)$$

and gives $H_*(O(q);\mathbb{F}_2)$ *the structure of a biassociative, bicommutative Hopf algebra.*

c. As an algebra

$$H_*(O(q);\mathbb{F}_2) = \mathbb{F}_2[e_1, e_2, \ldots, e_n, \ldots] \otimes E(h_1, h_2, \ldots, h_n, \ldots)$$

where the subscripts denote the dimensions of the generators and e_i *is the non-zero element of dimension* i *in* $H_i(O_1(q);\mathbb{F}_2) = \mathbb{F}_2$ *while* $h_i \in H_i(O_2(q);\mathbb{F}_2)$ *is the sum* $e_i + i_{\langle\beta\rangle}(e_i)$.

Proof. (a) has already been discussed in part. To see that the homology of the limit is the limit of homology groups is standard since homology is carried by chains with compact support.

To see (b) note that the inclusion $\mu_{n,m}: O_n(q) \times O_m(q) \longrightarrow O_{n+m}(q)$ has images all orthogonal matrices of the form $\begin{pmatrix} A & 0 \\ 0 & B \end{pmatrix}$ with A $n \times n$ and B $m \times m$. But the element $\begin{pmatrix} 0 & I_m \\ I_n & 0 \end{pmatrix} \in O_{n+m}(q)$ now acts to congugate $\mu_{n,m}$ and $\mu_{m,n}T$ where

$$T\colon\ O_n(q) \times O_m(q) \longrightarrow O_m(q) \times O_n(q)$$

is the interchange. Consequently, this pairing is commutative. Also, a similar conjugation shows that the multiplication fits correctly with inclusion so that, on the level of homology groups it passes to the limit groups and defines the desired algebra structure on $H_*(O(q);\mathbb{F}_2)$.

Finally we verify (c). Here, note that from our description of

$$H^*(D_{2^l};\mathbb{F}_2) = \mathbb{F}_2[a_1, b_1, w_2]/(ab = 0)$$

we have $\mathrm{res}^*(w)$ is the same in both conjugacy classes of subgroups $(\mathbb{Z}/2)^2 \subset D_{2^l}$. From this, if we set $f_i = i_{\langle\beta\rangle}(e_i)$, then a^i is dual to e_i, b^i is dual to f_i and both e_i^2 and f_i^2 are dual to w_i. Consequently we have $h_i^2 = 0$. On the other hand it is clear that the e_i, h_j generate $H_*(O(q);\mathbb{F}_2)$ and are independent. Hence by Hopf's theorem that commutative associative Hopf algebras over a field are tensor products of monogenic algebras it (c) follows. □

Examples 5.11 $H^*(\mathrm{O}_2(q);\mathbb{F}_2) = H^*(D_{2^l};\mathbb{F}_2) = \mathbb{F}_2[a+b,w](b)$ where $b^2 = (a+b)b$ gives the extension data and the Poincaré series is

$$P_{\mathrm{O}_2(q)}(x) = \frac{1+x}{(1-x)(1-x^2)}.$$

Now we describe $H^*(\mathrm{O}_3(q);\mathbb{F}_2)$. There are two conjugacy classes of $(\mathbb{Z}/2)^3$ in $O(3)$, the first associated to the form $\langle 1\rangle\perp\langle 1\rangle\perp\langle 1\rangle$ and the second to the form $\langle\beta\rangle\perp\langle\beta\rangle\perp\langle 1\rangle$. The first contributes a polynomial algebra $H^*((\mathbb{Z}/2)^3;\mathbb{F}_2)^{S_3} = \mathbb{F}_2[\sigma_1,\sigma_2,\sigma_3]$ and the second contributes $\mathbb{F}_2[\sigma_1,\sigma_2]\sigma_1 \otimes \mathbb{F}_2[\sigma_1]$. This can be seen since the terms in homology are all of the form $f_i f_j \otimes e_s$ with $0 \le i < j$. Consequently its Poincare series is

$$\begin{aligned} P_{\mathrm{O}_3(q)}(x) &= \frac{1}{(1-x)(1-x^2)(1-x^3)} + \frac{x}{(1-x)^2(1-x^2)} \\ &= \frac{(1+x)(1+x^2)}{(1-x)(1-x^2)(1-x^3)}. \end{aligned}$$

Finally, we consider $\mathrm{O}_4(q)$. Here there are three conjugacy classes of subgroups $(\mathbb{Z}/2)^4$ with contributions respectively $\mathbb{F}_2[\sigma_1,\sigma_2,\sigma_3,\sigma_4]$, elements dual to $f_i f_j \otimes e_r e_s$ with $0 \le i < j$ and duals to elements of the form $f_i f_j f_k f_l$ with $0 \le i < j < k < l$. Consequently the Poincaré series is

$$\begin{aligned} P_{\mathrm{O}_4(q)}(x) &= \frac{1+x^6}{(1-x)(1-x^2)(1-x^3)(1-x^4)} + \frac{x}{(1-x)^2(1-x^2)^2} \\ &= \frac{(1+x)(1+x^2)(1+x^3)}{(1-x)(1-x^2)(1-x^3)(1-x^4)}. \end{aligned}$$

The Cohomology Groups $H^*(\mathrm{O}_m(q);\mathbb{F}_2)$

The restriction map

5.12 $$H^*(\mathrm{O}_m(q);\mathbb{F}_2) \longrightarrow \coprod_{0 \le i < \frac{m}{2}} H^*((\mathbb{Z}/2)^{2i})^{S_{2i}} \otimes H^*((\mathbb{Z}/2)^{m-2i})^{S_{m-2i}}$$

is onto those elements which evaluate equally on monomials of the form

$$\bar{e}_{i_1} \otimes \cdots \otimes \bar{e}_{i_{2l-1}} \otimes \bar{e}_{i_{2l-1}} \otimes e_{i_{2l+1}} \otimes \cdots$$

and

$$\bar{e}_{i_1} \otimes \cdots \otimes \bar{e}_{i_{2l-2}} \otimes e_{i_{2l-1}} \otimes e_{i_{2l-1}} \otimes \cdots .$$

Examples of such elements are easily constructed. Note that

$$H^*((\mathbb{Z}/2)^m)^{S_r \times S_{m-r}} = \mathbb{F}_2[\bar{w}_1,\ldots,\bar{w}_r] \otimes \mathbb{F}_2[w_1,\ldots,w_{m-r}]$$

where $\bar{w}_i$ is the ith symmetric monomial in the first r coordinates and w_j is the jth symmetric monomial in the last $m-r$ coordinates. Then set

$$\begin{aligned} x_{2k} &= \sqcup \sum \bar{w}_{2j} \otimes w_{2(k-s)}, \\ x_{2k-1} &= \sqcup \sum_{p+q=k} \bar{w}_{2p} \otimes w_{2q-1}, \\ \bar{x}_{2k-1} &= \sqcup \sum_{p+q=k} \bar{w}_{2p-1} \otimes w_{2q}. \end{aligned}$$

This notation is meant to indicate that the restriction of the element to each of the conjugacy classes of elementary 2-groups has the indicated form. These elements satisfy the constraint conditions and consequently are in the image of $H^*(\mathrm{O}^{\pm}_{2s}(q);\mathbb{F}_2)$ or $H^*(\mathrm{O}_{2s+1}(q);\mathbb{F}_2)$. Indeed one has the following theorem.

Theorem 5.13 *$H^*(\mathrm{O}_{2n+1}(q);\mathbb{F}_2)$ is the subring of 6.8 generated by the elements x_k, $\bar{x}_k$. Similar generation results hold in the remaining cases. In all cases the Poincaré series of $H^*(\mathrm{O}^{\epsilon}_m(q);\mathbb{F}_2)$ is*

$$\frac{(1+x)(1+x^2)\cdots(1+x^{m-1})}{(1-x)(1-x^2)\cdots(1-x^m)} .$$

Further details, including the relations between the generators are explained in [FP]; pp. 264-277.

6. The Groups $H^*(\mathrm{Sp}_{2n}(q);\mathbb{F}_2)$

The symplectic groups differ from the orthogonal groups we have been considering in one important respect. It turns out that $H^*(\mathrm{Sp}_{2n}(q);\mathbb{F}_2)$ is not detected on elementary 2-subgroups. But there is an important similarity. Using the inclusion $H^n \subset H^n \perp H$ we obtain inclusions $\mathrm{Sp}_{2n}(q) \subset \mathrm{Sp}_{2n+2}(q)$, and passing to limits as in the orthogonal case we obtain a group $\mathrm{Sp}(q) = \lim_{\rightarrow}(\mathrm{Sp}_{2n}(q))$ together with a bicommutative, biassociative, Hopf algebra structure on $H_*(\mathrm{Sp}(q);\mathbb{F}_2)$. The Hopf algebra structure here forces the structure of the groups $H_*(\mathrm{Sp}(q);\mathbb{F}_2)$, and by restriction, the groups $H^*(\mathrm{Sp}_{2n}(q);\mathbb{F}_2)$ as well.

We have already seen that $\mathrm{Sp}_2(q) = \mathrm{SL}_2(q)$ and we have for all odd q the result $H^*(\mathrm{SL}_2(q);\mathbb{F}_2) = E(e_3) \otimes \mathbb{F}_2[b_4]$, [Q3]. $\mathrm{Sp}_{2n}(q)$ contains the subgroup $\mathrm{Sp}_2(q) \wr \mathcal{S}_n = \mathrm{SL}_2(q) \wr \mathcal{S}_n$ and this subgroup always contains a Sylow 2-subgroup for q odd. It follows that the detecting groups for $H^*(\mathrm{Sp}_{2n}(q);\mathbb{F}_2)$ have the form $\mathrm{SL}_2(q)^r \times (\mathbb{Z}/2)^s$ with $r+s \leq n$. Moreover, following the lines of argument in §5 we have that each of these groups is conjugate to a subgroup of $\mathrm{SL}_2(q)^n \subset \mathrm{SL}_2(q) \wr \mathcal{S}_n$. In particular

Lemma 6.1 *The two inclusions*

$$\mathrm{Sp}_{2n}(q) \times \mathbb{Z}/2 \xrightarrow{\Delta \times \mathrm{id}} \mathrm{Sp}_{2n}(q) \wr \mathbb{Z}/2 \longrightarrow \mathrm{Sp}_{4n}(q)$$

and

$$\mathrm{Sp}_{2n}(1) \times \mathbb{Z}/2 \xrightarrow{\mu} \mathrm{Sp}_{2n}(q)^2 \subset \mathrm{Sp}_{2n}(q) \wr \mathbb{Z}/2 \subset \mathrm{Sp}_{4n}(q)$$

are conjugate in $\mathrm{Sp}_{4n}(q)$. *Here* $\mu(\theta, T^{\epsilon}) = (\theta, \theta\tau^{\epsilon})$ *where* τ *is the non-trivial element in the center of* $\mathrm{Sp}_{2n}(q)$.

Proof. The ± 1 eigenspaces of T give a splitting of $\mathbb{F}^{4n}$ into $\mathbb{F}^{2n} \perp \mathbb{F}^{2n}$ and the action of $\Delta(\mathrm{Sp}_{2n}(q))$ is diagonal on these two subspaces. On the other hand T acts as $+1$ on the first summand and -1 on the second. In view of the fact that $\mathrm{Sp}_{2n}(q) \wr \mathbb{Z}/2$ is exactly the stabilizer of such an orthogonal splitting, the result follows. □

Lemma 6.2 *The restriction map*

$$H^*(\mathrm{Sp}_{2n}(q); \mathbb{F}_2) \longrightarrow H^*(\mathrm{SL}_2(q)^n; \mathbb{F}_2) = E(e_1, \ldots, e_n) \otimes \mathbb{F}_2[b_1, \ldots, b_n]$$

with $\dim(e_i) = 3$, $\dim(b_i) = 4$, $1 \le i \le n$ *is injective in cohomology.*

Dually, this gives

Corollary 6.3 *In homology the product map*

$$H_*(\mathrm{SL}_2(q); \mathbb{F}_2)^{\otimes^n} \longrightarrow H_*(\mathrm{Sp}_{2n}(q); \mathbb{F}_2)$$

is surjective. In particular $H_*(\mathrm{Sp}_{2n}(q); \mathbb{F}_2)$ *is a quotient of the subset of elements in* $\mathbb{F}_2[b_4, b_8, \ldots,] \otimes \mathbb{F}_2[e_3, e_7, \ldots, e_{4i-1}, \ldots]$ *consisting of products of* n *or less terms where the subscripts denote the dimensions of the generators.*

(The mapping $H_*(\mathrm{SL}_2(q); \mathbb{F}_2)^n \longrightarrow H_*(\mathrm{Sp}_{2n}(q); \mathbb{F}_2)$ factors through the $\mathcal{S}_n$ invariants and they have the description given in (6.3).)

Lemma 6.4 *The elements* e_{4i-1}^2 *are zero in* $H_*(\mathrm{Sp}_4(q); \mathbb{F}_2)$, *so it follows that there is a surjection of the products of* n *or less terms in the algebra* $\mathbb{F}_2[b_4, b_8, \ldots, b_{4i}, \ldots] \otimes E(e_3, \ldots, e_{4i-1}, \ldots)$ *onto* $H_*(\mathrm{Sp}_{2n}(q); \mathbb{F}_2)$.

Proof. (6.1) shows that the two inclusions $\mu: \mathrm{SL}_2(q) \times \mathbb{Z}/2 \longrightarrow \mathrm{SL}_2(q) \wr \mathbb{Z}/2$ and $\Delta \times \mathrm{id}: \mathrm{SL}_2(q) \times \mathbb{Z}/2 \longrightarrow \mathrm{SL}_2(q) \wr \mathbb{Z}/2$ are conjugate in $\mathrm{Sp}_4(q)$. Consequently,

$$\theta \in \mathrm{im}(H^*(\mathrm{Sp}_4(q); \mathbb{Z}/2) \subset H^*(\mathrm{SL}_2(q) \wr \mathbb{Z}/2; \mathbb{Z}/2)$$

only if $\mu^*(\theta) = (\Delta \times 1)^*(\theta)$.

In particular consider the elements $\Gamma(e_3 b^i)$. We have $\mu^*(\Gamma(e_3 b^i) = \mu^*(e_3 b^i \otimes e_3 b^i) = 0$, but

$$(\Delta \times \mathrm{id})^*(\Gamma(e_3 b^i)) = e_3 b^i \otimes e^{4i+3} + \text{ higher terms}$$

which is certainly not zero. It follows that $\Gamma(e_3 b^i)$ is not in the image of $H^*(\mathrm{Sp}_4(q); \mathbb{F}_2)$ so dually $e_{4i+3} \otimes e_{4i+3}$ maps to zero in $H_*(\mathrm{Sp}_4(q); \mathbb{F}_2)$ and $e_{4i+3}^2 = 0$ as asserted. □

We now turn to the proof of the main theorem

Theorem 6.5

$$H_*(\mathrm{Sp}(q);\mathbb{F}_2) = \mathbb{F}_2[b_4,\dots,b_{4n},\dots]\otimes E(e_3,\dots,e_{4n-1},\dots)\,,$$

and $H_*(\mathrm{Sp}_{2m}(q);\mathbb{F}_2)$ injects into $H_*(\mathrm{Sp}(q);\mathbb{F}_2)$ as the suspace spanned by all monomials $b_4^{i_1}\cdots b_{4l}^{i_l}e_3^{\epsilon_1}\dots e_{4s-1}^{\epsilon_s}$ with $\sum i_j+\sum\epsilon_j\le m$.

Proof. The main step in the proof of (6.5) is the following preliminary result.

Lemma 6.6

a. *There is a unique conjugacy class of maximal elementary subgroups $(\mathbb{Z}/2)^n\subset\mathrm{Sp}_{2n}(q)$ for each n and any elementary 2-group $(\mathbb{Z}/2)^s$ in $\mathrm{Sp}_{2n}(q)$ is contained in a subgroup $(\mathbb{Z}/2)^n$.*
b. *$(\mathbb{Z}/2)^n\times(\mathbb{Z}/2)^n\subset\mathrm{Sp}_{2n}(q)\wr\mathbb{Z}/2\subset\mathrm{Sp}_{4n}(q)$ is a closed system for $n>1$.*
c. *For $n=1$ there are two conjugacy classes of subgroups $(\mathbb{Z}/2)^2\subset\mathrm{Sp}_2(q)\wr\mathbb{Z}/2$. The first V_I is the center of $\mathrm{Sp}_2(q)^2$ and is normal in $\mathrm{Sp}_2(q)\wr\mathbb{Z}/2$ while the second $V_{II}\subset\mathrm{Sp}_2(q)\times\mathbb{Z}/2\xrightarrow{\Delta\times\mathrm{id}}\mathrm{Sp}_2(q)\wr\mathbb{Z}/2$.*

Proof (a) Given $(\mathbb{Z}/2)^w\subset\mathrm{Sp}_{2n}(q)$ we argue as in (5.3) using the ±1 eigenspaces for generators of $(\mathbb{Z}/2)^w$ to show that if $w<n$ then there is a subspace $V\subset\mathbb{F}_q^{2n}$ with the restriction of the symplectic form non-singular on V and V is a -1 eignenspace for t_1, but a $+1$ eigenspace for the other generators, and $\dim(V)\ge4$. This allows us to split V non-trivially as the orthogonal sum

$$V=\langle S\rangle\perp\cdots\perp\langle S\rangle$$

and extend $(\mathbb{Z}/2)^w$ to a subgroup $(\mathbb{Z}/2)^{w+1}\subset\mathrm{Sp}_{2n}(q)$. This process stops only when $w=n$. In particular, every $(\mathbb{Z}/2)^n\subset\mathrm{Sp}_{2n}(q)$ is associated uniquely with a splitting of the symplectic form

$$\mathbb{F}_q^{2n}=\langle S\rangle\perp\cdots\perp\langle S\rangle\,,$$

and from this it is direct that any two such subgroups are conjugate in $\mathrm{Sp}_{2n}(q)$.

(b). Let $W=(\mathbb{Z}/2)^w\subset\mathrm{Sp}_{2n}(q)\wr\mathbb{Z}/2$ and consider the projection $\pi\colon\mathrm{Sp}_{2n}(q)\wr\mathbb{Z}/2\to\mathbb{Z}/2$. If $\mathrm{im}(\pi|W)\ne\{1\}$ then there is a $t\in W$ with t of the form (θ,θ',T). Since

$$1=t^2=(\theta,\theta',T)(\theta,\theta',T)=(\theta\theta',\theta'\theta,1)$$

it follows that $t=(\theta,\theta^{-1},T)$ for some $\theta\in\mathrm{Sp}_{2n}(q)$. Thus, when we conjugate W by $(\theta^{-1},1,1)$, t is conjugated to $(1,1,T)$, so we can assume $(1,1,T)\in W$.

On the other hand, the centralizer of $(1,1,T)$ in $\mathrm{Sp}_{2n}(q)\wr\mathbb{Z}/2$ is

$$\Delta(\mathrm{Sp}_{2n}(q))\times\mathbb{Z}/2$$

which, by (a), has exactly one conjugacy class of elementary 2-groups $(\mathbb{Z}/2)^{n+1}$. Since $n \geq 2$ it follows that $n+1 < 2n$ and we assumed $w = 2n$. Thus, $W \subset \ker(\pi) = \mathrm{Sp}_{2n}(q) \times \mathrm{Sp}_{2n}(q)$ and (b) follows from (a).
(c). The proof of (b) actually shows that for all n, $\mathrm{Sp}_{2n}(q) \wr \mathbb{Z}/2$ contains exactly 2 conjugacy classes of elementary 2-subgroups, the first of the form $(\mathbb{Z}/2)^{2n} \subset \ker(\pi)$ and the second of the form $(\mathbb{Z}/2)^{n+1}$ which is not in the kernel of π and which is conjugate to a subgroup of $\Delta(\mathrm{Sp}_{2n}(q)) \times \mathbb{Z}/2$. (c) is a special case of this. □

Note that we can regard $\mathrm{Sp}_2(q) \wr \mathbb{Z}/2$ as the normalizer of a $\mathbb{Z}/2 \times \mathbb{Z}/2 \subset \mathrm{Sp}_4(q)$ and the $(\mathbb{Z}/2)^2$ is contained in $\mathrm{Sp}_2(q) \wr \mathbb{Z}/2$ as the normal subgroup V_I.

Lemma 6.7 *$N_{\mathrm{Sp}_4(q)}(V_{II}) \cap N_{Sq_4(q)}(V_I)$ is a split extension of the centralizer of V_{II}, $\Delta(\mathrm{Sp}_2(q)) \times \mathbb{Z}/2$ in $N_{\mathrm{Sp}_4(q)}(V_I)$ by a copy of $\mathbb{Z}/2$,*

$$N_{\mathrm{Sp}_4(q)}(V_{II}) \cap N_{\mathrm{Sp}_4(q)}(V_I) = (\Delta(\mathrm{Sp}_4(q)) \times \mathbb{Z}/2) \times_T \langle (1,-1,1) \rangle \, .$$

Proof. The Weyl group $W_{\mathrm{Sp}_4(q)}(V_I) = \mathbb{Z}/2 \subset \mathrm{GL}_2(2) = \mathcal{S}_3$. $(1,-1,1)$ normalizes V_{II} and acts non-trivially on it so is contained in the intersection. Hence the intersection above is the extension of the centralizer of V_{II} in $\mathrm{Sp}_2(q) \wr \mathbb{Z}/2$ by $(1,-1,1)$ as asserted. □

We now consider the double coset decomposition

$$\mathrm{Sp}_4(q) \;=\; \sqcup_1^m (\mathrm{Sp}_2(q) \wr \mathbb{Z}/2) g_i (\mathrm{Sp}_2(q) \wr \mathbb{Z}/2) \, .$$

For certain g_i we have the composite

$$\begin{array}{ccc}
V_{II} & \longrightarrow & \mathrm{Sp}_2(q) \wr \mathbb{Z}/2 \\
\big\downarrow{\scriptstyle i} & & \big\downarrow{\scriptstyle c_{g_i}} \\
\mathrm{Sp}_2(q) \wr \mathbb{Z}/2 \cap g_i^{-1} \mathrm{Sp}_2(q) \wr \mathbb{Z}/2 g_i & \hookrightarrow & g_i^{-1} \mathrm{Sp}_2(q) \wr \mathbb{Z}/2 g_i \\
\big\downarrow{\scriptstyle c_{g_i}} & & \big\downarrow{\scriptstyle i} \\
g_i \mathrm{Sp}_2(q) \wr \mathbb{Z}/2 g_i^{-1} \cap \mathrm{Sp}_2(q) \wr \mathbb{Z}/2 & \hookrightarrow & \mathrm{Sp}_2(q) \wr \mathbb{Z}/2
\end{array}$$

and the resulting image of V_{II} will be a copy of V_{II}, hence by modifying g_i it will be V_{II}, except in the one case covered in (6.1) where it will be V_I. Let $g_1 = 1$ and g_2 give the coset where this interchange happens.

Consider the composite map

$$E\colon\ H^*(\mathrm{Sp}_4(q) \wr \mathbb{Z}/2; \mathbb{F}_2) \xrightarrow{\mathrm{tr}} H^*(\mathrm{Sp}_4(q); \mathbb{F}_2) \xrightarrow{\mathrm{res}} H^*(\mathrm{Sp}_2(q) \wr \mathbb{Z}/2; \mathbb{F}_2) \xrightarrow{\mathrm{res}} H^*(V_I; \mathbb{F}_2).$$

By applying the double coset formula to the first two maps E can be rewritten as a sum of maps of the form

$$H^*(\mathrm{Sp}_2(q)\wr\mathbb{Z}/2;\mathbb{F}_2)\xrightarrow{\text{res}}H^*(L_i;\mathbb{F}_2)\xrightarrow{c_{g_i}}H^*(L_i';\mathbb{F}_2)$$
$$\xrightarrow{\text{tr}}H^*(\mathrm{Sp}_2(q)\wr\mathbb{Z}/2;\mathbb{F}_2)\xrightarrow{\text{res}}H^*(V_I;\mathbb{F}_2)$$

where $L_i = g_i\mathrm{Sp}_2(q)\wr\mathbb{Z}/2g_i^{-1}\cap\mathrm{Sp}_2(q)\wr\mathbb{Z}/2$. Again, each of the composites

$$H^*(L_i';\mathbb{F}_2)\xrightarrow{\text{tr}}H^*(\mathrm{Sp}_2(q)\wr\mathbb{Z}/2;\mathbb{F}_2)\xrightarrow{\text{res}}H^*(V_I;\mathbb{F}_2)$$

can be rewritten using the double coset formula as a sum of composites

$$\phi_{i,j}\colon\ H^*(L_i';\mathbb{F}_2)\xrightarrow{\text{res}}H^*(L_{i,j}';\mathbb{F}_2)\xrightarrow{c_j}H^*(L_{i,j}'';\mathbb{F}_2)\xrightarrow{\text{tr}}H^*(V_I;\mathbb{F}_2)$$

and $L_{i,j}' = g_j^{-1}L_i'g_j\cap V_I$. In particular $\phi_{i,j}\equiv 0$ if $L_{i,j}'\subsetneq V_I$, so the only terms which can enter non-trivially into the sum above are $\phi_{1,1} = \text{res}$ and the terms $\phi_{2,j}$.

In the case of $\phi_{2,j}$ we can choose the g_j giving the double coset decomposition $\mathrm{Sp}_2(q)\wr\mathbb{Z}/2 = \sqcup V_I g_j L_2' = \sqcup g_j L_2'$ so that $c_j^* = \text{id}$ for each j since L_2' is given by (6.7) and this group surjects onto $\text{im}(N_{Sq_4(q)}(V_{II})\longrightarrow\text{Aut}(V_{II}))$. It follows that the sum $\sum_j\phi_{2,j} = |\mathrm{Sp}_2(q)\wr\mathbb{Z}/2, L_2'|\text{res} = 0$ since the index of L_2' in $\mathrm{Sp}_2(q)\wr\mathbb{Z}/2$ is even.

Thus we have proved that E is just the restriction map

$$E = \text{res}\colon\ H^*(\mathrm{Sp}_2(q)\wr\mathbb{Z}/2;\mathbb{F}_2)\longrightarrow H^*(V_I;\mathbb{F}_2)$$

and the image of $H^*(\mathrm{Sp}_4(q);\mathbb{F}_2)$ in $H^*(V_I;\mathbb{F}_2)$ is $\mathbb{F}_2[b_4,b_8]$.

Now, to complete the proof of (6.5) we can proceed by induction, (6.4) and the result above providing the theorem for $\mathrm{Sp}_4(q)$. By Hopf's theorem on commutative Hopf algebras the only possible truncations on the polynomial parts would be relations of the form $b_{4j}^{2^l} = 0$ for some l. So it suffices to show that no such truncation can occur. Thus, assume there are no such truncations for $H^*(\mathrm{Sp}_{2^l}(q);\mathbb{F}_2)$, $l\ge 2$. But

$$(\mathbb{Z}/2)^{2^l}\times(\mathbb{Z}/2)^{2^l}\subset\mathrm{Sp}_{2^l}\wr\mathbb{Z}/2\subset\mathrm{Sp}_{2^{l+1}}(q)$$

is a closed system satisfying the hypothesis of the Cardenas-Kuhn theorem for $l\ge 2$ and the inductive step is direct. □

7. The Exceptional Chevalley Groups

The exceptional Chevellay groups are of two types. First there are the groups $G_2(q) \subset F_4(q) \subset E_6(q) \subset E_7(q) \subset E_8(q)$ which are analogues for finite field of the exceptional classical Lie groups. Then there are the groups corresponding to the "graph automorphisms" of the Dynkin diagrams. Classically, only the unitary groups have this type of description, but for finite fields (and other fields with similar automorphisms) there are further types. In particular the groups O_{2n}^- are associated to graph automorphisms. There are two further families of this type which were discovered by Steinberg, the first associated to the rotation through $2\pi/3$ of the Dynkin diagram D_4,

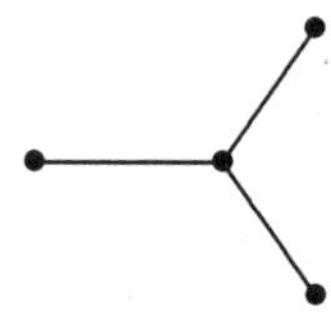

written ${}^3D_4(q)$, and called the "triality twisted" $D_4(q)$. The second is associated to the flip of the Dynkin diagram for $E_6(q)$,

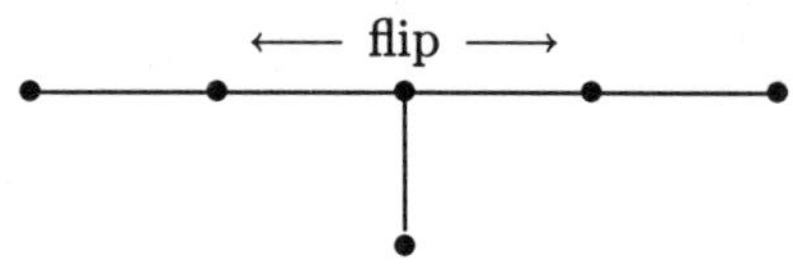

Then there are three further exceptional families, associated to graph automorphisms which are only valid at certain primes, the Suzuki groups $Su(2^{2m+1})$, and the two Ree families, ${}^2G_2(3^{2m+1})$, ${}^2F_4(2^{2m+1})$. The groups ${}^2G_2(3^{2m+1})$ have $\mathrm{Syl}_2({}^2G_2(3^{2m+1})) = (\mathbb{Z}/2)^3$ with Weyl group the semi-direct product $(\mathbb{Z}/7) \times_T \mathbb{Z}/3$. Consequently

$$H^*({}^2G_2(3^{2m+1}); \mathbb{F}_2) = \mathbb{F}_2[x, y, z]^{\mathbb{Z}/7 \times_T \mathbb{Z}/3}.$$

Aside from this and Quillen's general result mentioned in the introduction to this chapter not too much is known about most of these groups. However, recently, the situation for the groups ${}^3D_4(q)$ and ${}^2G_2(q)$ has begun to become much clearer, and there is related work by T. Hewett which promises to also clarify the situation for $F_4(q)$.

The groups ${}^3D_4(q)$ and $G_2(q)$ are simple for odd q and are characterized by a result of P. Fong and W.J. Wong, [FW],

Theorem 7.1 *Let G be a finite simple group with one conjugacy class of involutions. If the normalizer of an involution is the central product*

$$\mathrm{SL}_2(q) * \mathrm{SL}_2(q^v) \times_T \langle n \rangle$$

where v is 1 or 3 and n acts on each SL_2 as conjugation by $\begin{pmatrix} 0 & 1 \\ -\mu & 0 \end{pmatrix}$ *with μ a non-square in $\mathbb{F}_q$, then $G = G_2(q)$ if $v = 1$ and ${}^3D_4(q)$ if $v = 3$.*

It follows that $\mathrm{Syl}_2(G) \subset \mathrm{SL}_2(q) * \mathrm{SL}_2(q^v) \times_T \langle n \rangle$. From this $\mathrm{Syl}_2(G) = \mathcal{Q}_{2^m} * \mathcal{Q}_{2^m} \times_T \langle n \rangle$, and this group, in turn, is isomorphic to

$$(\mathbb{Z}/2^{m-1})^2 \times_T \langle y, v \rangle$$

where y acts to invert the elements of $(\mathbb{Z}/2^{m-1})$ while v interchanges them. In particular, this representation shows that G has 2-rank 3, so the largest 2-elementary subgroups are of the form $(\mathbb{Z}/2)^3$. For example, if a, b generate the two copies of $\mathbb{Z}/2^{m-1}$ then $I = \langle a^{2^{m-2}}, b^{2^{m-2}}, y \rangle$ and $II = \langle a^{2^{m-2}}, b^{2^{m-2}}, aby, \rangle$ give two distinct copies of $(\mathbb{Z}/2)^3 \subset \mathrm{Syl}_2(G)$.

For simplicity, in the rest of this section we assume that $q \equiv 1 \bmod (4)$.

The involution j which is centralized by $\mathrm{SL}_2(q) * \mathrm{SL}_2(q^v) \times_T \langle n \rangle$ is $(ab)^{2^{m-2}}$. With this notation results of [FW], [FM], show that G has 2 conjugacy classes of $(\mathbb{Z}/2)^2$'s, the first represented by $\langle j, a^{2^{m-2}} \rangle$, and the second by $\langle j, aby \rangle$ and 2 conjugacy classes of $(\mathbb{Z}/2)^3$'s, the first represented by I and the second by II. In particular, every $(\mathbb{Z}/2)^2$ is contained in a $(\mathbb{Z}/2)^3$. Further, in [FW] it is proved that the normalizer modulo the centralizer for each $(\mathbb{Z}/2)^2$ is $\mathrm{GL}_2(2) = \mathcal{S}_3$. Then in [FM] it is proved that the normalizer of I is given as a non-split extension

$$I \lhd E \longrightarrow \mathrm{GL}_3(2)$$

where $\mathrm{GL}_3(2)$ acts in the usual non-trivial way on I. (There are exactly two extensions of this type, one of which is split and one not split.) Finally, it is shown in [FM] that the normalizer of II is a non-split extension of the form

$$II \lhd W \longrightarrow \mathcal{S}_4$$

where $\mathcal{S}_4 \subset \mathrm{GL}_3(2)$ is the parabolic subgroup of all matrices in $\mathrm{GL}_3(2)$ of the form $\begin{pmatrix} A & B \\ 0 & 1 \end{pmatrix}$ with $A \in \mathrm{GL}_2(2)$. We write H_2 for the group

$$\begin{aligned} H_2 &= (\mathbb{Z}/4)^2 \times_T \mathbb{Z}/2 \\ &= \{a, b, y \mid a^4 = b^4 = y^2 = 1,\ ab = ba,\ ay = ya^{-1},\ by = by^{-1}\} \end{aligned}$$

and H for the extension $H_2 \times_T \mathbb{Z}/2 = \mathrm{Syl}_2(G_2(5))$.

Remark. It is also shown in [FM] that when we include $G(q) \subset G(q^2)$ then both the image of $I_{G(q)}$ and the image of $II_{G(q)}$ are conjugate to $I_{G(q^2)}$.

Remark. W also occurs as the normalizer of a $(\mathbb{Z}/2)^3$ in the Mathieu group M_{12}. It is discussed in Chapter VIII.

The data above gives a complete determination of the poset space for G in [FM]. The quotient of the poset space by the action of G is given below with the isotropy groups of the faces indicated.

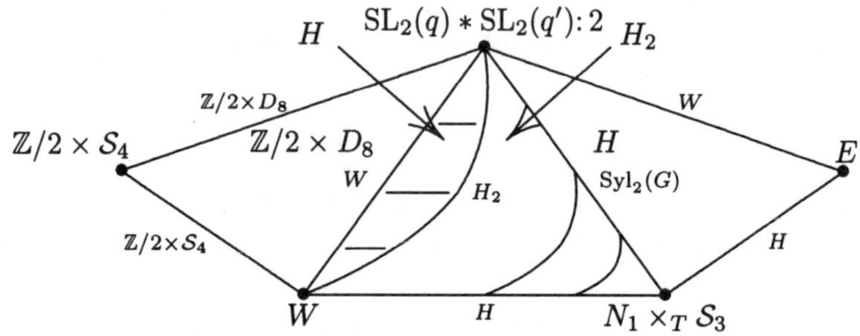

Here the labels in smaller type are the isotropy groups of the edges. Also, $N_1 = (\mathbb{Z}/(q-1) \times \mathbb{Z}/(q-1)) \times_T (\mathbb{Z}/2)$ since $q \equiv 1 \bmod (4)$.

There are two useful cancellation rules that can be used to simplify the diagram without changing the homology type of the associated classifying space. If a "free" edge, i.e. an edge incident on only one face, and the corresponding face have the same isotropy group then we can remove both edge and face. Likewise, if a free vertex, i.e. a vertex incident on only one edge, and the corresponding edge have the same isotropy group we can again remove both the edge and the vertex. By iterating these reductions the homotopy type of the resulting classifying space corresponds to the diagram

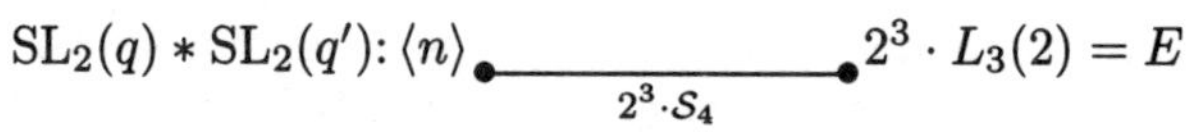

Thus we have the following theorem which reduces the determination of $H^*(G; \mathbb{F}_2)$ to a much simpler problem.

Theorem 7.2 *The intersection of E and $\mathrm{SL}_2(q) * \mathrm{SL}_2(q^v) \times_T \langle n \rangle$ in G is isomorphic to $W = N_G(II)$. Moreover the homomorphism*

$$\phi\colon\ \mathrm{SL}_2(q) * \mathrm{SL}_2(q^v) \times_T \langle n \rangle *_W E \longrightarrow G$$

is surjective and induces isomorphisms in cohomology with $(\mathbb{Z}/2)$ coefficients. Here $A *_B C$ is the free product of A and C, amalgamated over the common subgroup B.

These groups are studied as follows. First $\mathrm{Syl}_2(G) \subsetneq D_{2^n} \wr \mathbb{Z}/2$ as an index(2) subgroup, and from the Gysin sequence for this pair together with the fact that $H^*(D_{2^n} \wr \mathbb{Z}/2)$ is detected on elementary 2-groups it follows that $H^*(\mathrm{Syl}_2(G))$ is also detected on elementary 2-groups and its cohomology can be explicitly determined. In summary one has the results of Fong and Milgram, [FM], [M2],

Theorem 7.3

1. *$H^*(\mathrm{Syl}_2(G))$ is detected by restriction to its maximal 2-elementary subgroups $(\mathbb{Z}/2)^3$.*
2. *The Poincaré series for $\mathrm{Syl}_2(G)$ is independent of G and given as $\frac{1}{(1-x)^3}$.*
3. *$H^*(\mathrm{Syl}_2(G)) \cong \mathbb{F}_2[c_1, \Gamma_4, T_2](1, l_1, \gamma_2, \mu_3) \oplus \mathbb{F}_2[c_1, \Gamma_4, e_1](e_1, e_1\gamma_2)$ with the subscripts denoting the dimensions of the generators. The key multiplicative relation is $eT_2 = 0$.*

Since E contains representatives of both conjugacy classes of $(\mathbb{Z}/2)^3$ in G it follows that $H^*(G) \xrightarrow{\text{res}} H^*(E)$ is injective. Moreover, applying the Cardenas-Kuhn theorem to the closed system $I \lhd \mathrm{Syl}_2(G) \subset E$ we have

Lemma 7.4 image$(\text{res}^*: H^*(E) \to H^*(I)^{GL_3(2)})$ *is exactly* $H^*(I)^{GL_3(2)} = \mathbb{F}_2[d_4, d_6, d_7]$.

But the system $I \lhd E \subset G$ is a closed system as well, so we can again apply the Cardenas-Kuhn theorem and $\mathbb{F}_2[d_4, d_6, d_7]$ is the restriction to $H^*(I)$ of $H^*(G)$ as well.

Under the inclusion $G(q) \subset G(q^2)$ the groups I and II in $G(q)$ become conjugate in $G(q^2)$. (They are said to fuse.) It follows that

$$\text{image}(\text{res}^*:\ H^*(G) \to H^*(II))$$

contains $\mathbb{F}_2[d_4, d_6, d_7]$ as well. It remains to understand the remaining terms in the image of $H^*(G)$ in $H^*(II)$. In particular, it turns out that only the term ν_3 in dimension 3 must be determined by direct calculation as the generator in dimension 5 can be given as $Sq^2(\nu)$.

This was done in [M2] by completing analysis of $H^*(E)$ and its restrictions to the three non-conjugate $(\mathbb{Z}/2)^3$'s there, then applying Webb's theorem to $\mathcal{A}_2(G)$ to obtain the Poincaré series of $H^*(G)$, and finally applying an invariant argument to determine the exact class in 3. This gave a direct calculation without relying on Quillen's general result alluded to in the introduction.

If we use Quillen's result, then we only have to determine an element in $H^*(II)^{S_4}$. But a direct calculation (or a general theorem of T. Hewett [He]) shows that

$$H^*(II)^{S_4} = \mathbb{F}_2[\tilde{d}_2, \tilde{d}_3, D_4]\,.$$

Hence there is only one possible element in dimension 3, and the calculation is complete. The exact results, from [M2], are

Theorem 7.5

1. *Let $G = G_2(q)$ or ${}^3D_4(q)$ with $q \equiv 1 \bmod (4)$. Then the Poincaré series $P_G(x)$ is independent of q and equals*

$$\frac{(1+x^3)(1+x^5)(1+x^6)}{(1-x^4)(1-x^6)(1-x^7)}\,.$$

2. *$H^*(G)$ is free (and finitely generated) over a subpolynomial ring*

$$\mathbb{F}_2[d_4, d_6, d_7]$$

and if $q' = q^w$ then $G(q) \subset G(q')$ and the resulting restriction map takes the subpolynomial algebra for $G(q')$ isomorphically to the subpolynomial algebra for $G(q)$.

3. *As a ring $H^*(G)$ is generated by d_4, $d_6 = Sq^2(d_4)$, $d_7 = Sq^1(d_6)$, a generator β in dimension 3, and $f = Sq^2(\beta)$ in dimension 5. The relations are $\beta^4 = d_7 f + d_6\beta^2$, $f^2 = d_4\beta^2 + d_7\beta$, both of which are implied by the formula $Sq^4(f) = d_4 f + d_6\beta$.*

Remark. The Poincaré series in 7.5.1 is exactly the same as the Poincaré series of the sporadic group J_1 studied in II.6.9 and III.1.9. We will try to explain the relation between J_1 and $G_2(q)$ in Chapter IX where we use the plus construction. The group E studied above will play a crucial role in this analysis as the next remark will help to explain.

Remark 7.6 The group E also appears as a maximal finite subgroup of the topological Chevellay group G_2 of dimension 14 given as the automorphism group of the Cayley plane. G_2 is a subgroup of the orthogonal group $O(7)$. This embedding of E in was given by H.S.M. Coxeter, [Cox], and a particular representation is given in [CW], pg. 448, as the span of the permutation matrices $(1, 2, 3, 4, 5, 6, 7)$, $(1, 2, 4)(2, 6, 5)$, and the matrix $\delta_{(1,2,4,7)}(1, 2)(3, 6)$ where $\delta_{(a,b,c,d)}$ is the diagonal matrix in $O(7)$.

In fact, under this embedding I turns out to be a maximal torus in G_2 and E is its normalizer. Thus, by a general theorem of Borel, the inclusion induces an isomorphism

$$H^*(B_{G_2}; \mathbb{F}_2) \longrightarrow H^*(I; \mathbb{F}_2)^{GL_3(2)} .$$

An independent verification of this isomorphism is given in [M2] by using the explicit embedding to evaluate the composition

$$H^*(B_{O(7)}; \mathbb{F}_2) \longrightarrow H^*(B_{G_2}; \mathbb{F}_2) \longrightarrow H^*(I; \mathbb{F}_2)^{GL_3(2)}$$

on the Stiefel-Whitney classes (which generate $H^*(B_{O(7)}; \mathbb{F}_2)$, [MS]).

Remark. It is not hard to see that $H^*(G_2(\mathbb{F}_{p^\infty}); \mathbb{F}_2) = \varprojlim (H^*(G_2(p^n); \mathbb{F}_2)$ and that 7.4, 7.5 show that this limit is exactly $\mathbb{F}_2[d_4, d_6, d_7]$ which is, by 7.6 also equal to $H^*(G_2; \mathbb{F}_2)$. This result, due originally to Grothendieck is an aspect of his theory of Etale cohomology. A discussion of this remarkable theory lies outside the scope of this book however.

We refer the reader to the work of Kleinerman [Kl] for more on this topic.

Chapter VIII. Cohomology of Sporadic Simple Groups

0. Introduction

In this chapter we will describe recent progress towards understanding the cohomology of the sporadic simple groups. Briefly we recall that from the classification of finite simple groups, [Gor], it was shown that there exist 26 simple groups not belonging to infinite families (i.e. not of alternating or Lie type) and we study six of these groups here: four of the five Mathieu groups, the first Janko group, J_1, and the O'Nan group $O'N$.

Here are some of the reasons, (aside from pure curiosity), for understanding the cohomology of these groups.

As we will see in Chapter IX, we can add two and three dimensional cells to the classifying space of a perfect group to obtain a simply connected topological space, B_G^+, with the same homology as B_G, and the homotopy groups of these spaces are of basic importance in homotopy theory. The symmetric groups build up $Q(S^0)$, a process discussed in the introduction to Chapter VI and proved in IX.3. The general linear groups over a finite field similarly build the classifying spaces B_O and B_U as well as certain fibers of "Adams operations", $\Psi^k - 1$, known as $Im(J)$-spaces (IX.3 again). Indeed, in IX.3.2 we point out that these lead to a product splitting $Q(S^0) = Im(J) \times Coker(J)$ with the $Im(J)$ space completely understood.

It is natural to expect that the sporadic groups should play a role in the structure of the $Coker(J)$ space, though we are only beginning to understand some of the smaller sporadic groups in this framework. The group M_{11} which has 2-rank two has been studied by F. Cohen and the three 2-rank three sporadic groups M_{12}, J_1 and $O'N$ are currently being analyzed from this point of view. We only have partial information about $O'N$, but both J_1 and M_{12} are closely tied to the exceptional Lie group G_2, facts which are still slightly mysterious, but which we will explain in Chapter IX. Among the rank four sporadic groups we have determined the cohomology of M_{22} and M_{23} and the cohomology rings are discussed in §5. Neither one is Cohen-Macaulay so the rings are quite complex, but M_{23} turns out to be the first example of a finite group for which $H^i(G;\mathbb{Z}) = 0$ for $i \leq 4$. Indeed, it had been conjectured by C.Giffen [Gi] that the only finite group for which this is possible is the trivial group. Also, the $Coker(J)$ space is 5-connected with $H_6(Coker(J);\mathbb{Z}) = \mathbb{Z}/2$, and the natural inclusions $M_{23} \subset \mathcal{S}_{23} \subset \mathcal{S}_\infty$ induce an isomorphism of

$H_6(M_{23};\mathbb{Z})$ with the $\mathbb{Z}/2$ which generates $H_6(Coker(J);\mathbb{Z})$ in this dimension. Again, we expand on these remarks in §5 and Chapter IX.

Likewise, from the point of view of modular representations, the sporadic groups are key examples and it is apparent that cohomological invariants play a fundamental part in recent and ongoing research in this field. As was mentioned in our introduction to this book the ideas here are discussed in the books of Benson and Evens.

What we shall concentrate on is furnishing detailed calculations of the cohomology of a few of these groups. We will apply all the different techniques described in the previous chapters, hoping perhaps that this will serve as an additional justification and description of our methods. In particular, our calculation of $H^*(M_{11};\mathbb{F}_2)$ in §1 completely replaces the much more complex original calculation by Benson and Carlson. Also, our calculation of $H^*(M_{12},\mathbb{F}_2)$, where M_{12} is the Mathieu group of order 95,040, replaces the less natural and longer original calculation in [AMM2].

We use $\mathbb{F}_2$ coefficients throughout, so they are often suppressed.

1. The Cohomology of M_{11}

Let $G = M_{11}$, the first Mathieu group having order $7920 = 2^4\ 3^2\ 5\ 11$. M_{11} has 2-rank two with one conjugacy class of groups $(\mathbb{Z}/2)^2$ and one conjugacy class of involutions. From the Atlas, [Co], $N(2A) = 2\cdot\Sigma_4 = \mathrm{GL}_2(\mathbb{F}_3)$, (that is to say, the normalizer of an involution is the non-split extension of the form

$$\mathbb{Z}/2 \triangleleft N(2A) \longrightarrow S_4$$

isomorphic to $GL_3(\mathbb{F}_3)$), and we can also check that $N((\mathbb{Z}/2)^2) = \Sigma_4$. Thus the quotient $|A_2(M_{11}|/M_{11}$ has the form

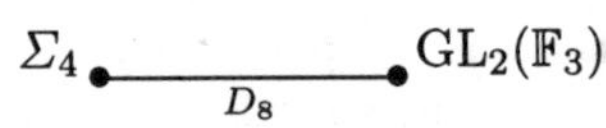

and we can apply V.3.3 to obtain the formula

$$H^*(M_{11}) \oplus H^*(D_8) = H^*(\mathrm{GL}_2(3)) \oplus H^*(\Sigma_4)\,.$$

We already know from IV.2.7 that $H^*(D_8;\mathbb{F}_2) = \mathbb{F}_2[\bar{x}_1,\bar{y}_1,w_2]/(\bar{x}\bar{y} = 0)$ which has Poincaré series

$$P_{D_8}(t) = \frac{2}{(1-x)(1-x^2)} - \frac{1}{1-x^2} = \frac{1}{(1-x)^2}\,.$$

Similarly, using VI.1.13, the Poincaré series for Σ_4 is $\frac{1+t+t^2+t^3}{(1-t^2)(1-t^3)}$. From VII.4.5 and III.4.2 the Poincaré series for $H^*(\mathrm{GL}_2(\mathbb{F}_3))$ is

$$\frac{(1+t)(1+t^3)}{(1-t^2)(1-t^4)} = \frac{1+t+t^2+t^3+t^4+t^5}{(1-t^3)(1-t^4)}$$

and so the Poincaré series for M_{11} is (as first computed in [We], compare [BC3])

$$\frac{1+t^5}{(1-t^3)(1-t^4)}.$$

The group $\mathrm{GL}_2(\mathbb{F}_3)$ contains a Sylow 2-subgroup of M_{11} and has its mod(2) cohomology detected on its elementary 2-subgroups. Consequently the same is true for M_{11}. More restrictively it follows from VII.4.4 that

$$H^*(M_{11};\mathbb{F}_2) \subseteq \mathbb{F}_2[x_1,x_2]^{\mathrm{GL}_2(\mathbb{F}_2)} = \mathbb{F}_2[d_2,d_3]\,,$$

(the Dickson algebra described in III.2.3) and since there is only one element in each of the dimensions 3, 4, and 5 in this ring we see that $H^*(M_{11}) \cong \mathbb{F}_2[d_3,d_2^2](1,d_2d_3)$. Hence we have

Theorem 1.2

$$H^*(M_{11}) \cong \mathbb{F}_2[v_3,v_4,v_5]/v_3^2v_4 + v_5^2 = 0$$

with Poincaré series $p(t) = \frac{1+t^5}{(1-t^3)(1-t^4)}$.

Remark 1.3 From the action of the Steenrod Algebra on the Dickson algebra, we have the following table giving the action of the Steenrod algebra on $H^*(M_{11})$,

gen.\\Sq	Sq^1	Sq^2	Sq^3	Sq^4
v_3	0	v_5	v_3^2	
v_4	0	v_3^2	0	v_4^2
v_5	v_3^2	0	0	$v_3^3 + v_4v_5$

2. The Cohomology of J_1

J_1 is the first Janko group, of order 175,560. It has a 2-Sylow subgroup isomorphic to $(\mathbb{Z}/2)^3$. Using this, we have already calculated its cohomology in II.6.9 using the invariant ring determined in III.1.9. For further remarks on $H^*(J_1)$ using a poset decomposition see V.2.11 and V.3. For convenience, we recall the result from II.6.9,

Theorem 2.1

$$H^*(J_1) \cong \mathbb{F}_2[x_3,y_4,z_7](\gamma_5,\mu_6)/ \begin{array}{r} \gamma^2 + y\mu + xz = 0 \\ \mu^2 + x^4 + x^2\mu + y^3 + \gamma z = 0. \end{array}$$

with Poincaré series

$$p(t) = \frac{(1+t^5)(1+t^6)}{(1-t^3)(1-t^4)(1-t^7)} = \frac{(1+x^3)(1+x^5)(1+x^6)}{(1-x^4)(1-x^6)(1-x^7)}\,.$$

The action of $\mathcal{A}(2)$ can be computed directly from the explicit form of the generators given in II.1.9, particularly x given in (II.1.10). We make the following remarks anticipating our discussion of the relation between J_1 and the exceptional group G_2 which we will give in Chapter IX. First $H^*(J_1;\mathbb{F}_2)$ is Cohen-Macaulay over the Dickson algebra $\mathbb{F}_2[d_4,d_6,d_7]$ and can be rewritten in the form

$$\mathbb{F}_2[d_4,d_6,d_7](1,x,Sq^2(x),x^2,xSq^2(x),x^3,x^2Sq^2(x),x^3Sq^2(x))\,.$$

In particular the quotient algebra by the ideal generated by the Dickson elements is

$$\begin{aligned} &H^*(J_1;\mathbb{F}_2)/(d_4,d_6,d_7)\\ &\qquad\cong \mathbb{F}_2[x,Sq^2(x)]/(x^4=(Sq^2(x))^2=0, x^2=Sq^1(Sq^2(x))) \end{aligned}$$

and this is exactly $H^*(G_2;\mathbb{F}_2)$, a "coincidence" which we will try to explain in Chapter IX. (We are indebted to F. Cohen for this description of $H^*(J_1;\mathbb{F}_2)$ and the geometric explanation which we give in Chapter IX.)

3. The Cohomology of M_{12}

In this section we study the Mathieu group M_{12}. It is considerably more complex than M_{11} and J_1 and we begin with a detailed analysis of the structure of its elementary 2-groups. We explicitly construct subgroups of M_{12} with respect to which each conjugacy class of maximal elementary 2-groups is weakly closed. Then we prove that $H^*(Syl_2(M_{12});\mathbb{F}_2)$ is detected by restriction to its elementary 2-subgroups and from that the determination of its cohomology is fairly direct. The discussion here is a modification, based on ideas in [FM], [M2], of the work of [AMM2] where $H^*(M_{12};\mathbb{F}_2)$ was first determined.

The Structure of the Mathieu Group M_{12}

The Mathieu group M_{12} is the subgroup of $\mathcal{S}_{12}$ generated by the following 6 elements as given for example in [Ha, pp. 79-80]:

$$\begin{aligned} u &= (1,2,3)(4,5,6)(7,6,9) \\ a &= (2,4,3,7)(5,6,9,8) \\ b &= (2,5,3,9)(4,8,7,6) \\ x &= (1,10)(4,5)(6,8)(7,9) \\ y &= (1,11)(4,6)(5,9)(7,8) \\ z &= (1,12)(4,7)(5,6)(8,9). \end{aligned}$$

It is 5-fold transitive and has order $2^6\ 3^3\ 5\ 11 = 95,040$. Note that $\langle a,b\rangle \cong \mathcal{Q}_8$ while $\langle x,y,z\rangle \cong \mathcal{S}_4$. When we conjugate a,b by x,y,z we obtain $xax = b$, $xbx = a$, $yay = ba$, $yby = b^{-1}$, $zaz = a^{-1}$, and $zbz = ba$, so $\mathcal{Q}_8$ is normal in the subgroup $W = \langle a,b,x,y,z\rangle$ which consequently has order $2^6\ 3 = 192$ and contains a 2-Sylow subgroup of M_{12}. In particular

$$H = Syl_2(M_{12}) = \langle a,b,d,k\rangle$$

where

$$\begin{aligned} d &= xyz &= (1,10,11,12)(4,8,7,6) \\ k &= xyxz &= (1,12)(5,9)(6,8)(10,11). \end{aligned}$$

Note that $bd = db$ so $\langle b,d\rangle = (\mathbb{Z}/4)^2$, and $k\theta k = \theta^{-1}$ for each $\theta \in \langle b,d\rangle$. Likewise, $ka = ak$, $ada^{-1} = b^{-1}d$, so setting $s = ad^2$ we have $s^2 = 1$, $\langle s,k\rangle = (\mathbb{Z}/2)^2$ and we can write H as a split extension $(\mathbb{Z}/4)^2 \times_T (\mathbb{Z}/2)^2$ where the twisting by s is given by $sds = b^{-1}d$, $sbs = b^{-1}$.

There are seven conjugacy classes of elements of order two in H, the central element $a^2 = (2,3)(4,7)(5,9)(6,8)$, one having two elements and representative

$$d^2 = (1,11)(4,7)(6,8)(10,12),$$

four with four elements in each conjugacy class having representatives

$$\begin{aligned} k &= (1,10)(4,7)(5,9)(11,12), \\ &\quad (1,12)(2,6)(3,8)(4,5)(7,9)(10,11), \\ &\quad (1,10)(2,5)(3,9)(4,6)(7,8)(11,12), \\ &\quad (1,11)(2,6)(3,8)(4,9)(5,7)(10,12), \end{aligned}$$

and finally one class with eight elements and representative z. The centralizers of the elements in the last 5 classes are given as follows, four copies of $D_8 \times \mathbb{Z}/2$ for the four classes with four elements and $(\mathbb{Z}/2)^3$ for the class with 8 elements. In $\mathcal{S}_{12}$ the involutions contained in M_{12} lie in two conjugacy classes, the class $\{4\}$ consisting of elements which are products of 4 disjoint involutions and $\{6\}$ consisting of those elements which are products of six disjoint involutions, so there are at least two distinct conjugacy classes of involutions in M_{12}. In fact, it is well known, [Co], that there are exactly two.

From the structure of the centralizers in H it is easy to see that every elementary 2-subgroup of H is contained in a $(\mathbb{Z}/2)^3$ and that there are exactly 8 distinct $(\mathbb{Z}/2)^3$'s, three of the form 4^7 (by which we mean that each non-identity element is in the class $\{4\}$), three of the form $4^3 6^4$, and finally two of the form $4^1 6^6$. These last two are conjugate in H and consequently weakly closed in $H \subset M_{12}$. In the other two classes two of the groups are conjugate and the third is normal in H.

Remark 3.1 These $(\mathbb{Z}/2)^3$'s are given as follows. The normal subgroup of type 4^7 is $V_1 = \langle b^2, d^2, k\rangle$ and the two other subgroups of this type are $V_2 = \langle b^2, d^2, bd^{-1}k\rangle$, $\langle b^2, d^2, d^{-1}k\rangle$. The normal subgroup of type $4^3 6^4$ is $V_3 = \langle b^2, d^2, bk\rangle$, and the two non-normal subgroups of this type are $V_4 = \langle b^2, k, s\rangle$, $\langle b^2, d^2 k, bs\rangle$, while the two subgroups of type $4^1 6^6$ are $V_5 = \langle b^2, bd^2 k, s\rangle$ and $\langle b^2, bk, ks\rangle$.

Set $H_{21} = \langle a, b, k, dkd^{-1}\rangle = \mathcal{Q}_8 \times_T K$ where $K \subset \mathcal{S}_4$ is the Klein group. Since k and dkd^{-1} act on $\mathcal{Q}_8$ by conjugation with a, ab respectively, it follows that $\langle ak, abdkd^{-1}\rangle \cong \mathcal{Q}_8$ commutes with $\langle a, b\rangle$ and H_{21} is the central product, $\mathcal{Q}_8 * \mathcal{Q}_8 \cong D_8 * D_8$, an extra special 2-group. H_{21} is also the subgroup of H which is spanned by the three groups of the form $4^3 6^4$, and the element of order three,

$$xy = (1, 10, 11)(4, 9, 8)(5, 6, 7) \tag{3.2}$$

which normalizes H_{21} permutes the three subgroups cyclically. Thus these groups form a single conjugacy class in M_{12} and are weakly closed in $W \subset M_{12}$. H_{21} also contains the normal 4^7 subgroup, and xy normalizes this group as well. Hence, W can be rewritten $(\mathbb{Z}/2)^3 \times_T \mathcal{S}_4$. Finally, H_{21} contains both of the $4^1 6^6$ subgroups, and each is normal in $\langle H_{21}, xy\rangle$.

The subgroup, $H_{22} \subset H$ which is spanned by the three subgroups of the form 4^7 also has order 32. It is $\langle b, d, k\rangle$ so $H_{22} \cong (\mathbb{Z}/4)^2 \times_T \mathbb{Z}/2$ with the element of order two acting to invert the elements in $(\mathbb{Z}/4)^2$. Consequently there are exactly four subgroups of the form $(\mathbb{Z}/2)^3 \subset H_{22}$, $\langle b^2, d^2, k\rangle$, $\langle b^2, d^2, bk\rangle$, $\langle b^2, d^2, dk\rangle$ and $\langle b^2, d^2, bdk\rangle$. The first, third, and fourth each have the form 4^7 and the second has the from $4^3 6^4$. Also, the element

$$T = (1, 4, 2)(3, 11, 7)(5, 10, 6)(8, 9, 12) \in M_{12} \tag{3.3}$$

normalizes H_{22} and cyclically permutes the three 4^7 subgroups while normalizing the $4^3 6^4$. Finally, the element $d^2 a = b^2 s \in H$ satisfies $d^2 a T d^2 a = T^{-1}$, so

$$W' = \langle H, T\rangle = H_{22} \times_T \mathcal{S}_3$$

and V_1 is weakly closed in $W' \subset M_{12}$. Summarizing the discussion above we have

Theorem 3.4

a. *There are precisely three conjugacy classes of groups $(\mathbb{Z}/2)^3$ contained in M_{12}. Under the inclusion $M_{12} \subset S_{12}$ they remain non-conjugate. The first, I, has the form 4^7, the second II, has the form $4^3 6^4$, and the third, III, has the form $4^1 6^6$.*
b. *There are subgroups W and W' of M_{12} described above so that $W \cap W' = H$ and $I \subset W'$, $II \subset W$ and $III \subset H$ are all weakly closed in M_{12}.*
c. *The Weyl group of I in M_{12} and the Weyl group of II in M_{12} are copies of S_4.*
d. *The Weyl group of III in M_{12} is $\mathcal{A}_4$, the Weyl group of III in W.*

(There isn't much to add to the discussion above to complete the proof. In the Atlas, [Co], it is noted that the centralizer of the involution of type $\{4\}$ is W, and the centralizer of an involution of type $\{6\}$ is $\mathbb{Z}/2 \times S_5$ and from this the result is direct.)

For later reference we give the explicit actions of xy and T on the eight $(\mathbb{Z}/2)^3$'s in H now. The notation is that the element in the nth position in the image group is the image of the nth element in the domain group. First the normalizing actions.

$$T\colon \langle b^2, d^2, bk\rangle \longrightarrow \langle d^2, b^2d^2, d^2bk\rangle, \qquad xy\colon \langle b^2, d^2, k\rangle \longrightarrow \langle b^2, k, kd^2\rangle\,. \tag{3.5}$$

The action of d^2xy on a representative group V_5 is

$$\langle b^2, d^2bk, s\rangle \longrightarrow \langle b^2, b^2d^2bks, d^2bk\rangle. \tag{3.6}$$

Next we give the action of xy on the three subgroups of type $4^3 6^4$

$$\langle b^2, d^2, bk\rangle \longrightarrow \langle b^2, k, b^2sk\rangle \longrightarrow \langle b^2, d^2k, bs\rangle, \tag{3.7}$$

and the acton of T on the three subgroups of type 4^7,

$$\langle b^2, d^2, k\rangle \longrightarrow \langle d^2, b^2d^2, bd^{-1}k\rangle \longrightarrow \langle b^2d^2, b^2, d^{-1}k\rangle. \tag{3.8}$$

Using the results above, and, for example, the detailed subgroup results in [AMM2] or [BR] we have the following picture of the poset space of 2–elementary abelian subgroups for M_{12}:

3.9 $|A_2(M_{12})|/M_{12}$

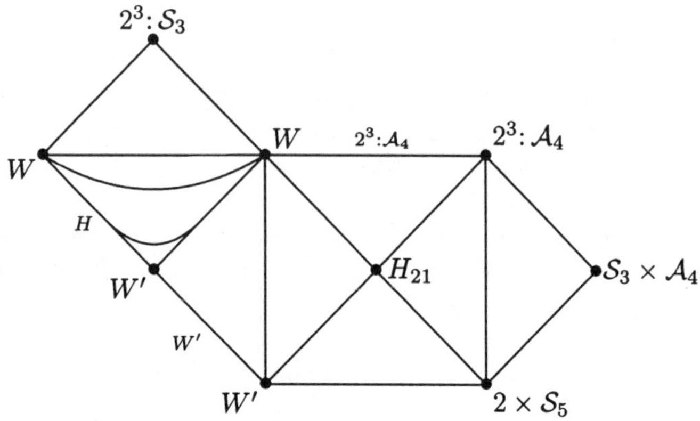

There are 9 vertices, 17 edges and 9 triangles in this orbit complex. We have only shown the isotropy information for the vertices and a few edges. The full details are given in [AMM2, p. 106]. In Webb's formula, V.3.3, most of the groups cancel out and we are left with only

(3.10) $$H^*(M_{12}) \oplus H^*(H) \cong H^*(W) \oplus H^*(W').$$

Additionally, a close analysis of the structure of the finite Chevellay groups $G_2(q)$ shows that for $q \equiv 3, 5 \bmod (8)$, $Syl_2(G_2(q)) \cong Syl_2(M_{12})$, and the configuration $W \cup_H W'$ is also contained in $G_2(q)$. See, e.g. [M2].

This completes our discussion of the subgroup structure of M_{12}. We now turn to the cohomology ring.

The Cohomology of M_{12}

Write $D_8 = \{x, y \mid x^2 = y^2 = (xy)^4 = 1\}$. Then $H_{22} \subset D_8 \times D_8$ with embedding given by $b \mapsto ((xy)^{-1}, xy)$, $d \mapsto (1, xy)$, and $y \mapsto (x, x)$, and this extends to a map $\pi: H \to D_8 \wr \mathbb{Z}/2$ by $\pi(s) = E$, the element which exchanges the two copies of D_8.

Theorem 3.11 *$H^*(H_{22}; \mathbb{F}_2)$ is detected by restriction to its four maximal elementary $(\mathbb{Z}/2)^3$ subgroups. In particular*

$$H^*(H_{22}; \mathbb{F}_2) = \mathbb{F}_2[w_1, w_2, c](1, x_1, x_2, x_1x_2)$$

with relations $x_1^2 = x_1c$, $x_2^2 = x_2c$ where the w's are two dimensional and the remaining generators are one dimensional.

Proof. Recall from IV.2.7 or IV.1.10 that $H^*(D_8; \mathbb{F}_2) = \mathbb{F}_2[\bar{x}, \bar{y}, w_2]/(\bar{x}\bar{y} = 0)$, where

$$H^1(D_8; \mathbb{F}_2) = Hom(D_8, \mathbb{F}_2^+)$$

with generators $\bar{x}(x) = 1$, $\bar{x}(y) = 0$, $\bar{y}(x) = 0$, $\bar{y}(y) = 1$. In $H^*(D_8 \times D_8; \mathbb{F}_2)$ write $x_1 = \bar{x} \otimes 1$, $x_2 = 1 \otimes \bar{x}$, $y_1 = \bar{y} \otimes 1$,$y_2 = 1 \otimes \bar{y}$, $w_1 = w \otimes 1$ and $w_2 = 1 \otimes w$. Then in the Gysin sequence for the inclusion $H_{22} \subset D_8 \times D_8$ we have that the map $\cup\chi\colon H^*(D_8 \times D_8; \mathbb{F}_2) \longrightarrow H^{*+1}(D_8 \times D_8; \mathbb{F}_2)$ is just $\cup(x_1 + x_1 + y_1 + y_2)$. We have

$$H^*(D_8 \times D_8; \mathbb{F}_2) = \mathbb{F}_2[x_1, y_1, x_2, y_2, w_1, w_2]/(x_1 y_1, x_2 y_2)$$

and this can be rewritten as

$$\mathbb{F}_2[\chi, x_1 + y_1, x_1, x_2, w_1, w_2]/(\chi x_2 = x_2^2 + x_2(x_1 + y_1), x_1(x_1 + y_1) = x_1^2).$$

In particular, multiplication by χ is injective, and from this the result is direct where c_1 is the image of $x_1 + y_1$. □

In IV.7.3 a special class, $\Gamma(x) \in H^{2i}(G \wr \mathbb{Z}/2; \mathbb{F}_2)$ is constructed for each $x \in H^i(G; \mathbb{F}_2)$. Moreover, these classes, their cup products with e^i where $e \in H^1(G \wr \mathbb{Z}/2; \mathbb{F}_2)$ is dual to E, and classes of the form $\mathrm{tr}(y)$, $y \in H^*(G \times G; \mathbb{F}_2)$ generate $H^*(G \wr \mathbb{Z}/2; \mathbb{F}_2)$. Also, in the Gysin sequence for the inclusion

$$(*) \qquad \pi\colon H \hookrightarrow D_8 \wr \mathbb{Z}/2$$

the map $H^*(D_8 \wr \mathbb{Z}/2; \mathbb{F}_2) \xrightarrow{\chi} H^{*+1}(D_8 \wr \mathbb{Z}/2; \mathbb{F}_2)$ is multiplication by $\mathrm{tr}(x_1 + y_1)$ which restricts to $H^*(D_8 \times D_8; \mathbb{F}_2)$ as $x_1 + x_2 + y_1 + y_2$. We can now state

Corollary 3.12

a. *$H^*(H; \mathbb{F}_2)$ is detected by restriction to the cohomology of its 5 conjugacy classes of maximal elementary two groups.*
b. *Up to extensions,*

$$\begin{aligned} H^*(H; \mathbb{F}_2) &= \mathbb{F}_2[c, \pi^*\mathrm{tr}(w_1), \pi^*\Gamma(w)](1, \pi^*\mathrm{tr}(x_1), \pi^*\Gamma(\bar{x}), \pi^*\mathrm{tr}(x_1 w_2)) \\ &\quad \oplus e_1 \mathbb{F}_2[e_1, c, \pi^*\Gamma(w)](1, \pi^*\Gamma(\bar{x})) \end{aligned}$$

where e_1 is one dimensional and dual to E.

Proof. We begin by consdering the Gysin sequence for the inclusion $(*)$. The kernel of $\cup\chi$ is $(e) = e\mathbb{F}_2[\Gamma(\bar{x}), \Gamma(\bar{y}), \Gamma(w), e]/(\Gamma(\bar{x})\Gamma(\bar{y}))$, but note that the square of any element in (e) is again a non-zero element of (e). Now, the transfer, $\mathrm{tr}\colon H^*(H; \mathbb{F}_2) \longrightarrow H^*(D_8 \wr \mathbb{Z}/2; \mathbb{F}_2)$, while it does not commute with cup products, does commute with squares (since they are stable cohomology operations), so there are no possible nilpotents in the cokernel of the restriction map $H^*(D_8 \wr \mathbb{Z}/2; \mathbb{F}_2) \longrightarrow H^*(H; \mathbb{F}_2)$.

Next we need $H^*(D_8 \wr \mathbb{Z}/2)/(\cup\chi)$. We have an exact sequence

$$0 \longrightarrow (e) \longrightarrow H^*(D_8 \wr \mathbb{Z}/2; \mathbb{F}_2) \longrightarrow H^*(D_8 \times D_8; \mathbb{F}_2)^{\mathbb{Z}/2} \longrightarrow 0.$$

When we cup this sequence with χ we obtain an exact sequence

$$0 \longrightarrow (e) \longrightarrow H^*(D_8 \wr \mathbb{Z}/2; \mathbb{F}_2)/(\chi) \\ \longrightarrow H^*(D_8 \times D_8)^{\mathbb{Z}/2}/(\chi H^*(D_8 \times D_8)^{\mathbb{Z}/2}) \longrightarrow 0,$$

and since the right hand quotient is easily seen to have no nilpotent elements the first statement follows.

We now use the spectral sequence of the extension $H_{22} \triangleleft H \xrightarrow{\pi} \mathbb{Z}/2$ with E_2-term $H^*_T(\mathbb{Z}/2; H^*(H_{22}; \mathbb{F}_2))$. Here the action of $\mathbb{Z}/2$ on $H^*(H_{22}; \mathbb{F}_2)$ is given by $T(c) = c$ since $c = x_1 + y_1 \sim x_2 + y_2$, $T(x_1) = x_2$, $T(w_1) = w_2$. Consequently we can calculate the E_2-term explicitly as

$$\begin{aligned} E_2 &= \mathbb{F}_2[c, w_1 + w_2, w_1 w_2](1, x_1 + x_2, x_1 x_2, x_1 w_2 + x_2 w_1) \\ &\qquad \oplus e\mathbb{F}_2[e, c, w_1 w_2](1, x_1 x_2). \end{aligned}$$

But each of the generators above, except c is in the image of π^* and hence is an infinite cycle, while c is also an infinite cycle, since $H^1(H; \mathbb{F}_2) = Hom(H, \mathbb{F}_2^+) = (\mathbb{Z}/2)^3$, and the spectral sequence collapses. This proves the second statement. □

From 3.12 it follows that the restriction images of all the generators to the five conjugacy classes of $(\mathbb{Z}/2)^3$'s are explicit. They can be determined as follows. From IV.7.1 and IV.7.2 we obtain the restrictions of $\Gamma(\theta)$ to the basic subgroups $K \times K$, $S \times K$, $K \times S$, $S \times S$, $\langle \Delta(K), E \rangle$ and $\langle \Delta(S), E \rangle$, and the image of tr is zero when restricted to these last two groups. But also, under π we obtain the following inclusions where $A = (xy)^2$:

$$\text{(3.13)} \quad \begin{array}{lcl} V_1 = \langle b^2, d^2, k \rangle & \mapsto & \langle A_1 + A_2, A_2, x_1 + x_2 \rangle \subset K \times K \\ V_2 = \langle b^2, d^2, bd^{-1}k \rangle & \mapsto & \langle A_1 + A_2, A_2, y_1 + x_2 \rangle \subset S \times K \\ V_3 = \langle b^2, d^2, bk \rangle & \mapsto & \langle A_1 + A_2, A_2, y_1 + y_2 + A_2 \rangle \subset S \times S \\ V_4 = \langle b^2, k, s \rangle & \mapsto & \langle A_1 + A_2, x_1 + x_2, E \rangle = \langle \Delta(K), E \rangle \\ V_5 = \langle b^2, bd^2k, s \rangle & \mapsto & \langle A_1 + A_2, y_1 + y_2, E \rangle = \langle \Delta(S), E \rangle. \end{array}$$

For definiteness, assume that w is given so that $\text{res}^*(w) = a^2 + ah$ in both $H^*(K)$ and $H^*(S)$ where a is dual to A and h is dual to x in K, y in S. In the each of the groups V_i write λ as the dual to b^2, τ is dual to the second generator, and h is dual to the third. Also, write $n = \tau^2 + \tau h$, $v = \lambda^4 + \lambda^2(\tau^2 + h^2 + \tau h) + \lambda(\tau^2 h + \tau h^2)$. Then we have the following table for the restrictions to the five $(\mathbb{Z}/2)^3$'s in 3.13.

(3.14)

gen.\group	V_1	V_2	V_3	V_4	V_5
c	h	h	h	τ	τ
$\pi^*tr(x_1)$	0	h	0	0	0
e	0	0	0	h	h
$\pi^*\Gamma(\bar{x})$	h^2	0	0	$\tau^2+\tau h$	0
$\pi^*tr(w_1)$	n	n	n	0	0
$\pi^*tr(x_1w_2)$	$\tau^2 h+\tau h^2$	$\lambda^2 h+\lambda h^2$	0	0	0
$\pi^*\Gamma(w)$	v	v	v	v	v

From this the detailed structure of $H^*(H;\mathbb{F}_2)$ can be easily obtained.

However, we now have a simple criterion for determining the cohomology of M_{12}, W, and W' directly from the table above by using the actions of T and (xy) detailed in 3.5-3.8, which lead to the following maps in cohomology:

(3.15)

Map	On Elements
$(xy)^*: H^*(V_1)\longrightarrow H^*(V_1)$	$h\mapsto h+\tau, \tau\mapsto h, \lambda\mapsto\lambda$
$T^*: H^*(V_2)\longrightarrow H^*(V_1)$	$h\mapsto h, \tau\mapsto\lambda+\tau, \lambda\mapsto\tau$
$T^*: H^*(V_3)\longrightarrow H^*(V_3)$	$h\mapsto h, \tau\mapsto h+\tau+\lambda, \lambda\mapsto\tau$
$(xy)^*: H^*(V_4)\longrightarrow H^*(V_3)$	$h\mapsto h, \tau\mapsto\tau+h, \lambda\mapsto\lambda+h$
$(d^2xy)^*: H^*(V_5)\longrightarrow H^*(V_5)$	$h\mapsto\tau, \tau\mapsto h+\tau, \lambda\mapsto\lambda+\tau$

Now the stability conditions for elements to be in $H^*(W)$, $H^*(W')$ and $H^*(M_{12})$ are easily written down.

Theorem 3.16

a. *$\alpha\in H^*(H;\mathbb{F}_2)$ is contained in the image of* res*: $H^*(W;\mathbb{F}_2)\longrightarrow H^*(H;\mathbb{F}_2)$ *if and only if* res$^*(\alpha)\in H^*(V_1)^{\mathbb{Z}/3}$*, and also in $H^*(V_5)^{\mathbb{Z}/3}$, while the map above from $H^*(V_4)$ to $H^*(V_3)$ stabilizes res$^*(\alpha)$.*

b. *$\alpha\in$ res$^*(H^*(W';\mathbb{F}_2))$ if and only if res$^*(\alpha)\in H^*(V_3)^{\mathbb{Z}/3}$ and the map from $H^*(V_2)$ to $H^*(V_1)$ stabilizes α.*

c. *$\alpha\in$ res$^*(H^*(M_{12}))$ if and only if the conditions in both (a.) and (b.) are satisfied, i.e., if and only if $\alpha\in H^*(W)\cap H^*(W')\subset H^*(H)$.*

Example 3.17 $\pi^*\Gamma(w)+c^4+e^4+(ce)^2+\pi^*tr(w_1)^2$ restricts to the Dickson element d_4 at each of the V_i. From this and the fact that $d_6=Sq^2(d_4)$, $d_7=Sq^1(d_6)$ it follows that $H^*(M_{12};\mathbb{F}_2)$ contains a copy of $\mathbb{F}_2[d_4,d_6,d_7]$. In fact it turns out that $H^*(M_{12};\mathbb{F}_2)$ is actually Cohen-Macaulay, that is to say, free and finitely generated, over this subalgebra.

Example 3.18 V_5 is weakly closed in $H\subset M_{12}$ and, from 3.14, the image of

$$res^*: H^*(H;\mathbb{F}_2)\longrightarrow H^*(V_5;\mathbb{F}_2)$$

is $\mathbb{F}_2[h,\tau,v_4]$ in $H^*(V_5)$. From 3.15 the action of $(d^2xy)^*$ fixes v_4 and acts on $\mathbb{F}_2[h,\tau]$ in the same way $\mathbb{Z}/3$ acts in III.1.3. Consequently, $H^*(V_5)^{\mathbb{Z}/3} = \mathbb{F}_2[h^2+h\tau+\tau^2, h^2\tau+h\tau^2, d_4](1, h^3+h^2\tau+\tau^3)$ is the image of restriction from $H^*(M_{12};\mathbb{F}_2)$. Thus, besides the copy of the Dickson algebra there is one two dimensional generator α and there are two three dimensional generators, $Sq^1(\alpha)$ and l_3 in $H^*(M_{12};\mathbb{F}_2)$. They are constructed as follows: $\alpha = (\pi^*tr(x_1))^2+\pi^*\Gamma(\bar{x})+e^2+c^2+ec$ which restricts to $(0,0,h^2,h^2,h^2+\tau h+\tau^2)$ and $l_3 = c^3+e^3+\pi^*tr(x_1)^3+(c+e)\pi^*\Gamma(\bar{x})+e^2c$ which restricts to $(0,0,h^3,h^3,h^3+h^2\tau+\tau^3)$.

Example 3.19 The map $T^{-*}(xy)^*T^*\colon H^*(V_2)\longrightarrow H^*(V_2)$ is given on elements by $h\mapsto h+\lambda$, $\tau\mapsto h+\lambda+\tau$, $\lambda\mapsto h$, so $\pi^*tr(x_1w_2)$ is stable for T^* and is also $\mathbb{Z}/3$ invariant in both $H^*(V_2)$, $H^*(V_1)$. Consequently, since it restricts to 0 in the remaining groups it is in the image from $H^*(M_{12})$. This gives us a third independent generator $m_3\in H^3(M_{12})$, and a generator $Sq^2(m_3)\in H^5(M_{12})$.

The remaining details of the determination of $H^*(M_{12};\mathbb{F}_2)$ are direct and simplified considerably by the weak closure conditions of 3.4 as 3.18 shows. We leave them to the reader and content ourselves with quoting the result from [AMM2].

Theorem 3.20 *$H^*(M_{12};\mathbb{F}_2)$ has the form $\mathbb{F}_2[\alpha_2,x_3,y_3,z_3,d_4,\gamma_5,d_6,d_7]/\mathcal{R}$ where the d_i are described above and $\mathcal{R}$ is the relation set*

$$\begin{array}{ll}
\alpha(x+y+z)=0 & x^3=\alpha^3x+\alpha d_4x+xd_6\\
xy=\alpha^3+x^2+y^2 & xz=\alpha^3+y^2\\
x^2y=\alpha^3z+\alpha d_4z+yd_6+\alpha d_7 & yz=\alpha^3+x^2\\
d_7x=d_4x^2+\alpha^2x^2 & \alpha\gamma=\alpha^2y\\
d_7y=\alpha^2d_6+\alpha^2y^2+d_4x^2+d_4y^2 & y\gamma=\alpha y^2\\
d_7z=\gamma^2+\alpha^2d_6+\alpha^2x^2+d_4x^2+d_4z^2 & x\gamma=\alpha^4+\alpha z^2\\
 & d_7^2=z^3\gamma+\alpha^2d_4d_6+\alpha^5d_4+\\
z^4=\gamma d_7+x^4+\alpha^4d_4+z^2d_6 & \quad zd_4d_7+zd_6(\gamma+\alpha z)\\
 & \quad +d_4^2(\alpha^3+xz+yz).
\end{array}$$

The Poincaré series for $H^*(M_{12};\mathbb{F}_2)$ is

$$\frac{1+t^2+3t^3+t^4+3t^5+4t^6+2t^7+4t^8+3t^9+t^{10}+3t^{11}+t^{12}+t^{14}}{(1-t^4)(1-t^6)(1-t^7)}$$

and $H^*(M_{12};\mathbb{F}_2)$ is Cohen-Macaulay over $\mathbb{F}_2[d_4,d_6,d_7]$.

(Note that all the generators have been constructed in 3.17-3.19 but x, y and z are linear combinations of $Sq^1(\alpha)$, l, and m and not these generators themselves.)

As a test the reader should calculate the Poincaré series for $H^*(W;\mathbb{F}_2)$ and $H^*(W';\mathbb{F}_2)$. Applying the result of Webb's formula, 3.10, then gives the

Poincaré series in 3.20. This was a critical step in [AMM2], but here, using the weak closure conditions, it only serves the role of assuring us that we have made no numerical errors.

4. Discussion of $H^*(M_{12};\mathbb{F}_2)$

Given a situation such as that of H, W, and W', we can find a universal completion $\Gamma = W *_H W'$ which makes the diagram below commute,

$$\begin{array}{ccc} H & \hookrightarrow & W' \\ \downarrow & & \downarrow{\scriptstyle \phi_2} \\ W & \stackrel{\phi_1}{\longrightarrow} & \Gamma \end{array}$$

and such that any Γ' which satisfies this (generated by W and W') is a quotient of Γ. Γ is called the amalgamated product of W and W' over H. It is well known, (see [Se3]), that an amalgamated product as above will act on a tree with finite isotropy, and orbit space of the form

$$W \bullet\!\!\stackrel{H}{\overline{\qquad\qquad}}\!\!\bullet W'$$

In [Go], Goldschmidt analyzed the situation for actions on the cubic tree (the tree of valence 3) and obtained a classification of finite primitive amalgams of index (3, 3) (this refers to the indexes $[W\colon H]$, $[W'\colon H]$). He shows that M_{12} is one of 15 such amalgams, necessarily a quotient of the universal one Γ.

From this we deduce the existence of an extension

$$1 \longrightarrow \Gamma' \longrightarrow \Gamma \longrightarrow M_{12} \longrightarrow 1 \tag{4.1}$$

where Γ' is a free group (it has cohomological dimension 1). Using the formula for Euler characteristics in [Brown], we have, on the one hand

$$\chi(\Gamma) = \frac{1}{192} + \frac{1}{192} - \frac{1}{64} = -\frac{1}{192}$$

(amalgamated product), and also

$$\chi(\Gamma) = \frac{\chi(\Gamma')}{|M_{12}|}.$$

Hence $\chi(\Gamma') = 95,040(-\frac{1}{192}) = -495$ and it follows that $\Gamma' \cong *_1^{496}\mathbb{Z}$, the free group on 496 generators.

We can now state

Theorem 4.2 *The natural map $\Gamma \longrightarrow M_{12}$ induces an isomorphism*

$$H^*(M_{12};\mathbb{F}_2) \longrightarrow H^*(\Gamma;\mathbb{F}_2).$$

Proof. Consider the map

$$\mathrm{res}_H^W \oplus \mathrm{res}_H^{W'}\colon\ H^*(W) \oplus H^*(W') \longrightarrow H^*(H)\,.$$

Its kernel is clearly $\mathrm{im}(\mathrm{res}_H^W) \cap \mathrm{im}(\mathrm{res}_H^{W'}) \cong H^*(M_{12})$. On the other hand, (3.10) gives that $H^*(W) \oplus H^*(W') \cong H^*(M_{12}) \oplus H^*(H)$. Hence $\mathrm{res}_H^W \oplus \mathrm{res}_H^{W'}$ is onto. On the other hand, from the structure of the orbit space of the tree described at the beginning of this section there is a classifying space for $W *_H W'$ of the form $B_W \cup_{B_H} B_{w'}$, and, applying the Mayer-Vietoris sequence we have a long exact sequence

$$\cdots \longrightarrow H^i(\Gamma) \longrightarrow H^i(W) \oplus H^i(W') \longrightarrow H^i(H) \longrightarrow H^{i+1}(\Gamma) \longrightarrow \cdots.$$

As it comes from a Mayer-Vietoris sequence the same map as before arises, hence the sequence splits and

$$H^*(W) \oplus H^*(W') \cong H^*(\Gamma) \oplus H^*(H)\,.$$

Consequently, by rank considerations and the fact that the finite subgroups in Γ are mapped isomorphically into M_{12} under the projections the proof is complete. □

Corollary 4.3 *$H^1(\Gamma';\mathbb{F}_2)$ is an M_{12}-acyclic $\mathbb{F}_2(M_{12})$-module of rank 496 which is not projective.*

Proof. The proof follows from considering the spectral sequence over $\mathbb{F}_2$ associated to (4.4) below and the observation that 64 does not divide 496. □

This representation has radically different cohomological behavior at distinct primes dividing $|M_{12}|$. For example, at $p = 3$ we have a sequence

$$H^{p-2}(M_{12}; H^1(\Gamma';\mathbb{F}_3)) \longrightarrow H^p(M_{12};\mathbb{F}_3) \longrightarrow H^p(W;\mathbb{F}_3) \oplus H^p(W';\mathbb{F}_3)$$

and clearly the term on the left must be non-trivial. It appears, however, that this module restricted to $M_{11} \subseteq M_{12}$ is the same one associated to the poset space for M_{11}.

To complete our discussion on M_{12}, we will explain the nature of its Poincaré series. Recall from 4.1 that $H^*(M_{12};\mathbb{F}_2)$ is Cohen-Macaulay over the Dickson algebra $\mathbb{F}_2[d_4, d_6, d_7]$. For any finite group G with Cohen-Macaulay cohomology Carlson and Benson, [BC2], have shown that the Poincaré series must satisfy a functional equation which in our case is

$$(*) \qquad p_{M_{12}}(t) = (-t)^{rk\, M_{12}} p_{M_{12}}(t)\,.$$

The method they use is to construct a projective $\mathbb{Z}G$-complex P^* of dimension $\sum_{i=1}^{rk\,G}(n_i-1)$ (where the $\{n_i\}$ are the dimensions of the generators of a polynomial subalgebra over which the cohomology is free and finitely generated), having the chain homotopy type of $C^*\left(\prod_{i=1}^{rk\,G} S^{n_i-1}\right)$. This is done by using cohomological varieties [BC1].

Then they consider the spectral sequence

$$E_2^{p,q}=H^p(G,H^q(P^*))\Longrightarrow H^{p+q}((P^*)^G)\,.$$

Let $v_i\in H^{n_i-1}(P^*)$ be the cohomology generators; by construction they transgress to $\rho_i\in H^{n_i}(G)$ and we have

$$E_\infty^{*,0}\cong H^*((P^*)^G)\cong H^*(G)/(\rho_i)$$

and so if $q(t)=P.S.\,H^*(P^{*G})$, then

$$p_G(t)=q(t)\Big/\prod_{i=1}^{rkG}(1-t^{n_i})\,.$$

$(P^*)^G$ is the algebraic orbit cochain complex, and hence will also satisfy Poincaré Duality, from which $p_G(t)$ satisfies $(*)$. The construction of a geometric complex X satisfying this is more delicate, and obstructions certainly exist in the general case.

For M_{12} the existence of such a complex can be proved, [M2], by considering, besides the map $\varGamma\to M_{12}$ of 4.2, also a map $\varGamma\to G_2(\mathbb{F}_{3^\infty})$ constructed in [M2] as a consequence of the remark following (3.10). Taking plus constructions as described in Chapter IX, we obtain a fibering $B_\varGamma^+\to B_{G_2(\mathbb{F}_{3^\infty})}^+$ and the fiber is a (2-local) finite complex with the correct Poincaré series. On the other hand the (2-local) homotopy equivalence $B_\varGamma^+\to B_{M_{12}}^+$ gives the desired map on the fiber complex. We do not know if this fiber is a manifold or not though it is a (2-local) finite dimensional Poincaré duality complex.

On the other hand, there is a closely related complex which is a manifold. We now elaborate on this.

Let Y be the graph associated to the Tits Building of $L_3(\mathbb{F}_2)$ first described in Chapter V. We recall that one associates a vertex to each proper subgroup in $(\mathbb{F}_2)^3$ and an edge to any proper flag. This is a trivalent graph with a transitive $L_3(\mathbb{F}_2)$-action, having as orbit space the edge

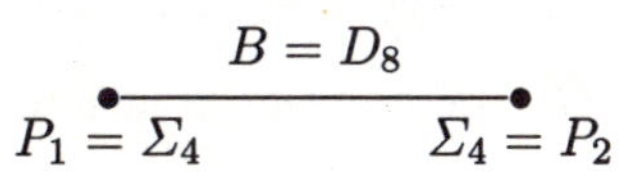

The Tits Building has the equivariant homotopy type of $A_2(L_3(\mathbb{F}_2))$, and we have

(4.4) $$\mathbb{Z}_{(2)}\oplus\mathbb{Z}_{(2)}[G/D_8]\cong\mathbb{Z}_{(2)}[G/\varSigma_4]\oplus\mathbb{Z}_{(2)}[G/\varSigma_4]\oplus P$$

where $P = H_1(Y, \mathbb{Z}_{(2)})$ is an 8-dimensional projective module, the so-called Steinberg representation.

The above also arises by considering the amalgamated product $\Gamma = \Sigma_4 \underset{D_8}{*} \Sigma_4$. The graph Y is a quotient of the cubic tree under a free normal subgroup $\Gamma' \subseteq \Gamma$, with quotient $L_3(\mathbb{F}_2)$. As for M_{12}, $H^*(\Gamma) \cong H^*(L_3(\mathbb{F}_2))$.

We consider the non-split extension E:

$$1 \to (\mathbb{Z}/2)^3 \to E \to L_3(\mathbb{F}_2) \to 1\,. \tag{4.5}$$

E is a group of order 1344 and it contains the subgroups W, W' which appear in M_{12}, realized as

$$\begin{aligned} 1 &\to (\mathbb{Z}/2)^3 \to W \to P_1 \to 1 \\ 1 &\to (\mathbb{Z}/2)^3 \to W' \to P_2 \to 1 \end{aligned} \tag{4.6}$$

and $\mathrm{Syl}_2(E) = \mathrm{Syl}_2(M_{12})$, realized as

$$1 \to (\mathbb{Z}/2)^3 \to H \to D_8 \to 1\,.$$

Denote $Q = (\mathbb{Z}/2)^3$, $G = L_3(\mathbb{F}_2)$ as before. Then

$$\begin{aligned} H^*(E) &\cong H^*(\mathrm{Hom}_E(F_*, \mathbb{F}_2)) \cong H^*(\mathrm{Hom}_Q(F_*, \mathbb{F}_2)^G) \\ &\cong H^*(G, \mathrm{Hom}_Q(F_*, \mathbb{F}_2)) \end{aligned}$$

where F_* is a free resolution of $\mathbb{Z}$ over $\mathbb{Z}E$. From this and Shapiro's formula, we deduce

$$\begin{aligned} H^*(E) \oplus H^*(H) \cong H^*(W) &\oplus H^*(W') \\ &\oplus H^*(G, \mathrm{Hom}_Q(F_*, \mathbb{F}_2) \otimes St)\,. \end{aligned}$$

Rearranging, we obtain

Theorem 4.7

$$H^*(E) \cong H^*(M_{12}) \oplus (H^*(Q) \otimes St)^{L_3(\mathbb{F}_2)}\,.$$

The group E is the normalizer of a $(\mathbb{Z}/2)^3$ in the compact Lie group G_2, which is a 14-dimensional manifold, with

$$H^*(B_{G_2}) \cong \mathbb{F}_2[d_4, d_6, d_7]\,. \qquad ([\mathrm{Bo2}])$$

Now E acts freely on G_2 and one can in fact show

Theorem 4.8

$$P.S.(H^*(G_2/E)) = P_E(t) \cdot (1 - t^4)(1 - t^6)(1 - t^7).$$

In [M2], $P_E(t)$ was determined to be

$$P_E(t) = \frac{1+t^2+3t^3+2t^4+4t^5+5t^6+4t^7+5t^8+4t^9+2t^{10}+3t^{11}+t^{12}+t^{14}}{(1-t^4)(1-t^6)(1-t^7)}.$$

The numerator represents the Poincaré series of the manifold G_2/E, and it clearly dominates our answer for $P_{M_{12}}(t)$, explaining the leading terms. As a corollary we obtain that the Poincaré series for $(H^*(Q) \otimes St)^{L_3(\mathbb{F}_2)}$ is

$$z(t) = \frac{t^4+t^5+t^6+2t^7+t^8+t^9+t^{10}}{(1-t^4)(1-t^6)(1-t^7)}.$$

Algebraically, the denominator is explained by the action of the Dickson algebra

$$\mathbb{F}_2[d_4, d_6, d_7] \cong H^*(Q)^{L_3(\mathbb{F}_2)}.$$

5. The Cohomology of Other Sporadic Simple Groups

The O'Nan Group $O'N$

The O'Nan group $O'N$ has order $460,815,505,920 = 2^9\ 3^4\ 5\ 7^3\ 11\ 19\ 31$, and in [AM3] we determine the poset space $|A_2(O'N)|/O'N$, obtaining the following picture:

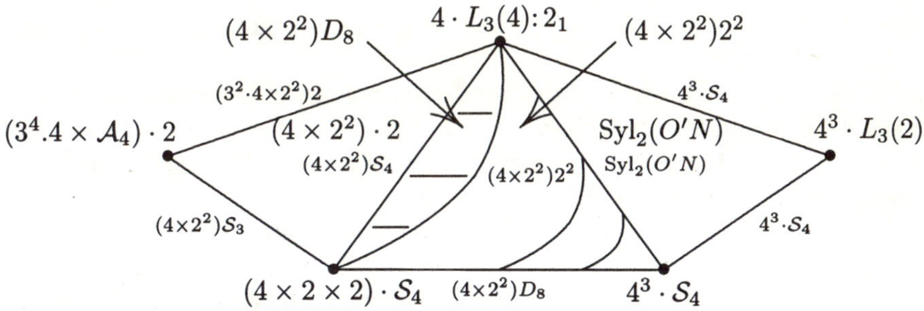

From this picture some easy cancellations give

5.1 $$H^*(O'N) \oplus H^*((\mathbb{Z}/4)^3 \cdot \Sigma_4) \cong H^*((\mathbb{Z}/4)^3 \cdot \mathrm{GL}_3(\mathbb{F}_2)) \oplus H^*(\mathbb{Z}/4 \cdot \mathrm{SL}_3(\mathbb{F}_4) \times_T \mathbb{Z}/2).$$

Our calculations show that the cohomology will be Cohen–Macaulay. Indeed, in this case the cohomology of $\mathrm{Syl}_2(O'N)$ is already Cohen–Macaulay, but is not detected by restriction to elementary 2-groups. We refer to [AM3] for complete details.

The Mathieu Group M_{22}

The third Mathieu group M_{22} of order $443,520 = 2^7\,3^2\,5\,7\,11$ is the subgroup of $\mathcal{S}_{22}$ generated by the three elements

$$
\begin{aligned}
x &= (1,19,7,9,12,11,15)(2,5,8,22,4,14,18)(6,17,21,20,10,16,13)\\
y &= (1,21,6,7,19)(3,22,11,12,4)(5,18,20,15,8)(9,13,16,10,14)\\
k &= (1,19,14,5,16,18,15,3)(2,12,20,9,13,7,17,10)(4,22)(6,21,11,8).
\end{aligned}
$$

There is a sporadic geometry described in [RSY] for M_{22} which also satisfies the hypotheses necessary for Webb's Theorem to apply. The associated complex has the following form, where $V_i = (\mathbb{Z}/2)^i$ and : denotes semi–direct product:

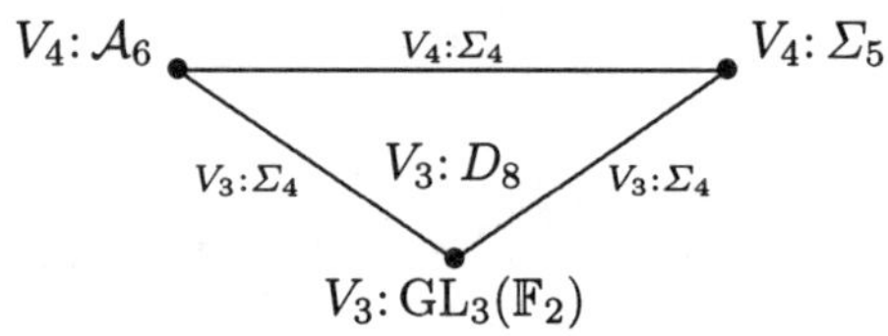

We apply V.3.3 to this picture and, after some cancellation and arguments of the type described from (4.5) on in §4, we have

5.2
$$
\begin{aligned}
H^*(M_{22}) \oplus H^*(\mathrm{Syl}_2(M_{22})) &\cong H^*(V_4\!:\Sigma_4) \oplus H^*(V_4\!:\Sigma_5)\oplus\\
&\qquad (H^*(V_3)\otimes St_1)^{\mathrm{GL}_3(\mathbb{F}_2)} \oplus (H^*(V_4)\otimes St_2)^{\mathcal{A}_6}
\end{aligned}
$$

where the Steinberg modules are those associated to $\mathrm{GL}_3(\mathbb{F}_2)$ and $\mathcal{A}_6$ respectively, thought of as alternating sums of the mod 2 homology of the poset spaces at $p = 2$.

However, it turns out that this is not the best way of determining $H^*(M_{22})$ because determining the cohomology of the components on the right hand side of 5.2 is comparable in difficulty to determining $H^*(M_{22})$ directly. Instead we first determine $H^*(Syl_2(M_{22}))$ and use detection on elementary subgroups, as in our analysis of M_{12}, to obtain enough information to understand the cohomology. There is one added complication here, though. It turns out that $H^*(Syl_2(M_{22});\mathbb{F}_2)$ is *not* entirely detected by restriction elementary abelian 2-subgroups.

A sylow subgroup of M_{22} is generated by the three elements k,

5.3
$$
\begin{aligned}
m &= (1,18)(2,12)(3,17)(5,20)(7,13)(9,15)(10,14)(16,19)\\
n &= (1,16)(2,13)(6,21)(7,18)(8,11)(12,19)(14,20)(15,17).
\end{aligned}
$$

The group structure of $\mathrm{Syl}_2(M_{22})$ has been studied in detail by Janko, [Ja]. He shows that there are exactly two subgroups isomorphic to $(\mathbb{Z}/2)^4$ contained in $\mathrm{Syl}_2(M_{22})$, E and F, with intersection $V_2 \cong (\mathbb{Z}/2)^2$, and each is normal in $\mathrm{Syl}_2(M_{22})$. A specific description of $\mathrm{Syl}_2(M_{22})$ can be given as follows. Let $\langle A,B,C,D\rangle = E$, then $\mathrm{Syl}_2(M_{22})$ has an index 2 normal subgroup, L, which is a split extension $L = E \times_T (\mathbb{Z}/2)^2$. Let α, β generate the $(\mathbb{Z}/2)^2$ then they satisfy $\alpha(A) = B$, $\alpha(B) = A$, $\alpha(C) = D$, $\alpha(D) = C$, while $\beta(A) = C$, $\beta(C) = A$, $\beta(B) = D$, $\beta(D) = B$.

$$\begin{array}{ccccc} & A & \leftrightarrow & B \\ \beta & \updownarrow & & \updownarrow \\ & C & \leftrightarrow & D \\ & & \alpha & \end{array}$$

Finally, $\mathrm{Syl}_2(M_{22})$ is a split extension $L \times_T (\mathbb{Z}/2)$, where the new generator e can be chosen so that $e(\alpha) = \beta$, $e(\beta) = \alpha$, and

$$\begin{aligned} e(A) &= BCD, \\ e(B) &= ABD, \\ e(C) &= ACD, \\ e(D) &= ABC. \end{aligned}$$

Then the group $F = \langle e, \alpha\beta, AD, BC \rangle$. Another group which is basic in $\mathrm{Syl}_2(M_{22})$ is the subgroup

$$LL = \langle \alpha, \beta, ABCD \rangle .$$

The normalizer of E in M_{22} is the semi-direct product $E \times_T \mathcal{A}_6$ and the normalizer of F is $F \times_T \mathcal{S}_5$. Finally, the remaining conjugacy class of extremal elementary 2-groups in M_{22} has LL as a representative, and the normalizer of LL in M_{22} is the split extension $LL \times_T GL_3(\mathbb{F}_2)$.

All three groups E, F, and LL are weakly closed in $\mathrm{Syl}_2(M_{22}) \subset M_{22}$. Also, in [AM1], the ring $H^*(\mathrm{Syl}_2(M_{22}))$ is determined and the image of restriction to these three subgroups can be easily read off from the results there. Thus the Cardenas-Kuhn theorem determines the image of the restriction map from $H^*(M_{22})$ in the cohomology of each of these three groups. The image in $H^*(E)$ is $H^*(E)^{\mathcal{A}_6}$, the image in $H^*(F)$ is $H^*(F)^{\mathcal{S}_5}$, and the image in $H^*(LL)$ is $\mathbb{F}_2[d_4, d_6, d_7] \cap \mathbb{F}_2[(\theta_4)^2, D_2, D_3]$. Here, if e, h, k are the one dimensional generators of $H^*(LL)$, then $D_2 = h^2 + hk + k^2$, $D_3 = h^2k + hk^2$ and $\theta_4 = e^4 + e^2 d_2 + e d_3$. This intersection is not the full Dickson algebra but

$$\mathbb{F}_2[d_4^2, d_6, d_7](1, d_4 d_6, d_4 d_7).$$

The algebra $H^*((\mathbb{Z}/2)^4; \mathbb{F}_2)^{\mathcal{A}_6}$ has been discussed at the end of Chapter III, and is the ring

$$\mathbb{F}_2[\bar{w}_3, \gamma_5, d_8, d_{12}](1, \gamma_9, b_{15}, \gamma_9 b_{15})$$

where $\bar{w}_3 = Sx_i^2 x_j$, $\gamma_5 = Sq^2(\bar{w}_3)$, $\gamma_9 = Sq^4\gamma_5$, but the $\mathcal{S}_5$ invariant subring is more complex. It is determined in [AM1], [AM4], and has the form

$$\mathbb{F}_2[\bar{w}_3, \gamma_5, d_8, d_{12}](1, n_6, n_8, \gamma_9, n_{10}, n_{12}, x_{12}, x_{14}, x_{15}, x_{16}, x_{18}, x_{24})$$

where $Sq^2(n_6) = n_8$, $Sq^4(n_6) = n_{10}$, $n_{12} = n_6^2$, $x_{12} = Sq^4(n_8)$ and $x_{14} = n_6 n_8$. Moreover, in the invariant subring

$$\begin{aligned} Sq^1(n_6) &= 0, \\ Sq^1(n_8) &= \bar{w}_3 n_6, \\ Sq^1(n_{10}) &= \bar{w}_3 n_8 + \gamma_5 n_6, \\ Sq^1(x_{12}) &= \bar{w}_3 n_{10} + \gamma_5 n_8 \\ Sq^1(x_{14}) &= \bar{w}_3 n_6^2. \end{aligned}$$

Consequently, $Sq^1(\bar{w}_3 x_{12} + \gamma_5 n_{10}) = \gamma_5^2 n_6$. The exact structure of the elements x_{15}, x_{16}, x_{18} and x_{24} are not known to us currently.

In any case, the image of $H^*(M_{22}; \mathbb{F}_2)$ in the direct sum

$$\mathbb{F}_2[a,b,c,d]^{\mathcal{A}_6} \oplus \mathbb{F}_2[a,b,c,d]^{S_5} \oplus \mathbb{F}_2[a,b,c]^{GL_3(2)}$$

will now be completely determined by specifying the "multiple image classes", i.e., those classes which must have non-trivial images in two or more of the summands.

It turns out that they are generated by $(\bar{w}_3, \bar{w}_3, 0)$, $(0, n_6, d_6)$, $(0, n_{10}, d_4 d_6)$ together with the polynomial ring $\mathbb{F}_2[d_8, d_{12}]$, where $d_8 \mapsto (d_8, d_8, d_4^2)$, $d_{12} \mapsto (d_{12}, d_{12}, d_6^2)$. In fact the above completely describes the multiple image classes when we note that $(\bar{w}_3^2, 0, 0)$, $(\gamma_5, 0, 0)$ and $(0, 0, d_7)$ are also in the restriction image. It is important to notice also that the multiple image property changes the Sq^1 operation on the elements which restrict respectively to $(0, n_6, d_6)$ and $(0, n_{10}, d_4 d_6)$, so in $H^*(M_{22})$ we have $Sq^1(n_6) = (0, 0, d_7)$ while $Sq^1(n_{10}) = (0, \bar{w}_3 n_8 + \gamma_5 n_6, d_4 d_7)$.

In summary, we can describe the non-nilpotent part of $H^*(M_{22})$ as the direct sum

$$H^*(V_4)^{\mathcal{A}_6} \oplus H^*(W_4)^{S_5} \oplus d_7 \mathbb{F}_2[d_4, d_6, d_7]$$

where the two copies of $\mathbb{F}_2[d_8, d_{12}](1, \bar{w}_3)$ in the first two rings are identified. The key technical step in this determination is to show that $(b_{15}, 0, 0)$ is in the image of the restriction map from $H^*(M_{22})$.

Finally, the radical, i.e., the interesection of the kernels of the three restriction maps above, is determined in [AM4] and shown to have the form

$$\mathbb{F}_2[d_8, d_{12}](a_2, a_7, a_{11}, a_{14})$$

where the mod (4) Bockstein $\beta_4(a_2) = \bar{w}_3$, while some higher Bockstein of a_7 is d_8, and a higher Bockstein of a_{11} is d_{12}. There are further higher Bocksteins which we do not completely understand at this time, but aside from that [AM4] gives a complete determination of $H^*(M_{22})$.

Remark. The cohomology ring structure of $H^*(M_{22})$ is given by specifying extension data in the radical. It turns out that this extension data is highly non-trivial so the exact sequence

$$\mathbb{F}_2[d_8, d_{12}(a_2, a_7, a_{11}, a_{14}) \longrightarrow H^*(M_{22}) \longrightarrow \mathcal{M}$$

where $\mathcal{M}$ is the image in $H^*(E) \oplus H^*(F) \oplus H^*(LL)$ described above does not split as rings even though it does split as $\mathbb{F}_2$-vector spaces. For example $\gamma_5(1)\gamma_5(2) = a_2 d_8$ is one of the many cup products which are non-zero in $H^*(M_{22})$ even though they are zero in $\mathcal{M}$.

Remark 5.4 The sporadic groups M_{23} and M^cL have the same Sylow 2-subgroup as M_{22} and each contains M_{22} as a subgroup. Thus the analysis above for M_{22} should apply to determine $H^*(M_{23})$ and $H^*(M^cL)$ as well and we discuss $H^*(M_{23}; \mathbb{F}_2)$ next. At this time, however, the details for M^cL have not yet been sorted out.

The Mathieu Group M_{23}

For M_{23} there is the double coset decomposition $M_{23} = M_{22} \sqcup M_{22}gM_{22}$. Here g satisfies the condition, $gM_{22}g^{-1} \cap M_{22} = M_{21} = L_3(4)$. Moreover there is an element $g \in M_{23} - M_{22}$ so $g^3 = 1$ and g normalizes $Syl_2(L_3(4))$, [Co]. Consequently, we can assume the second double coset is $M_{22}gM_{22}$ and $\theta \in H^*(M_{22})$ lies in $H^*(M_{23})$ if and only if $res^*(\theta) \in H^*(Syl_2(L_3(4)))^g$. Thus, the determination of the stable elements becomes quite straightforward. Indeed, both E and F remain non-conjugate in M_{23} but their Weyl groups become $\mathcal{A}_7$ and $(\mathbb{Z}/3 \times \mathcal{A}_5){:}\mathbb{Z}/2 \cong GL_2(\mathbb{F}_4) \times_T \mathbb{Z}/2$ respectively, while LL remains extremal. $H^*(E; \mathbb{F}_2)^{\mathcal{A}_7}$ is Cohen-Macaulay over the Dickson algebra $\mathbb{F}_2[d_8, d_{12}, d_{14}, d_{15}]$, is given at the end of Chapter III, and has Poincaré series

$$\frac{1 + x^{18} + x^{20} + x^{21} + x^{24} + x^{25} + x^{27} + x^{45}}{(1 - x^8)(1 - x^{12})(1 - x^{14})(1 - x^{15})}.$$

In particular, the multiple image classes (d_8, d_8, d_4^2), (d_{12}, d_{12}, d_6^2) remain, but the class $(\bar{w}_3, \bar{w}_3, 0)$ is no longer invariant. However, since $(0, 0, d_7)$ is still present in the restriction image, it follows that $(0, n_6, d_6)$ must still be in the image, and this accounts for all the multiple image classes. In particular, it is shown in [M5] that the Poincaré series for $H^*(F; \mathbb{F}_2)^{G_2(\mathbb{F}_4) \times_T \mathbb{Z}/2}$ is $p(x)/q(x)$ where

$$q(x) = (1 - x^{10})(1 - x^{12})(1 - x^{15})(1 - x^{24})$$

and $p(x)$ is a polynomial of degree 57 with terms of the form

$$1 + x^6 + 2x^8 + x^9 + x^{11} + 2x^{12} + x^{13} + 3x^{14} + \cdots .$$

This describes the image under restriction. The kernel of the restriction map is $\mathbb{F}_2[d_8, d_{12}](a_7, a_{11})$, and this gives a complete description of $H^*(M_{23}; \mathbb{F}_2)$. In particular note that $H^i(M_{23}; \mathbb{F}_2) = 0$ for $i \leq 5$, and $H^6(M_{23}; \mathbb{F}_2) = \mathbb{F}_2$ with restriction image $(0, n_6, d_6)$.

Using the fact that $H^*(\mathcal{S}_{23}; \mathbb{F}_2)$ is entirely detected on its 2-elementaries, we can study the restriction map

$$res^*: H^*(S_{23}; \mathbb{F}_2) \longrightarrow H^*(M_{23}; \mathbb{F}_2)$$

in dimension 6, and it is not hard to see that it is non-trivial.

Remark 5.5 The remaining Sylow p-subgroups of M_{23} and their normalizers are given as follows, $Syl_3(M_{23}) = (\mathbb{Z}/3)^2$ with Weyl group equal to the 2-Sylow subgroup of $GL_2(\mathbb{F}_3)$. The 5-Sylow subgroup is $\mathbb{Z}/5$ with Weyl group $\mathbb{Z}/4$. The 7-Sylow subgroup is $\mathbb{Z}/7$ with Weyl group $\mathbb{Z}/3$, the Weyl group for $\mathbb{Z}/11$ is $\mathbb{Z}/5$ and finally the Weyl group for $\mathbb{Z}/23$ is $\mathbb{Z}/11$. In each case it is direct to calculate the invariants, and only in the case of $\mathbb{Z}/7$ is the invariant subring less than 6-connected,

$$(\mathbb{F}_7[b] \otimes E(e))^{\mathbb{Z}/3} = \mathbb{F}_7[b^3] \otimes E(eb^2) .$$

So putting this together we have

Theorem 5.6 *$H_i(M_{23}; \mathbb{Z}) = 0$ for $i \leq 4$.*

This is the first finite group known for which this happens. Indeed, it had long been conjectured that if G is a finite group with $H_i(G; \mathbb{Z}) = 0$ for $i = 1, 2, 3$ then $G = \{1\}$.

Chapter IX. The Plus Construction and Applications

0. Preliminaries

Let G be a finite group. As we have seen, the classifying space BG has a very simple homotopy type as it is a $K(G,1)$. If G is perfect then $H_1(G;\mathbb{Z}) = 0$; suppose that we attach cells to BG to obtain a new, but simply-connected complex BG^+ with the same homology as before. Or equivalently so that the homotopy fiber of $BG \longrightarrow BG^+$ is acyclic, i.e. $H^i(\mathcal{F};\mathbb{Z}) = 0$ for all $i > 0$. The new complex will depend on G (as BG does) but the higher homotopy groups $\pi_i(BG^+)$ can be highly complicated invariants of G.

In this chapter we will describe a construction as above, due to Quillen, and known as the plus construction. In general, given a group G with a perfect normal subgroup N we obtain a homotopy fibration $X(N) \longrightarrow BG \longrightarrow BG^+$ with $\pi_1(X(N)) \cong N$, $X(N)$ acyclic. The main application of this is to afford a definition of the higher K-groups. For example, if $G = \mathrm{GL}(\mathbb{F}_q)$ and $N = E$, the subgroup generated by elementary matrices, then $\pi_i(BG^+) = K_i(\mathbb{F}_q)$ $i \geq 1$ *by definition.*

1. Definitions

We recall some notions from homotopy theory.

Definition 1.1
a. A space X is *acyclic* provided $\tilde{H}^*(X;\mathbb{Z}) = 0$.
b. A map $f\colon X_1 \longrightarrow X_2$ between path connected spaces is *acyclic* provided the homotopy fiber F_f of f is an acyclic space.

We note the following: if $f\colon X_1 \longrightarrow X_2$ is acyclic, then $f_*\colon \pi_1(X_1) \longrightarrow \pi_1(X_2)$ is an epimorphism with kernel a perfect normal subgroup.

We now provide a criterion for the acyclicity of a map. We follow the exposition given by Hausmann and Husemoller in [HH].

Proposition 1.2 *Let $f\colon X_1 \longrightarrow X_2$ be a map of connected spaces. Then f is acyclic if and only if, for any coefficient system L on X_2, the induced map*

$f_*: H_*(X_1; f^*(L)) \longrightarrow H_*(X_2; L)$ is an isomorphism, where $f^*(L)$ denotes the induced local coefficient system on X_1.

Proof. Recall that a local coefficient system L on X_2 is a module over $\pi_1(X_2)$ and $H_*(X_2; L) = H_*(C_*(\tilde{X}_2) \otimes_{\mathbb{Z}\pi_1(X_2)} L)$. Now consider the spectral sequence for the homotopy fibration $F \xrightarrow{i} X_1 \xrightarrow{f} X_2$ with

$$E^2_{p,q} = H_p(X_2; H_q(F; i^* f^* L)) \Rightarrow H_{p+q}(X_1; f^* L) \, .$$

Now $k^* f^* L$ is *trivial* on F, hence the spectral sequence only has one line if f is acyclic, and so the edge homomorphism

$$H_p(X_1; f^* L) \longrightarrow H_p(X_2; L)$$

induced by f is an isomorphism.

Conversely, assume f_* induces a homology isomorphism with any coefficent system L. In particular consider the coefficients $L = \mathbb{Z}\pi_1(X_2)$. In this case the isomorphism can be interpreted geometrically as follows. Take the free $\pi_1(X_2)$ bundle $\tilde{X}_2 \rightarrow X_2$ and pull it back to X_1 using f, i.e.

$$\begin{array}{ccc} X_3 & \xrightarrow{\tilde{f}} & \tilde{X}_2 \\ \Big\downarrow{\scriptstyle \pi'} & & \Big\downarrow{\scriptstyle \pi} \\ X_1 & \xrightarrow{f} & X_2 \end{array} \quad .$$

Then π, π' are fibrations with fiber $\pi_1(X_2)$. Comparing their respective spectral sequences, we see that f induces an isomorphism at the E^2-level by hypothesis

$$H_p(X_1; f^* \mathbb{Z}\pi_1(X_2)) \xrightarrow{f} H_p(X_2; \mathbb{Z}\pi_1(X_2))$$

and being filtration preserving, induces an isomorphism of the abutments

$$H_*(X_3) \xrightarrow{\tilde{f}_*} H_*(\tilde{X}_2) \, .$$

To prove acyclicity for f it suffices to prove it for $\tilde{f}$. From above we have that $\tilde{H}_0(F) = 0$. Now assume inductively that $\tilde{H}_j(F) = 0$, $j < n$. Consider the spectral sequence $E^2_{p,q} = H_p(\tilde{X}_2; H_q(F))$ where F is the homotopy fiber of $\tilde{f}$. Look at $E^r_{0,n}$; then, because the map from the total space to the base space is a homology isomorphism, $E^r_{0,n} = 0$ for $r > n+1$. However it can only be hit by

$$d_{n+1}\colon E^{n+1}_{n+1,0} \longrightarrow E^{n+1}_{0,n} \, ,$$

but $H_{n+1}(\tilde{X}_2)$ consists of permanent cocycles, so this is zero. We deduce that $\tilde{H}_n(F) = 0$ and so, inductively, we have proved that F is acyclic, so f is an acyclic map. □

2. Classification and Construction of Acyclic Maps

We begin this section by proving a proposition which allows us to compare acyclic maps.

Proposition 2.1 *Let $f_1: X \longrightarrow X_1$, and $f_2: X \longrightarrow X_2$ be maps between CW complexes so that f_1 is acyclic. Then there exists a map $h: X_1 \longrightarrow X_2$ with $hf_1 \simeq f_2$ if and only if $\ker(\pi_1(f_1)) \subseteq \ker(\pi_1(f_2))$ and h is unique up to homotopy. If f_2 is acyclic then h is acyclic, and h is a homotopy equivalence if and only if $\ker(\pi_1(f_1)) = \ker(\pi_1(f_2))$.*

Proof. Clearly, if h exists then $\pi_1(f_2) = \pi_1(h) \cdot \pi_1(f_1)$ so $\ker(\pi_1(f_1)) \subseteq \ker(\pi_1(f_2))$.

For the converse assume that f_1 is a cofibration and form the pushout diagram

$$\begin{array}{ccc} X & \xrightarrow{f_1} & X_1 \\ \Big\downarrow f_2 & & \Big\downarrow \phi_1 \\ X_2 & \xrightarrow{\phi_2} & X_1 \cup_X X_2 \end{array} .$$

Then $\pi_1(\phi_2): \pi_1(X_2) \longrightarrow \pi_1(X_1 \cup_X X_2) = \pi_1(X_1) *_{\pi_1(X)} \pi_1(X_2)$ and if

$$\ker(\pi_1(f_1)) \subseteq \ker(\pi_1(f_2))$$

it follows that $\pi_1(\phi_2)$ is an isomorphism. As ϕ_2 is also acyclic (which may be verified using (1.2)), we see that ϕ_2 must be a homotopy equivalence by Whitehead's theorem. Now assume χ is a homotopy inverse for ϕ_2 and let $h = \chi \cdot \phi_1$; then $h \cdot f_1 = \chi\phi_1 \cdot f_1 = \chi \cdot \phi_2 f_2 \simeq f_2$. Clearly h is uniquely determined up to homotopy. If f_2 is acyclic one can check that h must be acyclic. The rest is clear. □

We are now ready to construct acyclic maps.

Proposition 2.2 *Let X be a path-connected space and N a perfect normal subgroup of $\pi_1(X)$. Then there exists an acyclic map $f: X \longrightarrow X^+$ with $\ker(f) = N$. If X has the homotopy type of a CW-complex then so does X^+.*

Proof. We divide the argument into two steps. (I) First assume $\pi_1(X) = N$ is perfect. Let T_1 be a wedge of circles indexed by a set of generators for N and $\rho: T_1 \longrightarrow X$ a map so that $\pi_1(\rho)$ is surjective. Now form the cofiber $\gamma: X \longrightarrow X^*$ of ρ, i.e. attach a 2-cell for each circle. Clearly $\pi_1(X^*) = 1$ and the homology exact sequence yields exactness for

$$0 \longrightarrow H_q(X;\mathbb{Z}) \longrightarrow H_1(X^*;\mathbb{Z}) \longrightarrow 0 \qquad q \geq 3$$

$$0 \longrightarrow H_2(X) \longrightarrow H_2(X^*) \xrightarrow{t_1} H_1(T_1) \longrightarrow 0 .$$

Now using the Hurewicz theorem we have that $\pi_2(X^*) \cong H_2(X^*)$. Hence we can take a wedge T_2 of 2-spheres and a map $\mu: T_2 \longrightarrow X^*$ so that the composition

$$H_2(T_2) \longrightarrow H_2(X^*) \longrightarrow H_1(T_1)$$

is an isomorphism. Next, let $T_2 \longrightarrow X^* \longrightarrow X^+$ be a cofibration; as before we have exact sequences

$$0 \longrightarrow H_q(X^*) \longrightarrow H_q(X^+) \longrightarrow 0 \qquad q \geq 4,\ q = 1$$

$$0 \longrightarrow H_3(X^*) \longrightarrow H_3(X^+) \xrightarrow{t_2} H_2(T_2) \xrightarrow{\mu_*} H_2(X^*) \longrightarrow H_2(X^+) \longrightarrow 0\,.$$

Let f be the composition $X \longrightarrow X^* \longrightarrow X^+$; we claim it is a homology isomorphism. This is clear in dimensions $q \geq 4$, or $q = 1$. In dimension 2 we have that $\mu_*: H_2(T_2) \longrightarrow H_2(X^*)$ is monic by the construction. It follows that $H_3(X^*) \longrightarrow H_3(X^+)$ is an isomorphism and so is $H_3(X) \longrightarrow H_3(X^+)$. Now we have the following diagram with exact rows and columns:

$$\begin{array}{ccccccccc}
 & & & & 0 & & & & \\
 & & & & \uparrow & & & & \\
 & & & & H_1(T_1) & & & & \\
 & & & \nearrow\cong & \uparrow t_{1*} & & & & \\
0 & \longrightarrow & H_2(T_2) & \xrightarrow{\mu_*} & H_2(X^*) & \xrightarrow{r} & H_2(X^+) & \longrightarrow & 0 \\
 & & & & \uparrow s & H_2(f)\nearrow & & & \\
 & & & & H_2(X) & & & & \\
 & & & & \uparrow & & & & \\
 & & & & 0 & & & &
\end{array}$$

which is (vertically) split by $\epsilon: H_1(T_1) \longrightarrow H_2(X^*)$. Assume

$$\begin{aligned}
H_2(f)(x) = 0 \ &\Rightarrow\ r \cdot s(x) = 0 \\
&\Rightarrow\ s(x) = \mu_*(w) \\
&\Rightarrow\ 0 = t_{1*}s(x) = t_{1*}\mu_*(w) = v(w).
\end{aligned}$$

As v is an equivalence we conclude that $s(x) = w = 0$ and, as s is injective, $x = 0$ so $H_2(f)$ is injective.

Now let $z \in H_2(X^+)$, $z = r(y)$, $t_{1*}(y) \neq 0$. Then $t_{1*}(y) = v(w) = t_{1*} \cdot \mu_*(w)$, and $r(y - \mu_*(w)) = r(y) = z$ with $t_{1*}(x - \mu_*(w)) = 0$, i.e. $y - \mu_*(w) \in \operatorname{im}(s)$. Hence $H_2(f)$ is onto and we have shown that it is, in fact, an isomorphism. As X^+ is simply connected every local coefficient system on it is trivial. Hence $H_*(f)$ is an isomorphism for *for all* coefficients and therefore f is acyclic with $\ker \pi_1(f) = \pi_1(X) = N$.

(II) Now let $N \subseteq \pi_1(X)$ be a proper, perfect, normal subgroup, and denoted by $g: \tilde{X} \longrightarrow X$ the covering corresponding to N. Using (I) we construct an acyclic map $f_0: \tilde{X} \longrightarrow X_0$ with X_0 simply connected. We now change it up to homotopy into a cofibration and for the pushout diagram

$$\begin{array}{ccc} \tilde{X} & \xrightarrow{f_0} & X_0 \\ \downarrow & & \downarrow \\ X & \xrightarrow{f} & X \cup_{\tilde{X}} X_0 = X^+ \end{array} .$$

Once again, using (1.2), the fact that f_0 is an acyclic cofibration implies that f is also acyclic. Also, $\pi_1(X \cup_{\tilde{X}} X_0) = \pi_1(X)/\pi_1(\tilde{X}) \cong \pi_1(X)/N$ as X_0 is simply connected. Hence $\pi_1(f)$ is an epimorphism with kernel N, and the proof is complete. □

The previous two results can be combined to prove the following classification theorem.

Theorem 2.3 *Let X be a path-connected space with the homotopy type of a CW complex. The correspondence which assigns to an acyclic map $f: X \longrightarrow Y$ the subgroup $\ker\pi_1(f) \subseteq \pi_1(X)$ induces a bijection*

$$\left\{\begin{matrix} \textit{equivalence classes of} \\ \textit{acyclic maps on } X \end{matrix}\right\} \leftrightarrow \left\{\begin{matrix} \textit{set of normal, perfect} \\ \textit{subgroups of } \pi_1(X) \end{matrix}\right\} .$$

The space X^+ corresponding to a perfect normal subgroup $N \lhd \pi_1(X)$ is known as the *plus construction* of X with respect to N. If $N = \pi_1(X)$ it is simply called the *plus construction on* X. The importance of the construction described here will be illustrated by examples in the following section.

3. Examples and Applications

The Infinite Symmetric Group

From work of Dyer and Lashof [DL], there is an identification $Q(S^0)_0 \simeq (\Omega BC(S^0))_0$, where $C(S^0)$ denotes the monoid $\coprod_{n\geq 0} B\Sigma_n$, using the inclusion $\Sigma_n \times \Sigma_m \hookrightarrow \Sigma_{n+m}$ to give the multiplication. There is a natural map

$$B\Sigma_\infty \longrightarrow Q(S^0)_0 ,$$

which induces an isomorphism in homology. Now $\pi_1(Q(S^0)) \cong \mathbb{Z}/2 \cong \pi_1(B\Sigma_\infty)/\mathcal{A}_\infty$. Hence, taking the plus construction with respect to the maximal, normal, perfect subgroup $\mathcal{A}_\infty$ we deduce that

$$B\Sigma_\infty^+ \cong Q(S^0)_0 \ ,$$

i.e. $\pi_i(B\Sigma_\infty^+) \cong \pi_i^s(S^0)$, the ith stable homotopy group of the spheres. Now, if $G \hookrightarrow \Sigma_\infty$ is a perfect group, we have a natural map $BG^+ \longrightarrow B\Sigma_\infty^+$. In particular we have for $\pi_i(B\mathcal{A}_n^+) \xrightarrow{\beta_i^n} \pi_i(B\Sigma_\infty^+)$ that [H]

1. β_i^n is an isomorphism for $2 \le i < (n-1)/3$ or for $2 \le i < (n+1)/2$ and $n \equiv 2 \bmod (3)$.
2. When we invert 3, β_i^n is an isomorphism for $2 \le i < (n+1)/2$ except if $i = 3$ and $n = 6$.
3. β_i^n is an epimorphism with kernel isomorphic to $\mathbb{Z}/3$ if $n = 3i$ or $n = 3i+1$.

The General Linear Group Over a Finite Field

Let $\mathbb{F}_q$ denote the field with q elements and take $G = \mathrm{GL}(\mathbb{F}_q)$, $N = E$, the commutator subgroup. Then by definition

$$\pi_i(BG_+) = K_i(\mathbb{F}_q) \ ,$$

the higher K-groups of $\mathbb{F}_q$. Using the homology calculations we described for the general linear groups Quillen proved that, in fact, there is a homotopy equivalence $B\mathrm{GL}(\mathbb{F}_q)^+ \cong F\phi^q$ where $F\phi^q$ is described as the homotopy fiber of a map $BU \longrightarrow BU$.

Precisely, let ϕ^q denote the element in $[BU, BU]$ which represents the corresponding Adams operation in K-theory. Then $F\phi^q$ is the homotopy fiber corresponding to the operation $1 - \phi^q \colon BU \longrightarrow BU$. From this Quillen [Q3] calculated the higher K-groups of $\mathbb{F}_q$ as

$$K_n(\mathbb{F}_q) \ \cong \ \begin{cases} \mathbb{Z}/(q^i - 1) & n = 2i-1, \\ 0 & n = 2i. \end{cases}$$

This leads to information about $Q(S^0)$ and the stable homotopy groups of spheres as follows. Since the homotopy groups of $BGL(\mathbb{F}_p)^+$ are all finite a standard theorem in homotopy theory tells us that, up to homotopy type we have a product decomposition

$$BGL(\mathbb{F}_p)^+ \simeq \prod_{l \text{ prime}} BGL(\mathbb{F}_p)_l^+$$

where $\pi_i(BGL(\mathbb{F}_p)_l^+) = \mathrm{Syl}_l(\pi_i(BGL(\mathbb{F}_p)^+)$. It turns out that various of these $BGL(\mathbb{F}_p)_l^+$ are factors of $Q(S^0)$. In order to describe which ones we need a definition.

Definition 3.1 Given an odd prime q we say that p is *adopted* to q if $p \equiv 1 \bmod (q)$ but $p \not\equiv 1 \bmod (q^2)$.

Note that p is adopted to only a finite number of primes since p adopted to q implies that $p > q$. For example 19 is not adopted to any prime but both 7 and 13 are adopted to 3, while 11 is only adopted to 5 and 23 is only adopted to 11. It is well known that given q there is some p which is adopted to q.

We have

Theorem 3.2 *Let p_1 and p_2 be adapted to q, then the $BGL(\mathbb{F}_{p_i})_q^+$, $i = 1, 2$, are homotopy equivalent, and each is a factor of $Q(S^0)$. That is to say $Q(S^0) \cong V_q \times BGL(\mathbb{F}_{p_i})_q^+$.*

(The idea of the proof is to consider the injections

$$GL_n(\mathbb{F}_p) \xrightarrow{\text{reg}} S_{p^n} \xrightarrow{p} GL_{p^n}(\mathbb{F}_p)$$

where p is the inclusion as permutations of coordinates. From the determination in (VII.4) of $H^*(GL_n(\mathbb{F}_p); \mathbb{F}_q)$ we see that the cohomology calculation for the composition $(p \cdot reg)^*$ is determined by restricting to the maximal torus. A direct calculation then shows that the map surjects through a range which increases with n. Consequently it is an isomorphism through that range and thus a homotopy equivalence through that range when restricted to the the q piece. It follows on passing to limits that $Q(S^0)_q$ splits in the desired way.)

Remark. This process does not work to obtain a splitting at 2. The relevant space here is $BSO(\mathbb{F}_p)_2^+$, for $p \equiv 3 \bmod (8)$, and the proof of splitting is quite a bit more complex.

The Binary Icosahedral Group

Let G denote the binary icosahedral group. It is a group of order 120 which can be thought of as a double cover of $\mathcal{A}_5$. Classically it has been known to act freely on S^3 since it is a finite subgroup of the group $SU(2) \cong S^3$ which can also be thought of as the unit quaternions Sp(1) or Spin(3). Indeed, thinking of it as Spin(3) it double covers SO(3) and the conjugacy classes of finite subgroups of SO(3) are well known. In particular $\mathcal{A}_5$, the symmetry group of the icosahedron is a subgroup. Then the binary icosahedral group in Spin(3) is the inverse image of $\mathcal{A}_5$ under the double covering map. We have that $H_1(\mathcal{A}_5; \mathbb{Z}) = 0$ since $\mathcal{A}_5$ is simple. Also, we have that $H_2(\mathcal{A}_5; \mathbb{Z}) = \mathbb{Z}/2$ so it has a unique maximal central extension $\mathbb{Z}/2 \to \tilde{\mathcal{A}}_5 \to \mathcal{A}_5$ which is non-split. On the other hand it is not hard to show that the Sylow 2-subgoup of the binary icosahedral group is the quaternion group Q_8, so the extension above describes G. In particular it follows that $H_2(G; \mathbb{Z}) = H_1(G; \mathbb{Z}) = 0$. Now, by construction G acts freely on the unit sphere S^3, since it is a subgroup. Consequently the quotient manifold $S^3/G = M^3$ has $H_1(M^3; \mathbb{Z}) = H_1(G; \mathbb{Z})$, $H_2(M^3; \mathbb{Z}) = H_2(G; \mathbb{Z})$ and $H_3(M_3; \mathbb{Z}) = \mathbb{Z}$ since the action preserves orientation and the quotient is a compact oriented manifold. Thus M^3 is an example of a homology 3-sphere. It was originally discovered by Poincaré and is known as the Poincaré sphere. Taking a cellular decomposition of M^3 and lifting to the universal cover S^3 we obtain a G-free cellular decomposition of S^3 which gives us an exact sequence

$$\mathbb{Z} \xrightarrow{\partial_3} \mathcal{C}_3(S^3) \longrightarrow \mathcal{C}_2(S^3) \longrightarrow \mathcal{C}_1(S^3) \longrightarrow \mathcal{C}_0(S^3) \xrightarrow{\epsilon} \mathbb{Z}\,,$$

where the $\mathcal{C}_i(S^3)$ are finitely generated free $\mathbb{Z}(G)$ modules. When we paste copies of this exact sequence together using $\partial_3 \cdot \epsilon$ we obtain a long exact resolution of $\mathbb{Z}$ over $\mathbb{Z}(G)$. From this (since the map $H_3(S^3) \to H_3(M^3)$ is just multiplication by $deg(G) = 120$), it follows that

$$H_j(G;\mathbb{Z}) = \begin{cases} \mathbb{Z} & j = 0, \\ \mathbb{Z}/(120) & j \equiv 3 \bmod (4), \\ 0 & \text{otherwise.} \end{cases}$$

Now consider the classifying map $M^3 \xrightarrow{f} BG$ and the induced map on plus constructions

$$\begin{array}{ccccc} S^3 & \longrightarrow & M^3 & \longrightarrow & BG \\ \Big\downarrow \chi & & \Big\downarrow & & \Big\downarrow \\ F & \xrightarrow{j} & M^+ & \longrightarrow & BG^+ \end{array} .$$

As the maps between the plus constructions induce homology isomorphisms and G acts trivially on $H_*(S^3;\mathbb{Z})$, we can apply the comparison theorem to conclude that χ is a homology isomorphism. Since $H_i(G;\mathbb{Z}) = 0$, $i = 1, 2$, we see that BG^+ is 2-connected, hence F is simply connected and χ is a homotopy equivalence. Clearly $M^+ \simeq S^3$ and it follows that

$$\Omega(BG^+) \simeq Fiber(j) \simeq F_{120}\,,$$

where F_{120} is the homotopy fiber of the map of degree 120 from S^3 to itself. Thus we have an exact sequence

$$\xrightarrow{\partial} \pi_i(F_{120}) \longrightarrow \pi_i(S^3) \xrightarrow{\times 120} \pi_i(S^3) \xrightarrow{\partial} \pi_{i-1}(F_{120}) \longrightarrow \cdots .$$

It follows that there is an exact sequence

$$0 \longrightarrow \pi_i(S^3)/(120 \cdot \pi_i(S^3)) \longrightarrow \pi_i(BG^+) \longrightarrow \pi_{i-1}(S^3)_{120} \longrightarrow 0$$

for each $i \geq 2$, while $\pi_1(BG^+) = 0$. Here $\pi_i(S^3)_{120}$ denotes the subgroup of $\pi_i(S^3)$ consisting of elements whose order divides 120. This analysis is due to J.C. Hausmann [H]. In fact Hausmann has proved that if H is a perfect group with $H_2(H;\mathbb{Z}) = 0$ then $\pi_n(BG^+)$ for $n \geq 5$ is in one to one correspondence with the set of topological homology spheres with fundamental group H up to an appropriate notion of cobordism.

Remark. From the work of P. Selick, [Sel], F.R. Cohen, J.C. Moore, and J.A. Neisendorfer, [CMN], I.M. James, [Jam], and F.R. Cohen, [Coh], it follows that multiplication by the integer 120 on the 2, 3, and 5 torsion of $\pi_*(S^3)$ is trivial. Thus the homotopy groups of BG^+ split into two copies of $\pi_*(S^3)_{120}$ with a dimension shift. (We thank one of the referees for pointing this out to us.)

The Mathieu Group M_{12}

We have the map from the amalgamated product $W *_H W'$, to M_{12} given in VIII.4.1, which, from VIII.4.2 induces isomorphisms in mod (2) homology. Taking plus constructions gives us a map

$$B^+_{W*_H W'} \longrightarrow B^+_{M_{12}}$$

which is now a 2-local homotopy equivalence. On the other hand, from [FM], [M2], there is also a homomorphism $W *_H W' \rightarrow G_2(q)$ – which is injective on H and an isomorphism to $Syl_2(G_2(q))$ if $q \cong 3, 5 \bmod (8)$ – for any $q \not\cong 0 \bmod (2)$, and these homomorphisms fit together to give a homomorphism $W *_H W' \rightarrow G_2(p^\infty)$. For $p \cong 3, 5 \bmod (8)$ this gives a map

$$e_p \colon B^+_{W*_H W'} \longrightarrow B^+_{G_2(p^\infty)} .$$

VII.7.6 shows that $H^*(B^+_{G_2(p^\infty)}; \mathbb{F}_2) \cong \mathbb{F}_2[d_4, d_6, d_7]$ and that

$$e_p^* \colon \mathbb{F}_2[d_4, d_6, d_7] \rightarrow H^*(M_{12}; \mathbb{F}_2)$$

is an injection onto the same subalgebra in $H^*(M_{12}; \mathbb{F}_2)$, (the subalgebra which restricts to the Dickson algebra in each of the three conjugacy classes of $(\mathbb{Z}/2)^3$'s). Consequently, when we pass to the homotopy fibration, the fiber at the prime 2 is 14 dimensional with Poincaré series the numerator in the Poincaré series for $H^*(M_{12}; \mathbb{F}_2)$,

$$1 + t^2 + 3t^3 + t^4 + 3t^5 + 4t^6 + 2t^7 + 4t^8 + 3t^9 + t^{10} + 3t^{11} + t^{12} + t^{14} .$$

It would be very interesting to have a good geometric realization of this fiber. In particular, if it were the homotopy type of a closed parallelizable manifold of dimension 14, this would be very useful.

The Group J_1

There is also a close connection between $B^+_{J_1}$ and B_{G_2}. Here, if

$$LJ = (\mathbb{Z}/2)^3 \colon (\mathbb{Z}/7 \times_T \mathbb{Z}/3)$$

with the action induced by regarding $(\mathbb{Z}/2)^3$ as the additive subgroup of the field $\mathbb{F}_8$, then $LJ \subset E \subset G_2$ where E is the non-split extension $2^3 \cdot L_3(2)$ discussed in VII.7 (see VII.7.6 in particular).

If we could pass to plus constructions (with $\pi_1(B)$ equal to $\mathbb{Z}/3$), there would be a 2-equivalence

$$\pi_J \colon B^+_{(\mathbb{Z}/2)^3 \times_T (\mathbb{Z}/7 \times_T \mathbb{Z}/3)} \longrightarrow B^+_{J_1}$$

induced from the inclusion of the left hand subgroup as the normalizer of $Syl_2(J_1)$ in J_1. Of course, since $(\mathbb{Z}/2)^2 \times_T \mathbb{Z}/7$ is not perfect there are difficulties with this step, but what we can do is to kill the fundamental group by adding a two dimensional cell to kill the element of order three. A direct check shows that the resulting space has $H_2(B_{(\mathbb{Z}/2)^3 \times_T (\mathbb{Z}/7 \times_T \mathbb{Z}/3)} \cup e^2; \mathbb{Z}) = \mathbb{Z}$ and this is the second homotopy group, so we can kill this $\mathbb{Z}$ by adding a single three cell and we have a space with homology unchanged at 2 and 7, but the $\mathbb{Z}/3$ in dimension 1 is gone.

The inclusion of $(\mathbb{Z}/2)^3 \times_T (\mathbb{Z}/7 \times_T \mathbb{Z}/3)$ into G_2 induces a map of classifying spaces

$$e_J \colon B_{(\mathbb{Z}/2)^3 \times_T (\mathbb{Z}/7 \times_T \mathbb{Z}/3)} \longrightarrow B_{G_2}$$

which injects $H^*(B_{G_2}; \mathbb{F}_2)$ as the Dickson algebra in

$$H^*(B_{(\mathbb{Z}/2)^3 \times_T (\mathbb{Z}/7 \times_T \mathbb{Z}/3)}; \mathbb{F}_2).$$

Since B_{G_2} is three connected e_J lifts uniquely to a map of the space above with a two and a three cell added. Similarly, since J_1 is simple and the multiplier is $\{1\}$, [Co], so $H_2(J_1; \mathbb{Z}) = 0$ we can lift π_J to the space with cells attached and this lifted map is an equivalence at the prime 2. Consequently, at 2 we can identify the two spaces and we have a map

$$e_J \circ \pi_J^{-1} \colon (B^+_{J_1})_2 \longrightarrow (B_{G_2})_2,$$

and the fiber in this composition, at 2, and in cohomology looks like the fiber of e_J which is just the fourteen dimensional closed compact manifold $G_2 / \left((\mathbb{Z}/2)^3 \times_T (\mathbb{Z}/7 \times_T \mathbb{Z}/3)\right)$. This manifold, from VIII.2, has cohomology ring $H^*(J_1; \mathbb{F}_2)/(d_4, d_6, d_7)$ which is isomorphic to $H^*(G_2; \mathbb{F}_2)$ as a module of the Steenrod algebra $\mathcal{A}(2)$.

The lift of π_J restricted to this fiber induces a surjection in cohomology which explains the numerator in the Poincaré series for $H^*(J_1; \mathbb{F}_2)$ given in VIII.2.1. Also, John Harper has shown that any simply connected CW complex with the mod(2) cohomology of G_2 as a module over $\mathcal{A}(2)$ must, in fact be homotopic (at 2) to G_2. Thus, $B^+_{J_1}$, at 2, fibers over B_{G_2} with fiber having the homotopy type of G_2. This fibering is exotic, and we thank F. Cohen for describing it to us.

The Mathieu Group M_{23}

The embedding $M_{23} \subset \mathcal{S}_{23}$ given by letting it act as permutations on the 23 cosets of M_{22} is quite explicit. In particular, we have the commutative diagram

$$\begin{array}{ccc} M_{22} & \hookrightarrow & M_{23} \\ \downarrow & & \downarrow \\ \mathcal{S}_{22} & \hookrightarrow & \mathcal{S}_{23} \end{array}$$

where the embedding given by the left hand vertical arrow is given in VIII.5. Since $H^*(\mathcal{S}_n; \mathbb{F}_2)$ is determined by restriction to 2-elementaries, we can gain a great deal of information about the map on classifying spaces by restricting to the 2-elementaries E and F. Using the particular Sylow subgroup given in VIII.5.3 we find that E has four generators:

$$\begin{array}{cccc} (1,17)(3,18) & (2,15)(9,12) & (5,19)(16,20) & (7,10)(13,14) \\ (1,18)(3,17) & (2,12)(9,15) & (5,20)(16,19) & (7,13)(10,14) \end{array}$$

together with

$$(1,2)(3,9)(15,17)(12,18) \qquad (5,10(7,19)(13,16)(14,20)$$
$$(1,16)(17,20(3,5)(18,19)(2,13)(7,12)(9,10)(14,15).$$

These last two elements generate the intersection V_2 of E and F, and E is clearly contained in $(K \wr \mathbb{Z}/2) \wr \mathbb{Z}/2$ where $K \subset \mathcal{S}_4$ is the Klein group. The group F is generated by the last two elements together with two others, where the elements can now be written as follows.

$$\begin{array}{lllll} (1,13)(2,16) & (6,11)(8,21) & (3,5)(9,10) & (14,20)(15,17) & \\ (1,16)(2,13) & 6,21)(8,11) & & (14,20)(15,17) & (7,18)(12,19) \\ (1,2)(13,16) & & (3,9)(5,10) & (14,20)(15,17) & (7,19)(12,18) \\ (1,16)(2,13) & & (3,5)(9,10) & (14,15)(17,20) & (18,19)(7,12) \end{array}$$

and we see that $F \subset K \times K \times K \times K \times K$.

From the results of VI.1, VI.2, we see that the symmetric sum $Sd_3 \otimes d_3 \otimes 1 \otimes 1 \otimes 1$ in $H^*(K^5; \mathbb{F}_2)$ is in the image of restriction from $H^*(\mathcal{S}_{22}; \mathbb{F}_2)$, and it is a direct calculation to check that the image of this class under the map $H^*(K^5; \mathbb{F}_2) \longrightarrow H^*(F; \mathbb{F}_2)$ described above is non-zero. Consequently, using the splitting of 3.2 and the following remark, we can project $B^+_{M_{23}}$ to $V_2 = coker(J)$, which, from our knowledge of the stable homotopy of spheres, we know is 5-connected with $\pi_6(coker(J)) = \mathbb{Z}/2$, and it must be the case that the induced map $\pi_6(B^+_{M_{23}}) \longrightarrow \pi_6(coker(J))$ is an isomorphism.

4. The Kan–Thurston Theorem

From the results outlined in the previous section we can deduce that there are certain interesting topological spaces which have the homology of a $K(\pi, 1)$. Among them are $\Omega^\infty \Sigma^\infty$, $F\Psi^q$, and certain spaces of homeomorphisms we have not discussed. This very naturally leads to the question of whether or not this is true for a large class of spaces.

This was settled in the affirmative by Kan and Thurston [KT]; in fact they proved that *every path connected space has the homology of a* $K(\pi, 1)$. In this section we will outline a proof of their result (due to Maunder [Mau]), which can be stated more precisely as follows:

Theorem 4.1 *For every path–connected space* X *with basepoint there exists a space* TX*, and a map*

$$TX \xrightarrow{tX} X$$

which is natural with respect to X *and has the following properties:*

1. *the map* tX *induces an isomorphism on (singular) homology and cohomology*

$$H_*(TX, A) \cong H_*(X, A), \qquad H^*(X, A) \cong H^*(TX, A)$$

   for every local coefficient system A *on* X*,*
2. $\pi_i(TX)$ *is trivial for* $i \neq 1$ *and* $\pi_1 tX$ *is onto,*
3. *the homotopy type of* X *is completely determined by the pair of groups* $G_X = \pi_1 TX$*, and* $P_X = \ker \pi_1 tX$ *(note* $P_X \subset G_X$ *perfect). In fact, we have that* $X \simeq K(G_X, 1)^+$*, where the plus construction is taken with respect to* P_X*.*

To prove this result, we will need (given any group G) to construct an *acyclic* group CG into which the original group embeds. To do this we will use ideas due to Baumslag, Dyer and Heller [BDH]. First we need

Definition 4.2 A supergroup M of a group B is called a *mitosis* of B if there exist elements s, d in M such that

1. $M = \langle B, s, d \rangle$
2. $b^d = bb^s$ for all $b \in B$, and
3. $[b', b^s] = 1$ for all $b, b' \in B$.

Definition 4.3 A group M is *mitotic* if it contains a mitosis of every one of its finitely generated subgroups.

Our goal will be to show that every mitotic group is acyclic, and that every group embeds in a mitotic group. We introduce some notation. Let $\kappa: B \times B \to M$ be the homomorphism $\kappa(b', b) = b'b^s$ and $\lambda: B \to B \times B$ defined by $\lambda(b) = (b, 1)$. Then if $\mu: B \to M$ is the injection, clearly $\mu = \kappa\lambda$.

Lemma 4.4 *Let* $\phi: A \to B$ *be a homomorphism,* M *a mitosis of* B*, and* $\mu: B \to M$ *the given injection. Let* $\mathbb{F}$ *be any field such that* $\phi_*: H_i(A, \mathbb{F}) \to H_i(B, \mathbb{F})$ *is 0 for* $i = 1, 2, \ldots, n-1$*. Then* $(\mu\phi)_*: H_i(A, \mathbb{F}) \to H_i(M, \mathbb{F})$ *is 0 for* $i = 1, 2, \ldots, n$.

Proof. Clearly we only need to verify the claim for $i = n$. By the Kunneth formula,

$$H_n(B \times B, \mathbb{F}) \cong \sum_{i+j=n} H_i(B, \mathbb{F}) \otimes H_j(B, \mathbb{F}) .$$

Now let $\lambda': A \to A \times A$, $\lambda'(a) = (a, 1)$ and $\rho': A \to A \times A$, $\rho'(a) = (1, a)$. Then we have that $\mu\phi = \kappa(\phi \times \phi)\lambda'$, hence if $\alpha \in H_n(A, \mathbb{F})$,

$$(\mu\phi)_*(\alpha) = \kappa_*(\phi_* \otimes 1). \qquad (a)$$

We also have that $c_d\mu\phi = \kappa(\phi \times \phi)\Delta_A$, where c_d is conjugation by d. Using the fact that inner automorphisms are trivial in homology, we obtain

$$(\mu\phi)_*(\alpha) = \kappa_*(\phi_*\alpha \otimes 1) + \kappa_*(1 \otimes \phi_*\alpha). \qquad (b)$$

Similarly, $c_s\mu\phi = \kappa(\phi \times \phi)\rho'$ and hence

$$(\mu\phi)_*(\alpha) = \kappa_*(1 \otimes \phi_*\alpha). \qquad (c)$$

Combining (a), (b) and (c) yields $(\mu\phi)_*\alpha = 0$, proving the lemma. □

We now prove

Theorem 4.5 *Mitotic groups are acyclic.*

Proof. Assume that G is mitotic. If $K \subset G$ is a finitely generated subgroup, then $K = K_0 \subset K_1 \subset K_2 \subset \ldots \subset G$, where each injection $K_i \subset K_{i+1}$ is a mitosis. Now note that given any mitosis $B \to M$, the induced map $H_1(B, \mathbb{Z}) \to H_1(M, \mathbb{Z})$ is zero. Hence we obtain, using (4.4), that $H_i(K, \mathbb{F}) \to H_i(K_n, \mathbb{F})$ is zero for any field $\mathbb{F}$, and $i = 1, \ldots, n-1$. From this we deduce that $H_i(K, \mathbb{F}) \to H_i(G, \mathbb{F})$ is zero for all $i > 0$. Now G is the directed colimit of its finitely generated subgroups, and the colimit of their inclusion maps is the identity 1_G. Homology commutes with directed colimits, from which we deduce that $H_i(G, \mathbb{F}) = 0$ for any field, $i > 0$ and hence G is acyclic. □

We introduce the notion of algebraically closed groups.

Definition 4.6 A group G is said to be *algebraically closed* if every finite set of equations

$$h_i(g_1, \ldots, g_n, x_1, \ldots, x_m) = 1, \;\; i = 1, \ldots, k ,$$

in the variables $x_1, \ldots, x_m$ and constants $g_1, \ldots, g_n \in G$ which has a solution in some supergroup of G, already has a solution in G.

Theorem 4.7 *Algebraically closed groups are mitotic.*

Proof. Suppose that G is algebraically closed and denote $A = \langle g_1, \ldots, g_n \rangle$ a finitely generated subgroup of G. Let $D = G \times G$ and let $\widehat{G} = G \times 1$,

$H = \Delta(G)$, $K = 1 \times G$ be the corresponding copies of G embedded in D. We construct the extensions

$$E = \langle D, t; t^{-1}(g,1)t = (g,g),\ g \in G \rangle$$

$$m(G) = \langle E, u; u^{-1}(g,1)u = (1,g),\ g \in G \rangle\ .$$

Then G embeds as $\widehat{G}$ in $m(G)$, and the finitely many equations

$$g_i^{x_i}(g_i g_i^{x_2})^{-1} = 1,\ \ [g_i, g_j^{x_2}] = 1$$

with $i, j = 1, \ldots, n$ have a solution $x_1 = t, x_2 = u$ in $m(G)$. Thus they have a solution $x_1 = d$, $x_2 = s$ in G itself. As a consequence of this the group $\langle A, d, s \rangle$ is a mitosis of A in G, and so G is mitotic. □

Using the fact that any infinite group embeds in an algebraically closed group of the same cardinality, we obtain

Theorem 4.8 *Every infinite group can be embedded in an acyclic group of the same cardinality.*

We have therefore proved that given *any* group G there exists a group CG containing G such that CG is acyclic.

We will now give the proof of Theorem 4.1, following Maunder.

Proof. The first step is to prove the existence of TX satisfying (i) and (ii) when $X = L$, a connected simplicial complex with ordered vertices.

We proceed inductively: suppose that for each such L with at most $N-1$ simplexes, $t\colon TL \to L$ has been constructed satisfying (i) and (ii) and that this construction is natural for simplicial maps of L that are strictly order–preserving on each simplex. Assume also that, for each connected subcomplex $M \subset L$, $TM = t^{-1}M$, and that $\pi_1(TM) \to \pi_1(TL)$ is 1–1. Note that because every connected 1–dimensional complex is a $K(\pi, 1)$, we may start the induction by taking t to be the identity.

Let K be obtained from L by attaching an n–simplex $(n \geq 2)$ σ to $\partial\sigma \subset L$. Then $T(\partial\sigma) \subset T(L)$ and if $f\colon \sigma \to \Delta^n$ is the (unique) order–preserving simplicial homeomorphism to the standard n–simplex, the corresponding map $Tf\colon T(\partial\sigma) \to T(\partial\Delta^n)$ is a homeomorphism, and $T(\partial\Delta^n)$ is a $K(\pi,1)$. Now let $g\colon T(\partial\Delta^n) \to K(C\pi, 1)$ be a map realizing the embedding $\pi \hookrightarrow C\pi$, $C\pi$ acyclic. We take the mapping cylinder of the composition $gT(f)\colon T(\partial\sigma) \to K(C\pi, 1)$, and attach it to $T(L)$ along $T(\partial\sigma) \subset TL$; this will be TK. To extend t to $TK \to K$, we do it as usual on mapping cylinder coordinates (x, t) and by mapping $K(C\pi, 1)$ to the barycenter $\hat{\sigma}$ of σ.

The construction can be verified to be natural for simplicial maps that are strictly order–preserving on each simplex. Using the Mayer–Vietoris sequences for K, TK and the 5–lemma, it follows that $t\colon TK \to K$ induces isomorphisms of homology and cohomology for any coefficient system. Now note that

$$\pi_1(TK) \cong \pi_1(TL) *_\pi C\pi$$

and so the inclusions of TL, $K(C\pi, 1)$, $T(\partial\sigma)$ in TK induce monomorphisms of π_1—hence $\widetilde{T}K$ (the universal cover) contains multiple copies of the acyclic universal covers of all three. Using a lifted Mayer–Viaetoris sequence, this implies $\widetilde{T}K$ is acyclic, hence that TK is aspherical. Also note that as

$$\pi_1(K) = \pi_1(L) *_{\pi_1(\partial\sigma)} \pi_1(\sigma) ,$$

the map $t_*: \pi_1(TK) \to \pi_1(K)$ is onto.

Using induction on N, one can construct TK for all finite (ordered) simplicial complexes K. This can be extended to infinite simplicial complexes by taking the direct limit over finite subcomplexes.

Now if X is a path–connected space, let $\mathcal{S}X$ be its singular complex, $|\mathcal{S}X|$ its geometric realization. Denote by $|\mathcal{S}X|''$ the second derived complex (considered as a Δ-set). Then we can take $TX = T(|\mathcal{S}X|'')$. This will be a natural construction satisfying the desired properties, as a continuous map of X gives rise to a simplicial map of $|\mathcal{S}X|''$ that is strictly order–preserving on each simplex. Then tX is the map

$$TX = T(|\mathcal{S}X|'') \to |\mathcal{S}X|'' \simeq |\mathcal{S}X| \simeq X$$

Part (iii) follows from (i) and (ii) and the results in §1. □

Remark. One can in fact prove that if K is a finite connected simplicial complex, TK may be taken to be finite and of the same dimension as K (see [Mau]).

Chapter X. The Schur Subgroup of the Brauer Group

0. Introduction

In this final chapter we apply the techniques of group cohomology to the representation theory of finite groups. Given G a finite group we know that $\mathbb{F}(G)$ is semi-simple for any field of characteristic zero. Consequently, from the Wedderburn theorems there is a decomposition

$$\mathbb{F}(G) = \sum M_{n_i}(D_i) \tag{0.1}$$

where the D_i run over central simple division algebras with center $\mathbb{K}_i$ a finite cyclotomic extension of $\mathbb{F}$. The question that we answer here is the determination of all the classes $\{D_i\} \in B(\mathbb{F})$ which arise in this way, that is to say, which division algebras occur in the simple components of the group ring of a finite group.

When G is a finite group all the centers, $Z(D_i)$, in the semi-simple expansion of $\mathbb{F}(G)$ are cyclotomic extensions of $\mathbb{F}$, i.e., subfields of $\mathbb{F}(\zeta_m)$, and m divides the order of G. Indeed, if $\mathbb{K} = \mathbb{Q}(\zeta_{|G|})$, then $\mathbb{K}(G)$ is a direct sum of matrix algebras over $\mathbb{K}$. Conversely, if a sufficient number of roots of unity are not present, then $\mathbb{K}(G)$ cannot split in this fashion, since it is already not going to be true for the cyclic subgroups of G. For this reason, when studying the division algebras which occur in (0.1), it is sufficient to assume that the field $\mathbb{F}$ is cyclotomic.

The division algebras which we discuss from here on are all assumed to be of finite dimension over their centers. The content of this chapter is unpublished work of Milgram in the mid 1970's. The question of identifying the possible division algebras which arise in (0.1) was first raised by Fields and discussed in his joint paper with I. N. Herstein, [FH]. Later, M. Benard, [Br], Benard and M. Schacher, [BeS], and especially G. Janusz, [J], and T. Yamada, [Y], did important work on the question.

From our point of view the question becomes the determination of explicit maps of cohomology groups with twisted coefficients under restriction and change of coefficients. Hence we feel it is a suitable example with which to conclude this book.

1. The Brauer Groups of Complete Local Fields

Valuations and Completions

Definition 1.1 A *non-archimedean valuation* on a division algebra D is a map $\varphi: \mathbb{F} \longrightarrow \mathbb{R}_+$ where $\mathbb{R}_+$ is the non-negative real numbers satisfying the following three conditions.

(1) $\varphi(a) = 0$ if and only if $a = 0$.
(2) $\varphi(ab) = \varphi(a)\varphi(b)$.
(3) $\varphi(a+b) \leq \text{Max}(\varphi(a), \varphi(b))$.

A non-archimedean valuation trivially satisfies the triangle inequality. The valuation is discrete if the *value group* $\{\varphi(a) \mid a \in \mathbb{F}, a \neq 0\}$ is an infinite cyclic group. We also assume there is some $a \in \mathbb{F}$ with $\varphi(a) \neq 0$ to avoid trivial cases.

Example. The standard example is the p-adic valuation on the rationals. Let $n/m = p^{\alpha}w$ with $\alpha \in \mathbb{Z}$ and $w = \frac{n'}{m'}$ where both m' and n' are prime to p. Then $\varphi_p(n/m) = p^{\alpha}$. We will discuss examples of valuations on non-commutative division algebras later.

Example. Let $\mathcal{D}$ be a Dedekind domain (an integral domain in which every ideal is uniquely a product of prime ideals), and $\mathcal{P} \subset \mathcal{D}$ a prime ideal. Let $\mathbb{Q}(\mathcal{D})$ be the quotient field, and suppose that $m/n \in \mathbb{Q}(\mathcal{D})$. Then, if $(m) = \prod \mathcal{P}_i^{j_i}$, $(n) = \prod \mathcal{P}_k^{j_k}$ we have

$$(m/n) = \prod \mathcal{P}_i^{j_i} \prod \mathcal{P}_k^{-j_k} = \mathcal{P}^{\alpha}\mathcal{B}$$

where $\mathcal{B}$ is a product of powers of primes distinct from $\mathcal{P}$. Then the $\mathcal{P}$-adic valuation on $\mathbb{Q}(\mathcal{D})$ is given by $\varphi_{\mathcal{P}}(m/n) = e^{\alpha}$ for some $0 < e < 1$. In the case where $\mathcal{D}/\mathcal{P}$ is a finite field, e is usually taken to be $1/|\mathcal{D}/\mathcal{P}|$.

The valuation φ gives rise to a topology on D by defining a basis for the open neighborhoods of 0 as the inverse images $N_{\epsilon}(0) = \varphi^{-1}((0, \epsilon))$, and a basic set of open neighborhoods of a are given as $a + N_{\epsilon}(0)$. Two valuations are *equivalent* if and only if they give rise to the same topology on D. Thus setting $\phi_{p,a}(n/m) = a^{\alpha}$ for $0 < a < 1$ and $\alpha(n/m)$ as above, gives an equivalent valuation.

Remark 1.2 Two valuations φ and φ' are equivalent if and only if there is an $e \in \mathbb{R}_+$ so that $\varphi'(d) = \varphi(d)^e$ for all $d \in D$. See e.g., [P], p. 321.

The *valuation ring* $\mathcal{O}_\varphi \subset D$ is the subset of D consisting of all those $a \in D$ with $\varphi(a) \leq 1$. That $\mathcal{O}_\varphi$ is, in fact, a subring follows from conditions (2) and (3) in the definition, the first showing that it is closed under products, and the second showing that it is closed under sums. There is a maximal ideal $\mathcal{P}_\varphi \subset \mathcal{O}_\varphi$ defined as those $a \in D$ with $\varphi(a) < 1$. Thus, in the case of φ_p, we have that $\mathcal{O}_\varphi$ consists of those fractions in $\mathbb{Q}$ of the form $p^\alpha w$ with w as above, $\alpha \geq 0$ and $\mathcal{P}_\varphi$ consists of those fractions with $\alpha > 0$. In the case where D is a field, $\mathbb{F}$, the quotient $\mathcal{O}_\varphi/\mathcal{P}_\varphi = \mathbb{F}_\varphi$ is a field, called *the residue class field of the pair* $(\mathbb{F}, \varphi)$.

Lemma 1.3 *Let φ be a valuation on the division algebra D. Let $\mathcal{O}_\varphi$ be the valuation ring and $\mathcal{P}_\varphi$ be the maximal ideal, then the quotient $\mathcal{O}_\varphi/\mathcal{P}_\varphi$ is a division algebra.*

Proof. If $a \in \mathcal{O}_\varphi - \mathcal{P}_\varphi$ then $\varphi(a) = 1$. Consequently, $\varphi(a^{-1}) = 1$ and $a^{-1} \in \mathcal{O}_\varphi - \mathcal{P}_\varphi$. Likewise, if a is a unit of $\mathcal{O}_\varphi$ then $\varphi(a) = 1$ and $a \in \mathcal{O}_\varphi - \mathcal{P}_\varphi$. Hence $\mathcal{P}_\varphi$ is a maximal ideal in $\mathcal{O}_\varphi$ and the quotient is a division algebra. □

Definition. Let φ be a non-archimedean valuation on the division algebra D. Then the *completion* of D with respect to φ, $\hat{D}_\varphi$, is the completion of D with respect to the topology associated to φ.

Recall that the completion adds equivalence classes of Cauchy sequences to D, and a Cauchy sequence is a sequence of elements of D so that for every $\epsilon > 0$ there is an $N(\epsilon) \in \mathbb{N}$ and $\varphi(a_i - a_j) < \epsilon$ for $i, j > N(\epsilon)$. It is direct to check that $\hat{D}_\varphi$ is again a division algebra and φ extends to to a discrete valuation on $\hat{D}_\varphi$.

Remark 1.4 If $a_1, \ldots, a_i, \ldots$ are a complete set of representatives for the cosets of $\mathcal{P}_\varphi \subset \mathcal{O}_\varphi$, with $a_1 = 0$ representing the 0-coset, $a_2 = 1$ representing the coset of 1, then every element in the completion can be represented by a unique power series

$$s(a) = \sum a_{k_i} \pi^i$$

where $\pi \in \mathcal{P}_\varphi$ is any element which satisfies the condition $\varphi(\pi) \geq \varphi(\lambda)$ for all $\lambda \in \mathcal{P}_\varphi$. Such a π is called a *uniformizing parameter* for $\hat{D}_\varphi$.

Example. The ring $\hat{\mathbb{Z}}_p$. For the valuation φ_p on $\mathbb{Q}$ we consider the completion restricted to the valuation ring $\mathcal{O}_\varphi$. Write u for an element with $\varphi_p(u) = 1$. These are, as we have seen in the proof of (1.3), exactly the units of $\mathcal{O}_\varphi$. Then an element in the completion can be written as a formal power series

$$\hat{a} = \sum_{i=0}^{\infty} u_i p^i$$

where $0 \leq u_i < p$.

Definition. A discrete non-archimedean valuation on D is called *finite* if the quotient $\hat{\mathcal{O}}_p/\hat{\mathcal{P}}_p = \mathbf{D}_\varphi$ is finite.

If $\mathbb{K}$ is a finite extension of $\mathbb{Q}$ and φ is any non-archimedean valuation on $\mathbb{K}$ induced from a prime ideal of the ring of integers $\mathcal{O}_\mathbb{K}$ then the valuation is finite.

Remark 1.5

1. Let p be an odd prime. Then $\hat{\mathbb{Q}}_p^\bullet \cong \mathbb{Z} \times \hat{\mathbb{Z}}_p^+ \times (\mathbb{Z}/(p-1))$, where the $\mathbb{Z}$ is generated by the elements p^i, $i \in \mathbb{Z}$, the copy of the additive completion $\hat{\mathbb{Z}}_p^+$ is identified with the units with leading term $a_0 = 1$, and the $(p-1)^{st}$ roots of unity represent the units with leading terms $\neq 1$.
2. Let $p = 2$. Then $\hat{\mathbb{Q}}_2^\bullet \cong \mathbb{Z} \times \hat{\mathbb{Z}}_2^+ \times \mathbb{Z}/2$ where the $\mathbb{Z}$ and $\hat{\mathbb{Z}}_2^+$ are as above, but $\mathbb{Z}/2$ consists of the elements ± 1.
3. Let $\mathcal{O}$ be the ring of integers in a degree n extension, $\mathbb{K}$ of $\mathbb{Q}$, and suppose $\mathcal{P}$ is a prime ideal of $\mathcal{O}$ lying over the prime $(p) \subset \mathbb{Z}$, then

$$\hat{\mathbb{K}}_\mathcal{P}^\bullet = \mathbb{Z} \times (\hat{\mathbb{Z}}_p^+)^n \times \mathbb{Z}/p^i \times \mathbb{Z}/(p^r - 1) \,.$$

   Here the copy of $\mathbb{Z}$ is generated by a uniformizing parameter, and the residue class field is $\mathbb{F}_{p^r}$.

There are three key techniques used to verify (1.5.1)–(1.5.3). The first is Hensel's lemma, which allows us to lift the roots of unity in the residue class field to roots of unity in $\hat{\mathbb{K}}_\mathcal{P}$ by successive approximation. The second is the exponential map exp: $\pi^i\hat{\mathcal{O}}_\mathcal{P} \rightarrow \hat{\mathcal{O}}_\mathcal{P}^\bullet$ defined by the power series

$$exp(a) = 1 + \sum_1^\infty \frac{a^n}{n!} \,,$$

which maps a suitable power of the the maximal ideal in the (additive) valuation ring into the (multiplicative) units in the valuation ring. The third is the logarithm map which maps the units of the form $1 + \pi^l\hat{\mathcal{O}}_\mathcal{P}$ into the (additive) valuation ring,

$$log(1-b) = \sum_1^\infty \frac{b^n}{n} \,.$$

For details see for example [CF] or [La2].

The Brauer Groups of Complete Fields with Finite Valuations

If $D \subset D'$ is an extension of D to a division algebra D' with D' finite dimensional over the center of D, then φ extends uniquely to a function φ' on D' satisfying (1.1.1) and (1.1.2). Let $\mathbb{F}$ be the center of D and $\mathbb{K}$ be the center of D'. Then $\mathbb{F} \subset \mathbb{K}$ is a finite extension. Define $m = rs$ where $r^2 = \dim_{\mathbb{K}}(D')$ and $s = \dim_{\mathbb{F}}(\mathbb{K})$. Then the formula is

$$\varphi'(a) \;=\; \varphi(N_{\mathbb{K}/\mathbb{F}}(\varphi(a)))^{1/m}$$

where $N_{\mathbb{K}/\mathbb{F}}$ is the norm map and $\varphi\colon D' \longrightarrow \mathbb{K}$ is the reduced norm on D'.

Clearly φ' satisfies $\varphi'(ab) = \varphi'(a)\varphi'(b)$ and $\varphi'(a) = 0$ if and only if $a = 0$. Moreover, for $d \in \mathbb{F}$ we have $\varphi(d) = d^r$, $N_{\mathbb{K}/\mathbb{F}}(d) = d^s$ Hence, $\varphi'(d) = \varphi(d)$ by the uniqueness of mth roots in $\mathbb{R}^+$. The uniqueness of this extension is considerably more subtle. For details, see e.g., [P], pp. 305 - 330, especially p.329.

In general the extension above does not satisfy condition (1.1.3), and so, is not a valuation on D'. However, here is one case in which it does.

Theorem 1.6 *Let D be a central simple divison algebra with center a field $\mathbb{F}$, complete with respect to a valuation φ. Suppose, also that $|\mathbb{F}_\varphi|$ is finite. Then the extension of φ to D described above is a valuation on $\mathcal{D}$.*

(This is a special case of the lemma on page 329 of [P]. See also Proposition 1 on page 182 of [Se4].)

Corollary 1.7 *Let D satisfy the conditions of (1.6). Then*

1. *$\mathcal{O}_\varphi/\mathcal{P}_\varphi$ is a finite field, an extension of $\mathbb{F}_\varphi$ of degree n where $n^2 =$* $\dim_{\mathbb{F}}(D)$.
2. *There is a maximal field $\mathbb{K} = \mathbb{F}(\zeta_{p^{nr}-1}) \subset D$ where $|\mathcal{O}_\varphi(\mathbb{F})/\mathcal{P}| = p^r$.*

Proof. Let $\pi_D \in \mathcal{P}_\varphi$ be a uniformizing parameter. Let $\pi_{\mathbb{F}}$ be a uniformizing parameter for $\mathbb{F}$. Then there is an $e \geq 1$ so that $\varphi(\pi_D)^e = \varphi(\pi_{\mathbb{F}})$. Thus $\pi_{\mathbb{F}}\mathcal{O}_\varphi = \pi_D^e\mathcal{O}_\varphi$. But $\pi_{\mathbb{F}}\mathcal{O}_\varphi \cong \{\mathcal{O}_\varphi(\mathbb{F})/(\pi)\}^{n^2}$, while we have an alternate calculation of $|\mathcal{O}_\varphi/(\pi_D^e)| = |\mathcal{O}_\varphi/(\pi_D)|^e$ from the form of the power series expansion in (1.4). Hence we have $|\mathcal{O}_\varphi/\mathcal{P}| = |\mathcal{O}_\varphi(\mathbb{F})/\mathcal{P}|^{n^2/e}$.

The next step is to choose an element $k \in D$ so that the image of k in $\mathcal{O}_\varphi/\mathcal{P}_\varphi$ is a multiplicative generator. Since $k^{p^s-1} = 1 \bmod \mathcal{P}_\varphi$ where $s = rn^2/e$, we can use Hensel's lemma to replace k by an element, $k' \in D$, for which $(k')^{p^s-1} = 1$. If we use this root of unity k' we find that $\mathbb{K} = \mathbb{F}(\zeta_{p^s-1}) \subset D$. Hence $e \geq n$, since $\dim_{\mathbb{F}}(\mathbb{K}) = s \leq n$. On the other hand $\mathbb{F}(\pi_\varphi)$ is an extension of $\mathbb{F}$ of degree at least e, so $e \leq n$. It follows that $e = n$. □

Corollary 1.8 *Let D satisfy the conditions of (1.5), and $\mathbb{K}$ be the field of (1.7.2). Then* $\mathrm{Gal}(\mathbb{K}/\mathbb{F}) = \mathbb{Z}/n$. *Moreover, let g be the "Frobenius" generator* *of* Gal($\mathbb{K}/\mathbb{F}$), $\zeta \mapsto \zeta^{p^r}$, *then there is an element* $\pi \in D$ *with* $\pi k \pi^{-1} = g[k]$ *for all* $k \in \mathbb{K}$. *Also* $g^n = \pi_{\mathbb{F}}^t$ *where* $(t, n) = 1$ *and* $\pi_{\mathbb{F}} \in \mathbb{F}$ *is a uniformizing parameter for* $\mathbb{F}$. t *is well defined and depends only on* D.

Proof. For the most part this follows from the Noether-Skolem theorem. The rest follows from (1.7). □

In particular t/n is a well defined invariant of D in $\mathbb{Q}/\mathbb{Z}$. By construction, it follows that every division algebra D over a complete field $\mathbb{F}$ with finite $\mathbb{F}_\varphi$ is cyclic, of the form

$$\mathbb{A}(\mathbb{K}, g, \pi^t)\ ,$$

with $\mathbb{K}$ as above. Indeed, it is easily checked, using the Noether-Skolem theorem that t/n is independent of the explicit maximal subfield $\mathbb{F}(\zeta_{p^{rn}-1}) = \mathbb{K} \subset D$ which is selected.

The following result is standard and summarizes the most important facts about simple algebras over a field $\mathbb{F}$ as above.

Theorem 1.9 *Let* $\mathbb{F}$ *be a finite extension of* $\hat{\mathbb{Q}}_p$.
1. *An algebra of the form* $\mathbb{A}(\mathbb{K}, g, \pi^t)$, *as above, is a division algebra precisely when* $t \cong u \bmod \dim_{\mathbb{F}}(\mathbb{K})$ *where* u *is a unit.*
2. *If* $(t, n) = s > 1$ *then* $\mathbb{A}(\mathbb{K}, g, \pi^t) = M_s(D)$ *where* D *is the division algebra* $\mathbb{A}(\mathbb{K}', g, t/s)$ *and* $\dim_{\mathbb{F}}(\mathbb{K}') = n/s$.
3. *The Brauer group* $B(\mathbb{F}) \cong \mathbb{Q}/\mathbb{Z} = \lim_{n \mapsto \infty} H^2(\mathbb{Z}/n; \mathbb{F}(\zeta_{p^{rn}-1})^\bullet)$.

The following result is a special case of Proposition 7 on page 193 of [Se4].

Proposition 1.10 *Let* $\mathbb{F}$ *be a finite extension of* $\hat{\mathbb{Q}}_p$, *and suppose that* $\mathbb{K}$ *is a degree* n *extension of* $\mathbb{F}$. *Then, in the identification of* $B(\mathbb{F})$ *and* $B(\mathbb{K})$ *with* $\mathbb{Q}/\mathbb{Z}$ *in (1.9), the following diagram commutes*

$$\begin{array}{ccc} B(\mathbb{F}) & \xrightarrow{B(i)} & B(\mathbb{K}) \\ \big\downarrow{=} & & \big\downarrow{=} \\ \mathbb{Q}/\mathbb{Z} & \xrightarrow{n} & \mathbb{Z}/\mathbb{Z} \end{array}$$

where $B(i)$ *is the map of Brauer groups induced by* $\mathbb{A} \mapsto \mathbb{A} \otimes_{\mathbb{F}} \mathbb{K}$.

Corollary 1.11 *Suppose* $\mathbb{F}$ *is as in (1.10) and suppose* $\mathbb{K}$ *is ramified of degree* n *over* $\mathbb{F}$ *and has* Gal($\mathbb{K}/\mathbb{F}$) $= \mathbb{Z}/n$ *with* n *prime to* p. *Suppose also that a primitive* n^{th} *root of unity,* ζ_n *is contained in* $\mathbb{F}$. *Then the algebras* $\mathbb{A}(\mathbb{K}, T, \zeta_n^i)$, $0 < i < n$, *are central simple* $\mathbb{F}$*-algebras and generate the kernel of* $B(i)\colon B(\mathbb{F}) \longrightarrow B(\mathbb{K})$.

Proof. There is a short exact sequence in cohomology

$$0 \longrightarrow H^2(\mathbb{Z}/n; \mathbb{K}^\bullet) \longrightarrow B(\mathbb{F}) \longrightarrow B(\mathbb{K}) \longrightarrow 0 ,$$

and it suffices to check that the generators of $H^2(\mathbb{Z}/n; \mathbb{K}^\bullet)$ are the maps $(e_2 \mapsto \zeta_n^i)$ for the resolution of $\mathbb{Z}(\mathbb{Z}/n)$ given in (II.3.8). But $\delta_2 \colon \mathbb{K}^\bullet \to \mathbb{K}^\bullet$ is $\theta \mapsto T(\theta)/\theta$ where T generates $\mathbb{Z}/n$, so all cocycles are the elments in $\mathbb{F}^\bullet$. Likewise, the coboundary map is $\theta \mapsto N(\theta)$. So it suffices to observe that the ζ_n^i are not in the image of the norm. To see this consider the induced norm on $\mathcal{O}_{\mathbb{K}}/\mathcal{P}_{\mathbb{K}}$. Since the extension is totally ramified, we have

$$\mathcal{O}_{\mathbb{F}}/\mathcal{P}_{\mathbb{F}} = \mathcal{O}_{\mathbb{K}}/\mathcal{P}_{\mathbb{K}}$$

and the map induced by the norm is just $x \mapsto x^n$. (1.11) follows. □

2. The Brauer Group and the Schur Subgroup for Finite Extensions of $\mathbb{Q}$

The Brauer Group of a Finite Extension of $\mathbb{Q}$

Let $\mathbb{F}$ be a finite extension of $\mathbb{Q}$. Then $\mathcal{O}_{\mathbb{F}}$, the ring of integers in $\mathbb{F}$ is the set of all elements in $\mathbb{F}$ which are roots of a non-trivial monic polynomial with integer entries. It is a Dedekind ring. Let $\mathcal{P}$ be a prime ideal of $\mathcal{O}_{\mathbb{F}}$. Then there is a unique prime ideal of $\mathbb{Z}$, (p), so that $\mathcal{O}_{\mathbb{F}} \cap \mathbb{Z} = (p)$, and $\mathcal{P}$ is said to lie over (p) or be an extension of (p). More generally, if we have $\mathbb{Q} \subset \mathbb{L} \subset \mathbb{F}$, then $\mathcal{O}_{\mathbb{L}} \cap \mathcal{P}$ is a unique prime ideal, $\mathcal{P}_{\mathbb{L}}$, in $\mathcal{O}_{\mathbb{L}}$ and $\mathcal{P}$ is an extension of $\mathcal{P}_{\mathbb{L}}$ and lies over it. Likewise, given any prime ideal $\mathcal{P}_{\mathbb{L}}$ there is at least one prime ideal of $\mathcal{O}_{\mathbb{F}}$ lying over it.

Example 2.1 Let $\mathbb{F} = \mathbb{Q}(\zeta_n)$ and let (p) be a prime of $\mathbb{Z}$. This extension has dimension $\phi(n)$ where $\phi(n)$ is the Euler ϕ-function. Suppose that $(p, n) = 1$. Then there is a finite extension of the finite field $\mathbb{F}_p$ which contains a primitive n^{th} root of unity. The smallest such field has order p^r where r is the smallest positive integer so that $p^r \cong 1 \bmod (n)$. Then r divides $\phi(n)$ and the number of distinct primes lying over (p) is $\phi(n)/r$. These primes are obtained as follows. The Galois group $\mathbb{Z}/r$ of $\mathbb{F}_{p^r}$ over $\mathbb{F}_p$ acts on the primitive n^{th} roots of unity to break them into $\phi(n)/r$ orbits, each of length r. Consider the polynomial $p_i(x) = \prod_1^r (x - \lambda_j)$ where λ_j runs over the ith orbit. This polynomial has coefficients in $\mathbb{F}_p$ and hence lifts to a polynomial $\tilde{p}_i(x) \in \mathbb{Z}[x]$. Then the prime ideals have the form $(p, \tilde{p}_i(\zeta_n))$, for a fixed primitive n^{th} root ζ_n.

In particular, consider the primes over (2) in $\mathcal{O}_{\mathbb{Q}(\zeta_7)}$. Since $\mathbb{F}_8$ contains six primitive seventh roots of unity, they split into two orbits there $\{\zeta_7, \zeta_7^2, \zeta_7^4\}$

and $\{\zeta_7^3, \zeta_7^6, \zeta_7^5\}$, and the two polynomials are x^3+x+1, x^3+x^2+1, so the ideals are $(2, \zeta_7^3+\zeta_7+1)$, $(2, \zeta_7^3+\zeta_7^2+1)$.

For each prime $\mathcal{P} \subset \mathcal{O}_{\mathbb{F}}$ there is a valuation $\varphi_{\mathcal{P}}$, the associated completion of $\mathbb{F}$ at the valuation, $\hat{\mathbb{F}}_{\mathcal{P}}$, and the inclusion $i_{\mathcal{P}}: \mathbb{F} \subset \hat{\mathbb{F}}_{\mathcal{P}}$. This inclusion induces a map of Brauer groups $B(\mathbb{F}) \longrightarrow B(\hat{\mathbb{F}}_{\mathcal{P}})$ obtained on the one hand by tensoring $\mathbb{A} \mapsto \mathbb{A} \otimes_{\mathbb{F}} \hat{\mathbb{F}}_{\mathcal{P}}$, and on the other via coefficient mappings of second cohomology groups.

Also, for $\mathbb{F}$ as above there are k embeddings of $\mathbb{F}$ into $\mathbb{C}$ and r embeddings of $\mathbb{F}$ into $\mathbb{R}$, where $2k + r = n$, the dimension of $\mathbb{F}$ over $\mathbb{Q}$. The Brauer group of $\mathbb{C}$ is 0, but the Brauer group of $\mathbb{R}$ is $\mathbb{Z}/2$ with generator $\mathbb{H}$, the quaternions. Consequently, the r real embeddings, e_i, give r homomorphisms, $e_{i,*}: B(\mathbb{F}) \longrightarrow \mathbb{Z}/2$.

Taken together we get a homomorphism

$$E\colon\ B(\mathbb{F}) \longrightarrow \coprod_1^r B(\mathbb{R}) \oplus \coprod_{\mathcal{P}} B(\hat{\mathbb{F}}_{\mathcal{P}}) \;=\; \coprod_1^r \mathbb{Z}/2 \oplus \coprod_{\mathcal{P}} \mathbb{Q}/\mathbb{Z}\,.$$

(A priori, one would have expected that the image of E was in the *direct product* over all $\mathcal{P}$, but, in fact, a major theorem here asserts that the image of a central simple $\mathbb{F}$-division algebra is non-trivial at only a finite number of primes.) The main result here, and one of the main results of class field theory is that E is an injection with cokernel a single copy of $\mathbb{Q}/\mathbb{Z}$. Indeed, $B(\mathbb{F})$ can be identified with the kernel of the *sum* homomorphism

$$\coprod_1^r \mathbb{Z}/2 \oplus \coprod_{\mathcal{P}} \mathbb{Q}/\mathbb{Z} \longrightarrow \mathbb{Q}/\mathbb{Z}, \qquad \coprod \theta_i \mapsto \sum \theta_i\,.$$

Example 2.2 In the case $\mathbb{F} = \mathbb{Q}$ and D is the usual quaternion algebra

$$\mathbb{Q}(\mathbb{Q}(i), g, g^2 = -1)\,,$$

where g acts on $\mathbb{Q}(i)$ as conjugation, then D has image $1/2$ at the single real embedding, and image $1/2$ in $B(\hat{\mathbb{Q}}_2)$.

For general fields the determination of the Brauer group is much more complex. There are a number of special circumstances in the case of finite extensions of $\mathbb{Q}$ which make things simpler. We now quote some of these properties.

Theorem 2.3 *Let $\mathbb{F}$ be a finite extension of $\mathbb{Q}$ or $\hat{\mathbb{Q}}_p$, and suppose that $\mathbb{A}$ is a central simple $\mathbb{F}$-algebra of dimension n^2 over $\mathbb{F}$. Then there is a cyclic extension $\mathbb{K}$ of $\mathbb{F}$ of degree n, and, if T is a generator for* $\mathrm{Gal}(\mathbb{K}/\mathbb{F}) = \mathbb{Z}/n$ *we have that $\mathbb{A}$ is the cyclic algebra $\mathbb{A}(\mathbb{K}, T, T^n = \lambda)$ for some $\lambda \in \mathbb{F}^\bullet$.*

On the other hand, by class field theory, every *abelian* extension of finite extensions of $\mathbb{Q}$ or $\hat{\mathbb{Q}}_p$ is cyclotomic. Thus, for the field $\mathbb{K}$ above, we have that

$$\mathbb{K} \subset \mathbb{F}_{cycl} = \lim_{n \mapsto \infty} (\mathbb{F}(\zeta_n)) ,$$

and $\lim_n H^2(\mathrm{Gal}(\mathbb{F}(\zeta_n)/\mathbb{F}); \mathbb{F}(\zeta_n)^\bullet) \longrightarrow H^2(\mathbb{F}) = B(\mathbb{F})$ must be surjective. In fact, this map is actually an isomorphism (see e.g., [Se4]). Thus, for these fields the Brauer group is given strictly in terms of cyclotomic extensions.

The Schur Subgroup of the Brauer Group

Let G be a finite group, then the group ring $\mathbb{F}(G)$ is semi-simple for any $\mathbb{F}$ of characteristic zero, or if $char(\mathbb{F})$ is prime to the order of G. Hence we can write

$$\mathbb{F}(G) = \coprod M_{n_i}(D_i)$$

where $\mathbb{F}$ is a subfield of finite index in the center $Z(D_i)$, and D_i is a division algebra of finite dimension over the field $Z(D_i)$. Of course, when $|\mathbb{F}| < \infty$, the division algebra D_i must be a finite field, and the group ring is a sum of matrix algebras over fields. However, when $\mathbb{F}$ has characteristic zero the structure of the resulting division algebras, D_i, can be quite complex.

Example 2.4 Consider the split extension $\mathbb{Z}/7 \times_T \mathbb{Z}/9$, where the action of $\mathbb{Z}/9$ is given by $g(t) = t^2$, $t \in \mathbb{Z}/7$, where g generates the $\mathbb{Z}/9$. Then there is a cyclic algebra direct summand,

$$\mathbb{A}(\mathbb{Q}(\zeta_{21}), g, g^3 = \zeta_3) \subset \mathbb{Q}(G) ,$$

where $g[\zeta_{21}] = \alpha^{16}$. Thus, $g[\zeta_{21}^7] = \zeta_{21}^7$, while $g[\zeta_{21}^3] = \zeta_{21}^6$, so the center is $\mathbb{F} = \mathbb{Q}(\zeta_3, \sqrt{-7})$.
We now verify that this algebra has order 3 in the Brauer group of $\mathbb{F}$.

We work at the prime (7). There are two primes over (7) in the ring of integers, $\mathcal{O}(\mathbb{F})$, $\mathcal{P}_1 = (\sqrt{-7}, \zeta - 2)$ and $\mathcal{P}_2 = (\sqrt{-7}, \zeta^2 - 2)$, over (7), each ramified of degree 2 over $(7) \subset \mathbb{Z}$ and each with quotient isomorphic to $\mathbb{F}_7$.

Moreover, in $\mathcal{O}(\mathbb{Q}(\zeta_{21}))$ there is a unique prime over each of these $\mathcal{P}_i$, each ramified of degree 3. In particular, each quotient by each of the primes above is a copy of $\mathbb{F}_7$.

Choose one of these primes, say $\mathcal{P}_1 = \mathcal{P}$, and complete at it. Then the completion of $\mathbb{Q}(\zeta_{21})$ at the prime over $\mathcal{P}$ is still ramified of degree 3, and is Galois over $\hat{\mathbb{F}}_{\mathcal{P}}$ with Galois group $\mathbb{Z}/3$. Then (1.11) applies and the algebra is a division algebra at each prime, with invariants 1/3 and 2/3 respectively.

Definition 2.5 Let $\mathbb{F}$ have characteristic zero. Then the subset of the Brauer group of $\mathbb{F}$, $B(\mathbb{F})$, which consists of division algebras which occur as D_i in the expansion, (0.1), with center $\mathbb{F}$, as G runs over all finite groups, is called the *Schur subgroup* of $B(\mathbb{F})$. It is denoted $S(\mathbb{F})$.

Remark 2.6 In particular, the terminology implies that $S(\mathbb{F})$ *is* a subgroup of $B(\mathbb{F})$. Indeed, if $M_{n_i}(D_1)$ is a summand of $\mathbb{F}(G_1)$, $M_{m_i}(D_2)$ is a summand of $\mathbb{F}(G_2)$, then $M_{n_i m_i}(D_1 \otimes_{\mathbb{F}} D_2)$ is a summand of $\mathbb{F}(G_1 \times G_2)$. Moreover, if we define G^{op} as G with the reversed multiplication $g_1 \cdot g_2 = g_2 g_1$, then G^{op} is a group, and, if $M_n(D)$ occurs as a summand of $\mathbb{F}(G)$ then $M_n(D)^{op}$ occurs as a summand of $\mathbb{F}(G^{op})$. But $M_n(D)^{op}$ represents the inverse of $\{D\}$ in the Brauer group of $\mathbb{F}$.

The Group $(\mathbb{Q}/\mathbb{Z})$ and its Aut Group

The discussion above shows two things that we should take account of. First, it is sufficient to study the Schur subgroup in the case where $\mathbb{F}$ is a cyclotomic extension of $\mathbb{Q}$, and second, we do not need to consider all finite extensions of $\mathbb{Q}$, but only the cyclotomic ones. The universal cyclotomic extension is

$$\mathbb{Q}_{cycl} = \varinjlim \mathbb{Q}(\zeta_n) ,$$

and in this subsection, in preparation for this analysis, we determine the structure of $G = \mathrm{Aut}(\mathbb{Q}/\mathbb{Z}) = \mathrm{Gal}(\mathbb{Q}_{cycl}/\mathbb{Q})$.

The Sylow p subgroup of $\mathbb{Q}/\mathbb{Z}$ is the set of all fractions m/p^i with denominator a power of p. It can also be described as the direct limit $(\mathbb{Q}/\mathbb{Z})_p = \varinjlim (\mathbb{Z}/p^n)$, so we have the sequence of inclusions

$$\mathbb{Z}/p \subset \mathbb{Z}/p^2 \subset \mathbb{Z}/p^3 \subset \cdots \subset (\mathbb{Q}/\mathbb{Z})_p .$$

There is only one subgroup isomorphic to $\mathbb{Z}/p^n$, the set of equivalence classes of rational fractions with denominator p^n, $\mathbb{Z}/p^n \cong \{m/p^n \mid 0 \le m \le p^n - 1\}$. Hence if $\alpha \in \mathrm{Aut}(\mathbb{Q}/\mathbb{Z})_p$ then α restricted to $(\mathbb{Z}/p^n)$ is an isomorphism $\mathbb{Z}/p^n \longrightarrow \mathbb{Z}/p^n$.

For m and n relatively prime we have

$$\frac{a}{mn} = \frac{c}{m} + \frac{d}{n}$$

where $a = cn + dm$, it follows that $\mathbb{Q}/\mathbb{Z} = \coprod (\mathbb{Q}/\mathbb{Z})_p$, so $\mathrm{Aut}(\mathbb{Q}/\mathbb{Z}) = \prod \mathrm{Aut}((\mathbb{Q}/\mathbb{Z})_p)$, and

$$\mathrm{Aut}((\mathbb{Q}/\mathbb{Z})_p) = \varprojlim (\mathrm{Aut}(\mathbb{Z}/p^n)) .$$

Recall from (I.1.4) that

$$\begin{cases} \mathrm{Aut}(\mathbb{Z}/p^n) = \mathbb{Z}/p-1 \times \mathbb{Z}/p^{n-1} & \text{if } p \text{ is odd,} \\ \mathrm{Aut}(\mathbb{Z}/2^n) = \mathbb{Z}/2 \times \mathbb{Z}/2^{n-2} & \text{otherwise.} \end{cases}$$

The generators of $\mathrm{Aut}(\mathbb{Z}/2^n)$ are multiplication by -1 and by 5. For odd primes the generator of order p^{n-1} is $1+p$ while the generator of $\mathbb{Z}/(p-1)$ is fairly hard to predict, but in the first few cases can be chosen as

p	3	5	7	11	13	17	19	23	29	31
generator	2	2	3	2	2	3	2	5	2	3.

Consequently we have

Proposition 2.7 $\mathrm{Aut}(\mathbb{Q}/\mathbb{Z}) \cong \prod_{p\ prime} U\hat{\mathbb{Z}}_p$, *the infinite direct product of the units in the valuation ring* $\hat{\mathbb{Z}}_p$ *as* p *runs over all primes, where* $U\hat{\mathbb{Z}}_p$ *is* $\mathrm{Aut}((\mathbb{Z}/\mathbb{Z})_p)$.

Proof. The inclusion $\mathbb{Z}/p^n \subset \mathbb{Z}/p^{n+1}$ defines a restriction map

$$r\colon \mathrm{Aut}(\mathbb{Z}/p^{n+1}) \longrightarrow \mathrm{Aut}(\mathbb{Z}/p^n)$$

where $r(m)$ is multiplication by m again. In particular r is surjective with kernel $\mathbb{Z}/p$. Now, choose an infinite sequence of integers $A = (a_1, a_2, a_3, \cdots a_n \cdots)$ so that $a_{i+1} \equiv a_i \pmod{p^i}$ for all i, and for definiteness assume $0 \leq a_i < p^i$ so we may write $a_{i+1} = a_i + jp^i$, with $0 \leq j < p$. Clearly such a sequence A defines an element in $\mathrm{Aut}((\mathbb{Q}/\mathbb{Z})_p)$ given explicitly as $\{m/p^n \longrightarrow a_n m/p^n\}$, and conversely any element in $\mathrm{Aut}((\mathbb{Q}/\mathbb{Z})_p)$ defines such a sequence.

Note that we may also give such a sequence as a (formal) power series

$$A = \alpha_1 + \alpha_2 p + \alpha_3 p^2 + \cdots + \alpha_{n+1} p^n + \cdots \quad 0 \leq \alpha_i < p \text{ for all } i\,.$$

Moreover, the condition that A denote an element in $\mathrm{Aut}((\mathbb{Q}/\mathbb{Z})_p)$ is precisely that $\alpha_1 \neq 0$, and this identifies $\mathrm{Aut}((\mathbb{Q}/\mathbb{Z})_p)$ with $U\hat{\mathbb{Z}}_p$ as desired. □

3. The Explicit Generators of the Schur Subgroup

Cyclotomic Algebras and the Brauer-Witt Theorem

Let G be a group together with a representation $\mathcal{E} = (\mathbb{Z}/n, G, \varphi)$, as in (I.1.5), of G as an extension of the form

$$1 \longrightarrow \mathbb{Z}/n \overset{\triangleleft}{\longrightarrow} G \longrightarrow V \longrightarrow 1$$

where $\varphi\colon V \longrightarrow \mathrm{Aut}(\mathbb{Z}/n) = \mathrm{Aut}(\mathbb{Q}(\zeta_n))$ is an injection. Then there is a simple algebra, $A(G, \mathcal{E})$, with center $\mathbb{F} = (\mathbb{Q}(\zeta_n))^V$ associated to G and the extension data $\mathcal{E}$, which is constructed from the $\mathbb{Q}(\zeta_n)$ vector space with basis the elements of V by choosing a section $\Theta\colon V \longrightarrow G$ and introducing the relations

$$\begin{aligned} v\lambda v^{-1} &= \lambda^{\varphi(v)} \qquad v \in V,\ \lambda \in \mathbb{Z}/n, \\ w^{-1}v^{-1}(vw) &= \Theta(w)^{-1}\Theta(v)^{-1}\Theta(vw) \in \mathbb{Z}/n \qquad \text{for all } v,\ w \text{ in } V. \end{aligned}$$

Similar algebras were discussed in (I.8), but the algebras above are distinguished by the fact that the factor set $\{\Theta(v) \mid v \in V\}$ are all roots of unity. Such algebras are called *cyclotomic algebras*.

A key result in the theory of finite groups, the Brauer-Witt theorem ([Y], pp. 20-34), states that as the pairs $(G, \mathcal{E})$ run over all extensions of the form above, then the classes of the algebras $A(G, \mathcal{E})$ in their respective Brauer groups generate the Schur subgroups $S(\mathbb{F})$. In order to put this into the context of cohomology we need the maximal cylcotomic extension of $\mathbb{F}$ and its Galois group.

The Galois Group of the Maximal Cyclotomic Extension of $\mathbb{F}$

Let $\mathbb{F} \subset \mathbb{Q}(\zeta_n)$ be a cyclotomic field and $G_{\mathbb{F}} = \mathrm{Gal}(\mathbb{Q}_{cycl}/\mathbb{F})$. This subgroup of $\mathrm{Aut}(\mathbb{Q}/\mathbb{Z})$ is fundamental in what follows so we discuss it in some detail now.

To begin consider the extension $\mathbb{Q} \hookrightarrow \mathbb{Q}(\zeta_n)$. We have an exact sequence

$$1 \longrightarrow \mathrm{Gal}(\mathbb{Q}_{cycl}/\mathbb{Q}(\zeta_n)) \longrightarrow \mathrm{Gal}(\mathbb{Q}_{cycl}/\mathbb{Q}) \longrightarrow \mathrm{Gal}(\mathbb{Q}(\zeta_n)/\mathbb{Q}) \longrightarrow 1$$

and $\mathrm{Gal}(\mathbb{Q}(\zeta_n)/\mathbb{Q}) = \mathrm{Aut}(\mathbb{Z}/n)$. Moreover, if we write $n = 2^r p_2^{i_2} \cdots p_s^{i_s}$ where the p_j are odd, then from (I.1.4)

$$\mathrm{Aut}(\mathbb{Z}/n) = (\mathbb{Z}/2 \times \mathbb{Z}/2^{r-2}) \times \prod_1^s \mathbb{Z}/(p-1) \times \mathbb{Z}/p^{i_j - 1}$$

for $r \geq 2$, and the first term is missing otherwise.

From (2.7) $\mathrm{Aut}(\mathbb{Q}/\mathbb{Z}) = \prod_p U\hat{\mathbb{Z}}_p$ where $U\hat{\mathbb{Z}}_p = \mathbb{Z}/(p-1) \times \hat{\mathbb{Z}}_p^+$ for p odd and is $\mathbb{Z}/2 \times \hat{\mathbb{Z}}_2^+$ for $p = 2$. In particular,

$$G_{\mathbb{Q}(\zeta_n)} = \prod_{p|n} \hat{\mathbb{Z}}_p^+ \times \prod_{(p,n)=1} U\hat{\mathbb{Z}}_p$$

where the copy of $\hat{\mathbb{Z}}_p^+$ for $p|n$ is really the ideal $p^{i_j-1}\hat{\mathbb{Z}}_p^+ \subset \hat{\mathbb{Z}}_p^+$. From this we get the form of the general case since $\mathbb{Q} \subset \mathbb{F} \subset \mathbb{Q}(\zeta_n)$ for some n: $\mathrm{Gal}(\mathbb{Q}_{cycl}/\mathbb{F}) = \prod_{(n,p)=1} U\hat{\mathbb{Z}}_p \times W(\mathbb{F})$ where $W(\mathbb{F}) \subset \prod_{p|n} U\hat{\mathbb{Z}}_p$ and

$$G_{\mathbb{F}}/G_{\mathbb{Q}(\zeta_n)} = \mathrm{Gal}(\mathbb{Z}(\zeta_n)/\mathbb{F}) .$$

Since every finite index subgroup of $\hat{\mathbb{Z}}_p^+$ is isomorphic to $\hat{\mathbb{Z}}_p^+$ we can write

$$G_{\mathbb{F}} = \left\{ W^{tor} \times \prod_{(p,n)=1} \mathbb{Z}/(p-1) \right\} \times \prod_{p\ prime} \hat{\mathbb{Z}}_p^+ . \tag{3.1}$$

We now consider the groups G_v for v a non-archimedean valuation on $\mathbb{F} \subset \mathbb{Q}(\zeta_n)$, where G_v is defined as $\mathrm{Gal}(\hat{\mathbb{Q}}_{p,cycl}/\hat{\mathbb{F}}_v)$.

Theorem 3.2 *Let v be a non-archimedean valuation over p in $\mathbb{F}$. Then*

$$G_v = V \times \hat{\mathbb{Z}}_p^+ \times \prod_{q\ prime} \hat{\mathbb{Z}}_q^+ ,$$

where G_v is independent of the particular prime over p in $\mathbb{F}$ and $V \subset \mathbb{Z}/(p-1)$ if p is odd while $V = \mathbb{Z}/2$ or $\{1\}$ for $p = 2$.

Proof. That fact that G_v is independent of the particular prime $\mathcal{P}$ over (p) used to define v is due to the fact that in a cyclotomic field all the primes over (p) are permuted transitively by $\mathrm{Gal}(\mathbb{F}/\mathbb{Q})$ and the Galois groups are all abelian so conjugate subgroups are actually equal.

We now determine G_v beginning with $G_p = \mathrm{Gal}(\hat{\mathbb{Q}}_{g,cycl}/\hat{\mathbb{Q}}_p)$. Write $n = p^j\theta$ with $(\theta, p) = 1$. Then, if p^s is the first power of p which is congruent to 1 mod (θ) we have that $\hat{\mathbb{Q}}_p(\zeta_n) = \hat{\mathbb{Q}}_p(\zeta_{p^j}, \zeta_{p^s-1})$ and $\mathrm{Gal}(\hat{\mathbb{Q}}_p(\zeta_n)/\hat{\mathbb{Q}}_p) = \mathrm{Aut}(\mathbb{Z}/p^j) \times \mathbb{Z}/s$. Moreover, the generator of the $\mathbb{Z}/s$ can be selected to act trivially on ζ_{p^j} and as $\zeta \mapsto \zeta^p$ on ζ_{p^s-1}. In any case, passing to limits we obtain

$$\begin{aligned} G_p &= \varprojlim \mathrm{Aut}(\mathbb{Z}/p^j) \times \varprojlim (\mathbb{Z}/s) \\ &= U\hat{\mathbb{Z}}_p \times \prod_{q\ prime} \hat{\mathbb{Z}}_q^+ \\ &= \mathbb{Z}/(p-1) \times \left(\hat{\mathbb{Z}}_p^+\right)^2 \times \prod_{q \neq p} \hat{\mathbb{Z}}_q^+, \end{aligned}$$

and $G_v = \mathrm{Gal}(\hat{\mathbb{Q}}_{p,cycl}/\hat{\mathbb{F}}_v)$ has finite index in G_p. This completes the proof. □

The Cohomological Reformulation of the Schur Subgroup

We extend the definition of the limit cohomology group used in (I.8) to define the Brauer group to more general coefficients and arrive at the notion of *continuous cohomology* with general coefficients.

The conditions for the definition of continuous cohomology are that we have a group G given as an inverse limit, $G = \varprojlim (G_i)$, and coefficients, A, given as a direct limit $A = \varinjlim (A_i)$ where the limits are taken over surjections $p: G_{i+1} \to G_i$, and and injections $i_j: A_j \to A_{j+1}$ so

$$A = \bigcup A_n, \qquad A_1 \subset A_2 \subset \cdots \subset A_i \subset \cdots .$$

The groups G and A must also satisfy the consistency condition that for each n, G_n acts on A_n and $pg(a) = g(i(a))$ for all n, $a \in A_n$ and $g \in G_{n+1}$. Then $H^*_{cont}(G; A)$ is defined as $\varinjlim (H^*(G_n; A_n)$. A simple argument shows that $H^*_{cont}(G; A) = \lim_n \lim_m H^*(G_{n+m}; A_n)$ where G_{n+m} acts on A_n through the iterate projection of G_{n+m} onto G_n.

We clearly have that if $A_n = B_n \oplus C_n$, $i_n = i_{n,A} \oplus i_{n,B}$ for each n, then

$$H^*_{cont}(G; B \oplus C) = H^*_{cont}(G; B) \oplus H^*_{cont}(G; C) .$$

Next suppose that we have a product $G \times H = \varprojlim G_n \times H_n$. Then we have

Lemma 3.3 *$H^*_{cont}(G \times H; A) = H^*_{cont}(G; H^*_{cont}(H; A))$ if G acts trivially on A. Otherwise, there is a spectral sequence converging to $H^*_{cont}(G \times H; A)$ with E_2-term $H^*_{cont}(G; H^*_{cont}(H; A))$.*

Proof. Assume that $G = \varprojlim(G_n)$, $H = \varprojlim(H_m)$. Then we can choose as our resolution of $G_n \times H_m$ the tensor product of a resolution of G_n with a resolution of H_m. Filtering this by dimension in the complex for G_n gives the spectral sequence. Then, taking E_2-terms at each level gives $H^*(G_n; H^*(H_{n+m}; A_n))$. In the case of trivial G-action $E_2 = E_\infty$ at each level and the coefficient group need not be A_n, so we can pass to limits fixing G_n but varying H_{n+m} and the coefficients. In general we can pass to limits separately over m and n. So we fix $A_n \subset A$, pass to limits over m so that $H^*(G_n; H^*_{cont}(H; A_n)) = H^*_{cont}(G_n \times H; A_n)$. Then pass to limits over n to obtain our E_2 term for the calculation of $H^*_{cont}(G \times H; A)$. □

Example. $H^*_{cont}(G_{\mathbb{F}}; \mathbb{Q}/\mathbb{Z})$ is given as the limit over all m so that $\mathbb{F} \subset \mathbb{Q}(\zeta_m)$ of the cohomology groups $H^*(V_m(\mathbb{F}); \mathbb{Z}/m)$ where $V_m(\mathbb{F}) = G_{\mathbb{F}}/G_{\mathbb{Q}(\zeta_m)}$ and the limit is taken by noting that if $m' = mn$ then $V_{nm}(\mathbb{F})$ surjects onto $V_m(\mathbb{F})$ and $\mathbb{Z}/m \hookrightarrow \mathbb{Z}/nm$ as a unique subgroup, so we have the composition

$$H^*(V_m(\mathbb{F}); \mathbb{Z}/m) \xrightarrow{\pi^*} H^*(V_{nm}(\mathbb{F}); \mathbb{Z}/m) \xrightarrow{i_*} H^*(V_{nm}; \mathbb{Z}/nm) .$$

Using this definition we have the following cohomological reformulation of the Brauer-Witt theorem.

Theorem 3.4 *Let $\mathbb{F}$ be a cyclotomic field. Then the Schur subgroup $S(\mathbb{F})$ of the Brauer group $B(\mathbb{F})$ is the subgroup of $H^2_{cont}(G_{\mathbb{F}}; \mathbb{Q}^\bullet_{cycl}) = B(\mathbb{F})$ given as the image under the coefficient map*

$$H^2_{cont}(G_{\mathbb{F}}; \mathbb{Q}/\mathbb{Z}) \longrightarrow H^2_{cont}(G_{\mathbb{F}}; \mathbb{Q}^\bullet_{cycl}) .$$

Proof. The algebra $A(G, \mathcal{E})$ is completely specified by its cocycle $\Psi \in C^2_\varphi(V; \mathbb{Q}(\zeta_n)^\bullet)$, where

$$\Psi(|v|w|) = \Theta(w)^{-1}\Theta(v)^{-1}\Theta(vw)$$

and the action of V on $\mathbb{Z}/n$ is given by φ, but this is exactly the image in $C^2_\varphi(V; \mathbb{Q}(\zeta_n)^\bullet)$ of the cocycle, Θ, corresponding to the extension $\mathbb{Z}/n \overset{\triangleleft}{\longrightarrow} G \longrightarrow V$ in $C^2_\varphi(V; \mathbb{Z}/n)$.

Thus we have shown that the group $S(\mathbb{F})$ is determined as the subgroup spanned by the images of the compositions

$$H^2_\varphi(V;\mathbb{Z}/n)\longrightarrow H^2_\varphi(V;\mathbb{Q}(\zeta_n)^\bullet)\longrightarrow B(\mathbb{F}) .$$

The last composition is obtained by identifying V with $\mathrm{Gal}(\mathbb{Q}(\zeta_n)/\mathbb{F})$ and then taking the composition

$$H^2(\mathrm{Gal}(\mathbb{Q}(\zeta_n)/\mathbb{F};\mathbb{Q}(\zeta_n)^\bullet)\xrightarrow{\pi^*}H^2(G_{\mathbb{F}};\mathbb{Q}(\zeta_n)^\bullet)\xrightarrow{i_*}H^2(G_{\mathbb{F}};\mathbb{Q}_{cycl})=B(\mathbb{F})$$

where $\pi: G_{\mathbb{F}}\longrightarrow V$ is the induced map of Galois group and $i: \mathbb{Q}(\zeta_n)\hookrightarrow \mathbb{Q}_{cycl}$ is the inclusion of coefficients. (Again this is discussed in I.8.)

These maps from $H^2_\varphi(V;\mathbb{Z}/n)$ to $B(\mathbb{F})$ all factor through the composition

$$H^2_\varphi(G_{\mathbb{F}};\mathbb{Z}/n)\longrightarrow H^2_{cont}(G_{\mathbb{F}};\mathbb{Q}/\mathbb{Z})\longrightarrow H^2_{cont}(G_{\mathbb{F}};\mathbb{Q}_{cycl})$$

and the theorem follows. □

Remark. Continuous cohomology can be treated in terms of *continuous cochains* and has been usefully looked at from this point of view. Further discussion can be found in [Se4].

Now, suppose that v is a valuation on $\mathbb{F}$ over the valuation p on $\mathbb{Q}$. Let $G_v\subset G_{\mathbb{F}}\subset \mathrm{Aut}(\mathbb{Q}/\mathbb{Z})$ be the Galois group of $\hat{\mathbb{Q}}_{p,cycl}$ over $\hat{\mathbb{F}}_v$. Then we have, as above, that the Schur subgroup of $B(\hat{\mathbb{F}}_v)$ is given as the image of the coefficient map

$$H^2_{cont}(G_v;\mathbb{Q}/\mathbb{Z})\xrightarrow{i_*}H^2_{cont}(G_v;\hat{\mathbb{Q}}^\bullet_{cycl}), \tag{3.5}$$

and consequently we have

Corollary 3.6 *Let* $\mathrm{res}: B(\mathbb{F})\longrightarrow B(\hat{\mathbb{F}}_v)$ *be the map of Brauer groups induced by tensoring with* $\hat{\mathbb{F}}_v$*, then the map on Schur subgroups is given by the commutative diagram*

$$\begin{array}{ccc} H^*_{cont}(G_{\mathbb{F}};\mathbb{Q}/\mathbb{Z}) & \xrightarrow{i_*} & H^*_{cont}(G_{\mathbb{F}};\mathbb{Q}^\bullet_{cycl}) \\ \downarrow{\scriptstyle r_v^*} & & \downarrow{\scriptstyle r_v^*} \\ H^*_{cont}(G_v;\mathbb{Q}/\mathbb{Z}) & \xrightarrow{i_*} & H^2_{cont}(G_v;\hat{\mathbb{Q}}^\bullet_{p,cycl}) \end{array} .$$

In particular, since $\prod r_v^*: B(\mathbb{F})\rightarrow \coprod B(\hat{\mathbb{F}}_v)\times\coprod B(\mathbb{F}_{\infty,i})$ is injective, the determination of $S(\mathbb{F})$ breaks up into three steps:

1. the evaluation of the groups $H^2_{cont}(G_{\mathbb{F}};\mathbb{Q}/\mathbb{Z})$ and $H^2_{cont}(G_v;\mathbb{Q}/\mathbb{Z})$,
2. the evaluation of the coefficient maps

$$H^2_{cont}(G_v; \mathbb{Q}/\mathbb{Z}) \to H^2_{cont}(G_v; \hat{\mathbb{Q}}^{\bullet}_{p,cycl}),$$

3. the evaluation of the restriction maps

$$r_v^*: H^2_{cont}(G_{\mathbb{F}}; \mathbb{Q}/\mathbb{Z}) \to H^2_{cont}(G_v; \mathbb{Q}/\mathbb{Z}),$$

and the map at the infinite primes.

We answer these questions in order in the next two sections.

4. The Groups $H^*_{cont}(G_{\mathbb{F}}; \mathbb{Q}/\mathbb{Z})$ and $H^*_{cont}(G_v; \mathbb{Q}/\mathbb{Z})$

In this section we determine the cohomology groups with coefficients in $\mathbb{Q}/\mathbb{Z}$ of the subgroups of $\mathrm{Aut}(\mathbb{Q}/\mathbb{Z})$ associated to the cyclotomic fields $\mathbb{F}$ and $\hat{\mathbb{F}}_v$ discussed above. It turns out that they are isomorphic to untwisted cohomology groups with coefficients in the roots of unity contained in $\mathbb{F}$ or $\hat{\mathbb{F}}_v$, and rather than taking cohomology of the entire $G_{\mathbb{F}}$ or G_v with these coefficients we only need the torsion part.

The Cohomology Groups $H^*_{cont}(G_{\mathbb{F}}; \mathbb{Q}/\mathbb{Z})$

Let $G_{\mathbb{F}}$ be given in the notation of (3.1) where $W^{tor} \times \prod_{(p,n)=1} \mathbb{Z}/(p-1)$ is the torsion subgroup, $G_{\mathbb{F}}^{tor}$. Suppose, also that $\mathbb{Z}/m \subset \mathbb{Q}/\mathbb{Z}$ is the fixed set, $(\mathbb{Q}/\mathbb{Z})^{G_{\mathbb{F}}}$. Then we have

Theorem 4.1 *$H^*_{cont}(G_{\mathbb{F}}; \mathbb{Q}/\mathbb{Z}) = H^*_{cont}(G_{\mathbb{F}}^{tor}; \mathbb{Z}/m)$ where the action of $G_{\mathbb{F}}^{tor}$ is trivial on the coefficients $\mathbb{Z}/m$. In particular we have*

$$H^*_{cont}(G_{\mathbb{F}}^{tor}; \mathbb{Z}/m) = H^*(W^{tor}; \mathbb{Z}/m) \times H^*_{cont}\Big(\prod_{(p,n)=1} \mathbb{Z}/(p-1); \mathbb{Z}/m\Big).$$

Proof. We begin with some preliminary calculations for $A = \mathbb{Z}/p^i$ and $G \subset \mathrm{Aut}(A)$.

Lemma 4.2 *Let $A = \mathbb{Z}/p^i$ and suppose $G \subset \mathrm{Aut}(A)$.*

1. *Suppose p is odd, then $G = V \times \mathbb{Z}/p^r$ where $|V|$ divides $p-1$ and*
   a. *If $V = e$, then $H^*(G; \mathbb{Z}/p^i) = H^0(G; \mathbb{Z}/p^i) = (\mathbb{Z}/p^i)^G$.*
   b. *If $V \neq e$ then $H^*(G; \mathbb{Z}/p^i) = 0$.*
2. *Suppose that p is even.*
   a. *If $G = \mathbb{Z}/2 \times \mathbb{Z}/2^r$ with the first $\mathbb{Z}/2$ generated by $1 \leftrightarrow -1$ then*

$$H^j(G; \mathbb{Z}/2^i) = \begin{cases} \mathbb{Z}/2 & \text{generator } 2^{i-1} \text{ for } j \text{ even}, \\ \mathbb{Z}/2 & \text{generator } 1 \text{ for } j \text{ odd}. \end{cases}$$

b. If $G = \mathbb{Z}/2^r$ *and the generator is not* $1 \leftrightarrow -1$, *then*

$$H^j(G; \mathbb{Z}/2^i) = 0, \ j > 0, \ H^0(G; \mathbb{Z}/2^i) = inv_G(\mathbb{Z}/2^i) .$$

Proof of (4.2). For p odd G is cyclic, so we use the resolution in (II.3.8) and the formulae there for calculating cohomology groups. The things to note are that $\mathrm{Hom}_{\mathbb{Z}(\mathbb{Z}/n)}(\mathbb{Z}(\mathbb{Z}/n), A) \cong A$ and T, the generator of $\mathbb{Z}/n$ can be chosen so that it acts as multiplication by $u + p^i$ where $u \not\equiv 0 \bmod (p)$. If $u \not\equiv 1 \bmod (p)$, (which is the case when $|V| \neq 1$), then $T - 1$ is multiplication by a unit, so the cohmology trivializes. Otherwise $T - 1$ is multiplication by p^r but an easy calculation shows that $\sum_0^{p^{n-r}-1} T^i = p^{n-r}w$ where w is a unit $\mathrm{mod}(p^n)$ and the cohomology trivializes except in dimension 0.

As is usual here, the case $p = 2$ is slightly different. If G is cyclic we can use the resolution of (II.3.8) again, but if not then it is most convenient to first calculate cohomology with respect to $\mathbb{Z}/2^r$ and then the cohmology of the resulting complex with respect to $\mathbb{Z}/2$. □

We now return to the proof of (4.1). Write

$$H^*_{cont}(G_{\mathbb{F}}; \mathbb{Q}/\mathbb{Z}) = \coprod_p H^*_{cont}(G_{\mathbb{F}}; (\mathbb{Q}/\mathbb{Z})_p)$$

as p runs over all the primes of $\mathbb{Z}$. Fix p and consider the projection

$$\pi_p\colon \ G_{\mathbb{F}} \longrightarrow \mathrm{Aut}((\mathbb{Q}/\mathbb{Z})_p) = U\hat{\mathbb{Z}}_p .$$

π_p is surjective and split except at the primes which divide n. More generally, it is also possible that π_p is not surjective but its image still lifts to a direct summand of $G_{\mathbb{F}}$. In this case $G_{\mathbb{F}} = \mathrm{Ker}(\pi_p) \oplus U_p$ where $U_p \subseteq U\hat{\mathbb{Z}}_p$ and

$$H^*_{cont}(G_{\mathbb{F}}; (\mathbb{Q}/\mathbb{Z})_p) = H^*_{cont}(\mathrm{Ker}(\pi_p); H^*_{cont}(U_p; (\mathbb{Q}/\mathbb{Z})_p)) .$$

But (4.2) shows that $H^*_{cont}(U_p; (\mathbb{Q}/\mathbb{Z})_p) \cong 0$ if the torsion subgroup of U_p in non-empty. Hence, when π_p is surjective and split there is no contribution at that prime.

Assume now that p divides n. Then $W^{tor} = L_1 \oplus \mathbb{Z}/d$ where $L_1 \subset \mathrm{Ker}(\pi_p)$, so $G_{\mathbb{F}}^{tor} = L \oplus \mathbb{Z}/d$ with $L \subset \ker(\pi_p)$. As regards the $\hat{\mathbb{Z}}_p^+$ summands, only the $\hat{\mathbb{Z}}_q^+$ where $q = p$ or q divides $p - 1$ can fail to be in the kernel of π_p. Consequently, we can write

$$G_{\mathbb{F}} = G_{\mathbb{F}}(q) \oplus \left(\mathbb{Z}/d \oplus \hat{\mathbb{Z}}_p^+ \oplus \prod_{q|(p-1)} \hat{\mathbb{Z}}_q^+ \right)$$

where $G_{\mathbb{F}}(q)$ lies in $\ker(\pi_p)$, and we have

$$H^*_{cont}(G_{\mathbb{F}}; (\mathbb{Q}/\mathbb{Z})_p) = H^*_{cont}(G_{\mathbb{F}}(p); H^*_{cont}(\mathbb{Z}/d \oplus \hat{\mathbb{Z}}_p^+ \times \prod_{q|(p-1)} \hat{\mathbb{Z}}_q^+; (\mathbb{Q}/\mathbb{Z})_p)) .$$

The summand $\hat{\mathbb{Z}}_p^+ \subset \mathrm{Aut}((\mathbb{Q}/\mathbb{Z})_p)$ is the kernel of restriction to $\mathbb{Z}/p \subset (\mathbb{Q}/\mathbb{Z})_p$ and consequently, the coefficient cohomology groups can be obtained from the spectral sequence with E_2-term

$$\begin{aligned} H^*_{cont}(\mathbb{Z}/d \times \prod_{q|(p-1)} \hat{\mathbb{Z}}_q^+; H^*_{cont}(\hat{\mathbb{Z}}_p^+; (\mathbb{Q}/\mathbb{Z})_p)) \\ = H^*_{cont}(\mathbb{Z}/d \times \prod_{q|(p-1)} \hat{\mathbb{Z}}_q^+; \mathbb{Z}/p^i) \end{aligned}$$

where $\mathbb{Z}/p^i$ is the subgroup of $(\mathbb{Q}/\mathbb{Z})_p$ fixed under the action of $\hat{\mathbb{Z}}_p^+$. A second application of (4.2) shows that if $\mathbb{Z}/l \times \prod_{q|(p-1)} \hat{\mathbb{Z}}_q^+$ acts non-trivially on $\mathbb{Z}/p^i$, then the cohomology groups above are identically zero provided that p is odd. On the other hand, the only way they will act trivially is if a primitive root $\zeta_{p^i} \in \mathbb{F}$. We thus have that for p odd, the contribution of $(\mathbb{Q}/\mathbb{Z})_p$ is obtained as the untwisted cohomology groups

$$H^*_{cont}\left(G_{\mathbb{F}}^{tor}; H^*_{cont}\left(\prod_{q\neq p} \hat{\mathbb{Z}}_q^+; \mathbb{Z}/p^i\right)\right).$$

On the other hand, since $(q,p) = 1$, we have that the second term is just $\mathbb{Z}/p^i$ in degree 0 and the calculation gives the untwisted cohomology groups $H^*_{cont}(G_{\mathbb{F}}^{tor}; \mathbb{Z}/m)$, where $\mathbb{Z}/m$ is the cyclic group of *odd* roots of unity in $\mathbb{F}$.

Only the case $p = 2$ remains. Here $G_{\mathbb{F}} = G_{\mathbb{F}}(2) \times \left(\mathbb{Z}/2^l \times \hat{\mathbb{Z}}_2^+\right)$ where, if $l > 0$, the generator of $\mathbb{Z}/2^l$ acts on $(\mathbb{Q}/\mathbb{Z})_2$ by $x \leftrightarrow -x$, while the generator of the righthand copy of $\hat{\mathbb{Z}}_2^+$ acts via multiplication by $(-3)^{2^r}$ for an appropriate $r \geq 0$. Then

$$\begin{aligned} H^*_{cont}(G_{\mathbb{F}}; (\mathbb{Q}/\mathbb{Z})_2) &= H^*_{cont}(G_{\mathbb{F}}(2); H^*_{cont}(\mathbb{Z}/2^l \times \hat{\mathbb{Z}}_2^+; (\mathbb{Q}/\mathbb{Z})_2)) \\ &= H^*_{cont}(G_{\mathbb{F}}(2); H^*_{cont}(\mathbb{Z}/2^l; \mathbb{Z}/2^{r+2})). \end{aligned}$$

If $l > 0$ then $H^i_{cont}(\mathbb{Z}/2^l; \mathbb{Z}/2^{r+2}) = \mathbb{Z}/2$ which, in turn, is isomorphic to $H^i(\mathbb{Z}/2^l; \mathbb{Z}/2)$ for all $i \geq 0$. If $l = 0$ then a primitive 2^{r+2} root of unity is present in $\mathbb{F}$. But in each case the group is $H^*_{cont}(G_{\mathbb{F}}^{tor}; \mathbb{Z}/2^s)$ where $\mathbb{Z}/2^s$ is the cyclic group of *even* roots of unity in $\mathbb{F}$. This completes the proof. □

Remark 4.3 In the isomorphism above there is no difficulty except when the only even roots of unity in $\mathbb{F}$ are ± 1. In this case, while the groups are given correctly by the theorem, the cochains which represent them at the prime 2 come from the twisted cohomology of $\mathbb{Z}/2^l$ with coefficients in $\mathbb{Z}/2^{r+2}$, and this must be recalled when making explicit calculations. (See also (4.6).)

Remark. Note that $W^{tor} \subset G_{\mathbb{F}}^{tor}$ is given as a subgroup

$$W \subset \mathbb{Z}/2 \times \mathbb{Z}/(p_1 - 1) \times \cdots \times \mathbb{Z}/(p_r - 1),$$

and it is only at W^{tor} that tricky things can happen in cohomology. It is convenient to choose a basis $\alpha_1, \dots, \alpha_s$ for W so that the projection of α_i onto $\mathbb{Z}/2 = \{\pm 1\}$ is trivial for $i \geq 2$. Also, let λ_q represent a generator for $\mathbb{Z}/(q-1)$ when $(q, 2n) = 1$. Then the Kunneth theorem and (4.1) show that $H^2(G_{\mathbb{F}};\mathbb{Q}/\mathbb{Z})$ has generators of the form

$$(4.4)\quad \begin{cases} \beta(\alpha_i) & \text{the Bockstein of } \langle\alpha_i\rangle\text{, order g.c.d (order}(\alpha_i), m), \\ \beta(\lambda_q) & \text{the Bockstein of } \langle\lambda_q\rangle\text{, order g.c.d(order}(\lambda_q), m), \\ \langle\alpha_i\rangle \otimes \langle\lambda_q\rangle & i \geq 2, \\ \langle\alpha_i\rangle \otimes \langle\alpha_j\rangle & i > j \geq 2, \\ \langle\lambda_p\rangle \otimes \langle\lambda_q\rangle & p \text{ and } q \text{ prime to } 2n. \end{cases}$$

Additionally, if there is an element $\alpha_1 \in G_{\mathbb{F}}^{tor}$ which acts on $(\mathbb{Q}/\mathbb{Z})_2$ as $x \leftrightarrow -x$, then the further classes $\langle\alpha_1\rangle \otimes \langle\alpha_j\rangle$, $\langle\alpha_1\rangle \otimes \langle\lambda_q\rangle$, are also required.

More precisely, the terms $\langle * \rangle \otimes \langle * \rangle$ in (4.4) can be derived as follows. Let

$$4.B.1 \qquad \phi_{\alpha_i,\lambda_q}\colon G_{\mathbb{F}} \longrightarrow G_{\mathbb{F}}^{tor} \longrightarrow \mathbb{Z}/|\alpha_i| \times \mathbb{Z}/(q-1)$$

be the evident projection sending α_i to the first generator, λ_q to the second and all others to 0. Similarly for $\phi_{\lambda_q,\lambda_{q'}}$, etc. Then the $\phi_{*,*}$ induce restriction maps

$$\operatorname{res}(\alpha_i, \lambda_q)\colon\ H^*(\mathbb{Z}/|\alpha_i| \times \mathbb{Z}/(q-1); (\mathbb{Q}/\mathbb{Z})^{G_{\mathbb{F}}} \longrightarrow H^*(G_{\mathbb{F}};\mathbb{Q}/\mathbb{Z})$$

and $\langle\alpha_i\rangle \otimes \langle\lambda_q\rangle$ is in the image of $\operatorname{res}(\alpha_i, \lambda_q)$, etc. Again, the classes involving α_1 are slightly different. Here the map is

$$H^*(\mathbb{Z}/|\alpha_1| \times \mathbb{Z}/(q-1); \mathbb{Z}/2^{r_2}) \longrightarrow H^*(G_{\mathbb{F}};\mathbb{Q}/\mathbb{Z})$$

where the generator of $\mathbb{Z}/|\alpha_1|$ acts as multiplication by -1 on $\mathbb{Z}/2^{r_2+1}$, and this map carries a class of $H^*(\mathbb{Z}/|\alpha_1| \times \mathbb{Z}/(q-1); \mathbb{Z}/2^{r_2})$ onto $\langle\alpha_1\rangle \otimes \langle\lambda_q\rangle$.

The Local Cohomology with $\mathbb{Q}/\mathbb{Z}$ Coefficients

To this point we have determined the cohomology for the *global* situation. It remains to determine the cohomology groups for the various completions of $\mathbb{F}$.

Theorem 4.5 *Let v be a non-archimedean valuation over p in $\mathbb{F}$. We have*
1. *If p is odd then $H^2_{cont}(G_v;\mathbb{Q}/\mathbb{Z}) = \mathbb{Z}/|V|$.*
2. *If $p = 2$ then $V = \mathbb{Z}/2$ or $V = \{1\}$. When $V = \{1\}$ we have $H^2_{cont}(G_v;\mathbb{Q}/\mathbb{Z}) = 0$, but if $V = \mathbb{Z}/2$ then $H^2_{cont}(G_v;\mathbb{Q}/\mathbb{Z}) = \mathbb{Z}/2 \oplus \mathbb{Z}/2$.*

Proof. To begin consider the homomorphism $\pi\colon G_v \to \varprojlim \operatorname{Aut}(\mathbb{Z}/p^j) \to \hat{\mathbb{Z}}_p^+$. The subgroup $\operatorname{Ker}(\pi)$ is a split summand of G_v with complementary summand $\hat{\mathbb{Z}}_p^+$. Hence we have

$$H^*_{cont}(G_v; (\mathbb{Q}/\mathbb{Z})_p) = H^*_{cont}(\mathrm{Ker}(\pi); \mathbb{Z}/p^r)$$

where $\mathbb{Z}/p^r$ is the subgroup of invariants under the action of the complementary copy of $\hat{\mathbb{Z}}_p^+$ on $(\mathbb{Q}/\mathbb{Z})_p$. We can write $\mathrm{Ker}(\pi) = \hat{\mathbb{Z}}_p^+ \times (V \times \prod_{q \neq p} \hat{\mathbb{Z}}_q^+)$ and so we have a spectral sequence with E_2-term

$$H^*_{cont}(\hat{\mathbb{Z}}_p^+; H^*_{cont}(V \times \prod_{q \neq p} \hat{\mathbb{Z}}_q^+; \mathbb{Z}/p^r))$$

converging to $H^*_{cont}(G_v; (\mathbb{Q}/\mathbb{Z})_p)$. But unless $V = \{1\}$ and the $\hat{\mathbb{Z}}_q^+$ all act trivially on $\mathbb{Z}/p^r$ the second factor trivializes to 0. When $V = \{1\}$ and the action is trivial we obtain $H^*_{cont}(\hat{\mathbb{Z}}_p^+; \mathbb{Z}/p^r)$ where we can assume that the action is trivial. Passing to limits through the inverse system which gives rise to $\hat{\mathbb{Z}}_p^+$ we see that

$$H^i_{cont}(\hat{\mathbb{Z}}_p^+; \mathbb{Z}/p^r) = \begin{cases} \mathbb{Z}/p^r & i = 0 \text{ or } 1, \\ 0 & \text{otherwise,} \end{cases}$$

when the action is trivial.

It remains to study the situation for $(\mathbb{Q}/\mathbb{Z})_q$ with $q \neq p$. We assume that p is odd, and we have that $\mathbb{Z}/(p^s - 1) \subset \mathbb{F}^\bullet$ is the subgroup of roots of unity of order prime to p in $\mathbb{F}$. Then we need to determine $\varprojlim (H^*(V \times \hat{\mathbb{Z}}_p^+ \times \mathbb{Z}/r; \mathbb{Z}/(p^{rs} - 1))$ where the limit is taken over r. Since the action of $\mathbb{Z}/r$ on $\mathbb{Z}/(p^{rs} - 1)$ is non-trivial, this becomes

$$\begin{aligned} H^*_{cont}(V \times \hat{\mathbb{Z}}_p^+; \mathbb{Z}/(p^s - 1)) &= H^*(V; \mathbb{Z}/(p^s - 1)) \\ &= \begin{cases} \mathbb{Z}/(p^s - 1) & * = 0, \\ \mathbb{Z}/|V| & * > 0. \end{cases} \end{aligned}$$

Finally, we consider the case $p = 2$. We may assume that $V = \mathbb{Z}/2$ since otherwise the calculation goes exactly as above and the result is zero. When we consider the coefficients $(\mathbb{Q}/\mathbb{Z})_q$ with q odd the calculation also goes as before, and, since $2^s - 1$ is prime to 2, the result is again zero. It remains to determine $H^*_{cont}(G_v; (\mathbb{Q}/\mathbb{Z})_2)$. As before let $\pi: G_v \to \mathrm{Aut}((\mathbb{Q}/\mathbb{Z})_2)$, then $\mathrm{Ker}(\pi) = \prod_{q \text{ prime}} \hat{\mathbb{Z}}_q^+$, and the image of π is $\mathbb{Z}/2 \times \hat{\mathbb{Z}}_2^+$. Thus

$$\begin{aligned} H^*_{cont}(G_v; (\mathbb{Q}/\mathbb{Z})) &= H^*_{cont}(\mathrm{Ker}(\pi); H^*_{cont}(\mathbb{Z}/2 \times \hat{\mathbb{Z}}_2^+; (\mathbb{Q}/\mathbb{Z})_2)) \\ &= H^*_{cont}(\mathrm{Ker}(\pi); H^*(\mathbb{Z}/2; \mathbb{Z}/2^r)) \\ &= H^*_{cont}(\hat{\mathbb{Z}}_2^+; \mathbb{Z}/2) \otimes H^*(\mathbb{Z}/2; \mathbb{Z}/2^r) \\ &= H^*_{cont}(\hat{\mathbb{Z}}_2^+; \mathbb{Z}/2) \otimes \mathbb{F}_2[e], \end{aligned}$$

where e is one dimensional. Now we have already seen that

$$H^i_{cont}(\hat{\mathbb{Z}}_2^+; \mathbb{Z}/2) = \begin{cases} \mathbb{Z}/2 & i = 0, 1, \\ 0 & \text{otherwise,} \end{cases}$$

so we have that $H^2(G_v; (\mathbb{Q}/\mathbb{Z})) = \mathbb{Z}/2 \oplus \mathbb{Z}/2$ as asserted. (4.5) follows. □

Remark 4.6 As was the case for (4.1), the explicit cocycle representatives for the two copies of $(\mathbb{Z}/2)$ in (4.5) are somewhat deceptive. The explicit representatives can be given as follows. First, since $V = \mathbb{Z}/2$, if we take the resolution of $\mathbb{Z}/2$ in (II.3¿8), then $\left(e_2 \longrightarrow \frac{1}{2}\right) \in \mathrm{Hom}(\mathbb{Z}(\mathbb{Z}/2), \mathbb{Q}/\mathbb{Z})$ represents the first generator. For the second generator, there are two components needed, V and the copy of $\hat{\mathbb{Z}}_2^+$ which acts trivially on $(\mathbb{Q}/\mathbb{Z})_2$. A (continuous) resolution of $\hat{\mathbb{Z}}_2^+$ has two generators $\bar{e}_0$ in dimension zero and $\bar{e}_1$ in dimension one, with $\partial(\bar{e}_1) = (T-1)\bar{e}_0$. Here T is a (continuous) generator for $\hat{\mathbb{Z}}_2^+$. Tensoring the two resolutions we obtain a representative for the second generator as

$$\left(e_1^V \otimes \bar{e}_1 \longrightarrow \frac{1}{2^r}\right).$$

The Explicit Form of the Evaluation Maps at the Finite Valuations

Let v be a finite odd prime and $\phi \in G_v$ be the Frobenius element. Suppose also that ω generates the torsion subgroup V. Write

$$\phi = \prod(\mu_{v,i}\alpha_i) \times \left(\prod(\tau_{v,q}\lambda_q) \times (tor\ free)\right)$$

in G_v and similarly

$$\begin{cases} \omega = \lambda_q & \text{if } v \text{ lies over } q \\ \omega = \coprod \gamma_{v,i}\alpha_i & \text{otherwise.} \end{cases}$$

Let g be the generator $g\colon e_2 \longrightarrow \frac{1}{p^s-1}$ of $H^2(G_v;\mathbb{Q}/\mathbb{Z})$. Also, denote by $|w|$ the order of the class w. Then we have

Theorem 4.7 *Suppose v lies over an odd prime q not dividing n, then*
1. $\beta(\lambda_q) \longrightarrow g$, $\beta(\alpha_i) \longrightarrow 0$, $\beta(\lambda_{q'}) \longrightarrow 0$ *for* $q' \neq q$,
2. $\langle\lambda_{q'}\rangle \otimes \langle\lambda_{q''}\rangle \longrightarrow 0$ *unless* q' *or* $q'' = q$, *and then*

$$\langle\lambda_q\rangle \otimes \langle\lambda_{q'}\rangle \longrightarrow \left[\frac{\tau_{v,q'}(q-1)}{|\langle\lambda_q\rangle \otimes \langle\lambda_{q'}\rangle|}\right] g\,,$$

3. $\langle\lambda_{q'}\rangle \otimes \langle\alpha_i\rangle \longrightarrow \begin{cases} 0 & q' \neq q \\ \left[\frac{\mu_{v,i}(q-1)}{|\langle\lambda_{q'}\rangle \otimes \langle\alpha_i\rangle|}\right] g & \textit{otherwise.} \end{cases}$

Theorem 4.8 *If v lies over an odd prime p which divides n, then*
1. $\beta(\alpha_i) \mapsto \frac{|\omega|}{|\alpha_i|}\gamma_{v,i}g$, $\beta(\lambda_q) \longrightarrow 0$,
2. $\langle\alpha_r\rangle \otimes \langle\alpha_s\rangle \mapsto \left[\frac{|V|}{2}\frac{p^i-1}{|\langle\alpha_r\rangle\otimes\langle\alpha_s\rangle|}\gamma_{v,i}\gamma_{v,s} + (\gamma_{v,r}\mu_{v,s} - \gamma_{v,s}\mu_{v,r})\frac{|V|}{|\langle\alpha_r\rangle\otimes\langle\alpha_s\rangle|}\right] g$ *where p^i is the residue class degree of $\mathbb{F}_v$.*

3. $\langle\lambda_i\rangle \otimes \langle\lambda_j\rangle \mapsto 0$,

4. $\langle\alpha_r\rangle \otimes \langle\lambda_j\rangle \mapsto \left[\frac{\gamma_{v,r}\tau_{v,q}|V|}{|\langle\alpha_r\rangle\otimes\langle\lambda_j\rangle|}\right] g$.

It remains to describe the maps on the classes involving α_1, $\langle\alpha_1\rangle \otimes \langle\alpha_i\rangle$, and $\langle\alpha_1\rangle \otimes \langle\lambda_q\rangle$.

Theorem 4.9

1. *Let q be an odd prime prime to n. Assume $\phi^{(q)}$ acts on $(\mathbb{Q}/\mathbb{Z})_2$ as multiplication by $(-1+\mu 2^{r_2})$. Then*

$$\langle\alpha_1\rangle \otimes \langle\lambda_q\rangle \mapsto \mu\left(\frac{q-1}{2}\right) g\,.$$

2. *Let p be an odd prime dividing n, and assume ϕ^p acts on $(\mathbb{Q}/\mathbb{Z})_2$ as multiplication by $(-1+\nu 2^{r_2})$, then*

$$\langle\alpha_1\rangle \otimes \langle\alpha_i\rangle \mapsto \gamma_{v,i}\nu\left(\frac{|V|}{2}\right) g\,.$$

(Here $\mathbb{Z}/2^{r_2}$ is the invariant subgroup of $(\mathbb{Q}/\mathbb{Z})_2$ under the action of the torsion free part of G_v. That is, it is determined by the projection $G_v \longrightarrow \hat{\mathbb{Z}}_2^+ \subset U_2$.)

At the prime 2 we need the *non-ramified Frobenius.* This is the class which acts as the identity on $m/2^i$ and acts as multiplication by 2^s on m/n with n odd.

Theorem 4.10 *Let v by dyadic and write the non-ramified Frobenius at v as $\coprod \omega_i\alpha_i \times \prod \omega_q\lambda_q \times (tor\ free)$, then*

$$\begin{aligned} \langle\alpha_1\rangle \otimes \langle\alpha_i\rangle &\mapsto w_i \times (second\ factor), \\ \langle\alpha_1\rangle \otimes \langle\lambda_q\rangle &\mapsto \omega_q \times (second\ factor), \\ \beta(\alpha_1) &\mapsto (first\ generator), \end{aligned}$$

and the remaining terms map to 0.

The map to the infinite primes will be given in (5.6).

5. The Explicit Structure of the Schur Subgroup, $S(\mathbb{F})$

At this point what remains is the determination of the two maps r_v^* and $i_*\colon H^2_{cont}(G_v;\mathbb{Q}/\mathbb{Z}) \longrightarrow H^2_{cont}(G_v;\hat{\mathbb{Q}}^\bullet_{p,cycl})$.

The Map $H^*_{cont}(G_v;\mathbb{Q}/\mathbb{Z})\to H^2_{cont}(G_v;\hat{\mathbb{Q}}^\bullet_{p,cycl})$.

From (4.5.1) we have $H^2_{cont}(G_v;\mathbb{Q}/\mathbb{Z}) = \mathbb{Z}/|V|$ for v over an odd prime (p), and the group is $\mathbb{Z}/2\oplus\mathbb{Z}/2$ if v lies over (2) with $V\neq\{1\}$. In the case v over (2) explicit generators are given in (4.6). We now determine the map

$$H^2_{cont}(G_v;\mathbb{Q}/\mathbb{Z})\longrightarrow B(\hat{\mathbb{F}}_v) = \mathbb{Q}/\mathbb{Z}$$

given by the coefficient map $i_v^*\colon H^2_{cont}(G_v;\mathbb{Q}/\mathbb{Z})\to H^2_{cont}(G_v;\hat{\mathbb{Q}}^\bullet_{p,cycl})$. The result is

Theorem 5.1

1. *i_v^* is injective for all odd p.*
2. *For $p=2$ the first $\mathbb{Z}/2$ injects only if the degree of $\hat{\mathbb{F}}_v$ over $\hat{\mathbb{Q}}_2$ is odd, but the second $\mathbb{Z}/2$ always injects.*

Proof. We start with the proof of (5.1.1). Consider the ramified extension $\hat{\mathbb{F}}_\omega = \hat{\mathbb{F}}_v(\zeta_p)$, and note that $\mathrm{Gal}(\hat{\mathbb{F}}_\omega/\hat{\mathbb{F}}_v) = V$. Assume that $\mathbb{Z}/(p^r-1)$ is the subgroup of the roots of unity in $\hat{\mathbb{F}}_\omega$ having order prime to p, and let $\zeta_{p^r-1}=\zeta$ be a generator. We have

Lemma 5.2 *The generator of $H^2(V;\hat{\mathbb{F}}^\bullet_\omega)$ can be taken as $(e_2^v\to\zeta)$.*

Proof of (5.2). $V=\mathbb{Z}/n$ is cyclic and $\hat{\mathbb{F}}_\omega/\hat{\mathbb{F}}_v$ is totally ramified so (5.2) is a direct application of (1.11). □

Now (5.1.1) follows from the fact that inflation is an injection and the fact that $e_2^v\to\frac{1}{p^r-1})$ generates $H^2(G_v;\mathbb{Q}/\mathbb{Z})$.

We now turn to the proof of (5.1.2).

Lemma 5.3 *The image of $\left(e_2^v\to\frac{1}{2}\right)$ in $B(\hat{\mathbb{F}}_v)$ is $\frac{1}{2}$ if and only if deg$(\hat{\mathbb{F}}_v)$ over $\hat{\mathbb{Q}}_2$ is odd.*

Proof of (5.3). We first verify directly that for $\hat{\mathbb{F}}_v = \hat{\mathbb{Q}}_2$ the image of the first class is $\frac{1}{2}\in B(\hat{\mathbb{Q}}_2)$. Also, write U_2 for $\mathrm{Aut}((\mathbb{Q}/\mathbb{Z})_2) = \mathbb{Z}/2\times\hat{\mathbb{Z}}_2^+$. Then, passing to H^2's we have

$$\text{res:}\ H^2(U_2\times\textstyle\prod\hat{\mathbb{Z}}_p^+;\mathbb{Q}/\mathbb{Z})\longrightarrow H^2(G_v;\mathbb{Q}/\mathbb{Z})$$

takes $\left(e_2^v\to\frac{1}{2}\right)$ to the same class for $\hat{\mathbb{F}}_v$. But in the commutative diagram

$$\begin{array}{ccc}
H^2(U_2 \times \prod \hat{\mathbb{Z}}_p^+; \mathbb{Q}/\mathbb{Z}) & \xrightarrow{\text{res}} & H^2(G_v; \mathbb{Q}/\mathbb{Z}) \\
\downarrow & & \downarrow \\
H^2(U_2 \times \prod \hat{\mathbb{Z}}_p^+; \lim\limits_{n\longrightarrow\infty} \mathbb{Q}_2(\zeta_n)) & \xrightarrow{\text{res}} & H^2(G_v; \lim\limits_{n\longrightarrow\infty} \mathbb{Q}_2(\zeta)) \\
\Big\downarrow{=} & & \downarrow \\
B(\mathbb{Q}_2) & \xrightarrow{\text{res}} & B(\hat{\mathbb{F}}_v)
\end{array}$$

the map in the bottom line is multiplication by $deg(\hat{\mathbb{F}}_v)$ over $\hat{\mathbb{Q}}_2$ according to (1.10). □

It remains to consider the second class in (4.6). The situation is this, we have an embedding

$$G_v \hookrightarrow U_2 \times \hat{\mathbb{Z}}_2^+ \times \prod_{p \neq 2} \hat{\mathbb{Z}}_p^+$$

corresponding to the inclusion $\hat{\mathbb{Q}}_2 \subset \hat{\mathbb{F}}_v$. Let g be the infinite generator of U_2, (g: $\frac{1}{2^s} \longrightarrow \frac{-3}{2^s}$), and ϕ the (2-adic) Frobenius, the element which acts as $x \mapsto x^2$ on $(\mathbb{Q}/\mathbb{Z})_2$ and is the identity on the rest of $(\mathbb{Q}/\mathbb{Z})$. Then the embedding sends the generators of the piece $(\mathbb{Z}/2 \times \hat{\mathbb{Z}}_2^+ \times \hat{\mathbb{Z}}_2^+)$ onto $\mathbb{Z}/2 = \langle \tau \rangle$ producted with a finite index subgroup of $(\hat{\mathbb{Z}}_2^+)^2$. Such a subgroup has one of three forms:

$$\begin{aligned}
&2^{r-2}g + \lambda 2^{r-2+s}\phi,\ 2^{r-2+s+l}\phi,\ \lambda \text{ a unit},\ l, s \geq 1,\\
&2^s\phi + \lambda 2^{s+l}g,\ 2^{s+l+r-2}g,\ \lambda \text{ a unit},\ l \geq 1,\ r \geq 3,\\
&2^{r-2}g,\ 2^s\phi,\ r \geq 2,\ s \geq 0.
\end{aligned}$$

However, in the second case it is convenient to rewrite the basis as

$$(2^{s+l}g + \lambda' 2^s\phi,\ 2^{s+r-2}\phi)\,,$$

so that all three cases have the form $2^{r-2}g + \lambda 2^s\phi$, $2^{s+l}\phi$. In particular $\hat{\mathbb{F}}_v$ is contained in $\hat{\mathbb{Q}}_2(\zeta_{2^{r+l}}, \zeta_n)$ where $n = t2^{s+l} - 1$ for some odd t, as a subfield of index 2^l. But the odd roots of unity which are present in $\hat{\mathbb{F}}_v$ form a copy of the cyclic group $\mathbb{Z}/(2^{t2^s} - 1)$ so this extension is non-ramified except for the action of τ which gives rise to ramification of degree two.

We now break the process of determining the cohomology map into two steps. First we show that there is a corresponding non-zero class in a maximal totally ramified extension, and then, using the known behavior of the cohomology restriction map when the coefficients lie in a complete local field the result will follow.

Consider the totally ramified extension corresponding to

$$\hat{\mathbb{Z}}_2^+ \times \prod_{p \neq 2} \hat{\mathbb{Z}}_p^+ \subset G_v$$

where the generator of the $\hat{\mathbb{Z}}_2^+$ corresponds to $2^{r+l-2}g$, that is the extension $\hat{\mathbb{F}}_\omega$ of $\hat{\mathbb{F}}_v$ for which the Galois group of $\hat{\mathbb{Q}}_{p,cycl}$ over $\hat{\mathbb{F}}_\omega$ is the subgroup above. Clearly

$$\hat{\mathbb{F}}_\omega = \lim_{n\longrightarrow\infty} \hat{\mathbb{Q}}_2(\zeta_{2^{r+l}}, \zeta_{2^{ts^s+n}-1})$$

and $\text{Gal}(\hat{\mathbb{F}}_\omega/\hat{\mathbb{F}}_v) = \mathbb{Z}/2 \times \hat{\mathbb{Z}}_2^+$ where the generator τ acts as $\zeta_{2^i} \leftrightarrow \zeta_{2^i}^{-1}$, while it fixes $\zeta_{2^{t2^s+n}-1}$. h acts, ignoring a possible unit, as

$$\zeta_{2^i} \longrightarrow \zeta_{2^i}^{(-3)^{2^{r-2}}}, \qquad \zeta_{2^{t2^s+n}-1} \longrightarrow \left(\zeta_{2^{t2^s+n}-1}\right)^{2^s}.$$

We calculate as before, obtaining

$$H^2(\text{Gal}(\hat{\mathbb{F}}_\omega/\hat{\mathbb{F}}_v); \mathbb{Z}/2^{r+l} \times \lim_{n\longrightarrow\infty} (\mathbb{Z}/(2^{t2^{s+n}} - 1)) = \mathbb{Z}/2 \oplus \mathbb{Z}/2,$$

and the second generator is represented by the cochain

$$c = e_1^\tau \otimes e_1^h \longrightarrow \frac{1}{2^{r+l}}.$$

Now consider the diagram

$$(5.4)\qquad \begin{array}{ccc} H^2(\text{Gal}(\hat{\mathbb{F}}_\omega/\hat{\mathbb{F}}_v); \mathbb{Z}/2^{r+l}) & \xrightarrow{i_\omega} & H^2(\text{Gal}(\hat{\mathbb{F}}_\omega/\hat{\mathbb{F}}_v); \hat{\mathbb{F}}_\omega^\bullet) \\ \Big\downarrow \text{res} & & \Big\downarrow \text{res} \\ H^2(G_v; \mathbb{Q}/\mathbb{Z}) & \xrightarrow{i_v} & H^2(G_v; \lim_n \hat{\mathbb{Q}}_2(\zeta_n)) \end{array}.$$

We claim first that $\text{res}(\{c\})$ is the class of the second cochain in (4.6). Indeed, a simple chain approximation shows that under the restriction

$$e_1^v \otimes e_1^\phi \longrightarrow \frac{(h^{2^l} - 1}{h - 1} e_1^\gamma \otimes e_1^h$$

and

$$\frac{h^{2^l} - 1}{h - 1}\left(\frac{1}{2^{r+l}}\right) = \frac{1}{2^r}.$$

Since g is multiplication by -3 it follows that $r > 1$, and we now show that the map i_ω when restricted to $\{c\}$ is non-trivial. To see this, restrict to some extension

$$\hat{\mathbb{F}}'_\omega = \hat{\mathbb{Q}}_2(\zeta_{2^r}, \zeta_{2^{t2^s+n}-1}) \supset \hat{\mathbb{F}}_v$$

with Galois group $\mathbb{Z}/2 \times \mathbb{Z}/2^n$ where $n \geq l + r$. Let $\hat{\mathbb{F}}''_\omega$ be the invariant subfield under τ in $\hat{\mathbb{F}}'_\omega$, then $\hat{\mathbb{F}}''_\omega$ is the maximal unramified extension of $\hat{\mathbb{F}}_v$ in $\hat{\mathbb{F}}'_\omega$.

A uniformizing parameter for $\hat{\mathbb{F}}'_\omega$ is $\zeta_{2^{r+l}} - 1 = \pi$, and one for $\hat{\mathbb{F}}''_\omega$ is $(1 + \tau)\pi = \theta\pi^2$ for some unit $\theta \in \hat{\mathbb{F}}'_\omega$. The inflation map

$$\mathbb{Z}/2^n = H^2(\mathbb{Z}/2^n; \hat{\mathbb{F}}''^{\bullet}_\omega) \longrightarrow H^2(\mathbb{Z}/2 \times \mathbb{Z}/2^n; \hat{\mathbb{F}}'^{\bullet}_\omega)$$

is injective, taking the generator to

$$\{e_0^\tau \otimes e_2^h \longrightarrow \theta\bar{\theta}\pi^2\}$$

where $\bar{\theta} \in \hat{\mathbb{F}}'_\omega$ is another unit and $\theta\bar{\theta}\pi^2 \in \hat{\mathbb{F}}_v$. Thus the class of the cocycle above has order 2^n in $H^2(\mathbb{Z}/2 \times \mathbb{Z}/2^n; \hat{\mathbb{F}}'^{\bullet}_\omega)$.

Lemma 5.5 *For $n > l + r$ we have that the norm map $N^h(\pi) \in \hat{\mathbb{F}}_v$.*

Proof of (5.5). $\tau N^h(\pi) = N^h(\tau(\pi)) = N^h(-\zeta_{2^{r+l}}\pi) = N^h(\zeta_{2^{r+l}})N^h(\pi)$. But, if

$$\bar{N}^h(\zeta_{2^{r+l}}) = \zeta_{2^{r+l}}^{1+(-3)^{2^{r-2}}+\cdots+(-3)^{2^{r-2}(2^l-1)}} = \zeta_{2^r}\ ,$$

then $N^h(\zeta_{2^{r+l}}) = [\bar{N}^h(\zeta_{2^{r+l}})]^{2^{n-l}} = 1$, so $N^h(\pi)$ is invariant and so belongs to $\hat{\mathbb{F}}_v$. □

Now we can complete the proof of (5.1). Calculating explicitly,

$$\delta[-(e_0^\tau \otimes e_2^h \longrightarrow N^h(\pi))] \text{ is } (e_1^\tau \otimes e_1^h \longrightarrow \zeta_{2^{r+l}}) \sim (e_0^\tau \otimes e_2^h \longrightarrow N^h(\pi))\ .$$

But $N^h(\pi) = \mu\pi^{2^n}$ for μ some unit in $\hat{\mathbb{F}}'_\omega$, and hence, from (5.5) we have

$$N^h(\pi) \ = \ (\pi^2\theta\bar{\theta})^{2^{n-1}} w$$

where $w \in \hat{\mathbb{F}}_v$ is a unit, and hence the norm of something in $\hat{\mathbb{F}}''_\omega$, so we have the equivalence of cocycles,

$$(e_0^\tau \otimes e_2^h \longrightarrow N^h(\pi)) \sim 2^{n-1}\{e_0^\tau \otimes e_2^h \longrightarrow \theta\bar{\theta}\pi^2\}$$

which is the element of order 2 in $H^2(\mathbb{Z}/2 \times \mathbb{Z}/2^n; \hat{\mathbb{F}}'^{\bullet}_\omega)$. This completes the proof of (5.1). □

The Invariants at the Infinite Real Primes

Since $\mathbb{F}$ is a cyclotomic field it has a real valuation if and only if complex conjugation is an element of $G_{\mathbb{F}}$. But this is the case if and only if $\tau: x \leftrightarrow -x \in G_{\mathbb{F}}$.

If $\tau \in G_{\mathbb{F}}$ then $(\mathbb{Q}/\mathbb{Z})^{G_{\mathbb{F}}} = \mathbb{Z}/2$ and $H^2_{cont}(G_{\mathbb{F}}^{tor}; \mathbb{Z}/2)$ is determined in (4.4).

We now give the evaluation result for infinite real places.

Theorem 5.6 *Write*

$$\tau \ = \ \prod \mu_{\infty,i}\alpha_i \times \prod \left(\frac{q-1}{2}\right) \lambda_q\ .$$

Then, in the map $r^\infty\colon H^2_{cont}(G^{tor}_{\mathbb{F}};\mathbb{Z}/2)\to H^2(\mathbb{Z}/2;\mathbb{Z}/2)=B(\mathbb{R})$ *we have*

1. $r^\infty(\beta(\lambda_q))\neq 0$
   $r^\infty(\beta(\alpha_i))\neq 0$ *only if* $\mu_{\infty,i}\not\equiv 0 \bmod 2$.
2. $r^\infty\langle\alpha_1\rangle\otimes\langle *\rangle = 0.$
3. $r^\infty\langle\alpha_i\rangle\otimes\langle\alpha_j\rangle = \mu_{\infty,i}\mu_{\infty,j}(e^2),$
   $r^\infty\langle\alpha_i\rangle\otimes\langle\lambda_q\rangle = \frac{q-1}{2}\mu_{\infty,i}(e^2),$
   $r^\infty\langle\lambda_q\rangle\otimes\langle\lambda_{q'}\rangle = \frac{q-1}{2}\cdot\frac{q'-1}{2}(e^2)$

otherwise. In particular, the evaluation is the same at each infinite prime.

Proof. We consider the composite

$$\mathbb{Z}/2\xrightarrow{\ i\ }G_{\mathbb{F}}\xrightarrow{\ \phi_{\alpha_i,\lambda_q}\ }\mathbb{Z}/|\alpha_i|\times\mathbb{Z}/(q-1)\ ,$$

where $i(T)=\tau$, and we construct a chain approximation using the diagonal approximation $e_i\mapsto\sum e_j\otimes T^j e_{i-j}$ of (IV.3.6) and the easily derived chain approximation for the map of groups $i\colon\mathbb{Z}/2\to\mathbb{Z}/2n$ sending T to a^n. Using the resolutions of (II.3.8) we obtain that a chain approximation is

$$\begin{aligned} e_{2i}&\longrightarrow e_{2i}\\ e_{2i+1}&\longrightarrow\frac{a^n-1}{a-1}e_{2i+1}\,. \end{aligned}$$

Combining these we get a chain approximation for $\phi_{\alpha_i,\lambda_q}\cdot i$ as

5.7
$$\begin{aligned} e_1\ &\longrightarrow\ \frac{\alpha_i^{\mu_{\infty,i}}-1}{\alpha_i-1}\lambda_q^{(q-1)/2}e_1^{\alpha_i}\otimes e_0^{\lambda_q}\\ &\qquad+\frac{\lambda_q^{(q-1)/2}-1}{\lambda_q-1}e_0^{\alpha_i}\otimes e_1^{\lambda_q}\\ e_2\ &\longrightarrow\ \lambda_q^{(q-1)/2}e_2\otimes e_0+\frac{(\alpha_i^{\mu_{\infty,i}}-1)(1-\lambda_q^{(q-1)/2})}{(\alpha_i-1)(\lambda_q-1)}e_1\otimes e_1\\ &\qquad+e_0\otimes e_2. \end{aligned}$$

From this (5.6.1) and (5.6.3) follow directly. To get (5.6.2) note that a chain for $\langle\alpha_1\rangle\otimes\langle\lambda_q\rangle$ is given by the formula

5.8
$$c\ =\ \left(e_1^{\alpha_1}\otimes e_1^{\lambda_q}\to\frac{1}{2^{r_2+1}}\right)-\left(e_0^{\alpha_1}\otimes e_2^{\lambda_q}\to\frac{q-1}{2^{r_2+2}}\right).$$

Now note that in this case $2\alpha_1=e$, and use (5.7) to show that $\phi_{\alpha_1,\lambda_q}\cdot i)^{\#}(c)=0$.

This completes the proof of (5.6) and forms a good model for the remaining calculations at the finite primes. □

The Remaining Local Maps

The techniques above work equally well at the finite places. Indeed there are only a few things to be careful about. The following lemma is the key to calculating the images of the mixed terms $\langle\lambda_q, \lambda_{q'}\rangle$, $\langle\alpha_i, \lambda_q\rangle$ etc. at the v lying over odd primes. The notation is that established in (4.7) through (4.10).

Lemma 5.9 *Let $\mathbb{F}_v$ be the completion of $\mathbb{F}$ at v and $\bar{F}_v$ be its residue field which has order p^s. Let ψ be the Frobenius map, $\psi(\zeta_n) = \zeta_n^{p^s}$ for $(n,p) = 1$. Then, in $H^2(V \times \prod_q \hat{\mathbb{Z}}_q^+; \mathbb{Z}/p^s - 1)$ the mixed class $\langle\psi\rangle \otimes \langle v\rangle$ maps to $-g$.*

Proof. We construct a specific cohomology between the two classes:

$$\delta\left(e_0^\psi \otimes e_1^v \longrightarrow \frac{1}{|V|(p^s-1)}\right) = \left(e_1^\psi \otimes e_1^v \longrightarrow \frac{1}{|V|}\right) + \left(e_0^\psi \otimes e_2^v \longrightarrow \frac{1}{p^s-1}\right). \quad \square$$

From (5.9) theorems (4.7) and (4.8) follow easily but (4.9) is slightly more delicate. From (5.8) we have the expression for the required cocycle. We consider the composition

5.10 $$G_v \longrightarrow G_{\mathbb{F}} \longrightarrow \mathbb{Z}/2 \times \mathbb{Z}/(q-1)$$

where the $\mathbb{Z}/2$ represents the action of α_1 on $(\mathbb{Q}/\mathbb{Z})_2$.

Assume that

$$\begin{array}{lcl} \psi & \longrightarrow & (u_{v,1}\alpha_1) \times \theta_{v,q}\lambda_q \\ v & \longrightarrow & \lambda_q \end{array}$$

and, in order for there to be anything interesting, clearly $u_{v,1}$ must be odd. So we assume this also. Then the chain of (5.8) maps back to

$$-\left(e_1^v \otimes e_1^\psi \longrightarrow \frac{1}{2^{r_2}}\right) - \left(e_2^v \otimes e_0^\psi \longrightarrow \frac{q-1}{2^{r_2+1}}\right).$$

Now, ψ acts on $\frac{1}{2^s}$ by multiplying it by $-1 + \lambda 2^{r_2}$ and we have

$$\begin{aligned} \delta\left(e_1^v \otimes e_0^\psi \longrightarrow \frac{1}{2^{r_2+1}}\right) &= -\left(e_1^v \otimes e_1^\psi \longrightarrow -\frac{1}{2^{r_2}} + \frac{\lambda}{2}\right) \\ &\quad + \left(e_2^v \otimes e_0^\psi \longrightarrow \frac{q-1}{2^{r_2+1}}\right) \\ &= \left(e_1^v \otimes e_1^\psi \longrightarrow \frac{\lambda}{2}\right) + \left(e_1^v \otimes e_1^\psi \longrightarrow \frac{1}{2^{r_2}}\right) \\ &\quad + \left(e_2^v \otimes e_0^\psi \longrightarrow \frac{q-1}{2^{r_2+1}}\right) \end{aligned}$$

and (4.9.1) follows from (5.9). (4.9.2) follows similarly.

The considerations leading to (4.10) are essentially the same. Here the important ψ is the generator of the $\hat{\mathbb{Z}}_2^+ \subset \hat{\mathbb{Z}}_2^+ \times \hat{\mathbb{Z}}_2^+$ which is the kernel of the projection $\pi: G_v \to U_2$, since the second generator of $H^2(G_v; \mathbb{Q}/\mathbb{Z})$ is

$$\left(e_1^v \otimes e_1^\psi \to \frac{1}{2^{r_2+1}} \right) .$$

References

[Ad1] J.F. Adams, Structure and applications of the Steenrod algebra, Comm. Math. Helv.**32** (1958), 180–214.

[Ad2] J.F. Adams, On the non-existence of elements of Hopf invariant One, Ann. Math.**72** (1960), 20–104.

[A1] A. Adem, Cohomological non-vanishing for modules over p-groups, J. Algebra **141** No.2 (1991), 376–381.

[A2] A. Adem, Cohomological restrictions on finite group actions, J. Pure and Applied Algebra **54** (1988), 117–139.

[AB] A. Adem, W. Browder, The free rank of symmetry of $(S^n)^k$, Inv. Math.**92** (1988), 431–440.

[AM1] A. Adem, R.J. Milgram, $\mathcal{A}_5$–invariants, the cohomology of $L_3(4)$ and related extensions, Proc. Lond. Math. Soc. **66**(1993), 187–224.

[AM2] A. Adem, R.J. Milgram, Invariants and cohomology of groups, Bol. Soc. Math. Mex., (to appear).

[AM3] A. Adem, R.J. Milgram, The geometry and cohomology of the O'Nan Group $O'N$, in preparation.

[AM4] A. Adem, R.J. Milgram, The cohomology of the Mathieu Group M_{22}, Topology, (to appear).

[AMM1] A. Adem, J. Maginnis, R.J. Milgram, Symmetric invariants and cohomology of groups, Math. Ann. **287** (1990), 391–411.

[AMM2] A. Adem, J. Maginnis, R.J. Milgram, The geometry and cohomology of the Mathieu group M_{12}, J. Algebra **139** (1991), 90–133.

[Adem] J. Adem, The relations on Steenrod powers of cohomology classes, *Algebraic Geometry and Topology*, Princeton (1957), 191–238.

[AP] C. Allday, V. Puppe, *Cohomological Methods in Transformation Groups,* Cambridge Studies in Advanced Mathematics 32, Cambridge University Press (1993).

[AE] J. Alperin, L. Evens, Representations, resolutions and Quillen's dimension theorem, J.Pure & Applied Algebra **22** (1981), 1–9.

[ANT] Artin, Nesbitt, Thrall, *Rings With Minimal Condition*, U. of MIchigan Press, (1944).

[BP] M. Barratt, S. Priddy, On the homology of non-connected monoids and their associated groups, Comm. Math. Helv. **47** (1972), 1–14.

[BDH] G. Baumslag, M. Dyer, A. Heller, The topology of discrete groups, J.Pure & Applied Algebra **16** (1980), 1–49.

[BM] H. Behr, J. Mennicke, A presentation of the group $PSL(2,p)$, Can. J. Math **20** (1968), 1432–1438.

[Be] M. Benard, The Schur subgroup I, J. of Algebra **22** (1972), 374–377.

[BeS] M. Benard, M. Schacher, The Schur subgroup II, J. of Algebra **22** (1972), 378–385.

[B] H. Bender, On groups with abelian 2-Sylow subgroups, Math. Zeit. **117** (1970), 164–176.

[Be] D. Benson, *Representations and Cohomology II: Cohomology of Groups and Modules*, Cambridge Studies in Advanced Mathematics, Cambridge University Press (1991).

[BC1] D. Benson, J. Carlson, Complexity and multiple complexes, Math. Zeit. **195** (1987), 221–238.

[BC2] D. Benson, J. Carlson, Projective resolutions and Poincaré duality complexes, to appear.

[BC3] D. Benson, J. Carlson, Diagrammatic methods for modular representations and cohomology, Comm. Algebra **15** (1987), 53–121.

[BCR] D. Benson, J. Carlson, G. Robinson, On the vanishing of group cohomology, J. Algebra **131** (1990), 40–73.

[Bo1] A. Borel, *Seminar on Transformation Groups*, Ann. of Math. Studies **46** Princeton U. Press, (1960).

[Bo2] A. Borel, La cohomologie mod 2 de certains espaces homogènes, Comm. Math. Helv. **27** (1953), 165–197.

[BHMM] C. Boyer, J. Hurtubise, B.M. Mann, R.J. Milgram, The topology of instanton moduli spaces, I: The Atiyah-Jones conjecture, Ann. of Math., (to appear).

[Bre1] G. Bredon, *Introduction to Compact Transformation Groups*, Academic Press New York, (1972).

[Bre2] G. Bredon, *Equivariant Cohomology Theories*, Lecture Notes in Mathematics, **34** (1967), Springer Berlin.

[Brow] W. Browder, Cohomology and group actions, Inv. Math. **71** (1983), 599–608.

[Brown] K. Brown, *Cohomology of Groups*, Springer-Verlag Graduate Texts in Mathematics **87**, (1982).

[BR] F. Buekenhout, S. Rees, The subgroup structure of the Mathieu group, Math. of Computation, **50**(1988), 595-605.

[Card] H. Cardenas, El álgebra de cohomologiá del grupo simétrico de grado p^2, Boletín Sociedad Matemática Mexicana **10** (1965), 1–30.

[C] J. Carlson, The varieties and cohomology ring of a module, J. Algebra **85** (1983), 104–143.

[Ca] G. Carlsson, On the rank of abelian groups acting freely on $(S^n)^k$, Inv. Math. **69** (1982), 393–404.

[Car] H. Cartan, Sur l'iteration des operations de Steenrod, Comm. Math. Helv. **29** (1955), 40–58.

[CE] H. Cartan, S. Eilenberg, *Homological Algebra*, Princeton University Press, Princeton NJ (1956).

[Carter] R. Carter, *Simple Groups of Lie Type*, Wiley, New York (1972).

[CF] J.W.S. Cassels, A. Fröhlich, *Algebraic Number Theory*, Academic Press, (1967).

[CS] T. Chang, T. Skjelbred, Group actions on Poincaré duality spaces, Bull. A.M.S., **78** (1972), 1024–1026.

[Ch] G.R. Chapman, Generators and relations for the cohomology ring of J_1, in LMS Lecture Notes **71** (1982), Cambridge University Press.

[CW] A.M. Cohen, D.B. Wales, Finite subgroups of $G_2(\mathbb{C})$, Comm. in Algebra, **11** (1983), 441-459.

[Coh] F.R. Cohen, Two-primary analogues of Selick's theorem and the Kahn-Priddy theorem for the 3-sphere, Topology, **23** (1984), 401–421.

[C^2M^2] F.R. Cohen, R.L. Cohen, B.M. Mann, R.J. Milgram, The topology of rational functions and divisors of surfaces, Acta Math., **166** (1991), 163–221.

[CMN] F.R. Cohen, J.C. Moore, J.A. Neisendorfer, The double suspension and exponents of the homotopy groups of spheres, Ann. Math., **118** (1979), 549–565.

[Co] J.H. Conway, R.T. Curtis, S.P. Norton, R.A. Wilson, *Atlas of Finite Groups*, Clarendon Press, Oxford (1985).

[Cox] H.S.M. Coxeter, Integral Cayley numbers, Duke J., **13**(1946), 561-578.

[DM] J.F. Davis, R.J. Milgram, *A Survey of the Spherical Space Form Problem*, Mathematical Reports, **2** Part 2, Harwood Press, (1985).

[DM2] J.J. Davis, R.J. Milgram, Semicharacteristics, bordism, and free group actions, Trans. A.M.S., **312** (1989), 53–83.

[D] L. Dickson, The alternating group on eight letters and the quaternary linear congruence group modulo two, Math. Ann. **54** (1901), 564–569.

[Do] A. Dold, 'Homology of symmetric products and other functors of complexes, Ann. Math., **68** (1958), 54–80.

[DT] A. Dold, R. Thom, Quasifaserungen und unendliche symmetrische Produkte, Ann. Math. **67** (1958), 239–281.

[DW] W.G. Dwyer, C.W. Wilkerson, A new finite loop space at the prime 2, J.A.M.S., **6**(1993), 37–64.

[DL] E. Dyer, R. Lashof, Homology of iterated loop spaces, Am. J. Math. **84** (1962), 35–88.

[E] B. Eckmann, Cohomology of groups and transfer, Ann. Math. **58** (1953), 481–493.

[EM] S. Eilenberg, S. MacLane, Cohomology theory in abstract groups I & II, Ann. Math. **48** (1947), 51–78 and 326–341.

[Ev1] L. Evens, The cohomology ring of a finite group, Trans. A.M.S. **101** (1961), 224–239.

[Ev2] L. Evens, *Cohomology of Groups*, Oxford University Press (1992).

[FP] S. Fiedorowicz, S. Priddy, *Homology of Classical Groups over Finite Fields and their Associated Infinite Loop Spaces*, Lecture Notes in Mathematics, **674**, Springer-Verlag (1978).

[FH] K.L. Fields, I.N. Herstein, On the Schur subgroup of the Brauer group, J. of Algebra, **20** (1972), 70–71.

[FM] P. Fong, R.J. Milgram, On the geometry and cohomology of the simple groups $G_2(q)$ and ${}^3D_4(q)$, preprint, Stanford University (1990).

[FW] P. Fong, W. Wong, A Characterization of the finite simple groups $PSp(4,q)$, $G_2(q)$, $D_4^2(q)$ I, Nagoya J. Math. **36** (1969), 143–184.

[G] W. Gaschütz, Kohomologische Trivialitäten un aussere automorphismen von p-gruppen, Math. Zeit. **88** (1965), 432–433.

[Gi] C. Giffen, List of Problems, Contemporary Mathematics 19 (1983).

[Go] D. Goldschmidt, Automorphisms of trivalent graphs, Ann. Math. **111** (1980), 377–406.

[Gor] D. Gorenstein, *Finite Simple Groups: An Introduction to Their Classification*, University Series in Mathematics, Plenum Press (1982).

[Gr1] K. Gruenberg, *Relation Modules of Finite Groups*, AMS Regional Conf. Series **25**, (1976).

[Gr2] K. Gruenberg, *Cohomological Topics in Group Theory*, Lecture Notes in Mathematics **143**, Springer-Verlag (1970).

[Ha] M. Hall Jr., *The Theory of Groups*, Macmillian, (1959).

[H] J.C. Hausmann, Manifolds with a Given Homology and Fundamental Group, Comm. Math. Helv. **53** (1978), 113–134.

[HH] J.C. Hausmann, D. Husemoller, Acyclic maps, L'Enseignement Math.**25** (1979), 53–75.

[He] T. Hewett, Thesis, Stanford University (1991).

[HS] G. Hochschild, J.P. Serre, Cohomology of group extensions, Trans. A.M.S. **74** (1953), 110–134.

[HiS] P. Hilton, U. Stammbach, *A Course in Homological Algebra*, Springer–Verlag (1971).

[Hu] Hungerford, *Algebra*, Graduate Texts in Mathematics **73**, Springer-Verlag (1974).

[Jam] I.M. James, On the suspension sequence, Ann. Math., **65** (1957), 74–107.

[Ja] Z. Janko, A characterization of the Mathieu simple groups, J. Algebra, **9**(1968), 1–19.

[J] G.J. Janusz, The Schur group of an algebraic number field, Ann. of Math., **103** (1976), 253-281.

[KP] D. Kahn, S. Priddy, On the transfer in the homology of symmetric groups, Math. Proc. Camb. Phil. Soc. **83** (1978), 91–101.

[KT] D. Kan, W. Thurston, Every connected space has the cohomology of a $K(\pi, 1)$, Topology **15** (1976), 253–258.

[Kl] S.N. Kleinerman, The Cohomology of Chevalley Groups of Exceptional Lie type, AMS Memoir (1982).

[K] N. Kuhn, Chevalley group theory and the transfer in the homology of the symmetric groups, Topology **24** (1985), 247–264.

[Ku] T. Kuo, On the exponent of $H^n(G, \mathbb{Z})$, J. Algebra **7** (1967), 160–167.

[La] S. Lang, *Algebra*, Addison-Wesley (1971).

[La2] S. Lang, *Algebraic Number Theory*, Addison-Wesley, (1973).

[Li] A. Liulevicius, A theorem in homological algebra and stable homotopy of projective spaces, Trans. A.M.S., **109**(1963), 540–552.

[L] R. Lyndon, The cohomology theory of group extensions, Duke Math. J. **15** (1948), 271–292.

[MM] I. Madsen, R.J. Milgram, *The Classifying Spaces for Surgery and Cobordism of Manifolds*, Ann. Math. Studies **92**, Vol.92 Princeton University Press, (1979).

[Ma] B.M. Mann, The cohomology of the symmetric groups, Trans. A.M.S. **242** (1978),

[MM1] B.M. Mann, R.J. Milgram, Some space of holomorphic maps to complex Grassmann manifolds, J. Diff. Geom., **33** (1991), 301–324.

[MM2] B.M. Mann, R.J. Milgram, On the moduli space of $SU(n)$ monopoles and holomorphic maps to flag manifolds, J. Diff. Geom., **38**(1993), 39–103.

[Mau] C.R.F. Maunder, A short proof of a theorem of Kan and Thurston, Bull. London Math. Soc. **13** (1981), 325–327.

[Mac] S. Maclane, The Milgram bar construction as a tensor product of functors, *Saunders Maclane Selected Papers*, 435-453, Springer-Verlag, (1979).

[M1] R.J. Milgram, The mod 2 spherical characteristic classes, Ann. Math. **92** (1970), 238–261.

[M2] R.J. Milgram, On the geometry and cohomology of the simple groups $G_2(q)$ and ${}^3D_4(q)$, II, Preprint, Stanford University, (1989).

[M3] R.J. Milgram, The homology of symmetric products, Trans. A.M.S. **138** (1969), 251–265.

[M4] R.J. Milgram, The bar construction and abelian H-spaces, Ill. J. Math. **11** (1967), 242-250.

[M5] R.J. Milgram, The cohomology of the Mathieu group M_{23}, (preprint, Stanford University, 1993).

[Mi] J. Milnor, The Steenrod algebra and its dual, Ann. Math. **67** (1958), 150–171.

[Mi2] J. Milnor, Groups which act on S^n without fixed points, Amer. J. Math., **79** (1957), 623–630.

[MS] J. Milnor, J. Stasheff, *Characteristic Classes*, Ann. Math. Studies **76**, Princeton University Press, (1974).

[Mu] H. Mui, Modular invariant theory and cohomology algebras of the symmetric groups, J. Fac. Sci. Tokyo **22** (1975), 310–369.

[Na] M. Nakaoka, Homology of the infinite symmetric group, Ann. Math. **73** (1961), 229–257.

[N] T. Nakayama, On modules of trivial cohomology over a finite group I, Ill. J. Math. **1** (1957), 36–43.

[P] R. Pierce, *Associative Algebras*, Graduate Texts in Mathematics, **83** Springer-Verlag (1980).

[Q1] D. Quillen, The spectrum of an equivariant cohomology ring I & II, Ann. Math. **94** (1971), 549–602.

[Q2] D. Quillen, Cohomology of groups, ICM Proceedings, Nice 1970, Gauthier Villars (1971), Vol.II, 47–51.

[Q3] D. Quillen, On the cohomology and K-theory of the general linear groups over a finite field, Ann. Math. **96** (1972), 552–586.

[Q4] D. Quillen, The mod 2 cohomology rings of extra-special 2-groups and the spinor groups, Math. Ann. **194** (1971), 197–212.

[Q5] D. Quillen, Homotopy properties of the poset of non-trivial p-subgroups of a group, Adv. Math. **28** (1978), 101–128.

[Q6] D. Quillen, The Adams conjecture, Topology **10** (1971), 67–80.

[QV] D. Quillen, B. Venkov, Cohomology of finite groups and elementary abelian subgroups, Topology **11** (1972), 317–318.

[RS] M. Richardson, P.A. Smith, Periodic transformations of complexes, Ann. Math. **39** (1938), 611–633.

[R] D.S. Rim, Modules Over finite groups, Ann. Math. **69** (1959), 700–712.

[RSY] A. Ryba, S. Smith, S. Yoshiara, Some projective modules determined by sporadic geometries, UIC preprint (1988).

[S] G. Segal, The topology of rational functions, Acta Math., **143** (1979), 39–72.

[Sel] P. Selick, Odd primary torsion in $\pi_k(S^3)$, Topology, **17** (1978), 407–412.

[Se1] J.P. Serre, Cohomologie modulo 2 des complexes d' Eilenberg–MacLane, Comm. Math. Helv. **27** (1953), 198–132.

[Se2] J.P. Serre, Sur la dimension cohomologique des groupes profinis, Topology **3** (1965), 413–420.

[Se3] J.P. Serre, *Trees*, Springer-Verlag (1980).

[Se4] J.P. Serre, *Local Fields*, Graduate Texts in Mathematics, **67**, Springer-Verlag, (1979).

[Sn] V. Snaith, A stable decomposition for $\Omega^n \Sigma^n X$, J. London Math. Soc. **7** (1974), 577–583.

[Sp] E. Spanier, *Algebraic Topology*, Springer Verlag (1989).

[St] R. Stanley, Invariant theory, Bull. A.M.S. (1979)

[Ste] N. Steenrod, Homology groups of symmetric groups and reduced power operations, Proc. Nat. Acad. Sci. **39** (1953), 213–217.

[Ste2] N. Steenrod, Milgram's classifying space of a topological group, Topology **7** (1968), 319-368.

[SE] N. Steenrod, D. Epstein, *Cohomology Operations*, Ann. Math. Studies **50**, Princeton University Press (1962).

[S1] R.G. Swan, The p-period of a finite group, Ill. J. Math. **4** (1960), 341–346.

[S2] R.G. Swan, Minimal resolutions for finite groups, Topology **4** (1965), 193–208.

[S3] R.G. Swan, A new method in fixed-point theory, Comm. Math. Helv. **34**, (1960).

[TY] M. Tezuka, N. Yagita, The cohomology of subgroups of $GL_n(\mathbb{F}_q)$, Contemporary Math. **19** (1983), 379–396.

[Th] C. Thomas, *Characteristic Classes and the Cohomology of Finite Groups*, Cambridge Studies in Advanced Mathematics 9, Cambridge University Press (1986).

[V] B. Venkov, Cohomology algebras for some classifying spaces, Dokl. Akad. Nauk. SSR **127** (1959), 943–944.

[Wa] C.T.C. Wall, Resolutions for extensions of groups, Proc. Camb. Phil. Soc., **57**(1961), 251–255.

[We] P. Webb, A local method in group cohomology, Comm. Math. Helv. **62** (1987), 135–167.

[W] C. Wilkerson, A primer on Dickson invariants, in Contemporary Math. **19** (1983), 421–434.

[Wo] J. Wolf, *Spaces of Constant Curvature*, Publish or Perish, Inc. (1977).

[Y] T. Yamada, *The Schur subgroup of the Brauer Group*, Lecture Notes in Mathematics **397**, Springer-Verlag, (1974).

Index

Grundlehren der mathematischen Wissenschaften

A Series of Comprehensive Studies in Mathematics

A Selection

204. Popov: Hyperstability of Control Systems
205. Nikol'skiĭ: Approximation of Functions of Several Variables and Imbedding Theorems
206. André: Homologie des Algébres Commutatives
207. Donoghue: Monotone Matrix Functions and Analytic Continuation
208. Lacey: The Isometric Theory of Classical Banach Spaces
209. Ringel: Map Color Theorem
210. Gihman/Skorohod: The Theory of Stochastic Processes I
211. Comfort/Negrepontis: The Theory of Ultrafilters
212. Switzer: Algebraic Topology – Homotopy and Homology
215. Schaefer: Banach Lattices and Positive Operators
217. Stenström: Rings of Quotients
218. Gihman/Skorohod: The Theory of Stochastic Procrsses II
219. Duvant/Lions: Inequalities in Mechanics and Physics
220. Kirillov: Elements of the Theory of Representations
221. Mumford: Algebraic Geometry I: Complex Projective Varieties
222. Lang: Introduction to Modular Forms
223. Bergh/Löfström: Interpolation Spaces. An Introduction
224. Gilbarg/Trudinger: Elliptic Partial Differential Equations of Second Order
225. Schütte: Proof Theory
226. Karoubi: K-Theory. An Introduction
227. Grauert/Remmert: Theorie der Steinschen Räume
228. Segal/Kunze: Integrals and Operators
229. Hasse: Number Theory
230. Klingenberg: Lectures on Closed Geodesics
231. Lang: Elliptic Curves: Diophantine Analysis
232. Gihman/Skorohod: The Theory of Stochastic Processes III
233. Stroock/Varadhan: Multidimensional Diffusion Processes
234. Aigner: Combinatorial Theory
235. Dynkin/Yushkevich: Controlled Markov Processes
236. Grauert/Remmert: Theory of Stein Spaces
237. Köthe: Topological Vector Spaces II
238. Graham/McGehee: Essays in Commutative Harmonic Analysis
239. Elliott: Proabilistic Number Theory I
240. Elliott: Proabilistic Number Theory II
241. Rudin: Function Theory in the Unit Ball of C^n
242. Huppert/Blackburn: Finite Groups II
243. Huppert/Blackburn: Finite Groups III
244. Kubert/Lang: Modular Units
245. Cornfeld/Fomin/Sinai: Ergodic Theory
246. Naimark/Stern: Theory of Group Representations
247. Suzuki: Group Theory I
248. Suzuki: Group Theory II
249. Chung: Lectures from Markov Processes to Brownian Motion
250. Arnold: Geometrical Methods in the Theory of Ordinary Differential Equations
251. Chow/Hale: Methods of Bifurcation Theory
252. Aubin: Nonlinear Analysis on Manifolds. Monge-Ampère Equations
253. Dwork: Lectures on ρ-adic Differential Equations
254. Freitag: Siegelsche Modulfunktionen
255. Lang: Complex Multiplication
256. Hörmander: The Analysis of Linear Partial Differential Operators I

257. Hörmander: The Analysis of Linear Partial Differential Operators II
258. Smoller: Shock Waves and Reaction-Diffusion Equations
259. Duren: Univalent Functions
260. Freidlin/Wentzell: Random Perturbations of Dynamical Systems
261. Bosch/Güntzer/Remmert: Non Archimedian Analysis – A System Approach to Rigid Analytic Geometry
262. Doob: Classical Potential Theory and Its Probabilistic Counterpart
263. Krasnosel'skiĭ/Zabreĭko: Geometrical Methods of Nonlinear Analysis
264. Aubin/Cellina: Differential Inclusions
265. Grauert/Remmert: Coherent Analytic Sheaves
266. de Rham: Differentiable Manifolds
267. Arbarello/Cornalba/Griffiths/Harris: Geometry of Algebraic Curves, Vol. I
268. Arbarello/Cornalba/Griffiths/Harris: Geometry of Algebraic Curves, Vol. II
269. Schapira: Microdifferential Systems in the Complex Domain
270. Scharlau: Quadratic and Hermitian Forms
271. Ellis: Entropy, Large Deviations, and Statistical Mechanics
272. Elliott: Arithmetic Functions and Integer Products
273. Nikol'skiĭ: Treatise on the Shift Operator
274. Hörmander: The Analysis of Linear Partial Differential Operators III
275. Hörmander: The Analysis of Linear Partial Differential Operators IV
276. Liggett: Interacting Particle Systems
277. Fulton/Lang: Riemann-Roch Algebra
278. Barr/Wells: Toposes, Triples and Theories
279. Bishop/Bridges: Constructive Analysis
280. Neukirch: Class Field Theory
281. Chandrasekharan: Elliptic Functions
282. Lelong/Gruman: Entire Functions of Several Complex Variables
283. Kodaira: Complex Manifolds and Deformation of Complex Structures
284. Finn: Equilibrium Capillary Surfaces
285. Burago/Zalgaller: Geometric Inequalities
286. Andrianov: Quadratic Forms and Hecke Operators
287. Maskit: Kleinian Groups
288. Jacod/Shiryaev: Limit Theorems for Stochastic Processes
289. Manin: Gauge Field Theory and Complex Geometry
290. Conway/Sloane: Sphere Packings, Lattices and Groups
291. Hahn/O'Meara: The Classical Groups and K-Theory
292. Kashiwara/Schapira: Sheaves on Manifolds
293. Revuz/Yor: Continuous Martingales and Brownian Motion
294. Knus: Quadratic and Hermitian Forms over Rings
295. Dierkes/Hildebrandt/Küster/Wohlrab: Minimal Surfaces I
296. Dierkes/Hildebrandt/Küster/Wohlrab: Minimal Surfaces II
297. Pastur/Figotin: Spectra of Random and Almost-Periodic Operators
298. Berline/Getzler/Vergne: Heat Kernels and Dirac Operators
299. Pommerenke: Boundary Behaviour of Conformal Maps
300. Orlik/Terao: Arrangements of Hyperplanes
301. Loday: Cyclic Homology
303. Lange/Birkenhake: Complex Abelian Varieties
303. DeVore/Lorentz: Constructive Approximation
304. Lorentz/v. Golitschek/Makovoz: Constructive Approximation. Advanced Problems
305. Hiriart-Urruty/Lemaréchal: Convex Analysis and Minimization Algorithms I. Fundamentals
306. Hiriart-Urruty/Lemaréchal: Convex Analysis and Minimization Algorithms II. Advanced Theory and Bundle Methods
307. Schwarz: Quantum Field Theory and Topology
308. Schwarz: Topology for Physicists
309. Adem/Milgram: Cohomology of Finite Groups
310. Giaquinta/Hildebrandt: Calculus of Variations, Vol. I: The Lagrangian Formalism
311. Giaquinta/Hildebrandt: Calculus of Variations, Vol. II: The Hamiltonian Formalism

Springer-Verlag and the Environment

We at Springer-Verlag firmly believe that an international science publisher has a special obligation to the environment, and our corporate policies consistently reflect this conviction.

We also expect our business partners – paper mills, printers, packaging manufacturers, etc. – to commit themselves to using environmentally friendly materials and production processes.

The paper in this book is made from low- or no-chlorine pulp and is acid free, in conformance with international standards for paper permanency.